STUDENT SOLUTIONS MANUAL

Gloria Langer
Laurel Technical Services

INTRODUCTORY ALGEBRA FOR COLLEGE STUDENTS

SECOND EDITION

ROBERT BLITZER

PRENTICE HALL, Upper Saddle River, NJ 07458

Acquisitions Editor: ***Karin Wagner***
Editorial Assistant: ***Audra Walsh***
Production Editor: ***Dawn Blayer***
Special Projects Manager: ***Barbara A. Murray***
Supplement Cover Manager: ***Paul Gourhan***
Production Coordinator: ***Alan Fischer***

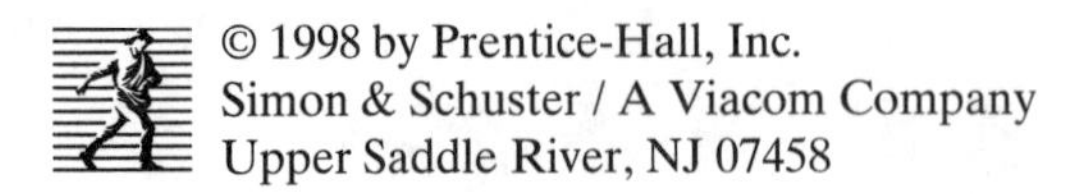

Simon & Schuster / A Viacom Company
Upper Saddle River, NJ 07458

Printed in the United States of America

10 9 8 7 6 5

ISBN 0-13-860594-7

Prentice-Hall International (UK) Limited, *London*
Prentice-Hall of Australia Pty. Limited, *Sydney*
Prentice-Hall Canada, Inc., *Toronto*
Prentice-Hall Hispanoamericana, S.A., *Mexico*
Prentice-Hall of India Private Limited, *New Delhi*
Prentice-Hall of Japan, Inc., *Tokyo*
Simon & Schuster Asia Pte. Ltd., *Singapore*
Editora Prentice-Hall do Brasil, Ltda., *Rio de Janeiro*

CONTENTS

Chapter 1

Problem Set 1.1

1. $\frac{10}{15}=\frac{2\cdot\cancel{5}}{3\cdot\cancel{5}}=\frac{2}{3}$

3. $\frac{15}{18}=\frac{\cancel{3}\cdot 5}{\cancel{3}\cdot 6}=\frac{5}{6}$

5. $\frac{35}{50}=\frac{\cancel{5}\cdot 7}{\cancel{5}\cdot 10}=\frac{7}{10}$

7. $\frac{7}{56}=\frac{1\cdot\cancel{7}}{\cancel{7}\cdot 8}=\frac{1}{8}$

9. $\frac{3}{8}\cdot\frac{7}{11}=\frac{3\cdot 7}{8\cdot 11}=\frac{21}{88}$

11. $\frac{1}{10}\cdot\frac{5}{12}=\frac{1\cdot 5}{10\cdot 12}=\frac{5}{120}=\frac{1\cdot\cancel{5}}{\cancel{5}\cdot 24}=\frac{1}{24}$

13. $\frac{2}{3}\cdot\frac{9}{4}=\frac{2\cdot 9}{3\cdot 4}=\frac{18}{12}=\frac{3\cdot\cancel{6}}{2\cdot\cancel{6}}=\frac{3}{2}$

15. $3\frac{2}{5}\cdot 2\frac{7}{8}=\frac{17}{5}\cdot\frac{23}{8}$
$=\frac{17\cdot 23}{5\cdot 8}$
$=\frac{391}{40}$
$=9\frac{31}{40}$

17. $\frac{5}{4}\div\frac{3}{8}=\frac{5}{4}\cdot\frac{8}{3}$
$=\frac{5\cdot 8}{4\cdot 3}$
$=\frac{40}{12}$
$=\frac{\cancel{4}\cdot 10}{\cancel{4}\cdot 3}$
$=\frac{10}{3}$

19. $\frac{18}{5}\div 2=\frac{18}{5}\cdot\frac{1}{2}$
$=\frac{18\cdot 1}{5\cdot 2}$
$=\frac{18}{10}$
$=\frac{\cancel{2}\cdot 9}{\cancel{2}\cdot 5}$
$=\frac{9}{5}$

21. $2\frac{1}{3}\div 1\frac{1}{6}=\frac{7}{3}\div\frac{7}{6}$
$=\frac{7}{3}\cdot\frac{6}{7}$
$=\frac{\cancel{7}\cdot 6}{3\cdot\cancel{7}}$
$=\frac{6}{3}=2$

23. $1\frac{4}{5}\div\frac{3}{20}=\frac{9}{5}\cdot\frac{20}{3}$
$=\frac{9\cdot 20}{5\cdot 3}$
$=\frac{180}{15}$
$=\frac{\cancel{15}\cdot 12}{\cancel{15}}=12$

25. $\frac{2}{11}+\frac{3}{11}=\frac{2+3}{11}=\frac{5}{11}$

27. $\frac{5}{6}-\frac{1}{6}=\frac{5-1}{6}$
$=\frac{4}{6}=\frac{\cancel{2}\cdot 2}{\cancel{2}\cdot 3}=\frac{2}{3}$

29. $\frac{7}{12}+\frac{1}{12}=\frac{7+1}{12}$
$=\frac{8}{12}=\frac{2\cdot\cancel{4}}{3\cdot\cancel{4}}=\frac{2}{3}$

31. $\frac{1}{2}+\frac{1}{5}=\frac{1}{2}\cdot\frac{5}{5}+\frac{1}{5}\cdot\frac{2}{2}$
$=\frac{5}{10}+\frac{2}{10}$
$=\frac{5+2}{10}=\frac{7}{10}$

33. $\frac{3}{4}+\frac{3}{20}=\frac{3}{4}\cdot\frac{5}{5}+\frac{3}{20}$
$=\frac{15}{20}+\frac{3}{20}$
$=\frac{15+3}{20}$
$=\frac{18}{20}=\frac{\cancel{2}\cdot 9}{\cancel{2}\cdot 10}=\frac{9}{10}$

35. $\frac{7}{8}+\frac{1}{6}=\frac{7}{8}\cdot\frac{3}{3}+\frac{1}{6}\cdot\frac{4}{4}$
$=\frac{21}{24}+\frac{4}{24}$
$=\frac{21+4}{24}$
$=\frac{25}{24}=1\frac{1}{24}$

37. $\frac{17}{25}+\frac{4}{15}=\frac{17}{25}\cdot\frac{3}{3}+\frac{4}{15}\cdot\frac{5}{5}$
$=\frac{51}{75}+\frac{20}{75}$
$=\frac{71}{75}$

39. $\frac{13}{18}-\frac{2}{9}=\frac{13}{18}-\frac{2}{9}\cdot\frac{2}{2}$
$=\frac{13}{18}-\frac{4}{18}$
$=\frac{13-4}{18}$
$=-\frac{9}{18}=\frac{1}{2}$

41. $\frac{4}{3}-\frac{3}{4}=\frac{4}{3}\cdot\frac{4}{4}-\frac{3}{4}\cdot\frac{3}{3}$
$=\frac{16}{12}-\frac{9}{12}$
$=\frac{16-9}{12}=\frac{7}{12}$

43. $\frac{7}{10}-\frac{3}{16}=\frac{7}{10}\cdot\frac{8}{8}-\frac{3}{16}\cdot\frac{5}{5}$
$=\frac{56}{80}-\frac{15}{80}$
$=\frac{56-15}{80}$
$=\frac{41}{80}$

45. $1\frac{5}{6}+3\frac{3}{8}=\frac{11}{6}+\frac{27}{8}$
$=\frac{11}{6}\cdot\frac{4}{4}+\frac{27}{8}\cdot\frac{3}{3}$
$=\frac{44}{24}+\frac{81}{24}$
$=\frac{125}{24}=5\frac{5}{24}$

47. $6\frac{3}{5}-3\frac{1}{2}=\frac{33}{5}-\frac{7}{2}$
$=\frac{33}{5}\cdot\frac{2}{2}-\frac{7}{2}\cdot\frac{5}{5}$
$=\frac{66}{10}-\frac{35}{10}$
$=\frac{31}{10}=3\frac{1}{10}$

49. $4\frac{1}{8}-\frac{1}{2}-\frac{3}{4}=\frac{33}{8}-\frac{1}{2}-\frac{3}{4}$
$=\frac{33}{8}-\frac{1}{2}\cdot\frac{4}{4}-\frac{3}{4}\cdot\frac{2}{2}$
$=\frac{33}{8}-\frac{4}{8}-\frac{6}{8}$
$=\frac{33-4-6}{8}$
$=\frac{23}{8}=2\frac{7}{8}$

	Fraction	Decimal	Percent
51.	$\frac{19}{20}$	0.95	95%
53.	$\frac{7}{20}$	0.35	35%
55.	$\frac{1}{50}$	0.02	2%
57.	$\frac{1}{200}$	0.005	0.5%

59. $0.26(400)=104$

61. $\frac{1}{2}\cdot\frac{3}{4}=\frac{1\cdot 3}{2\cdot 4}=\frac{3}{8}$
$\frac{3}{8}$ cup of sugar

63. $3\frac{3}{4}\div 3=\frac{15}{4}\cdot\frac{1}{3}$
$=\frac{\cancel{3}\cdot 5}{\cancel{3}\cdot 4}$
$=\frac{5}{4}$
$=1\frac{1}{4}$
$1\frac{1}{4}$ acres each

65. $\frac{3}{4}+\frac{2}{5}=\frac{3}{4}\cdot\frac{5}{5}+\frac{2}{5}\cdot\frac{4}{4}$
$=\frac{15}{20}+\frac{8}{20}$
$=\frac{15+8}{20}$
$=\frac{23}{20}=1\frac{3}{20}$

The total distance covered is $1\frac{3}{20}$ miles.

$\frac{3}{4}-\frac{2}{5}=\frac{15}{20}-\frac{8}{20}$
$=\frac{15-8}{20}$
$=\frac{7}{20}$

Walked $\frac{7}{20}$ mile farther than jogged.

67. a. $38\frac{1}{2}-\frac{1}{4}+\frac{5}{8}=\frac{77}{2}-\frac{1}{4}+\frac{5}{8}$
$=\frac{77}{2}\cdot\frac{4}{4}-\frac{1}{4}\cdot\frac{2}{2}+\frac{5}{8}$
$=\frac{308}{8}-\frac{2}{8}+\frac{5}{8}$
$=\frac{308-2+5}{8}$
$=\frac{311}{8}=38\frac{7}{8}$

The value is $\$38\frac{7}{8}$.

b. $20\cdot 38\frac{7}{8}=\frac{20}{1}\cdot\frac{311}{8}$
$=\frac{20\cdot 311}{1\cdot 8}$
$=\frac{6220}{8}$
$=\frac{\cancel{4}\cdot 1555}{2\cdot\cancel{4}}$
$=777\frac{1}{2}$

The total value is \$777.50.

c. Gain of $-\frac{1}{4}+\frac{5}{8}=-\frac{2}{8}+\frac{5}{8}=\frac{3}{8}$

Profit of $20\cdot\frac{3}{8}=\frac{20\cdot 3}{1\cdot 8}=\frac{60}{8}=7\frac{1}{2}$

Money earned = \$7.50

69. $0.15(60)=9$
\$9 tip

71. \$3,502.50 + 0.28(32,350 – 23,350)
= 3,502.50 + 0.28(9000)
= 3,502.50 + 2520
= 6022.50
\$6022.50 tax

73. d is true

75. d is true

77–83. Answers may vary.

85. $\left(\frac{2}{3}+\frac{5}{9}\right)\div\left(\frac{1}{4}+\frac{1}{12}\right)$
$=\left(\frac{2}{3}\cdot\frac{3}{3}+\frac{5}{9}\right)\div\left(\frac{1}{4}\cdot\frac{3}{3}+\frac{1}{12}\right)$
$=\left(\frac{6}{9}+\frac{5}{9}\right)\div\left(\frac{3}{12}+\frac{1}{12}\right)$
$=\frac{11}{9}\div\frac{4}{12}$
$=\frac{11}{9}\cdot\frac{12}{4}$
$=\frac{11\cdot 12}{9\cdot 4}$
$=\frac{132}{36}=\frac{11\cdot\cancel{12}}{3\cdot\cancel{12}}=\frac{11}{3}$

87. To date, the team has won 0.6(70) = 42 games. There are a total of 70 + 40 = 110 games. In order to win 70% of the total games, they would have to win 0.7(110) = 77 games. So they would have to win 77 – 42 = 35 of the remaining 40 games.

89. $\frac{1\cancel{9}}{\cancel{9}5}=\frac{1}{5}$

Problem Set 1.2

1. {1, 2, 3}

3. {0, 1, 2, 3, 4, 5}

5. {–2, –1, 0, 1, ...}

7. {–6, –5, –4, –3, ...}

9. {7}

11. $\left\{-\frac{3}{4}\right\}$

13. {0}

15. $\left\{\frac{2}{3},\ 1\right\}$

17. $\{\pi\}$

19. a. Natural numbers: $\{\sqrt{100}\}$ since $\sqrt{100} = 10$.

b. Whole numbers: $\{0, \sqrt{100}\}$
The whole numbers consist of the natural numbers and 0.

c. Integers: $\{-9, 0, \sqrt{100}\}$.

d. Rational numbers:
$\{-9, -\frac{4}{5}, 0, 0.25, 5\frac{1}{8}, 9.2, \sqrt{100}\}$
Each of the numbers can be expressed as the quotient of two integers.
$-9 = \frac{-9}{1}$, $0 = \frac{0}{1}$, $0.25 = \frac{1}{4}$,
$5\frac{1}{8} = \frac{41}{8}$, $9.2 = 9\frac{1}{5} = \frac{46}{5}$,
$\sqrt{100} = 10 = \frac{10}{1}$.

e. Irrational numbers: $\{\sqrt{3}, e\}$. In decimal form neither number terminates nor has a repeating pattern.

f. Real numbers:
$\{-9, -\frac{4}{5}, 0, 0.25, \sqrt{3}, e, 5\frac{1}{8}, 9.2, \sqrt{100}\}$

21. a. Natural numbers: $\{\sqrt{49}\}$ since $\sqrt{49} = 7$

b. Whole numbers: $\{0, \sqrt{49}\}$

c. Integers: $\{-7, 0, \sqrt{49}\}$

d. Rational numbers:
$\{-7, -0.\bar{6}, 0, \sqrt{49}\}$

e. Irrational numbers: $\{\sqrt{50}\}$

f. Real numbers:
$\{-7, -0.\bar{6}, 0, \sqrt{49}, \sqrt{50}\}$

23. $\frac{1}{2} < 2$; $\frac{1}{2}$ is to the left of 2 so $\frac{1}{2}$ is less than 2.

$\frac{1}{2}$ 2
−2 −1 0 1 2

25. $3 > -\frac{5}{2}$; $-\frac{5}{2}$ is to the left of 3 so $-\frac{5}{2} < 3$ or $3 > -\frac{5}{2}$.

$-\frac{5}{2}$ 3
−3 −2 −1 0 1 2 3

27. $-4 > -6$; -6 is to the left of -4 so $-6 < -4$ or $-4 > -6$.

−6 −4
−7 −6 −5 −4 −3

29. $-2.5 < 1.5$; -2.5 is to the left of 1.5 so $-2.5 < 1.5$.

−2.5 1.5
−3 −2 −1 0 1 2 3

31. $-\frac{3}{4} > -\frac{5}{4}$, $-\frac{5}{4}$ is to the left of $-\frac{3}{4}$ so $-\frac{5}{4} < -\frac{3}{4}$ or $-\frac{3}{4} > -\frac{5}{4}$.

$-\frac{5}{4}$ $-\frac{3}{4}$
−2 −1 0

33. $-4.5 < 3$; -4.5 is to the left of 3 so $-4.5 < 3$.

−4.5 3
−5 −4 −3 −2 −1 0 1 2 3 4

35. $\sqrt{2} < 1.5$; $\sqrt{2}$ is to the left of 1.5 so $\sqrt{2} < 1.5$.

$\sqrt{2}$
1.0 1.25 1.5

37. $0.\bar{3} > 0.3$;
0.3 is to the left of $0.\bar{3}$ so $0.3 < 0.\bar{3}$ or

$0.\bar{3} > 0.3.$

$0.\bar{3}$

0.3 0.35 0.4

39. $-\pi > -3.5$; -3.5 is to the left of $-\pi \approx -3.14$ so $-3.5 < -\pi$ or $-\pi > -3.5$.

$-\pi$

-3.5 -3.25 -3.0

41. Opposite of 6; -6

43. Opposite of -7; $-(-7) = 7$

45. Opposite of $\frac{2}{3}$; $-\frac{2}{3}$

47. Opposite of $-\sqrt{5}$; $-(-\sqrt{5}) = \sqrt{5}$

49. $|6| = 6$ because the distance between 6 and 0 on the number line is 6.

51. $|-7| = 7$ because the distance between -7 and 0 on the number line is 7.

53. $\left|\frac{2}{3}\right| = \frac{2}{3}$ because the distance between $\frac{2}{3}$ and 0 on the number line is $\frac{2}{3}$.

55. $\left|-\sqrt{13}\right| = \sqrt{13}$ because the distance between $-\sqrt{13}$ and 0 on the number line is $\sqrt{13}$.

57. 20; 20° above zero

59. -8.5; A loss of 8.5 pounds

61. 3000 A deposit of $3000.00

63. 3.7; A surplus of 3.7 billion dollars

65. c is true; $-\frac{1}{2}$ is an example of a rational number which is not positive.

67. d is true; $\frac{7}{3} = 2\frac{1}{3}$ and the whole numbers less than $\frac{7}{3}$ are $\{0, 1, 2\}$.

69. b is true; $|-4| = 4,\ 4 > 3$.

71. c is false; $|0| = 0,\ |-4| = 4$ and $0 < 4$.

73. $-\sqrt{12} \approx -3.464$, -4 and -3

75. $2 - 3\sqrt{5} \approx -4.708$, -5 and -4

77–85. Answers may vary.

87. Group Activity

Review Problems

88. $\frac{1}{4} \div \frac{1}{2} = \frac{1}{4} \cdot \frac{2}{1} = \frac{2}{4} = \frac{1}{2}$

89. $\frac{3}{4} - \frac{1}{5} = \frac{3}{4} \cdot \frac{5}{5} - \frac{1}{5} \cdot \frac{4}{4} = \frac{15}{20} - \frac{4}{20} = \frac{11}{20}$

90. $\frac{8}{5} \cdot \frac{11}{12} = \frac{2 \cdot 4 \cdot 11}{5 \cdot 3 \cdot 4} = \frac{22}{15}$

Problem Set 1.3

1. a. $(0.09 - 0.05)(5{,}720{,}000{,}000)$
$= (0.04)(5{,}720{,}000{,}000)$
$= 228{,}800{,}000$

b. $100\% - 12\% = 88\%$
Subtract 12% from 100%. Add all of the percents except Africa's percent. The first method is faster.

c. Asia $= 0.59(5{,}720{,}000{,}000)$
$= 3{,}374{,}800{,}000$
Oceania $= 0.01(5{,}720{,}000{,}000)$
$= 57{,}200{,}000$
Africa $= 0.12(5{,}720{,}000{,}000)$
$= 686{,}400{,}000$
Latin America $= 0.09(5{,}720{,}000{,}000)$
$= 514{,}800{,}000$
Europe $= 0.09(5{,}720{,}000{,}000)$
$= 514{,}800{,}000$

N. America = 0.05(5,720,000,000)
= 286,000,000
former USSR = 0.05(5,720,000,000)
= 286,000,000

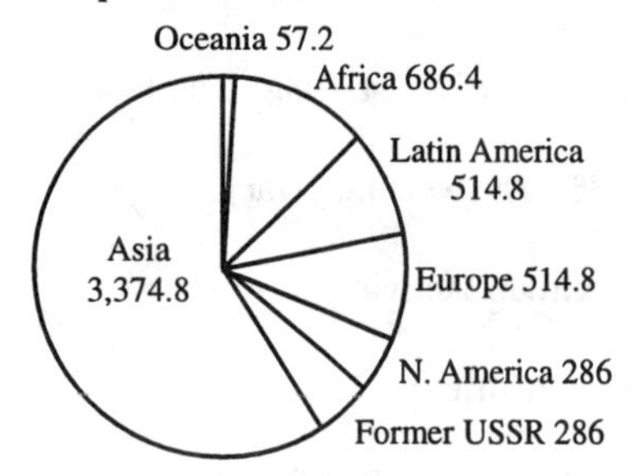

d. No; we are not given the area of the land for each region

e. No; the percentages for all regions may change by 2050.

3. a. 85.8 + 26 + 10.4 + 10.4 + 18.2 + 10.4 + 2.6 + 18.2 + 49.4 + 13 + 15.6 = 260
U.S. population = 260 million

b. $\frac{13}{260} = 0.05, 5\%$

c. 10%

5. a. 1994

b. 1992, 1993

c. These patients are staying in the hospital for shorter periods of time and a greater percentage are being treated as outpatients, as time goes on.

7. a. No; the graph shows that U.S. spends more of its GDP on health care than in other rich countries, but it does not explicitly show that the U.S. spends more.

b. 13.4% – 5.3% = 8.1%

c. Each is independent for each year.

9. a. 1980; 55,000,000

b. 1990

11. Quadrant I

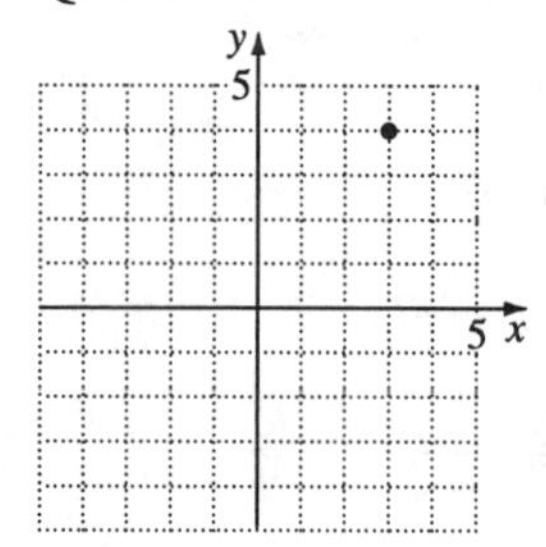

13. Quadrant II

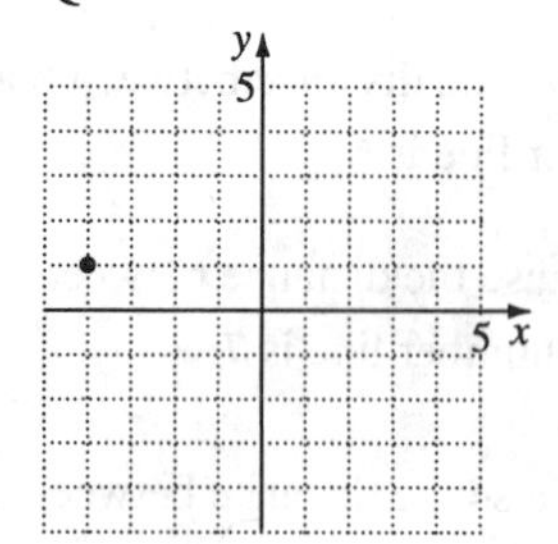

15. Quadrant III

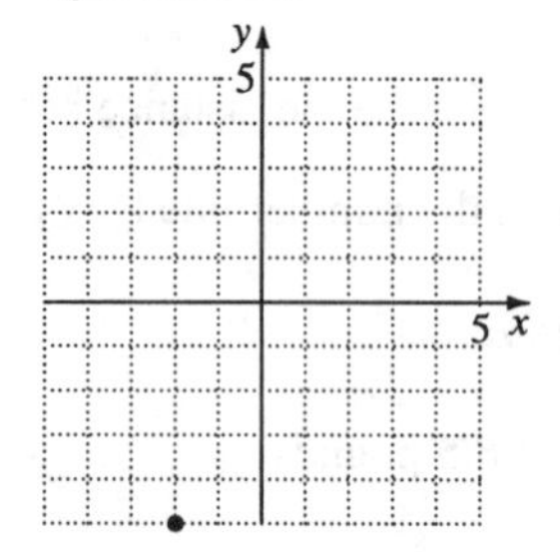

17. Quadrant IV

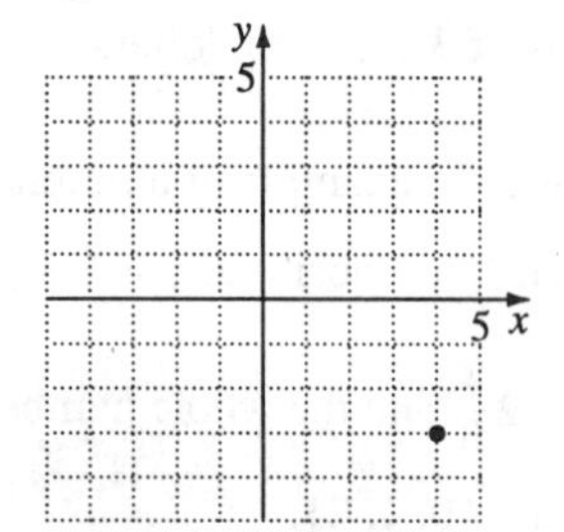

19.

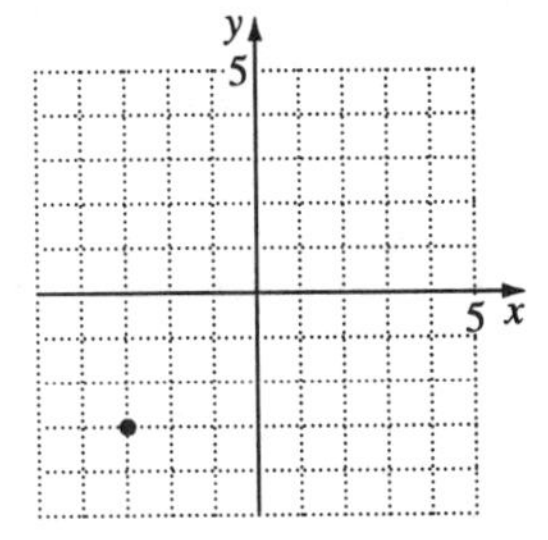

21.

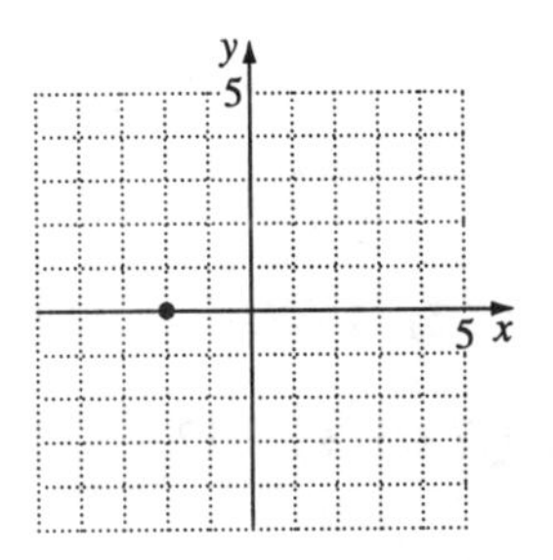

23.

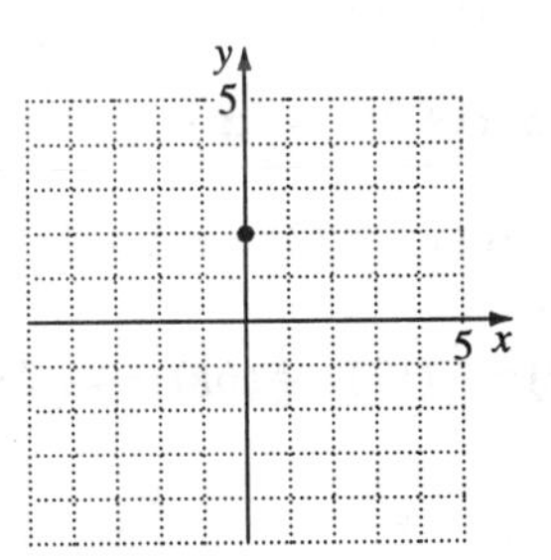

25.

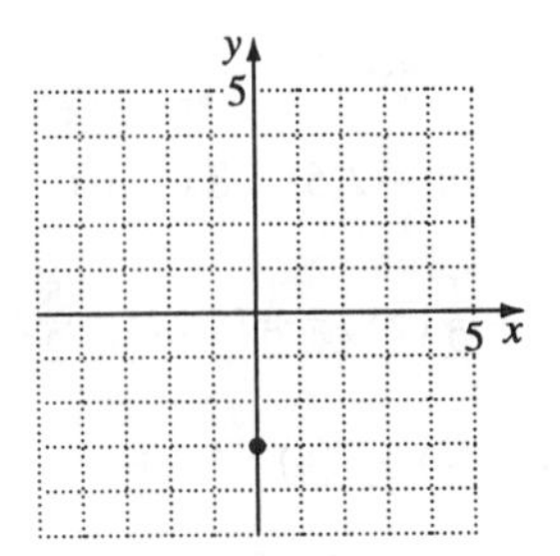

27.

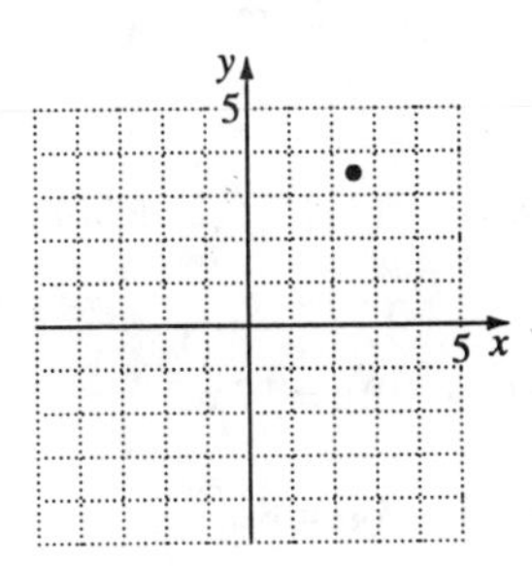

29.

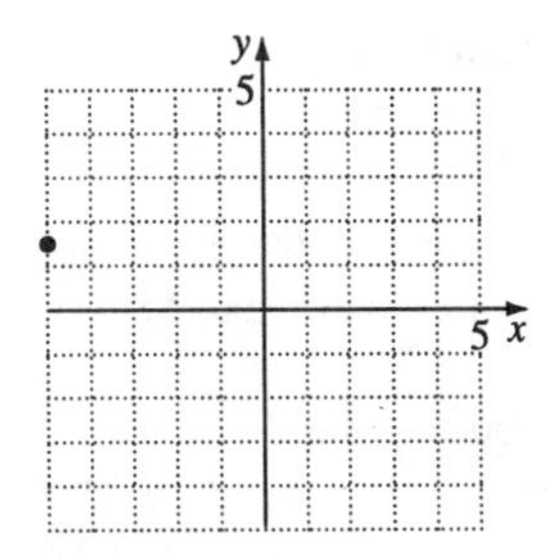

31.

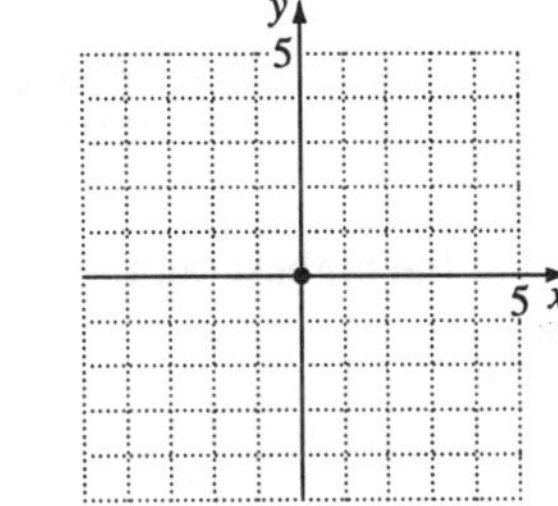

33.

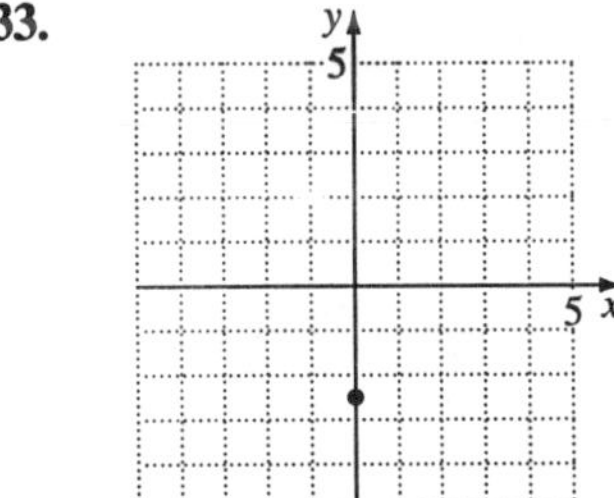

35. (5, 2)

37. (–6, 5)

39. (–2, –3)

41. (5, –3)

43. **a.** (1970, 1.5%), (1975, 4%), (1980, 15%), (1985, 21%), (1990, 30%)

b. In each year (x-value) the percent of total degrees given in dentistry (y-value) were given to women.

c. No, it is unlikely that the trend will continue.

d. Answers may vary.

45. a. $A(0, 8000)$
If the price is not increased, the manufacturer will make $8000.

b. $B(2, 11{,}000)$
If the price is increased by $2, the manufacturer will make $11,000.

c. No; the profit peaks at a $4 increase and then declines.

d. $9; $12,000; the highest point is (4, 12,000).

47. c is true

49–55. Answers may vary.

57. a. C

b. A

c. B

59. Group Activity

Review Problems

60. $|x| = 4, x = -4, 4, \{-4, 4\}$

61. $\frac{1}{3} > 0.33$

62. $|-5| - |-2| = 5 - 2 = 3$

Problem Set 1.4

1. $x + 7 = 7 + x$

3. $x + 4y = 4y + x$

5. $4x + 7y = 7y + 4x$

7. $4(x + 6) = 4(6 + x)$

9. $x \cdot 7 = 7 \cdot x$

11. $6 + xy = 6 + yx$

13. $4(b + 5) = (b + 5) \cdot 4$

15. $7 + (5 + x) = (7 + 5) + x = 12 + x$

17. $7(4x) = (7 \cdot 4)x = 28x$

19. $3(x + 5) = 3(x) + 3(5) = 3x + 15$

21. $8(2x + 3) = 8(2x) + 8(3) = 16x + 24$

23. $\frac{1}{3}(12 + 6r) = \frac{1}{3}(12) + \frac{1}{3}(6r) = 4 + 2r$

25. $5(x + y) = 5x + 5y$

27. $3(x - 2) = 3(x) - 3(2) = 3x - 6$

29. $2(4x - 5) = 2(4x) - 2(5) = 8x - 10$

31. $\frac{1}{2}(5x - 12) = \frac{1}{2}(5x) + \frac{1}{2}(-12) = \frac{5}{2}x - 6$

33. $(2x + 7)4 = 2x(4) + 7(4) = 8x + 28$

35. $6(x + 3 + 2y) = 6(x) + 6(3) + 6(2y)$
$= 6x + 18 + 12y$

37. $5(3x - 2 + 4y) = 5(3x) - 5(2) + 5(4y)$
$= 15x - 10 + 20y$

39. $7x + 10x = (7 + 10)x = 17x$

41. $11a - 3a = (11 - 3)a = 8a$

43. $3 + (x + 11) = (3 + 11) + x = 14 + x$

45. $5y - 3 + 6y = (5y + 6y) - 3 = 11y - 3$

47. $2x + 5 + 7x - 4 = (2x + 7x) + (5 - 4) = 9x + 1$

49. $11a + 12 - 3a - 2 = (11a - 3a) + (12 - 2)$
$= 8a + 10$

51. $5(3x + 2) - 4 = 15x + 10 - 4 = 15x + 6$

53. $12 + 5(3x - 2) = 12 + 15x - 10$
$= 15x + (12 - 10) = 15x + 2$

55. $7(3a + 2b) + 5(4a - 2b)$
$= 21a + 14b + 20a - 10b$
$= 21a + 20a + 14b - 10b = 41a + 4b$

57. **a.** $16{,}020 + 1527x$
$= 16{,}020 + 1527(8)$
$= 16{,}020 + 12{,}216 = \$28{,}236$
In 1988, the average salary was \$28,236.

b. $x = 1991 - 1980$
$x = 11$
$16{,}020 + 1527(11)$
$= 16{,}020 + 16{,}797 = \$32{,}817$

c. $16{,}020 + 1527x = 1527x + 16{,}020$

59. $\frac{DA + D}{24} = \frac{200(12) + 200}{24}$
$= \frac{2400 + 200}{24} = \frac{2600}{24} = 108.\overline{3}$ mg
$\frac{D(A+1)}{24} = \frac{200(12+1)}{24}$
$= \frac{200(13)}{24} = \frac{2600}{24} = 108.\overline{3}$ mg
The second form is easier to use.

61. $4x(y + 3)$; $4xy + 12x$

63. c is true

65–67. Answers may vary.

69. Commutative

71. Commutative

73. Answers may vary.

75. **a.** Yes;
$a * b = a + b + 3$
$b * a = b + a + 3 = a + b + 3$
$a * b = b * a$

b. Yes;
$(a * b) * c = (a + b + 3) * c$
$= (a + b + 3) + c + 3$
$= a + b + c + 6$
$a * (b * c) = a * (b + c + 3)$
$= a + (b + c + 3) + 3$
$= a + b + c + 6$
$(a * b) * c = a * (b * c)$

77. No; Example: Let $a = 1$, $b = 2$, $c = 3$ then
$a + bc = 1 + 6 = 7$
$a(b + c) = 1(2 + 3) = 5$
$7 \neq 5$

Review Problems

79. **a.** Natural numbers $\left\{\frac{18}{3}, \sqrt{81}\right\}$;
$\frac{18}{3} = 6$, $\sqrt{81} = 9$

b. Whole numbers $\left\{\frac{18}{3}, \sqrt{81}\right\}$

c. Integers $\left\{-23, \frac{18}{3}, \sqrt{81}\right\}$

d. Rational numbers
$\left\{-23, \frac{17}{3}, \frac{18}{3}, \sqrt{81}\right\}$

e. Irrational numbers
$\left\{\frac{5\pi}{3}, \sqrt{83}\right\}$

f. Real numbers
$\left\{-23, \frac{17}{3}, \frac{18}{3}, \frac{5\pi}{3}, \sqrt{81}, \sqrt{83}\right\}$

80. White and Hispanic

81. $\frac{2}{3}+\frac{4}{5}=\frac{2}{3}\cdot\frac{5}{5}+\frac{4}{5}\cdot\frac{3}{3}=\frac{10}{15}+\frac{12}{15}=\frac{22}{15}$

Problem Set 1.5

1. $-8+3=-5$

3. $-10+2=-8$

5. $-6+6=0$

7. $-4+(-5)=-9$

9. $-7+(-5)=-\left(|-7|+|-5|\right)=-(7+5)=-12$

11. $12+(-8)=+\left(|12|-|-8|\right)=+(12-8)=4$

13. $6+(-9)=-\left(|-9|-|6|\right)=-(9-6)=-3$

15. $-9+(+4)=-\left(|-9|-|+4|\right)=-(9-4)=-5$

17. $-0.4+(-0.9)=-1.3$

19. $-3.6+2.1=-1.5$

21. $-9+(-9)=-18$

23. $9+(-9)=0$

25. $-\frac{7}{10}+\left(-\frac{3}{10}\right)=-\frac{10}{10}=-1$

27. $\frac{9}{10}+\left(-\frac{3}{5}\right)=\frac{9}{10}+\left(-\frac{6}{10}\right)=\frac{3}{10}$

29. $-\frac{5}{8}+\frac{3}{4}=-\frac{5}{8}+\frac{6}{8}=\frac{1}{8}$

31. $-\frac{3}{7}+\left(-\frac{4}{5}\right)=-\frac{15}{35}-\frac{28}{35}=-\frac{43}{35}$

33. $3\frac{1}{2}+\left(-4\frac{1}{4}\right)=-\left(\left|-4\frac{1}{4}\right|-\left|3\frac{1}{2}\right|\right)$
$=-\left(3\frac{5}{4}-3\frac{2}{4}\right)=-\frac{3}{4}$

35. $-8.74-8.74=-17.48$

37. $85+(-15)+(-20)+12$
$=(85+12)+(-15-20)$
$=97-35=62$

39. $-45+\left(-\frac{3}{7}\right)+25+\left(-\frac{4}{7}\right)$
$=(-45+25)+\left(-\frac{3}{7}-\frac{4}{7}\right)$
$=-20+(-1)=-21$

41. $3.5+(-45)+(-8.4)+72$
$=(3.5-8.4)+(72-45)$
$=-4.9+27=22.1$

43. $-8x+5x=(-8+5)x=-3x$

45. $7x+(-5y)+(-9x)+2y$
$[7x+(-9x)]+[(-5y)+2y]$
$=(7-9)x+(-5+2)y=-2x-3y$

47. $7(3a-5)+6(2-9a)$
$=21a-35+12-54a$
$=(21-54)a+(12-35)$
$=-33a-23$

49. $100+(-56)=44$°F

51. $-1312+712=-600$
The elevation of the person is 600 feet below sea level.

53. Temperature at 8:00 A.M. + rise 15°F by noon + fall 5°F by 4 P.M.
$=-7+15-5=3$
The temperature at 4:00 P.M. was 3°F.

55. Start at 27-yard line + 4-yard gain + 2-yard loss + 8-yard gain + 12-yard loss
$=27+4-2+8-12=39-14=25$
The location of the football at the end of the fourth play is at the 25-yard line.

57. Starting price + rose $1\frac{1}{2}$ points
+ fell $\frac{1}{4}$ point + fell $\frac{1}{2}$ point
$= 35\frac{1}{2} + 1\frac{1}{2} - \frac{1}{4} - \frac{1}{2} = 37 - \frac{3}{4} = 36\frac{1}{4}$
The final price for the stock is $36\frac{1}{4}$ per share.

59. a. $3.58 + (-0.19) + (-0.11) + (-0.08) + (-0.06) = 3.14$

b. $3.14 + 0.02 + (-0.02) + 0 + 0.05 = 3.19$

61. The sum of x and 18,000
18,000 more than x
x increased by 18,000
x plus 18,000
18,000 added to x

63. d is true; If $x > 0$, then $0 + (-x) = -x$.
The sum of zero and a negative number is always a negative number.

65. $3\sqrt{5} - 2\sqrt{7} - \sqrt{11} + 4\sqrt{3} = 5.0283$

67. $-2\sqrt{7} - \sqrt{5} \approx -7.53$
$-3\sqrt{3} - 2\sqrt{2} \approx -8.02$
$-2\sqrt{7} - \sqrt{5} > -3\sqrt{3} - 2\sqrt{2}$

69. Answers may vary.

71. $-18y + 11x + (-3y) + 3x = 7(2x - 3y)$

Review Problems

73. $\{x|x$ is an integer but not a natural number$\}$
$\{\ldots, -3, -2, -1, 0\}$

74. a. Natural numbers: $\left\{\sqrt{9}\right\}$

b. Whole numbers: $\left\{0, \sqrt{9}\right\}$

c. Integers: $\left\{-17, 0, \sqrt{9}\right\}$

d. Rational numbers:
$\left\{-17, -\frac{2}{3}, 0, \bar{3}, \sqrt{9}, 10\frac{1}{7}\right\}$

e. Irrational numbers:
$\left\{\sqrt{5}, \pi, \sqrt{7}\right\}$

f. Real numbers:
$\left\{-17, -\frac{2}{3}, 0, \bar{3}, \sqrt{5}, \pi, \sqrt{7}, \sqrt{9}, 10\frac{1}{7}\right\}$

75. a. 11%

b. 1930 to 1940

Problem Set 1.6

1. $13 - 8 = 13 + (-8) = 5$

3. $8 - 15 = 8 + (-15) = -7$

5. $4 - (-10) = 4 + 10 = 14$

7. $-6 - (-17) = -6 + 17 = 11$

9. $-12 - (-3) = -12 + 3 = -9$

11. $-11 - 17 = -28$

13. $\frac{1}{5} - \left(-\frac{3}{5}\right) = \frac{1}{5} + \frac{3}{5} = \frac{4}{5}$

15. $-\frac{4}{5} - \left(-\frac{1}{5}\right) = -\frac{4}{5} + \frac{1}{5} = -\frac{3}{5}$

17. $\frac{1}{2} - \left(-\frac{1}{4}\right) = \frac{1}{2} + \frac{1}{4} = \frac{3}{4}$

19. $-4.4 - 9.3 = -13.7$

21. $-7.2 - (-5.1) = -7.2 + 5.1 = -2.1$

23. $3.1 - (-6.03) = 3.1 + 6.03 = 9.13$

25. $13-2-(-8)=13-2+8=11+8=19$

27. $9-8+3-7=1+3-7=4-7=-3$

29. $-6-2+3-10$
$=-8+3-10=-5-10=-15$

31. $-10-(-5)+7-2=-10+5+7-2$
$=-5+7-2=2-2=0$

33. $-23-11-(-7)+(-25)$
$=-23-11+7+(-25)$
$=-34+7+(-25)$
$=-27+(-25)=-52$

35. $-823-146-50-(-832)$
$=-823-146-50+832$
$=-969-50+832$
$=-1019+832=-187$

37. $1-\frac{2}{3}-\left(-\frac{5}{6}\right)=1-\frac{2}{3}+\frac{5}{6}$
$=\frac{1}{3}+\frac{5}{6}=\frac{2}{6}+\frac{5}{6}=\frac{7}{6}=1\frac{1}{6}$

39. $-30-14+11-(-9)-(-6)+17$
$=-30-14+11+9+6+17$
$=-44+11+9+6+17$
$=-33+9+6+17$
$=-24+6+17$
$=-18+17$
$=-1$

41. $-0.16-5.2-(-0.87)$
$=-0.16-5.2+0.87$
$=-5.36+0.87=-4.49$

43. $-\frac{3}{4}-\frac{1}{4}-\left(-\frac{5}{8}\right)$
$=-\frac{3}{4}-\frac{1}{4}+\frac{5}{8}=-1+\frac{5}{8}=-\frac{3}{8}$

45. $-17x-(-23x)=-17x+23x=6x$

47. $7a-(-15a)+4a$
$=7a+15a+4a=26a$

49. $3-4y-(-9)+6y$
$=3-4y+9+6y$
$=3+9-4y+6y=12+2y$

51. $-6-(-7b)+7b+5b-(-13)$
$=-6+7b+7b+5b+13$
$=-6+13+7b+7b+5b$
$=7+19b$

53. $-13x-5y-(-9x)+7y$
$=-13x-5y+9x+7y$
$=-13x+9x-5y+7y$
$=-4x+2y$

55. Elevation of peak of Mount Whitney
– elevation of Death Valley
= 14,494 – (–282)
= 14,494 + 282
= 14,776
The peak of Mount Whitney is 14,776 feet above Death Valley.

57. $98-(-90)=98+90=188$
The range of temperatures is 188°F.

59. 1992(%) – 1982(%)
$26.7-22.1=4.6\%$
$40.7-39.0=1.7\%$
$13.5-11.9=1.6\%$
$10.6-9.6=1.0\%$
$4.7-4.2=0.5\%$
$3.3-3.0=0.3\%$
$12.5-13.0=-0.5\%$
$17.4-18.6=-1.2\%$
$34.5-39.0=-4.5\%$

61. x minus 8%
x decreased by 8%
The difference between x and 8%
8% less than x

63. $36-15=21;\ b-21\%$

65. $35-26=9;\ d+9\%$

67. Area of the second floor: $3600 - x$ square feet

69. c is true.
If x represents a positive number ($x > 0$), then the difference between 0 and a negative number ($-x$) is $0 - (-x) = 0 + x = x$ which is always a positive number.

71. $-5\sqrt{3} - (-2\sqrt{11}) - (-\sqrt{17}) + \sqrt{6} = 4.5456$

73. $-\sqrt{7} - (-\sqrt{3}) \approx -0.91$
$-\sqrt{11} - (-\sqrt{5}) \approx -1.08$
$-\sqrt{7} - (-\sqrt{3}) > -\sqrt{11} - (-\sqrt{5})$

75. Answers may vary.

77. Smallest possible difference: Dec 31 in year A and Jan 1 in year (A + 1)(the next year). The difference in birthdays is 1 day.

Largest possible difference: Jan 1 in year A and Dec 31 in year (A + 1)(the next year). The difference in birthdays is 1 day less than 2 years.

79. Group Activity

Review Problems

80. a. Natural numbers: $\{\sqrt{1}\}$

b. Whole numbers: $\{0, \sqrt{1}\}$

c. Integers: $\{-123, 0, \sqrt{1}\}$

d. Rational numbers:
$\left\{-123, -\frac{3}{9}, 0, 0.4\dot{5}, \sqrt{1}, 8\frac{1}{5}\right\}$

e. Irrational numbers: $\{\sqrt{7}, e\}$

f. Real numbers: $\left\{-123, -\frac{3}{9}, 0, 0.4\dot{5}, \sqrt{1}, \sqrt{7}, e, 8\frac{1}{5}\right\}$

81. First reading + increase of 4°F
+ decrease of 17°F + decrease of 2°F
$= 12 + 4 - 17 - 2$
$= -3$
The temperature of the final reading is –3°F.

82. $-\frac{1}{2} \ ? \ -\frac{1}{10}$
$-\frac{5}{10} < -\frac{1}{10}$

Problem Set 1.7

1. $6(-9) = -(6 \cdot 9) = -54$

3. $(-7)(-3) = +(7 \cdot 3) = 21$

5. $(-2)(6) = -12$

7. $(-13)(-1) = 13$

9. $0(-5) = 0$

11. $\frac{1}{2}(-14) = -7$

13. $\left(-\frac{3}{4}\right)(-20) = \frac{3 \cdot 20}{4 \cdot 1} = 15$

15. $-\frac{3}{5}\left(-\frac{4}{7}\right) = \frac{3 \cdot 4}{5 \cdot 7} = \frac{12}{35}$

17. $-\frac{7}{9} \cdot \frac{2}{3} = -\frac{7 \cdot 2}{9 \cdot 3} = -\frac{14}{27}$

19. $\left(\frac{4}{15}\right)\left(-1\frac{1}{4}\right) = \left(\frac{4}{15}\right)\left(-\frac{5}{4}\right) = -\frac{4 \cdot 5}{15 \cdot 4} = -\frac{1}{3}$

21. $(-4.1)(0.03) = -(4.1)(0.03) = -0.123$

23. $(-3.8)(-2.4) = +(3.8)(2.4) = 9.12$

25. $(3.08)(-0.25) = -0.77$

27. $(-5)(-2)(-3)(4)$
$= 10(-3)(4) = -30(4) = -120$

29. $-2(-3)(-4)(-1) = 6(-4)(-1) = -24(-1) = 24$

31. $-3\left(-\frac{1}{6}\right)(-50) = \frac{1}{2}(-50) = -25$

33. $(-3)(-1)(-2)\left(-\frac{1}{2}\right)(-4) = 3(-2)\left(-\frac{1}{2}\right)(-4)$
$= 3(-4) = -12$

35. $-\frac{1}{8}(-24)\left(-\frac{1}{2}\right)(-6) = 3\left(-\frac{1}{2}\right)(-6) = 3(3) = 9$

37. $(-5)(-5)(-5) = 25(-5) = -125$

39. $-7 - (-3)(2) = -7 - (-6) = -7 + 6 = -1$

41. $8(-7) - (-11)$
$= -56 - (-11) = -56 + 11 = -45$

43. $(-6)(4) - (-4)(2)$
$= -24 - (-8) = -24 + 8 = -16$

45. $7(-2)(-5) - (-11) = 70 + 11 = 81$

47. $-15 - (-3)(-4)(-2) = -15 + 24 = 9$

49. $(-4)(2)(-1) - (-5)(3)(-2) = 8 - 30 = -22$

51. $-6(-8 - 2) = -6(-10) = 60$

53. $6 - 4(2 - 10) = 6 - 4(-8) = 6 + 32 = 38$

55. $4(2 + 5) - 5(7 + 3)$
$= 4(7) - 5(10) = 28 - 50 = -22$

57. $2(8 - 10) - 3(-6 + 4)$
$= 2(-2) - 3(-2) = -4 + 6 = 2$

59. $(4 - 11)(6 - 10) = (-7)(-4) = 28$

61. $(-3 - 2)(-6 + 10) = (-5)(4) = -20$

63. $(-4 - 6)(-3) + 5$
$= (-10)(-3) + 5 = 30 + 5 = 35$

65. $-3(-5) + 7(-1) = 15 + (-7) = 8$

67. $4(3) - 5(-2) + 7(-3)$
$= 12 + 10 - 21 = 22 - 21 = 1$

69. $\frac{1}{2} - \left(-\frac{1}{2}\right)\left(\frac{1}{4}\right) = \frac{1}{2} - \left(-\frac{1}{8}\right)$
$= \frac{1}{2} + \frac{1}{8} = \frac{4}{8} + \frac{1}{8} = \frac{5}{8}$

71. $\left(-\frac{1}{2}\right)\left(\frac{1}{6}\right) - \left(-\frac{1}{3}\right)(3)$
$= -\frac{1}{12} - (-1)$
$= -\frac{1}{12} + 1$
$= -\frac{1}{12} + \frac{12}{12} = \frac{11}{12}$

73. $\frac{2}{3} - 2\left(\frac{7}{8} - \frac{5}{12}\right)$
$= \frac{2}{3} - 2\left(\frac{21}{24} - \frac{10}{24}\right)$
$= \frac{2}{3} - 2\left(\frac{11}{24}\right)$
$= \frac{2}{3} - \frac{11}{12} = \frac{8}{12} - \frac{11}{12} = -\frac{3}{12} = -\frac{1}{4}$

75. $-5(2x) = -10x$
$-5[2 \cdot (-3)] = -10(-3)$
$-5(-6) = 30$
$30 = 30$

77. $-4\left(-\frac{3}{4}y\right) = 3y$
$-4\left[-\frac{3}{4}(-3)\right] = 3(-3)$
$-4\left(\frac{9}{4}\right) = -9$
$-9 = -9$

79. $8x + x = 9x$
$8(-3) + (-3) = 9(-3)$
$-24 - 3 = -27$
$-27 = -27$

81. $-5x + x = -4x$
$-5(-3) + (-3) = -4(-3)$

$15 - 3 = 12$
$12 = 12$

83. $6b - 7b = -b$
$6(-3) - 7(-3) = -(-3)$
$-18 + 21 = 3$
$3 = 3$

85. $-y + 4y = 3y$
$-(-3) + 4(-3) = 3(-3)$
$3 - 12 = -9$
$-9 = -9$

87. $-4(2x - 3) = -8x + 12$
$-4[2(-3) - 3] = -8(-3) + 12$
$-4(-6 - 3) = 24 + 12$
$-4(-9) = 36$
$36 = 36$

89. $-3(-2x + 4) = 6x - 12$
$-3[-2(-3) + 4] = 6(-3) - 12$
$-3(6 + 4) = -18 - 12$
$-3(10) = -30$
$-30 = -30$

91. $-(2y - 5) = -2y + 5$
$-[2(-3) - 5] = -2(-3) + 5$
$-(-6 - 5) = 6 + 5$
$-(-11) = 11$
$11 = 11$

93. $4(2y - 3) - (7y + 2)$
$= 8y - 12 - 7y - 2$
$= y - 14$
$4[2(-3) - 3] - [7(-3) + 2] = -3 - 14$
$4(-6 - 3) - (-21 + 2) = -17$
$4(-9) - (-19) = -17$
$-36 + 19 = -17$
$-17 = -17$

95. $-5(-2x - 1) - 11x$
$= 10x + 5 - 11x = 5 - x$
$-5[-2(-3) - 1] - 11(-3) = 5 - (-3)$
$-5(6 - 1) + 33 = 5 + 3$
$-5(5) + 33 = 8$
$-25 + 33 = 8$
$8 = 8$

97. $3\left(\frac{1}{3}x + \frac{1}{3}\right) = x + 1$
$3\left[\frac{1}{3}(-3) + \frac{1}{3}\right] = -3 + 1$
$3\left(-1 + \frac{1}{3}\right) = -2$
$3\left(-\frac{2}{3}\right) = -2$
$-2 = -2$

99. $2\left(-\frac{1}{2}x + \frac{1}{2}\right) = -x + 1$
$2\left[-\frac{1}{2}(-3) + \frac{1}{2}\right] = -(-3) + 1$
$2\left(\frac{3}{2} + \frac{1}{2}\right) = 3 + 1$
$2\left(\frac{4}{2}\right) = 3 + 1$
$2(2) = 4$
$4 = 4$

101. $x = 1982 - 1980 = 2$
$1.271(2) - 1.4 = 1.142$ million

103. $x = 1986 - 1980 = 6$
$1.271(6) - 1.4 = 6.226$ million

105. $x = 1990 - 1980 = 10$
$1.271(10) - 1.4 = 11.31$ million

107. Answers may vary.

109. a. $-(0.04x - 30) - 5.7 - 0.031x$
$= -0.04x + 30 - 5.7 - 0.031x$
$= -0.04x - 0.031x + 30 - 5.7$
$= -0.071x + 24.3$

b. $x = 1980 - 1948 = 32$
$-0.071(32) + 24.3 = 22.028$
$22.028 \approx 22.03$
Very well

c. $x = 1960 - 1948 = 12$
$-0.071(12) + 24.3 = 23.448$
$23.448 \approx 24$
Fairly well

d. $x = 2000 - 1948 = 52$
$-0.071(52) + 24.3 = 20.608$
20.608 seconds

111. The value of one quarter times number of quarters
$25¢ \cdot x$
$25x$ cents

113. 70% of number of kilograms of the metal
$0.70 \cdot x$
$0.70x$ kilograms

115. 45 miles per hour times x hours
$45 \cdot x$
$45x$ miles

117. Cost of VCR + tax
$x + 0.12x$
1.12 dollars

119. Percent discount times dollar amount
$0.45 \cdot x$
$0.45x$ dollars

121. Price of item minus discount
$x - 0.45x$
$0.55x$ dollars

123. Weight of person + weight of each bag times the number of bags
$140 + 80 \cdot x$
$140 + 80x$ pounds

125. Months per year times salary per month
12 months $\cdot\ x$ dollars per month
$12x$ dollars

127. Fee per night plus fee per person times number of people
$20 + 5x$ dollars

129. Earnings per hour plus earnings per student times the number of students helped
$4.50 + 0.30x$ dollars

131. The consecutive odd integer that follows any odd integer represented by x: $x + 2$

133. c is true;
$$\left(-\frac{1}{2}\right)\left(-\frac{1}{2}\right) = +\left(\frac{1}{2}\right)\left(\frac{1}{2}\right) = +\left(\frac{1 \cdot 1}{2 \cdot 2}\right) = \frac{1}{4}$$

135. a is true

137. $y = 382.75x + 1742$

10,000

20

National Average

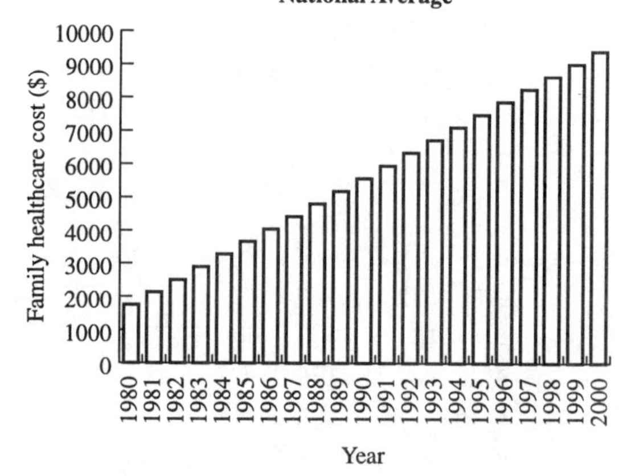

139. Cost for first minute plus cost for each additional minute times the number of additional minutes
(which is 1 less than the total number of minutes)

$15 + 5(x - 1)$ cents
$15 + 5x - 5$
$10 + 5x$ cents

141. $a + b = ab$ when a is a natural number and b is a mixed number, $1\frac{1}{c}$, where c is one less than the whole number a.
$b = 1\frac{1}{a-1}$ or $b = \frac{a}{a-1}$

Review Problems

142. a. Natural numbers: $\{1492\}$

b. Whole numbers: $\{0, 1492\}$

c. Integers: $\left\{-\sqrt{25},\ 0,\ 1492\right\}$

d. Rational numbers:
$\left\{-\sqrt{25},\ 0,\ \frac{17}{125},\ 1492\right\}$

e. Irrational numbers:
$\left\{-\sqrt{2},\ \frac{\pi}{2}\right\}$

f. Real numbers:
$\left\{-\sqrt{25},\ -\sqrt{2},\ 0,\ \frac{17}{125},\ \frac{\pi}{2},\ 1492\right\}$

143. $\{\, x \mid \sqrt{x}$ is irrational and x is a natural number between 2 and 10, not including 2 and not including 10$\}$:
$\left\{\sqrt{3},\ \sqrt{5},\ \sqrt{6},\ \sqrt{7},\ \sqrt{8}\right\}$

144. a. 62,000

b. Labrador Retrievers

Problem Set 1.8

1. $7^2 = 7 \cdot 7 = 49$

3. $4^3 = 4 \cdot 4 \cdot 4 = 64$

5. $(-4)^2 = (-4)(-4) = 16$

7. $(-4)^3 = (-4)(-4)(-4) = -64$

9. $(-2)^4 = (-2)(-2)(-2)(-2) = 16$

11. $-2^4 = -2 \cdot 2 \cdot 2 \cdot 2 = -16$

13. $2^6 = 2 \cdot 2 \cdot 2 \cdot 2 \cdot 2 \cdot 2 = 64$

15. $\left(\frac{2}{3}\right)^2 = \frac{2}{3} \cdot \frac{2}{3} = \frac{4}{9}$

17. $\left(-\frac{1}{3}\right)^3 = \left(-\frac{1}{3}\right)\left(-\frac{1}{3}\right)\left(-\frac{1}{3}\right) = -\frac{1}{27}$

19. $\left(-\frac{3}{4}\right)^3 = \left(-\frac{3}{4}\right)\left(-\frac{3}{4}\right)\left(-\frac{3}{4}\right) = -\frac{27}{64}$

21. $\left(-\frac{2}{3}\right)^4 = \left(-\frac{2}{3}\right)\left(-\frac{2}{3}\right)\left(-\frac{2}{3}\right)\left(-\frac{2}{3}\right)$
$= \frac{16}{81}$

23. $-\left(\frac{2}{3}\right)^4 = -\frac{2}{3} \cdot \frac{2}{3} \cdot \frac{2}{3} \cdot \frac{2}{3} = -\frac{16}{81}$

25. $-\left(-\frac{1}{2}\right)^3 = -\left(-\frac{1}{2}\right)\left(-\frac{1}{2}\right)\left(-\frac{1}{2}\right) = \frac{1}{8}$

27. $(-1)^{17} = (-1)(-1)(-1)\cdots(-1)(-1) = -1$

29. $-(-1)^{13} = -(-1)(-1)(-1)\cdots(-1)(-1)$
$= -(-1) = 1$

31. $-(-1)^{12} = -(-1)(-1)(-1)(-1)(-1)\cdots(-1)$
$= -1$

33. $(-1.2)^3 = (-1.2)(-1.2)(-1.2) = -1.728$

35. $\frac{1}{4^3} = \frac{1}{4 \cdot 4 \cdot 4} = \frac{1}{64}$

37. $\frac{-12}{4} = -12 \cdot \frac{1}{4} = -3$

39. $\frac{21}{-3} = 21\left(-\frac{1}{3}\right) = -7$

41. $\frac{-90}{-3} = -90\left(-\frac{1}{3}\right) = 30$

43. $\frac{0}{-7} = 0\left(-\frac{1}{7}\right) = 0$

45. $\frac{-7}{0}$ is undefined

47. $(-480) \div 24 = -480 \cdot \frac{1}{24} = -20$

49. $(465) \div (-15) = 465\left(-\frac{1}{15}\right) = -31$

51. $\frac{-15 \cdot 9}{0.003} = -5300$

53. $\frac{-8.25}{-0.05} = 165$

55. $4.06 \div (-0.7) = 4.06 \cdot \left(-\frac{1}{0.7}\right) = -5.8$

57. $-\frac{14}{9} \div \frac{7}{8} = -\frac{14}{9} \cdot \frac{8}{7}$
$= -\frac{16}{9}$ or $-1\frac{7}{9}$

59. $\frac{3}{8} \div \left(-\frac{3}{4}\right) = \frac{3}{8} \cdot \left(-\frac{4}{3}\right) = -\frac{1}{2}$

61. $-\frac{4}{3} \div \left(-\frac{16}{9}\right) = -\frac{4}{3} \cdot \left(-\frac{9}{16}\right)$
$= \frac{1}{3} \cdot \frac{9}{4} = \frac{3}{4}$

63. $0 \div \left(-\frac{3}{7}\right) = 0 \cdot \left(-\frac{7}{3}\right) = 0$

65. $-\frac{3}{7} \div 0 = -\frac{3}{7} \cdot \frac{1}{0}$ is undefined

67. $-\frac{5}{7} \div \left(-\frac{5}{7}\right) = -\frac{5}{7} \cdot \left(-\frac{7}{5}\right) = 1$

69. $-\frac{5}{7} \div \frac{5}{7} = -\frac{5}{7} \cdot \left(\frac{7}{5}\right) = -1$

71. $6 \div \left(-\frac{2}{5}\right) = 6 \cdot \left(-\frac{5}{2}\right) = -15$

73. $-1\frac{2}{3} \div \left(-\frac{2}{9}\right) = -\frac{5}{3} \cdot \left(-\frac{9}{2}\right)$
$= \frac{15}{2}$ or $7\frac{1}{2}$

75. $\frac{7}{12} \div (-7) = \frac{7}{12} \cdot \left(-\frac{1}{7}\right) = -\frac{1}{12}$

77. $\left(3 - 4\frac{1}{3}\right) \div \left(-\frac{2}{3} + \frac{5}{6}\right)$
$= \left(-1\frac{1}{3}\right) \div \left(\frac{1}{6}\right) = -\frac{4}{3} \cdot \frac{6}{1} = -8$

79. $6x^2 + 11x^2 = 17x^2$

81. $9x^3 - 4x^3 = 5x^3$

83. $7x^4 + x^4 = 8x^4$

85. $16x^2 - 17x^2 = -x^2$

87. $2x^2 + 2x^3$ cannot be simplified because exponents are not equal

89. $6x^2 - 6x^2 = 0$

91. $3x^2 + 4x^3 - 2x^2 - x^3$
$= 3x^2 - 2x^2 + 4x^3 - x^3$
$= x^2 + 3x^3$

93. $3x^2 = 3(17)^2 = 3(289) = 867$ grams

95. $200 - 16t^2$
$= 200 - 16(3)^2$
$= 200 - 16(9)$
$= 200 - 144 = 56$
56 feet

97. **a.** 1989: $x = 0$: $\frac{893.5}{0+14.2} \approx 62.9\%$
1990: $x = 1$: $\frac{893.5}{1+14.2} \approx 58.8\%$
1991: $x = 2$: $\frac{893.5}{2+14.2} \approx 55.2\%$
1992: $x = 3$: $\frac{893.5}{3+14.2} \approx 51.9\%$

b. It decreased.

c. $x = 2000 - 1989 = 11$
$\frac{893.5}{11+14.2} = \frac{893.5}{25.2} \approx 35.5\%$

99. $\frac{(-2)+(-7)+(-4)+1+(-13)+(-15)+(-26)+29+(-22)+(-10)+2}{11} = \frac{-67}{11} \approx -6.1\%$
On average, the percent change is –6.1%.

101. Total cost ÷ number of oranges
$\frac{5}{x}$ dollars

103. Height in inches ÷ 12 inches per foot
$\frac{c}{12}$ feet

105. Sum of items ÷ number of items
$\frac{12+x}{2}$

107. Length in centimeters
$\div$ 100 centimeters per meter
$\frac{x}{100}$ meters

109. b is true: Since a is negative and c is positive, then $a-c$ is negative. Since b is positive, then $\frac{a-c}{b}$ must be negative because a negative divided by a positive is a negative.

111. d is true, $(-4)^3 = -64,\ -4^3 = -64$

113. $\frac{600{,}000p}{100-p}$

$p = 90$: $\frac{600{,}000(90)}{100-90} = \$5{,}400{,}000$

$p = 95$: $\frac{600{,}000(95)}{100-95} = \$11{,}400{,}000$

$p = 98$: $\frac{600{,}000(98)}{100-98} = \$29{,}400{,}000$

$p = 99$: $\frac{600{,}000(99)}{100-99} = \$59{,}400{,}000$

$p = 99.9$: $\frac{600{,}000(99.9)}{100-99.9} = \$599{,}400{,}000$

$p = 99.999$:

$\frac{600{,}000(99.999)}{100-99.999} = \$59{,}999{,}400{,}000$

The cleanup cost soars upward as the percentage of polluting bacteria removed approaches 100%.

115. Answers may vary.

117. Total cost $\div$ number of calculators
$\frac{50}{x-1}$ dollars

119. Number of boxes $\div$ amount per worker
$x \text{ boxes} \div \left(\frac{1}{2} \text{ box / worker}\right) = 2x \text{ workers}$

121. t^2 meters

Review Problems

122. $\{x|x$ is a whole number less than 6 and a positive number$\}$
$\{1, 2, 3, 4, 5\}$

123. $12°F - (-16°F) = 12°F + 16°F = 28°F$

124. $\frac{5}{8} \cdot \frac{1}{3} = \frac{5 \cdot 1}{8 \cdot 3} = \frac{5}{24}$

Problem Set 1.9

1. $-45 \div 5 \cdot 3 = -9 \cdot 3 = -27$

3. $-3 + 5(1-4)^3 = -3 + 5(-3)^3$
$= -3 + 5(-27) = -3 - 135 = -138$

5. $16 - 2 \cdot 3 - 25 = 16 - 6 - 25 = 10 - 25 = -15$

7. $-12 \div 3 + 18 \div 9 = -4 + 2 = -2$

9. $6 + 12 - 12 \div 4 = 6 + 12 - 3 = 18 - 3 = 15$

11. $14 - 2 \cdot 5 - 20 = 14 - 10 - 20 = 4 - 20 = -16$

13. $(14-2) \cdot 5 - 20 = 12 \cdot 5 - 20 = 60 - 20 = 40$

15. $(-30) \div (-6)\left(-\frac{1}{3}\right) = -30 \div 2 = -15$

17. $\frac{10+8}{5^2 - 4^2} = \frac{18}{25-16} = \frac{18}{9} = 2$

19. $[2(6-2)]^2 = [2(4)]^2 = (8)^2 = 64$

21. $-8 + 4(3-5)^3 = -8 + 4(-2)^3$
$= -8 + 4(-8) = -8 - 32 = -40$

23. $36 - 24 \div 2^3 \cdot 3 - 1$
$= 36 - 24 \div 8 \cdot 3 - 1$
$= 36 - 3 \cdot 3 - 1 = 36 - 9 - 1 = 26$

25. $(15 - 3^3)^2 = (15-27)^2 = (-12)^2 = 144$

27. $16 - (-3)(-12) \div 9$
$= 16 - (36) \div 9 = 16 - 4 = 12$

29. $[7+3(2^3-1)]\div 21$
$=[7+3(8-1)]\div 21$
$=[7+3(7)]\div 21$
$=(7+21)\div 21$
$=28\div 21=28\cdot\frac{1}{21}=\frac{4}{3}$ or $1\frac{1}{3}$

31. $\frac{37+15\div(-3)}{16}=\frac{37+(-5)}{16}=\frac{32}{16}=2$

33. $\frac{3}{5}\left(\frac{2}{3}-\frac{3}{4}\right)=\frac{3}{5}\left(\frac{8-9}{12}\right)=\frac{3}{5}\left(-\frac{1}{12}\right)=-\frac{1}{20}$

35. $4(3-6)^2-2(3-4)=4(-3)^2-2(-1)$
$=4(9)+2=36+2=38$

37. $\frac{5(4-6)}{2}-\frac{27}{-3}=\frac{5(-2)}{2}+9=-5+9=4$

39. $5-5\div 5\cdot 5-5^2$
$=5-5\div 5\cdot 5-25$
$=5-1\cdot 5-25$
$=5-5-25=-25$

41. $\frac{4^2-3^2}{(4-3)^2}=\frac{16-9}{1^2}=\frac{7}{1}=7$

43. $3(-2)^3-5(-2)+4$
$=3(-8)+10+4$
$=-24+10+4=-10$

45. $[5+3(-2)]^7=(5-6)^7=(-1)^7=-1$

47. $\left(\frac{3}{2}\right)^2\div\left(-\frac{3}{4}\right)=\frac{9}{4}\cdot\left(-\frac{4}{3}\right)=-3$

49. $6(6-7)^3-9(3-6)^2$
$=6(-1)^3-9(-3)^2$
$=6(-1)-9(9)=-6-81=-87$

51. $\frac{(-11)(-4)+2(-7)}{7-(-3)}=\frac{44-14}{7+3}=\frac{30}{10}=3$

53. $-2^2+4[16\div(3-5)]$
$=-4+4[16\div(-2)]$
$=-4+4(-8)=-4-32=-36$

55. $24\div\frac{3^2}{8-5}-(-6)$
$=24\div\frac{9}{3}+6=24\div 3+6=8+6=14$

57. $5(x-3)+2=5x-15+2=5x-13$

59. $-3[5(x-3)+2]$
$=-3(5x-15+2)$
$=-3(5x-13)$
$=-15x+39$

61. $3[6-(y+1)]=3(6-y-1)$
$=3(5-y)=15-3y$

63. $7-4[3-(-4y-5)]$
$=7-4(3+4y+5)$
$=7-4(8+4y)$
$=7-32-16y$
$=-25-16y$

65. $3[6x-2y-4(-5x-y)]$
$=3(6x-2y+20x+4y)$
$=3(26x+2y)=78x+6y$

67. $12x-4[6x-8y-(2x-4y)]$
$=12x-4(6x-8y-2x+4y)$
$=12x-4(4x-4y)$
$=12x-16x+16y$
$=-4x+16y$

69. $H=0.8(200-A)$
$=0.8(200-150)$ substitute 150 for A
$=0.8(50)$
$=40$ The handicap is 40.
$150+40=190$. The final score is 190.

71. $M = \frac{2x}{1-x}$

Percentage of Contaminant Removed	$M = \frac{2x}{1-x}$	Required Amount of Money in Monetary Pool
50% ($x = 0.5$)	$M = \frac{2(0.5)}{1-0.5} = \frac{1}{0.5} = 2$	\$2 million
60% ($x = 0.6$)	$M = \frac{2(0.6)}{1-0.6} = \frac{1.2}{0.4} = 3$	\$3 million
70% ($x = 0.7$)	$M = \frac{2(0.7)}{1-0.7} = \frac{1.4}{0.3} = 4.\overline{6}$	\$4.$\overline{6}$ million
80% ($x = 0.8$)	$M = \frac{2(0.8)}{1-0.8} = \frac{1.6}{0.2} = 8$	\$8 million
90% ($x = 0.9$)	$M = \frac{2(0.9)}{1-0.9} = \frac{1.8}{0.1} = 18$	\$18 million
95% ($x = 0.95$)	$M = \frac{2(0.95)}{1-0.95} = \frac{1.9}{0.05} = 38$	\$38 million
99% ($x = 0.99$)	$M = \frac{2(0.99)}{1-0.99} = \frac{1.98}{0.01} = 198$	\$198 million

As the percent of contaminant gets closer and closer to 100% the cost increases rapidly.

73. $D = \frac{14,400}{x^2 + 10x}$

For $x = 10$,

$$D = \frac{14,400}{(10)^2 + 10(10)}$$
$$= \frac{14,400}{100 + 100}$$
$$= \frac{14,400}{200}$$

= 72 people at \$10/calculator

For $x = 15$,

$$D = \frac{14,400}{15^2 + 10(15)}$$
$$= \frac{14,400}{225 + 150}$$
$$= \frac{14,400}{375}$$

= 38.4 people at \$15/calculator

The difference is $72 - 38.4 = 33.6 \approx 34$.

34 more people are willing to purchase calculators at a price of \$10 than at a price of \$15.

75. $C = 469x - 1700$
$C = 469(5) - 1700 = 645$
$C = 469(6) - 1700 = 1114$
$C = 469(7) - 1700 = 1583$
$T = -82x + 1972$
$T = -82(5) + 1972 = 1562$
$T = -82(6) + 1972 = 1480$
$T = -82(7) + 1972 = 1398$
The sales of compact discs are increasing, while the sales of turntables are decreasing. The graphs match the results.

77. $y = 0.0002x^3 - 0.02x^2 + 3x + 151$
$x = 0,$
$y = 0.0002(0)^3 - 0.02(0)^2 + 3(0) + 151 =$
151 million
$x = 10,$
$y = 0.0002(10)^3 - 0.02(10)^2 + 3(10) + 151$
$= 179.2$ million
$x = 20,$
$y = 0.0002(20)^3 - 0.02(20)^2 + 3(20) + 151$
$= 204.6$ million
$x = 30,$
$y = 0.0002(30)^3 - 0.02(30)^2 + 3(30) + 151$
$= 228.4$ million
$x = 40,$
$y = 0.0002(40)^3 - 0.02(40)^2 + 3(40) + 151$
$= 251.8$ million
They are very close.

79. $C = \frac{8x + 5000}{x}$
$x = 100,\ C = \frac{8(100) + 5000}{100} = \$58$
$x = 200,\ C = \frac{8(200) + 5000}{200} = \$33$
$x = 300,\ C = \frac{8(300) + 5000}{300} = \$24.67$
$x = 400,\ C = \frac{8(400) + 5000}{400} = \$20.50$
$x = 500,\ C = \frac{8(500) + 5000}{500} = \$18$
Yes; the cost decreases.

81. d is true;
$\frac{|3-7| - 2^3}{(-2)(-3)} = \frac{|-4| - 8}{6} = \frac{4-8}{6} = -\frac{4}{6} = -\frac{2}{3}$
and $-\frac{1}{3} - \frac{1}{3} = -\frac{2}{3}$.

83. a. $y = 1.113x - 1.5239$
They are very close.

b.

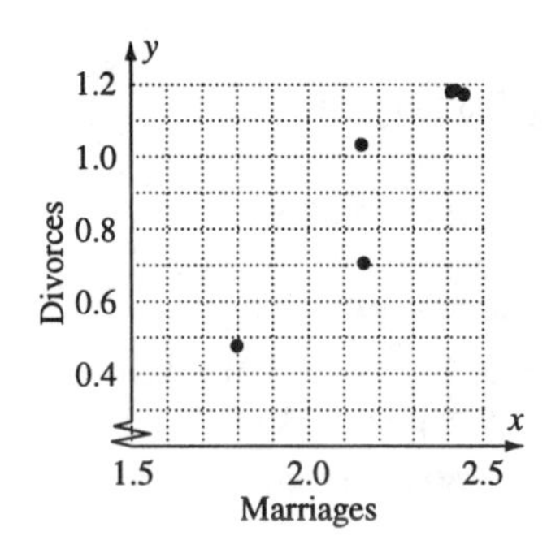

As the number of marriages increase, the number of divorces increase.

85. $A = P\left(1 + \frac{r}{n}\right)^{nt}$
$A = 5000\left(1 + \frac{0.07}{n}\right)^{10n}$

a. $A = 5000\left(1 + \frac{0.07}{1}\right)^{10 \cdot 1} = \$9835.76$

b. $A = 5000\left(1 + \frac{0.07}{2}\right)^{10 \cdot 2} = \$9948.94$

c. $A = 5000\left(1 + \frac{0.07}{4}\right)^{10 \cdot 4} = \$10,007.99$

d. $A = 5000\left(1 + \frac{0.07}{12}\right)^{10 \cdot 12} = \$10,048.31$

e. $A = 5000\left(1 + \frac{0.07}{365}\right)^{10 \cdot 365} = \$10,068.09$

87. Answers may vary.

89. $Q = ac = (-1)(-2) = 2 \qquad a = -1,\ c = -2$
$M = Q^2 - a^2 = 2^2 - (-1) = 3$
$P = \frac{1}{3}a^2M = \frac{1}{3}(-1)^2(3) = 1$

91. $1 \div 2 + 3(4 \times 5) = 60\frac{1}{2}$

Review Problems

93. $\frac{5}{12} + \frac{7}{9} = \frac{5}{12} \cdot \frac{3}{3} + \frac{7}{9} \cdot \frac{4}{4} = \frac{15}{36} + \frac{28}{36} = \frac{43}{36}$

94. $-68 - (-20) = \ 68 + 20 = -48°\text{F}$

95. $t = 2000 - 1980 = 20$
$P = 2(20)^2 + 22(20) + 320$
$P = 800 + 440 + 320$
$P = 1560$ thousand

Chapter 1 Review

1. $\frac{9}{10} \cdot \frac{18}{7} = \frac{9 \cdot 2 \cdot 9}{2 \cdot 5 \cdot 7} = \frac{81}{35}$

2. $\frac{15}{32} \div 5 = \frac{15}{32} \cdot \frac{1}{5} = \frac{3 \cdot 5}{32 \cdot 5} = \frac{3}{32}$

3. $\frac{7}{9} + \frac{5}{12} = \frac{7}{9} \cdot \frac{4}{4} + \frac{5}{12} \cdot \frac{3}{3}$
$= \frac{28}{36} + \frac{15}{36} = \frac{43}{36}$

4. $\frac{3}{4} - \frac{2}{15} = \frac{3}{4} \cdot \frac{15}{15} - \frac{2}{15} \cdot \frac{4}{4}$
$= \frac{45}{60} - \frac{8}{60} = \frac{37}{60}$

5. $5\frac{3}{4} - 3\frac{5}{8} = \frac{23}{4} - \frac{29}{8}$
$= \frac{23}{4} \cdot \frac{2}{2} - \frac{29}{8}$
$= \frac{46}{8} - \frac{29}{8}$
$= \frac{17}{8}$ or $2\frac{1}{8}$

6. $\frac{20}{36} = \frac{5}{9};\ 0.\overline{5};\ 55.56\%$

7. $\{x \mid x$ is a whole number that is less than $6\}$
$\{0, 1, 2, 3, 4, 5\}$

8. $\{x \mid x$ is an integer that is greater than $-3\}$
$\{-2, -1, 0, 1, 2, \ldots\}$

9. a. Natural numbers: $\{\sqrt{81}\}$

b. Whole numbers: $\{0,\ \sqrt{81}\}$

c. Integers: $\{-17,\ 0,\ \sqrt{81}\}$

d. Rational numbers:
$\{-17,\ -\frac{9}{13},\ 0,\ 0.75,\ 5\frac{1}{4},\ \sqrt{81}\}$

e. Irrational numbers:
$\{\sqrt{2},\ \pi\}$

f. Real numbers:
$\{-17,\ -\frac{9}{13},\ 0,\ 0.75,\ \sqrt{2},\ \pi,$
$5\frac{1}{4},\ \sqrt{81}\}$

10. $17 > 5$

11. $-|-3.2|\ \square\ -(-3.2)$
$-3.2 < 3.2$

12. $0 > -\frac{1}{3}$

13. $-\frac{1}{4} < -\frac{1}{5}$

14. $(0.11 - 0.02)(880$ million$) = 79.2$ million

15. a. No; The homeless are not represented.

b. $\frac{10.7}{25.5 + 10.7 + 6.1 + 64.3} = 10\%$

c. Multiple family = 24%
Single family detached = 60%
Single attached = 6%

16. a. 32%

b. Stroke

c. Accidents, Pneumonia/Influenza, Diabetes, AIDS, Suicide, and Liver Ailments

17. a. 23%

b. 0.32(400) = 128

c. Walking, Bicycling, Camping

18. a. 1981; 6 crimes per 1000 population

b. 1976, 1978, 1983, 1985

c. 1974, 1977, 1981, 1986, 1991, 1992

d. 1984, 5.1 crimes per 1000 population

e. 1988, 5.7 crimes per 1000 population

f. When unemployment is down, crime is up.

19. Quadrant IV

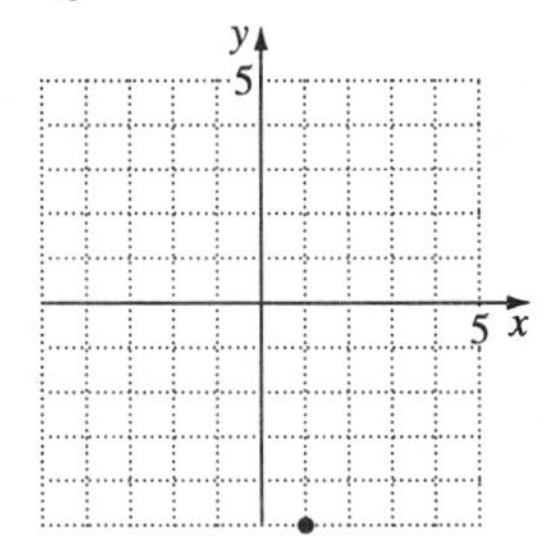

20. Quadrant IV

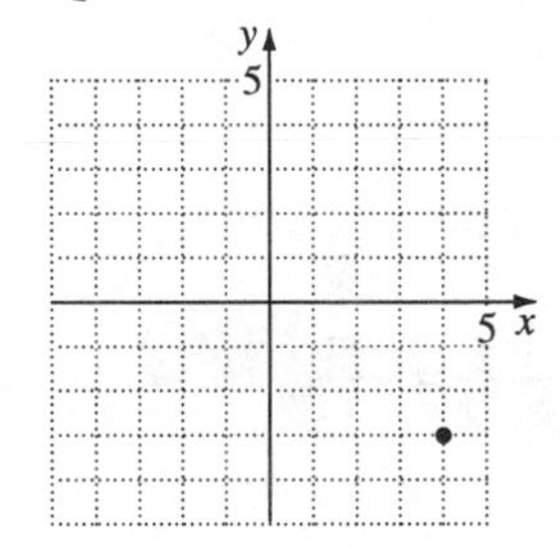

21. Quadrant I

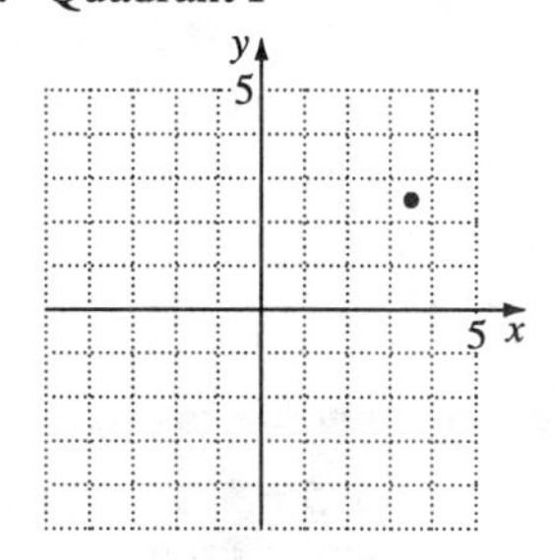

22. Quadrant II

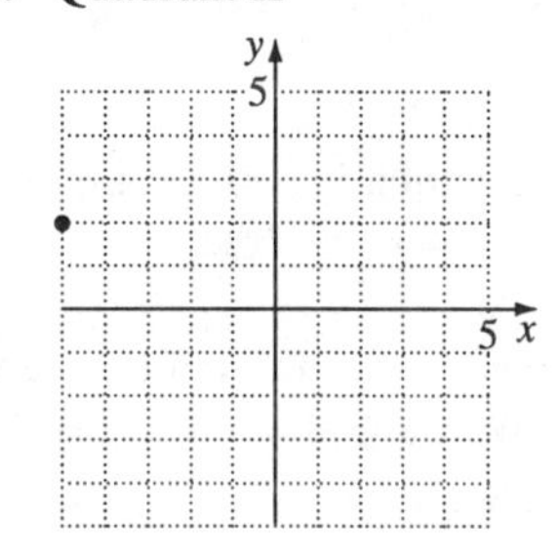

23. $A(5, 6)$
$B(-2, 0)$
$C(-5, 2)$
$D(-4, -2)$
$E(0, -5)$
$F(3, -1)$

24.

Except for two outliers, as wine consumption increases, deaths due to heart disease decreases.

25. $3x + 5 = 5 + 3x$

26. $t(1.24) + 313.6 = 1.24t + 313.6$

28. $-3(5x) = (-3 \cdot 5)x = -15x$

29. a. $100(2x + 6) - 41.5x - 25$
$= 200x + 600 - 41.5x - 25$
$= 200x - 41.5x + 600 - 25$
$= 158.5x + 575$

b. $x = 0$: $158.5(0) + 575 = \$575$ thousand
$x = 1$: $158.5(1) + 575 = \$733.5$ thousand
$x = 2$: $158.5(2) + 575 = \$892$ thousand
$158.5 thousand

30. $-1312 + 512 = -800$
The person's elevation is 800 feet below sea level.

31. Elevation of plane – elevation of submarine
$26{,}500 - (-650) = 26{,}500 + 650 = 27{,}150$
The difference in elevation is 27,150 feet.

32. $1821 + 5400 - 4330 - 1714 = 1177$

33. a. $34 - 8 = 26$; $a + 26$

b. $99 - 69 = 30$; $b - 30$

c. $\frac{95}{5} = 19$; $19c$

d. The sum of *d* and 16
16 more than *d*
16 added to *d*

e. *e* minus 38
e decreased by 38
38 less than *e*

34. a. $45 - (-5) = 45 + 5 = 50°\text{F}$

b. $-12 + 26 = 14°\text{F}$

c. $\frac{-12 + (-5) + 6 + 26 + 45 + 60}{6}$
$= \frac{120}{6} = 20°\text{F}$

35. $5x$

36. $2(x + 1) + 2(x + 2)$
$= 2x + 2 + 2x + 4 = 4x + 6$

37. $39.05 + 0.45x$

38. $150 + 10x$

39. $x - 0.35x = 0.65x$

40. $\frac{x}{12}$

41. $\frac{20}{x}$

42. $\frac{6}{x}$

43. $8 + (-11) = 8 - 11 = -3$

44. $-\frac{3}{4} + \frac{1}{5} = -\frac{15}{20} + \frac{4}{20} = -\frac{11}{20}$

45. $7 + (-5) + (-13) + 4$
$= 7 + 4 + (-5) + (-13)$
$= 11 + (-18) = -7$

46. $-7.8 + 4.1 + 13 + (-5.2)$
$= -7.8 + (-5.2) + 4.1 + 13$
$= -13 + 17.1 = 4.1$

47. $-9 - (-13) = -9 + 13 = 4$

48. $-7 - (-5) + 11 - 16 = -7 + 5 + 11 - 16 = -7$

49. $-\frac{3}{5} - \frac{9}{10} = -\frac{6}{10} - \frac{9}{10}$
$-\frac{15}{10} = -\frac{3}{2}$ or $-1\frac{1}{2}$

50. $-7(-12) = 84$

51. $-2.3(4.5) = -10.35$

52. $\frac{3}{5}\left(-\frac{5}{11}\right) = -\frac{3 \cdot \cancel{5}}{\cancel{5} \cdot 11} = -\frac{3}{11}$

53. $5(-3)(-2)(-4)$
$= -15(-2)(-4) = 30(-4) = -120$

54. $-3\left(-\frac{1}{6}\right)(40) = \frac{1}{2}(40) = 20$

55. $(-4)^2 = 16$

56. $(-2)^5 = -32$

57. $\left(-\frac{2}{3}\right)^2 = \frac{4}{9}$

58. $45 \div (-5) = -\frac{45}{5} = -9$

59. $-\frac{4}{5} \div \left(-\frac{2}{5}\right) = -\frac{4}{5} \cdot \left(-\frac{5}{2}\right) = \frac{20}{10} = 2$

60. $\frac{-25}{0.05} = -500$

61. $-40 \div 5 \cdot 2 = -8 \cdot 2 = -16$

62. $-3 + 4(4-7)^3$
$= -3 + 4(-3)^3$
$= -3 + 4(-27)$
$= -3 - 108 = -111$

63. $16 \div 4^2 - 2 = 16 \div 16 - 2 = 1 - 2 = -1$

64. $(16 \div 4)^2 - 2 = (4)^2 - 2 = 16 - 2 = 14$

65. $16 \div (4^2 - 2) = 16 \div (16 - 2)$
$= 16 \div 14 = \frac{16}{14} = \frac{8}{7}$ or $1\frac{1}{7}$

66. $(-10)(-6) - (-8)(4)$
$= 60 - (-32) = 60 + 32 = 92$

67. $-8[-4 - 5(-3)] = -8(-4 + 15) = -8(11) = -88$

68. $\frac{6(-10+3)}{2(-15) - 9(-3)} = \frac{6(-7)}{-30 + 27} = \frac{-42}{-3} = 14$

69. $8^2 - 36 \div 3^2 \cdot 4 - (-7)$
$= 64 - 36 \div 9 \cdot 4 + 7$
$= 64 - 4 \cdot 4 + 7$
$= 64 - 16 + 7 = 55$

70. $3 + (9 \div 3)^3 - 25 \div 5 \cdot 2 - (4-1)^3$
$= 3 + 3^3 - 5 \cdot 2 - 3^3$
$= 3 + 27 - 10 - 27 = -7$

71. $\frac{5}{12} - \frac{11}{12} \div \left(\frac{1}{6} - \frac{3}{8}\right)$
$= \frac{5}{12} - \frac{11}{12} \div \left(-\frac{5}{24}\right)$
$= \frac{5}{12} - \frac{11}{12} \cdot \left(-\frac{24}{5}\right)$
$= \frac{5}{12} + \frac{22}{5} = \frac{25 + 264}{60} = \frac{289}{60}$

72. $11x + 7y + (-13x) + (-6y)$
$= 11x + (-13x) + 7y + (-6y)$
$= -2x + y$

73. $-13a - 7b - (-5a) + 13b$
$= -13a - 7b + 5a + 13b$
$= -13a + 5a - 7b + 13b$
$= -8a + 6b$

74. $3(x + 5) - 7x$
$= 3x + 15 - 7x$
$= -4x + 15$

75. $-6(3x - 4) = -18x + 24$

76. $7(2y - 5) - (15y - 2)$
$= 14y - 35 - 15y + 2$
$= -y - 33$

77. $4(-5x - 1) - 3(6x - 1)$
$= -20x - 4 - 18x + 3$
$= -38x - 1$

78. $\frac{1}{6}(6x - 6) = x - 1$

79. $2[7x - 3(2x - 1)]$
$= 2(7x - 6x + 3)$
$= 2(x + 3) = 2x + 6$

80. $[3(x + 5) - 7] - [2(x - 1) + 5]$
$= (3x + 15 - 7) - (2x - 2 + 5)$
$= (3x + 8) - (2x + 3)$
$= 3x + 8 - 2x - 3 = x + 5$

81. $D = -1545.5x + 49{,}391$

1988: $x = 0$ $D = -1545.5(0) + 49{,}391 = 49{,}391$

1989: $x = 1$ $D = -1545.5(1) + 49{,}391 = 47{,}845.5$

1990: $x = 2$ $D = -1545.5(2) + 49{,}391 = 46{,}300$

The number of deaths is decreasing.

82. $t = \sqrt{\frac{h}{16}}$

$t = \sqrt{\frac{144}{16}} = \sqrt{9} = 3$

3 seconds

83. $D = 9.2t^2 - 46.7t + 480$

1980: $x = 10$ $D = 9.2(10)^2 - 46.7(10) + 480 = \$933$ billion

1990: $x = 20$ $D = 9.2(20)^2 - 46.7(20) + 480 = \$3226$ billion

They are very close.

84. **a.** $y = 12x - x^2$

x	0	1	2	3	4	5	6	7	8	9	10	11	12
y	0	11	20	27	32	35	36	35	32	27	20	11	0

b.

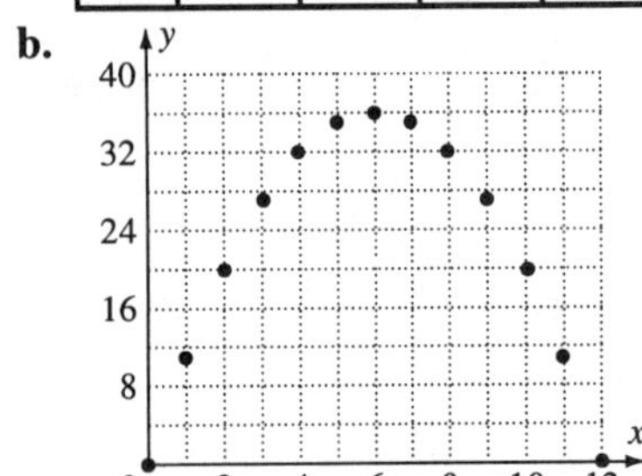

They are very similar.

85. $P = 1.7t + 230$

a. $t = 1990 - 1980 = 10$

$P = 1.7(10) + 230 = 247$ million

b. 2000: $t = 20$ $P = 1.7(20) + 230 = 264$ million

2010: $t = 30$ $P = 1.7(30) + 230 = 281$ million

2020: $t = 40$ $P = 1.7(40) + 230 = 298$ million

2030: $t = 50$ $P = 1.7(50) + 230 = 315$ million

2040: $t = 60$ $P = 1.7(60) + 230 = 332$ million

2050: $t = 70$ $P = 1.7(70) + 230 = 349$ million

c. Medium projection

86. $T = 3(A-20)^2 \div 50 + 10$

Time for 40-year-old:
$$\begin{aligned} T &= 3(40-20)^2 \div 50 + 10 \\ &= 3(20)^2 \div 50 + 10 \\ &= 3(400) \div 50 + 10 \\ &= 1{,}200 \div 50 + 10 \\ &= 24 + 10 = 34 \end{aligned}$$

Time for 30-year-old:
$$\begin{aligned} T &= 3(30-20)^2 \div 50 + 10 \\ &= 3(10)^2 \div 50 + 10 \\ &= 3(100) \div 50 + 10 \\ &= 300 \div 50 + 10 \\ &= 6 + 10 = 16 \end{aligned}$$

difference in time: 34 – 16 = 18

The difference in time is 18 seconds.

87. $N = -45t^4 + 446t^3 - 517t^2 + 2026t + 984$

1982: $t = 0$; $N = 984$

1983: $t = 1$; $N = 2894$

1984: $t = 2$; $N = 5816$

1985; $t = 3$; $N = 10{,}806$

1986; $t = 4$; $N = 17{,}840$

1987; $t = 5$; $N = 25{,}814$

1988; $t = 6$; $N = 32{,}544$

1989; $t = 7$; $N = 34{,}766$

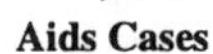

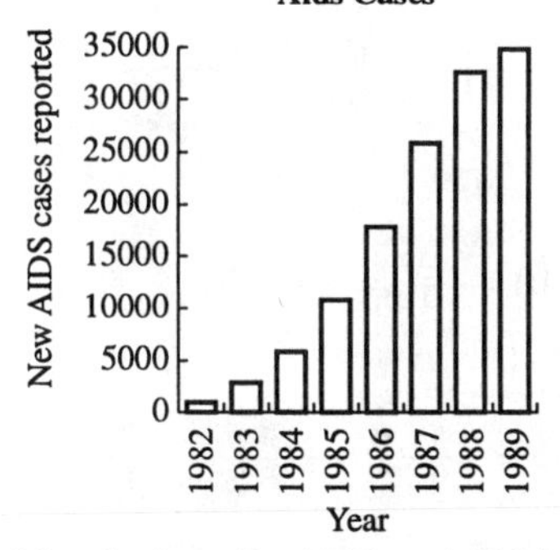

No; the bars for 1983 and 1984 in the text are low.

Chapter 1 Test

1. $\{-5, -4, -3, -2, -1\}$

2. Rational numbers can be written as the quotient of two integers.
$-7 = -\frac{7}{1}$, $-\frac{4}{5} = -\frac{4}{5}$, $0 = \frac{0}{1}$, $0.25 = \frac{1}{4}$,
$\sqrt{4} = 2 = \frac{2}{1}$, $\frac{22}{7} = \frac{22}{7}$
so -7, $-\frac{4}{5}$, 0, 0.25, $\sqrt{4}$, $\frac{22}{7}$ are the rational numbers of the set.

3. 16% of America's population in the year 2050 will be African-American.
0.16×390 million
= 62.4 million African-Americans

4. Total Revenue (in millions of dollars):
$135 + 120 + 50 + 45 + 45 + 40 + 30 + 35$
$= 500$
Computer products: $\frac{135}{500} = 0.27$
27% of the 1996 purchases on computer Web sites were for computer products.

5. Anxiety, headache, sprains/strains

6. **a.** 1991, 15 million crimes

   b. 1984

7. $(-5, -2)$

8. 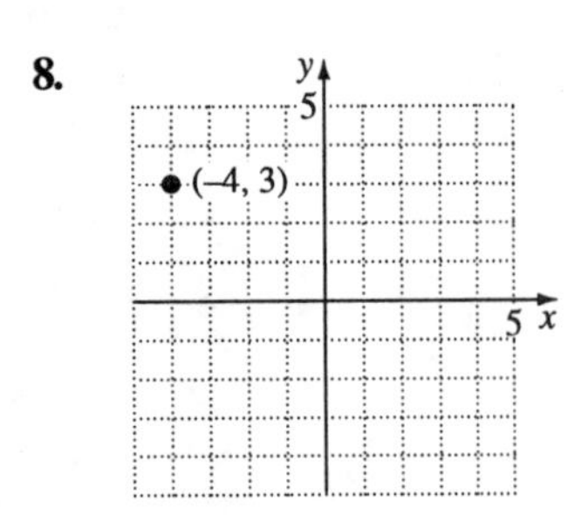

9. $8x^2 + 7x = 7x + 8x^2$

10. $x \cdot (5 \cdot y) = (x \cdot 5) \cdot y$

11. 16,200 feet above sea level: +16,200
830 feet below sea level: –830
Difference: $+16{,}200 - (-830)$
$= 16{,}200 + 830 = 17{,}030$ feet

12. Triple the product of the height and width
$3 \cdot (h \cdot w) = 3hw$

13. Perimeter = 2 (length) + 2(width)
$= 2(2x + 1) + 2(x)$
$= 4x + 2 + 2x$
$= 6x + 2$

14. $5.3 - (-9.2) = 5.3 + 9.2 = 14.5$

15. $-9 + 3 + (-11) + 6$
$= -6 + (-11) + 6$
$= -17 + 6 = -11$

16. $-\frac{2}{11} + \frac{5}{44} = \left(-\frac{2}{11} \cdot \frac{4}{4}\right) + \frac{5}{44} = \frac{-8}{44} + \frac{5}{44}$
$= \frac{-8+5}{44} = \frac{-3}{44} = -\frac{3}{44}$

17. $7.3 - 10.4 = -3.1$

18. $3(-17) = -3(17) = -51$

19. $\left(-\frac{3}{7}\right)\left(-\frac{7}{15}\right) = \left(\frac{3}{7}\right)\left(\frac{7}{15}\right) = \frac{\overset{1}{\cancel{3}} \cdot \overset{1}{\cancel{7}}}{\underset{1}{\cancel{7}} \cdot \underset{5}{\cancel{15}}} = \frac{1}{5}$

20. $-50 \div 10 = -\frac{\overset{5}{5\cancel{0}}}{\underset{1}{1\cancel{0}}} = -\frac{5}{1} = -5$

21. $-\frac{5}{9} \div \frac{2}{5} = -\frac{5}{9} \cdot \frac{5}{2} = -\frac{5 \cdot 5}{9 \cdot 2} = -\frac{25}{18}$

22. $-6 - (5 - 12) = -6 - (-7) = -6 + 7 = 1$

23. $(-3)(4) - (2)(6) = -12 - 12 = -24$

24. $\frac{3(-2) - 2(2)}{-2(8-3)} = \frac{-6-4}{-2(5)} = \frac{-10}{-10} = 1$

25. $(6-8)^2(5-7)^3 = (-2)^2(-2)^3 = (-2)^5$
$= (-2)(-2)(-2)(-2)(-2) = -32$

26. $11x - (7x - 4) = 11x - 7x + 4 = 4x + 4$

27. $9(-3x - 2) - 4(5x - 3)$
$= -27x - 18 - 20x + 12$
$= -47x - 6$

28. $6 - 2[3(x + 1) - 5] = 6 - 2[3x + 3 - 5]$
$= 6 - 2[3x - 2] = 6 - 6x + 4 = -6x + 10$

29. 1994 is ten years after 1984 so $t = 10$.
$S = 91t + 164 = 91(10) + 164$
$= 910 + 164 = 1074$
The 1994 average annual salary is \$1074 thousand.

30. 1970 is ten years after 1960 so $t = 10$.
$C = -3.1t^2 + 51.4t + 4024$
$= -3.1(10)^2 + 51.4(10) + 4024$
$= -3.1(100) + 514 + 4024$
$= -310 + 514 + 4024 = 4228$
The 1970 average annual consumption of cigarettes is 4228.

Chapter 2

Problem Set 2.1

1. $x-7=13$
$x-7+7=13+7$
$x+0=20$
$x=20$
Check: $x-7=13$
$20-7=13$
$13=13$ √
The solution is 20.

3. $z+5=-12$
$z+5-5=-12-5$
$z=-17$
Check: $z+5=-12$
$-17+5=-12$
$-12=-12$ √
The solution is –17.

5. $-3=x+14$
$-3-14=x+14-14$
$-17=x$
Check: $-3=-17+14$
$-3=-3$ √
The solution is –17.

7. $-18=y-5$
$-18+5=y-5+5$
$-13=y$
Check: $-18=-13-5$
$-18=-18$ √
The solution is –13.

9. $7+z=13$
$z=13-7$
$z=6$
Check: $7+6=13$
$13=13$ √
The solution is 6.

11. $-3+y=-17$
$y=-17+3$
$y=-14$
Check: $-3-14=-17$
$-17=-17$ √
The solution is –14.

13. $x+\frac{1}{3}=\frac{7}{3}$
$x=\frac{7}{3}-\frac{1}{3}$
$x=\frac{6}{3}$
$x=2$
Check: $2+\frac{1}{3}=\frac{7}{3}$
$\frac{6}{3}+\frac{1}{3}=\frac{7}{3}$
$\frac{7}{3}=\frac{7}{3}$ √
The solution is 2.

15. $t+\frac{5}{6}=-\frac{7}{12}$
$t=-\frac{7}{12}-\frac{5}{6}$
$t=-\frac{7}{12}-\frac{10}{12}$
$t=-\frac{17}{12}$
Check: $-\frac{17}{12}+\frac{5}{6}=-\frac{7}{12}$
$-\frac{17}{12}+\frac{10}{12}=-\frac{7}{12}$
$-\frac{7}{12}=-\frac{7}{12}$ √
The solution is $-\frac{17}{12}$.

17. $x-\frac{3}{4}=\frac{9}{2}$
$x=\frac{9}{2}+\frac{3}{4}$
$x=\frac{18}{4}+\frac{3}{4}$
The solution is $x=\frac{21}{4}$
Check: $\frac{21}{4}-\frac{3}{4}=\frac{9}{2}$

$\frac{18}{4} = \frac{9}{2}$
$\frac{9}{2} = \frac{9}{2}$ √
The solution is $\frac{21}{4}$.

19. $-\frac{1}{5} + y = -\frac{3}{4}$
$y = -\frac{3}{4} + \frac{1}{5}$
$y = -\frac{15}{20} + \frac{4}{20}$
$y = -\frac{11}{20}$
Check: $-\frac{1}{5} - \frac{11}{20} = -\frac{3}{4}$
$-\frac{4}{20} - \frac{11}{20} = -\frac{3}{4}$
$-\frac{15}{20} = -\frac{3}{4}$
$-\frac{3}{4} = -\frac{3}{4}$ √
The solution is $-\frac{11}{20}$.

21. $3.2 + x = 7.5$
$3.2 + x - 3.2 = 7.5 - 3.2$
$x = 4.3$
Check: $3.2 + 4.3 = 7.5$
$7.5 = 7.5$ √
The solution is 4.3.

23. $x - \frac{3}{4} = \frac{9}{2}$
$x - \frac{3}{4} + \frac{3}{4} = \frac{9}{2} + \frac{3}{4}$
$x = \frac{21}{4}$
Check: $\frac{21}{4} - \frac{3}{4} = \frac{9}{2}$
$\frac{18}{4} = \frac{9}{2}$
$\frac{9}{2} = \frac{9}{2}$ √
The solution is $\frac{21}{4}$.

25. $5 = -13 + y$
$5 + 13 = y$
$18 = y$
Check: $5 = -13 + 18$
$5 = 5$ √
The solution is 18.

27. $-\frac{3}{5} = -\frac{3}{2} + s$
$-\frac{3}{5} + \frac{3}{2} = s$
$-\frac{6}{10} + \frac{15}{10} = s$
$\frac{9}{10} = s$
Check: $-\frac{3}{5} = -\frac{3}{2} + \frac{9}{10}$
$-\frac{6}{10} = -\frac{15}{10} + \frac{9}{10}$
$-\frac{6}{10} = -\frac{6}{10}$ √
The solution is $\frac{9}{10}$.

29. $830 + y = 520$
$y = 520 - 830$
$y = -310$
Check: $830 - 310 = 520$
$520 = 520$ √
The solution is –310.

31. $r + 3.7 = 8$
$r = 8 - 3.7$
$r = 4.3$
Check: $4.3 + 3.7 = 8$
$8 = 8$ √
The solution is 4.3.

33. $3\frac{2}{5} + x = 5\frac{2}{5}$
$x = 5\frac{2}{5} - 3\frac{2}{5}$
$x = 2$
Check: $3\frac{2}{5} + 2 = 5\frac{2}{5}$
$5\frac{2}{5} = 5\frac{2}{5}$ √
The solution is 2.

35. $-3.7 + m = -3.7$
$m = -3.7 + 3.7$
$m = 0$
Check: $-3.7 + 0 = -3.7$

$-3.7 = -3.7$ √
The solution is 0.

37. $6y + 3 - 5y = 14$
$y + 3 = 14$
$y = 14 - 3$
$y = 11$
Check: $6(11) + 3 - 5(11) = 14$
$66 + 3 - 55 = 14$
$14 = 14$ √
The solution is 11.

39. $7 - 5x + 8 + 2x + 4x - 3 = 2 + 3 \cdot 5$
$x + 12 = 2 + 15$
$x = 17 - 12$
$x = 5$
Check: $7 - 5(5) + 8 + 2(5) + 4(5) - 3$
$= 2 + 3 \cdot 5$
$7 - 25 + 8 + 10 + 20 - 3 = 2 + 15$
$45 - 28 = 17$
$17 = 17$ √
The solution is 5.

41. $7y + 4 = 6y - 9$
$7y - 6y + 4 = -9$
$y = -9 - 4$
$y = -13$
Check: $7(-13) + 4 = 6(-13) - 9$
$-91 + 4 = -78 - 9$
$-87 = -87$ √
The solution is -13.

43. $18 - 7x = 12 - 6x$
$18 = 12 + x$
$6 = x$
Check: $18 - 7(6) = 12 - 6(6)$
$18 - 42 = 12 - 36$
$-24 = -24$ √
The solution is 6.

45. a. $C + M = S$
$325 + M = 650$
$M = 650 - 325$
$M = 325$
\$325

b. $C + M = S$
$M = S - C$

47. a. $F - \frac{9}{5}C = 32$
$F = 32 + \frac{9}{5}C$

b. $F = 32 + \frac{9}{5}(0) = 32$
$F = 32 + \frac{9}{5}(5) = 41$
$F = 32 + \frac{9}{5}(10) = 50$
$F = 32 + \frac{9}{5}(15) = 59$
$F = 32 + \frac{9}{5}(20) = 68$
32°F, 41°F, 50°F, 59°F, 68°F

c. (0,32), (5, 41), (10, 50), (15, 59), (20, 68)

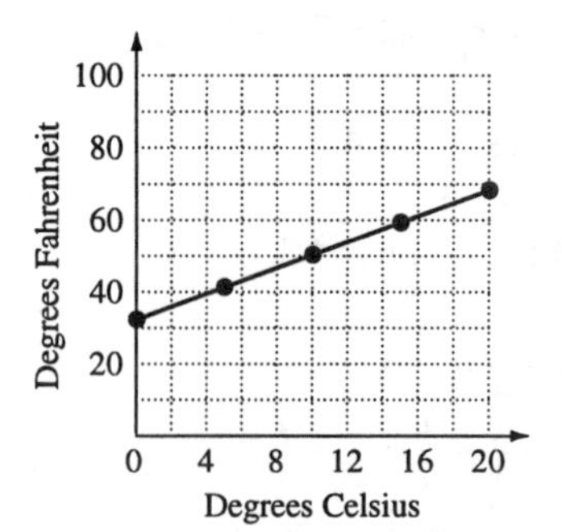

Linear relationship; as degrees Celsius increase, degrees Fahrenheit increase.

49. a. Let x = the salary in 1993.
$x + 68{,}425 - 115{,}100 = 1{,}073{,}579$
$x - 46{,}675 = 1{,}073{,}579$
$x = 1{,}073{,}579 + 46{,}675$
$x = 1{,}120{,}254$
\$1,120,254

b. The salary in 1993 plus the increase in salary from 1993 to 1994 (\$68,425) minus the decrease in salary from 1994 to 1995 ($115,100) equals the salary in 1995 ($1,073,579).

51. c is true.
$4 - x = -3x$
$4 - x + x = -3x + x$
$4 = -2x$
$-2 = x$

53. $17x - 0.12973 = 16x + 4.1$
$17x - 16x = 4.1 + 0.12973$
$x = 4.22973$

55–59. Answers may vary.

Review Problems

61. $-16 - (50 + 5^2) = -16 - (50 \div 25)$
$= -16 - 2 = -18$

62. $2(5 - 3y) - (2y - 4)$
$= 10 - 6y - 2y + 4$
$= -8y + 14$

63. $d = t^2 + 7t = 5^2 + 7(5) \qquad t = 5$ seconds
$= 25 + 35 = 60$
The shock wave travels 60 miles in 5 seconds.

Problem Set 2.2

1. $5x = 45$
$\frac{5x}{5} = \frac{45}{5}$
$x = 9$
Check: $5(9) = 45$
$45 = 45$ √
The solution is 9.

3. $7b = 56$
$\frac{7b}{7} = \frac{56}{7}$
$b = 8$
Check: $7(8) = 56$
$56 = 56$ √
The solution is 8.

5. $8r = -24$
$\frac{8r}{8} = -\frac{24}{8}$
$r = -3$
Check: $8(-3) = -24$
$-24 = -24$ √
The solution is -3.

7. $-3y = -15$
$\frac{-3y}{-3} = \frac{-15}{-3}$
$y = 5$
Check: $-3(5) = -15$
$-15 = -15$ √
The solution is 5.

9. $-8m = 2$
$\frac{-8m}{-8} = \frac{2}{-8}$
$m = -\frac{1}{4}$
Check: $-8\left(-\frac{1}{4}\right) = 2$
$2 = 2$ √
The solution is $-\frac{1}{4}$.

11. $7y = 0$
$\frac{7y}{7} = \frac{0}{7}$
$y = 0$
Check: $7(0) = 0$
$0 = 0$ √
The solution is 0.

13. $\frac{y}{3} = 4$
$3\left(\frac{y}{3}\right) = 3(4)$
$y = 12$
Check: $\frac{12}{3} = 4$
$4 = 4$ √
The solution is 12.

15. $-\frac{x}{5}=11$
$(-5)\left(-\frac{x}{5}\right)=(-5)(11)$
$x=-55$
Check: $-\frac{(-55)}{5}=11$
$11=11$ √
The solution –55.

17. $-\frac{x}{5}=-10$
$(-5)\left(-\frac{x}{5}\right)=(-5)(-10)$
$x=50$
Check: $-\frac{50}{5}=-10$
$-10=-10$ √
The solution is 50.

19. $\frac{2}{3}y=8$
$\frac{3}{2}\left(\frac{2}{3}y\right)=\frac{3}{2}(8)$
$y=12$
Check: $\frac{2}{3}(12)=8$
$8=8$ √
The solution is 12.

21. $-\frac{2}{5}a=\frac{6}{15}$
$-\frac{5}{2}\left(-\frac{2}{5}a\right)=-\frac{5}{2}\left(\frac{6}{15}\right)$
$a=-1$
Check: $-\frac{2}{5}(-1)=\frac{6}{15}$
$\frac{2}{5}=\frac{2}{5}$ √
The solution is –1.

23. $-\frac{7}{2}x=-21$
$-\frac{2}{7}\left(-\frac{7}{2}x\right)=-\frac{2}{7}(-21)$
$x=6$
Check: $-\frac{7}{2}(6)=-21$
$-21=-21$ √
The solution is 6.

25. $-r=7$
$(-1)(-r)=(-1)(7)$
$r=-7$
Check: $-(-7)=7$
$7=7$ √
The solution is –7.

27. $-15=-y$
$(-1)(-15)=(-1)(-y)$
$15=y$
Check: $-15=-(15)$
$-15=-15$ √
The solution is 15.

29. $-4y-2y=24$
$-6y=24$
$y=-4$
Check: $-4(-4)-2(-4)=24$
$16+8=24$
$24=24$ √
The solution is –4.

31. $5y+3y-4y=10+2$
$4y=12$
$y=3$
Check: $5(3)+3(3)-4(3)=10+2$
$15+9-12=12$
$12=12$ √
The solution is 3.

33. $-6-2=5y+3y-10y$
$-8=-2y$
$\frac{-8}{-2}=y$
$y=4$
Check: $-6-2=5(4)+3(4)-10(4)$
$-8=20+12-40$
$-8=-8$ √
The solution is 4.

35. $3y-2=9$
$3y=9+2$
$3y=11$
$y=\frac{11}{3}$

Check: $3\left(\frac{11}{3}\right)-2=9$
$11-2=9$
$9=9$ √
The solution is $\frac{11}{3}$.

37. $2a+1=7$
$2a=6$
$a=3$
Check: $2(3)+1=7$
$6+1=7$
$7=7$ √
The solution is 3.

39. $-2y+5=7$
$-2y=2$
$y=-1$
Check: $-2(-1)+5=7$
$2+5=7$
$7=7$ √
The solution is -1.

41. $-2y-5=7$
$-2y=7+5$
$-2y=12$
$y=-6$
Check: $-2(-6)-5=7$
$12-5=7$
$7=7$ √
The solution is -6.

43. $12=4m+3$
$12-3=4m$
$9=4m$
$\frac{9}{4}=m$
Check: $12=4\left(\frac{9}{4}\right)+3$
$12=9+3$
$12=12$ √
The solution is $\frac{9}{4}$.

45. $0.03x+21=27$
$0.03x=27-21$
$0.03x=6$
$x=\frac{6}{0.03}$
$x=200$
Check: $0.03(200)+21=27$
$6+21=27$
$27=27$ √
The solution is 200.

47. $-x-3=3$
$-x=3+3$
$-x=6$
$x=-6$
Check: $-(-6)-3=3$
$6-3=3$
$3=3$ √
The solution is -6.

49. $-x-\frac{1}{3}=\frac{2}{3}$
$-x=\frac{2}{3}+\frac{1}{3}$
$-x=1$
$x=-1$
Check: $-(-1)-\frac{1}{3}=\frac{2}{3}$
$1-\frac{1}{3}=\frac{2}{3}$
$\frac{2}{3}=\frac{2}{3}$ √
The solution is -1.

51. $6y=2y-12$
$6y-2y=-12$
$4y=-12$
$y=-3$
Check: $6(-3)=2(-3)-12$
$-18=-6-12$
$-18=-18$ √
The solution is -3.

53. $3x=-2x-15$
$3x+2x=-15$
$5x=-15$

$x = \frac{-15}{5}$
$x = -3$
Check: $3(-3) = -2(-3) - 15$
$-9 = 6 - 15$
$-9 = -9$ √
The solution is –3.

55. $-5y = -2y - 12$
$-5y + 2y = -12$
$-3y = -12$
$y = \frac{-12}{-3}$
$y = 4$
Check: $-5(4) = -2(4) - 12$
$-20 = -8 - 12$
$-20 = -20$ √
The solution is 4.

57. $8y + 4 = 2y - 5$
$8y - 2y + 4 = -5$
$6y + 4 = -5$
$6y = -5 - 4$
$6y = -9$
$y = -\frac{9}{6}$
$y = -\frac{3}{2}$
Check: $8\left(-\frac{3}{2}\right) + 4 = 2\left(-\frac{3}{2}\right) - 5$
$-12 + 4 = -3 - 5$
$-8 = -8$ √
The solution is $-\frac{3}{2}$.

59. $6x - 5 = x + 5$
$6x - x - 5 = 5$
$5x = 5 + 5$
$5x = 10$
$x = 2$
Check: $6(2) - 5 = 2 + 5$
$12 - 5 = 7$
$7 = 7$ √
The solution is 2.

61. $6x + 14 = 2x - 2$
$6x - 2x + 14 = -2$
$4x = -2 - 14$
$4x = -16$
$x = -4$
Check: $6(-4) + 14 = 2(-4) - 2$
$-24 + 14 = -8 - 2$
$-10 = -10$ √
The solution is –4.

63. $-3y - 1 = 5 - 2y$
$-3y + 2y - 1 = 5$
$-y = 5 + 1$
$-y = 6$
$y = -6$
Check: $-3(-6) - 1 = 5 - 2(-6)$
$18 - 1 = 5 + 12$
$17 = 17$ √
The solution is –6.

65. $\frac{1}{5}y - 4 = -6$
$5\left(\frac{1}{5}y - 4\right) = 5(-6)$
$5\left(\frac{1}{5}y\right) - 5(4) = -30$
$y - 20 = -30$
$y = -30 + 20$
$y = -10$
Check: $\frac{1}{5}(-10) - 4 = -6$
$-2 - 4 = -6$
$-6 = -6$ √
The solution is –10.

67. $\frac{2}{3}y - 5 = 7$
$3\left(\frac{2}{3}y - 5\right) = 3(7)$
$3\left(\frac{2}{3}y\right) - 3(5) = 21$
$2y - 15 = 21$
$2y = 21 + 15$
$2y = 36$
$y = 18$

Check: $\frac{2}{3}(18)-5=7$
$12-5=7$
$7=7$ √
The solution is 18.

69. $\frac{2}{3}x-\frac{3}{4}=\frac{5}{12}$
$12\left(\frac{2}{3}x-\frac{3}{4}\right)=12\left(\frac{5}{12}\right)$
LCM of 3, 4, 12 is 12.
$12\left(\frac{2}{3}x\right)-12\left(\frac{3}{4}\right)=5$
$8x-9=5$
$8x=5+9$
$8x=14$
$x=\frac{14}{8}$
$x=\frac{7}{4}$
Check: $\frac{2}{3}\left(\frac{7}{4}\right)-\frac{3}{4}=\frac{5}{12}$
$\frac{14}{12}-\frac{9}{12}=\frac{5}{12}$
$\frac{5}{12}=\frac{5}{12}$ √
The solution is $\frac{7}{4}$.

71. $\frac{1}{2}x+\frac{1}{12}=\frac{3}{8}$
$24\left(\frac{1}{2}x+\frac{1}{12}\right)=24\left(\frac{3}{8}\right)$
LCM of 2, 8, 12 is 24.
$24\left(\frac{1}{2}x\right)+24\left(\frac{1}{12}\right)=9$
$12x+2=9$
$12x=9-2$
$12x=7$
$x=\frac{7}{12}$
Check: $\frac{1}{2}\left(\frac{7}{12}\right)+\frac{1}{12}=\frac{3}{8}$
$\frac{7}{24}+\frac{2}{24}=\frac{9}{24}$
$\frac{9}{24}=\frac{9}{24}$ √
The solution is $\frac{7}{12}$.

73. $\frac{3}{7}-\frac{5}{8}x=\frac{1}{7}$
$56\left(\frac{3}{7}-\frac{5}{8}x\right)=56\left(\frac{1}{7}\right)$
LCM of 7, 8 is 56.
$56\left(\frac{3}{7}\right)-56\left(\frac{5}{8}x\right)=8$
$24-35x=8$
$-35x=8-24$
$-35x=-16$
$x=\frac{16}{35}$
Check: $\frac{3}{7}-\frac{5}{8}\left(\frac{16}{35}\right)=\frac{1}{7}$
$\frac{3}{7}-\frac{2}{7}=\frac{1}{7}$
$\frac{1}{7}=\frac{1}{7}$ √
The solution is $\frac{16}{35}$.

75. $\frac{1}{6}x-\frac{1}{8}x=\frac{1}{12}$
$24\left(\frac{1}{6}x-\frac{1}{8}x\right)=24\left(\frac{1}{12}\right)$
LCM of 6, 8, 12 is 24.
$24\left(\frac{1}{6}x\right)-24\left(\frac{1}{8}x\right)=2$
$4x-3x=2$
$x=2$
Check: $\frac{1}{6}(2)-\frac{1}{8}(2)=\frac{1}{12}$
$\frac{1}{3}-\frac{1}{4}=\frac{1}{12}$
$\frac{4}{12}-\frac{3}{12}=\frac{1}{12}$
$\frac{1}{12}=\frac{1}{12}$ √
The solution is 2.

77. $\frac{1}{3}x+\frac{2}{5}=\frac{1}{5}x-\frac{2}{5}$
$15\left(\frac{1}{3}x+\frac{2}{5}\right)=15\left(\frac{1}{5}x-\frac{2}{5}\right)$
LCM of 3, 5 is 15.
$15\left(\frac{1}{3}x\right)+15\left(\frac{2}{5}\right)=15\left(\frac{1}{5}x\right)-15\left(\frac{2}{5}\right)$
$5x+6=3x-6$
$5x-3x=-6-6$
$2x=-12$

$x = -6$

Check: $\frac{1}{3}(-6) + \frac{2}{5} = \frac{1}{5}(-6) - \frac{2}{5}$

$-\frac{6}{3} + \frac{2}{5} = -\frac{6}{5} - \frac{2}{5}$

$-\frac{24}{15} = -\frac{8}{5}$

$-\frac{8}{5} = -\frac{8}{5}$ ✓

The solution is –6.

79. a. $A = BH$

$\frac{A}{H} = \frac{BH}{H}$

$\frac{A}{H} = B$

b. $A = BH$

$40 = 10H$

$\frac{40}{10} = \frac{10H}{10}$

$4 = H$

4 inches

81. a. $M = \frac{1}{5}n$

$3 = \frac{1}{5}n$

$5(3) = 5\left(\frac{1}{5}n\right)$

$15 = n$

15 seconds

b. $M = \frac{1}{5}n$

$5(M) = 5\left(\frac{1}{5}n\right)$

$5M = n$

83. $M = \frac{A}{s}$

$M = \frac{A}{740}$

$2.03 = \frac{A}{740}$

$740(2.03) = A$

$1502.2 = A$

1502.2 mi/hr

85. $M = \frac{A}{s}$

$sM = s\left(\frac{A}{s}\right)$

$sM = A$

87. $3.7x - 19.46 = -39.74$

$3.7x - 19.46 + 19.46 = -39.74 + 19.46$

$3.7x = -20.28$

$\frac{3.7x}{3.7} = \frac{-20.28}{3.7}$

$x \approx -5.4811$

89–93. Answers may vary.

95. Group Activity.

Review Problems

96. $\frac{1}{3} - \left(-\frac{1}{4}\right) = \frac{1}{3} + \frac{1}{4} = \frac{7}{12}$

97. a. 1985

b. \$650

c. Answer may vary; possible answer: \$6000
Extend the horizontal and vertical axes. Continue the trend upward. Read across to the expenditures for year 2000.

98. $\frac{2}{3}(3y+9) + \frac{1}{2}(2y-6) + 8$

$= \frac{2}{3}(3y) + \frac{2}{3}(9) + \frac{1}{2}(2y) - \frac{1}{2}(6) + 8$

$= 2y + 6 + y - 3 + 8$

$= (2y + y) + (6 - 3 + 8) = 3y + 11$

Problem Set 2.3

The check for Problems 1–55 is left to the student.

1. $3x - 7x + 30 = 10 - 2x$

$-4x + 30 = 10 - 2x$

$-4x + 30 - 10 = 10 - 2x - 10$

$-4x + 20 = -2x$

$-4x + 20 + 4x = -2x + 4x$

$20 = 2x$
$10 = x$
The solution is 10.

3. $3x + 6 - x = 8 + 3x - 6$
$2x + 6 = 2 + 3x$
$2x + 6 - 2 = 2 + 3x - 2$
$2x + 4 = 3x$
$2x + 4 - 2x = 3x - 2x$
$4 = x$
The solution is 4.

5. $6y + 25 - 4y = 4y - 4 + y + 29$
$2y + 25 = 5y + 25$
$2y + 25 - 25 = 5y + 25 - 25$
$2y = 5y$
$2 \neq 5$
No solution.

7. $3(x - 2) = 12$
$3x - 6 = 12$
$3x - 6 + 6 = 12 + 6$
$3x = 18$
$x = 6$
The solution is 6.

9. $-2(y + 3) = -9$
$-2y - 6 = -9$
$-2y - 6 + 6 = -9 + 6$
$-2y = -3$
$y = \frac{3}{2}$
The solution is $\frac{3}{2}$.

11. $-2(y + 4) + 7 = 3$
$-2y - 8 + 7 = 3$
$-2y - 1 = 3$
$-2y - 1 + 1 = 3 + 1$
$-2y = 4$
$y = -2$
The solution is –2.

13. $6x - (3x + 10) = 14$
$6x - 3x - 10 = 14$
$3x - 10 = 14$
$3x - 10 + 10 = 14 + 10$
$3x = 24$
$x = 8$
The solution is 8.

15. $2(4 - 3x) = 2(2x + 5)$
$8 - 6x = 4x + 10$
$8 - 6x - 10 = 4x + 10 - 10$
$-2 - 6x = 4x$
$-2 - 6x + 6x = 4x + 6x$
$-2 = 10x$
$-\frac{2}{10} = x$
$x = -\frac{1}{5}$
$-\frac{1}{5}$

17. $3(2y + 3) = -3y - 9$
$6y + 9 = -3y - 9$
$6y + 9 - 9 = -3y - 9 - 9$
$6y = -3y - 18$
$6y + 3y = -3y - 18 + 3y$
$9y = -18$
$\frac{9y}{9} = \frac{-18}{9}$
$y = -2$
The solution is –2.

19. $3(y + 3) = -2(2y - 1)$
$3y + 9 = -4y + 2$
$3y + 9 - 9 = -4y + 2 - 9$
$3y = -4y - 7$
$3y + 4y = -4y - 7 + 4y$
$7y = -7$
$\frac{7y}{7} = \frac{-7}{7}$
$y = -1$
The solution is –1.

21. $8(y + 2) = 2(3y + 4)$
$8y + 16 = 6y + 8$
$8y + 16 - 16 = 6y + 8 - 16$
$8y = 6y - 8$

$8y - 6y = 6y - 8 - 6y$
$2y = -8$
$y = -4$
The solution is −4.

23. $3(y+1) = 7(y-2) - 3$
$3y + 3 = 7y - 14 - 3$
$3y + 3 = 7y - 17$
$3y + 3 - 3 = 7y - 17 - 3$
$3y = 7y - 20$
$3y - 7y = 7y - 20 - 7y$
$-4y = -20$
$\frac{-4y}{-4} = \frac{-20}{-4}$
$y = 5$
The solution is 5.

25. $5(2z - 8) - 2 = 5(z - 3) + 3$
$10z - 40 - 2 = 5z - 15 + 3$
$10z - 42 = 5z - 12$
$10z - 42 + 42 = 5z - 12 + 42$
$10z = 5z + 30$
$10z - 5z = 5z + 30 - 5z$
$5z = 30$
$\frac{5z}{5} = \frac{30}{5}$
$z = 6$
The solution is 6.

27. $17(x+3) = 13 + 4(x - 10)$
$17x + 51 = 13 + 4x - 40$
$17x + 51 = 4x - 27$
$17x + 51 - 51 = 4x - 27 - 51$
$17x = 4x - 78$
$17x - 4x = 4x - 78 - 4x$
$13x = -78$
$\frac{13x}{13} = \frac{-78}{13}$
$x = -6$
The solution is −6.

29. $6 = -4(1 - x) + 3(x + 1)$
$6 = -4 + 4x + 3x + 3$
$6 = -1 + 7x$
$6 + 1 = -1 + 7x + 1$
$7 = 7x$
$\frac{7}{7} = \frac{7x}{7}$
$1 = x$
The solution is 1.

31. $10(y+4) - 4(y-2) = 3(y-1) + 2(y-3)$
$10y + 40 - 4y + 8 = 3y - 3 + 2y - 6$
$6y + 48 = 5y - 9$
$6y + 48 - 48 = 5y - 9 - 48$
$6y - 5y = 5y - 57 - 5y$
$y = -57$
The solution is −57.

33. $9 - 6(2z + 1) = 3 - 7(z - 1)$
$9 - 12z - 6 = 3 - 7z + 7$
$3 - 12z = 10 - 7z$
$3 - 12z - 3 = 10 - 7z - 3$
$-12z = 7 - 7z$
$-5z = 7$
$\frac{-5z}{-5} = \frac{7}{-5}$
$z = -\frac{7}{5}$
The solution is $-\frac{7}{5}$.

35. $\frac{x}{3} + \frac{x}{2} = \frac{5}{6}$
$6\left(\frac{x}{3} + \frac{x}{2}\right) = 6\left(\frac{5}{6}\right)$
LCM = 6
$2x + 3x = 5$
$5x = 5$
$\frac{5x}{5} = \frac{5}{5}$
$x = 1$
The solution is 1.

37. $20 - \frac{z}{3} = \frac{z}{2}$
$6\left(20 - \frac{z}{3}\right) = 6\left(\frac{z}{2}\right)$
LCM = 6
$120 - 2z = 3z$
$120 - 2z + 2z = 3z + 2z$
$120 = 5z$

$\frac{120}{5} = \frac{5z}{5}$
$24 = z$
The solution is 24.

39. $\frac{3x}{4} - 3 = \frac{x}{2} + 2$
$4\left(\frac{3x}{4} - 3\right) = 4\left(\frac{x}{2} + 2\right)$
$\text{LCM} = 4$
$3x - 12 = 2x + 8$
$3x - 12 + 12 = 2x + 8 + 12$
$3x = 2x + 20$
$3x - 2x = 2x + 20 - 2x$
$x = 20$
The solution is 20.

41. $\frac{3x}{5} - x = \frac{x}{10} - \frac{5}{2}$
$10\left(\frac{3x}{5} - x\right) = 10\left(\frac{x}{10} - \frac{5}{2}\right)$
$\text{LCM} = 10$
$6x - 10x = x - 25$
$-4x = x - 25$
$-4x - x = x - 25 - x$
$-5x = -25$
$\frac{-5x}{-5} = \frac{-25}{-5}$
$x = 5$
The solution is 5.

43. $\frac{5z-1}{7} - \frac{3z-2}{5} = 1$
$35\left(\frac{5z-1}{7}\right) - 35\left(\frac{3z-2}{5}\right) = 35(1)$
$5(5z - 1) - 7(3z - 2) = 35$
$25z - 5 - 21z + 14 = 35$
$4z + 9 = 35$
$4z + 9 - 9 = 35 - 9$
$4z = 26$
$\frac{4z}{4} = \frac{26}{4}$
$z = \frac{13}{2}$
The solution is $\frac{13}{2}$.

45. $\frac{z-3}{4} - 1 = \frac{z}{2}$
$4\left(\frac{z-3}{4} - 1\right) = 4\left(\frac{z}{2}\right)$
$\text{LCM} = 4$
$z - 3 - 4 = 2z$
$z - 7 = 2z$
$z - 7 - z = 2z - z$
$-7 = z$
The solution is -7.

47. $\frac{2y-3}{9} + \frac{y-3}{2} = \frac{y+5}{6} - 1$
$18\left(\frac{2y-3}{9}\right) + 18\left(\frac{y-3}{2}\right) = 18\left(\frac{y+5}{6}\right) - 18(1)$
$2(2y - 3) + 9(y - 3) = 3(y + 5) - 18$
$4y - 6 + 9y - 27 = 3y + 15 - 18$
$13y - 33 = 3y - 3$
$13y - 33 + 33 = 3y - 3 + 33$
$13y = 3y + 30$
$13y - 3y = 3y + 30 - 3y$
$10y = 30$
$\frac{10y}{10} = \frac{30}{10}$
$y = 3$
The solution is 3.

49. $15.2 - 3.4x = 9.76$
$100(15.2 - 3.4x) = 100(9.76)$
$100(15.2) - 100(3.4x) = 100(9.76)$
$1520 - 340x = 976$
$1520 - 340x - 1520 = 976 - 1520$
$-340x = -544$
$\frac{-340x}{-340} = \frac{-544}{-340}$
$x = 1.6$
The solution is 1.6.

51. $2.24y - 9.28 = 5.74y + 5.42$
$100(2.24y - 9.28) = 100(5.74y + 5.42)$
$100(2.24y) - 100(9.28)$
$= 100(5.74y) + 100(5.42)$
$224y - 928 = 574y + 542$
$224y - 928 - 542 = 574y + 542 - 542$
$224y - 1470 = 574y$

$224y - 1470 - 224y = 574y - 224y$
$-1470 = 350y$
$\frac{-1470}{350} = \frac{350y}{350}$
$-4.2 = y$
The solution is –4.2.

53. $0.2x - 0.5 = 1.2x - 0.6$
$10(0.2x - 0.5) = 10(1.2x - 0.6)$
$10(0.2x) - 10(0.5) = 10(1.2x) - 10(0.6)$
$2x - 5 = 12x - 6$
$2x - 5 + 6 = 12x - 6 + 6$
$2x + 1 = 12x$
$2x + 1 - 2x = 12x - 2x$
$1 = 10x$
$\frac{1}{10} = \frac{10x}{10}$
$\frac{1}{10} = x$
The solution is $\frac{1}{10}$.

55. $3.2y - 2.2 = 4.9y + 5.9$
$10(3.2y - 2.2) = 10(4.9y + 5.9)$
$32y - 22 = 49y + 59$
$32y - 22 - 59 = 49y + 59 - 59$
$32y - 81 = 49y$
$32y - 81 - 32y = 49y - 32y$
$-81 = 17y$
$\frac{-81}{17} = \frac{17y}{17}$
$y = -\frac{18}{17} \approx -4.765$
The solution is –4.765

57. a. $\frac{c}{2} + 80 = 2F$
$\frac{c}{2} + 80 - 80 = 2F - 80$
$\frac{c}{2} = 2F - 80$
$2\left(\frac{c}{2}\right) = 2(2F - 80)$
$c = 4F - 160$

b. $c = 4(40) - 160 = 0$
$c = 4(60) - 160 = 80$
$c = 4(65) - 160 = 100$
$c = 4(70) - 160 = 120$
$c = 4(80) - 160 = 160$

c. (40, 0), (60, 80), (65, 100), (70, 120), (80, 160)

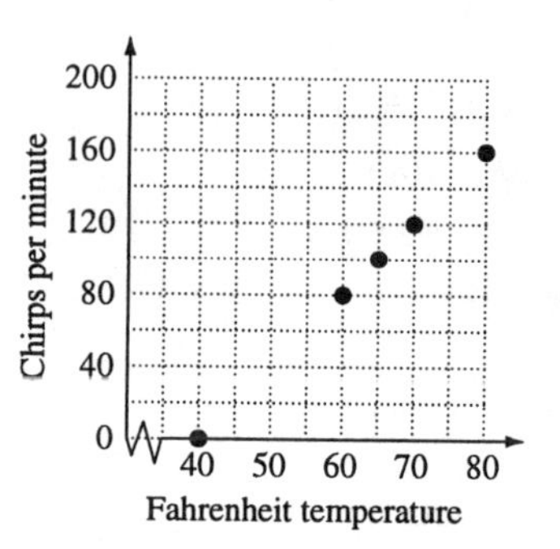

Linear relationship; when the temperature increases there are more chirps per minute.

59. c is true.
$2 - 3y = 11$
$-3y = 9$
$y = -3$
$y^2 + 2y - 3 = (-3)^2 + 2(-3) - 3$
$= 9 - 6 - 3 = 0$
Neither positive nor negative

61. $8.05x + 2.03x = 17.06 - 4.3$
$10.08x = 12.76$
$x \approx 1.265873$
The solution is 1.265873.

63. $3.7y - 15.1 = 9y - 6.2$
$3.7y - 9y = -6.2 + 15.1$
$-5.3y = 8.9$
$y \approx -1.679245$
The solution is –1.679245.

65. $19.25x - 63.1x = 14.9 - 52.04$
$-43.85x = -37.14$
$x \approx 0.846978$
The solution is 0.846978.

67. $-6.1x + 11.03 = 11x + 5.17$
$-17.1x = -5.86$
$x \approx 0.342690$
The solution is 0.342690.

69. Answers may vary.

71. $2(3x + 4) = 3x + 2[3(x - 1) + 2]$
$6x + 8 = 3x + 2(3x - 3 + 2)$
$6x + 8 = 3x + 2(3x - 1)$
$6x + 8 = 3x + 6x - 2$
$6x + 8 - 8 = 9x - 2 - 8$
$6x = 9x - 10$
$6x - 9x = 9x - 10 - 9x$
$-3x = -10$
$\frac{-3x}{-3} = \frac{-10}{-3}$
$x = \frac{10}{3}$
The solution is $\frac{10}{3}$.

73. $x(x-5) = 4x(x+2) - 3(x^2 + x - 7)$
$x^2 - 5x = 4x^2 + 8x - 3x^2 - 3x + 21$
$x^2 - 5x = x^2 + 5x + 21$
$x^2 - 5x - x^2 = x^2 + 5x + 21 - x^2$
$-5x = 5x + 21$
$-5x - 5x = 5x + 21 - 5x$
$-10x = 21$
$x = -\frac{21}{10}$
The solution is $-\frac{21}{10}$.

75. Group Activity.

Review Problems

76. $\frac{10-3x}{2} = \frac{10-3(-4)}{2} = \frac{10+12}{2} = \frac{22}{2}$
$= 11$

77. $3(2x - 5) - (x - 4) = 6x - 15 - x + 4$
$= 5x - 11$

78. $-10\frac{1}{2} < -10\frac{1}{4}$

Problem Set 2.4

1. $y = 10(x - 65) + 50$
$250 = 10(x - 65) + 50$
$250 = 10x - 650 + 50$
$250 = 10x - 600$
$250 + 600 = 10x - 600 + 600$
$850 = 10x$
$\frac{850}{x} = \frac{10x}{10}$
$85 = x$
85 mi/hr
There is a dot on the line in the graph where the vertical line representing 85 mi/hr intersects with the horizontal line representing \$250.

3. a. $E = 0.215t + 71.05$
$73.35 = 0.215t + 71.05$
$73.35 - 71.05 = 0.215t$
$2.3 = 0.215t$
$\frac{2.3}{0.215} = \frac{0.215t}{0.215}$
$10.698 \approx t$
$1950 + 10 = 1960$
The point above 1960 is approximately parallel to the life expectancy 73.35 years.

b. $E = 0.215t + 71.05$
$E - 71.05 = 0.215t + 71.05 - 71.05$
$E - 71.05 = 0.215t$
$\frac{E - 71.05}{0.215} = \frac{0.215t}{0.215}$
$t = \frac{E - 71.05}{0.215}$

c. $t = \frac{81.8 - 71.05}{0.215}$
$t = \frac{10.75}{0.215}$
$t = 50$
$1950 + 50 = 2000$
Place a point at the intersection of the lines representing Year = 2000 and 81.8 years.

5. $\frac{W}{4A} = P$

$\frac{W}{4(24)} = 28$

$\frac{W}{96} = 28$

$96\left(\frac{W}{96}\right) = 96(28)$

$W = 2688$

2688 lb

7. $T = t - \frac{h}{100}$ $h < 12,000$

$t = 30°C$, $h = ?$, $T = 0°C$

$0 = 30 - \frac{h}{100}$

$\frac{h}{100} = 30$

$h = 3000$

height: 3000 meters

9. $PB = A$

$P = 18\% = 0.18$, $B = 40$, $A = ?$

$0.18(40) = A$

$A = 7.2$

11. $PB = A$

$P = ?$, $B = 15$, $A = 3$

$P(15) = 3$

$P = \frac{1}{5} = 0.20 = 20\%$

$PB = A$

13. $PB = A$

$P = ?$, $B = 3$, $A = 15$

$P(3) = 15$

$P = 5 = 5.00 = 500\%$

$PB = A$

15. $PB = A$

$P = 60\% = 0.60$, $B = ?$, $A = 3$

$0.60B = 3$

$B = 5$

17. $PB = A$

$P = 18\% = 0.18$, $B = ?$, $A = \$2,970$

$0.18B = 2,970$

$B = 16,500$

Your salary is $16,500 for that year.

19. $PB = A$

$P = ?$, $B = \$60,000$, $A = \$7,500$

$P(60,000) = 7,500$

$P = 0.125 = 12.5\%$

12.5% of goal raised.

21. $A = LW$ for L

$\frac{A}{W} = L$

$L = \frac{A}{W}$

23. $A = \frac{1}{2}bh$ for b

$\frac{1}{2}bh = A$

$bh = 2A$

$b = \frac{2A}{h}$

25. $Prt = I$ for P

$P = \frac{I}{rt}$

27. $E = mc^2$ for m

$mc^2 = E$

$m = \frac{E}{c^2}$

29. $y = mx + b$ for m

$mx + b = y$

$mx = y - b$

$m = \frac{y-b}{x}$

31. $A = \frac{1}{2}(a+b)$ for a

$\frac{1}{2}(a+b) = A$

$a + b = 2A$

$a = 2A - b$

33. $S = P + Prt$ for r

$P + Prt = S$

$Prt = S - P$

$r = \frac{S-P}{Pt}$ or $\frac{S}{Pt} - \frac{1}{t}$

35. $I = \frac{E}{R}$
$IR = E$
$R = \frac{E}{I}$

37. $L = a + (n-1)d$ for n
$a + (n-1)d = L$
$a + nd - d = L$
$nd = L + d - a$
$n = \frac{L+d-a}{d}$ or $\frac{L-a}{d} + 1$

39. $3x + y = 6$
$3x + y - 3x = 6 - 3x$
$y = 6 - 3x$
$y = 6 - 3(2)$
$y = 6 - 6$
$y = 0$

41. $2x = 4y - 6$
$2x + 6 = 4y - 6 + 6$
$2x + 6 = 4y$
$\frac{2x+6}{4} = \frac{4y}{4}$
$\frac{x+3}{2} = y$
$y = \frac{10+3}{2}$
$y = \frac{13}{2}$

43. $2y = 6 - 5x$
$\frac{2y}{2} = \frac{6-5x}{2}$
$y = \frac{6-5x}{2}$
$y = \frac{6-5(-1)}{2}$
$y = \frac{6+5}{2}$
$y = \frac{11}{2}$

45. $-3x = 21 - 6y$
$-3x - 21 = 21 - 6y - 21$
$-3x - 21 = -6y$
$\frac{-3x-21}{-6} = \frac{-6y}{-6}$
$\frac{x+7}{2} = y$
$y = \frac{0+7}{2}$
$y = \frac{7}{2}$

47. $-12 = -x - 4y$
$-12 + x = -x - 4y + x$
$-12 + x = -4y$
$\frac{-12+x}{-4} = \frac{-4y}{-4}$
$\frac{-12+x}{-4} = y$
$y = \frac{-12-3}{-4}$
$y = \frac{-15}{-4}$
$y = \frac{15}{4}$

49. c is true.
a is not true;
$y + 3y = 4y$
$4y = 4y$, so $y = 1$ is a solution.
b is not true; $A = LW$ so $W = \frac{A}{L}$
c is true; $y - x = 7$, so $y = x + 7$
d is not true; $x - b = 6x - c$
$c - b = 5x$
$x = \frac{c-b}{5}$

51. $y = -0.358709x + 256.835$
$y = 3$ minutes $= 180$ seconds, $x = ?$
$180 = -0.358709x + 256.835$
$0.358709x = 76.835$
Using a calculator:
$x = 214.1987$
x years after 1900:
$1900 + 214.1987 = 2114.1987$
Sometime during the year 2114.

53. $F = \frac{9}{5}C + 32$
$\frac{9}{5}C = F - 32$
$C = \frac{5}{9}(F - 32)$

$C = \frac{5}{9}(98.6 - 32)$, $F = 98.6$
$C = \frac{5}{9}(66.6)$
$C = 37$, so 37°C
$K = C + 273$
$K = 37 + 273$
$K = 310$
The normal body temperature of 98.6°F corresponds to 310K.

Review Problems

55. 77%

56. 17-year-old boys

57. As boys and girls get older, a larger percent is sexually active.

Problem Set 2.5

1. Let x = the number.
$5x - 7 = 123$
$5x = 130$
$x = 26$
The number is 26.

3. Let x = the number.
$2x + 4 = 36$
$2x = 32$
$x = 16$
The number is 16.

5. Let x = the number.
$2(4 + x) = 36$
$8 + 2x = 36$
$2x = 28$
$x = 14$
The number is 14.
This problem is 2 times a quantity $(4 + x)$.
Problem 3 is 2 times x then add 4.

7. Let x = the smaller page number.
Then $x + 1$ = the larger page number.
$x + (x + 1) = 629$
$2x + 1 = 629$
$2x = 628$
$x = 314$
The smaller page number is 314. The larger page number is 314 + 1 = 315. The page numbers are 314 and 315.

9. Let x = short side of triangle.
Then $x + 1$ = long side of triangle and $x + 2$ = diagonal of triangle.
$x + (x + 1) + (x + 2) = 12$
$3x + 3 = 12$
$3x = 9$
$x = 3$
$x + 1 = 4$
$x + 2 = 5$
The diagonal has a length of 5 m.

11. Let x = the width.
Then $x + 2$ = the length.
$2x + 2(x + 2) = 332$
$2x + 2x + 4 = 332$
$4x + 4 = 332$
$4x = 328$
$x = 82$
$x + 2 = 84$
The playground's dimensions are 82 ft by 84 ft.

13. Let x = the percent living in the Midwest.
Then $x + 2$ = the percent living in the South.
$x + (x + 2) = 100 - 18 - 58$
$2x + 2 = 24$
$2x = 22$
$x = 11$
$x + 2 = 13$
The percents of Asian-Americans living in the Midwest and the South are 11% and 13%, respectively.

15. Let x = the length of the first piece. Then $2x$ = the length of the second piece, and $x + 4$ = the length of the third piece.
$x + (2x) + (x + 4) = 64$
$4x + 4 = 64$
$4x = 60$
$x = 15$
$2x = 30$
$x + 4 = 19$
The lengths are 15 m, 30 m, and 19 m.

17. Let x = the number of deaths from stroke. Then $4x + 140{,}900$ = the number of deaths from heart disease.
$x + (4x + 140{,}900) = 889{,}600$
$5x + 140{,}900 = 889{,}600$
$5x = 748{,}700$
$x = 149{,}740$
$4x + 140{,}900 = 739{,}860$
The numbers of deaths per year from heart disease and from stroke are 739,860 and 149,740, respectively.

19. Let x = the number in Russia. Then $2x - 4$ = the number in U.S. and $x - 2$ = the number in France.
$x + (2x - 4) + (x - 2) = 38$
$4x - 6 = 38$
$4x = 44$
$x = 11$
$2x - 4 = 18$
$x - 2 = 9$
The number of nuclear reactors no longer in service for U.S., Russia, and France are 18, 11, and 9, respectively. Germany, 9; UK, 7; Italy, 4; Armenia, Canada, Ukraine, 2; Spain and Slovakia, 1.

21. Falls: $0.22x = 1760$; $x = 8000$
There are 8000 Americans who suffer spinal cord injuries.
Acts of violence: $0.16(8000) = 1280$
Sports injuries: $0.13(8000) = 1040$
Other: $0.04(8000) = 320$
Vehicular accidents: $0.45(8000) = 3600$

23. Let p equal the price of the car. $0.12p$ equals the price reduction.
$p - 0.12p = 17{,}600$
$0.88p = 17{,}600$
$p = \frac{17{,}600}{0.88}$
$p = 20{,}000$
price of car, \$20,000

25. Let p equal the price of the car.
$0.065p$ = sales tax.
$p + 0.065p = 17{,}466$
$1.065p = 17{,}466$
$p = 16{,}400$
price of car, \$16,400

27. Let x equal the dealer's cost of the refrigerator.
$0.25x$ = markup
$x + 0.25x = 584$
$1.25x = 584$
$x = 467.20$
dealer's cost, \$467.20

29. Let x equal the original price of the sofa.
$\frac{2}{7}x$ = reduction
$x - \frac{2}{7}x = 235$
$\frac{5}{7}x = 235$
$x = \frac{7}{5}(235)$
$x = 329$
sofa's price, \$329

31. Let x equal the length of call. Cost of first minute is \$0.75. Cost of each additional minute is \$0.60.
$0.75 + 0.60(x - 1) = 12.15$
$0.75 + 0.60x - 0.60 = 12.15$
$0.60x + 0.15 = 12.15$
$0.60x = 12.00$

$x = 20$
length of call, 20 minutes

33. Let x equal the number of advertisements distributed.
$45 + 0.05x = 82.50$
$0.05x = 37.50$
$x = 750$
750 advertisements were distributed.

35. Let x = the number of years after 1939 when this will occur.
$1.44x + 280 = 366.4$
$1.44x = 86.4$
$x = 60$
$1939 + 60 = 1999$

37. Let x = the number of years for this to occur.

a. $30{,}000 + 150x = 36{,}000$
$150x = 6000$
$x = 40$
40 years
Electric Cost at 40 years
$5000 + 40(1100)$
$= 5000 + 44{,}000 = \$49{,}000$

b. $30{,}000 + 150x = 5000 + 1100x$
$25{,}000 = 950x$
$26.3 \approx x$
They will be the same after about 26.3 years.
Cost = $30{,}000 + 150(26.3) = 33{,}945$
It will cost about \$33,945.

c. The point is about (26.3, 33,945).
This represents after 26.3 years, the prices will be equal at about \$33,945.
This agrees with the answers in part (b).

39. a is true; $(x + 7) - x = 7$.

41–43. Answers may vary.

45. Let x equal the number of inhabitants per square mile in Canada. Canada is x. The United States is $9x - 1$. Australia is $x - 2$. England is $10(9x - 1) - 9$
$= 90x - 10 - 9 = 90x - 19$. England exceeds the sum of the others by 537.
$90x - 19 = x + (9x - 1) + (x - 2) + 537$
$90x - 19 = 11x + 534$
$79x = 553$
$x = 7$
Population density:
Canada: 7 inhabitants per square mile
United States: $9(7) - 1 = 63 - 1$
$= 62$ inhabitants per square mile
Australia:
$7 - 2 = 5$ inhabitants per square mile
England: $90(7) - 19 = 630 - 19$
$= 611$ inhabitants per square mile

47. Let x = amount of money you had originally.
amount of money spent: $\frac{1}{5}x$
amount left: $x - \frac{1}{5}x = \frac{4}{5}x$
amount of money lost:
$\frac{1}{3}\left(\frac{4}{5}x\right) = \frac{4}{15}x$
amount left: $\frac{4}{5}x - \frac{1}{3}\left(\frac{4}{5}x\right) = \frac{2}{3}\left(\frac{4}{5}x\right) = \frac{8}{15}x$
$\frac{8}{15}x = 96$
$x = 180$
original amount of money, \$180

49. Let x equal the number of oranges in each of the original three piles.
$3x$ = total number of oranges
$3x - 2$ = total after 2 are thrown away.
$\frac{3x-2}{2} = 32$
$3x - 2 = 64$
$3x = 66$
$x = 22$
There are 22 oranges in each of the original piles.

51. Group Activity.

Review Problems

52. **a.** \$27,500

b. Teachers

53. $4-2(2-y)$
$=4-4+2y=2y$

54. $4-\frac{3y}{2}=y-1$
$4-\frac{3y}{2}+1=y-1+1$
$5-\frac{3y}{2}=y$
$5-\frac{3y}{2}+\frac{3y}{2}=y+\frac{3y}{2}$
$5=\frac{2y}{2}+\frac{3y}{2}$
$5=\frac{5y}{2}$
$10=5y$
$2=y$

Problem Set 2.6

1. $x>-7$

a. $7>-7$; Yes

b. $0>-7$; Yes

c. $-7.2>-7$; No

3. $6-5y\geq 7$

a. $6-5\left(-\frac{1}{5}\right)\geq 7?$
$6+1\geq 7$?
$7\geq 7$; Yes

b. $6-5(0)\geq 7$?
$6-0\geq 7$?
$6\geq 7$; No

c. $6-5(-4)\geq 7$?
$6+20\geq 7$?
$26\geq 7$; Yes

5. $x>6$

4 5 6 7 8

7. $y<-4$

−6 −5 −4 −3 −2

9. $x\geq -3$

−5 −4 −3 −2 −1

11. $x\leq 4$

2 3 4 5 6

13. $-2<x\leq 5$

−3 −2 −1 0 1 2 3 4 5 6

15. $-1<x<4$

−3 −2 −1 0 1 2 3 4 5 6

17. $x>-2$

19. $x\geq 4$

21. $x\geq 3$

23. $x-3>2$
$x-3+3>2+3$
$x>5$
$\{x|x>5\}$

3 4 5 6 7

25. $x+4\leq 9$
$x+4-4\leq 9-4$
$x\leq 5$
$\{x|x\leq 5\}$

3 4 5 6 7

27. $y-3<0$
$y-3+3<0+3$
$y<3$

$\{y|y < 3\}$

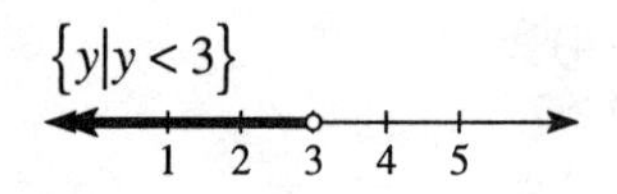

29. $3x + 4 \le 2x + 7$
$3x - 2x \le 7 - 4$
$x \le 3$
$\{x|x \le 3\}$

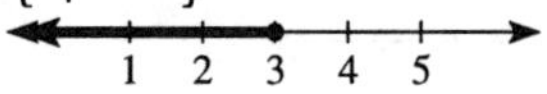

31. $5x - 9 < 4x + 7$
$5x - 4x < 7 + 9$
$x < 16$
$\{x|x < 16\}$

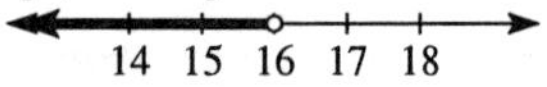

33. $7x - 7 > 6x - 3$
$7x - 6x > -3 + 7$
$x > 4$
$\{x|x > 4\}$

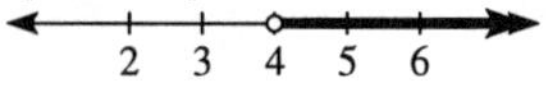

35. $x - \frac{2}{3} > \frac{1}{2}$
$x - \frac{2}{3} + \frac{2}{3} > \frac{1}{2} + \frac{2}{3}$
$x > \frac{3}{6} + \frac{4}{6}$
$x > \frac{7}{6}$
$\left\{x \middle| x > \frac{7}{6}\right\}$

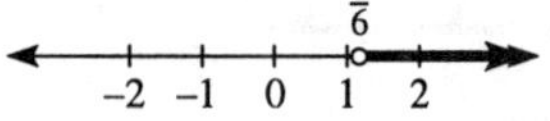

37. $y + \frac{7}{8} \le \frac{1}{2}$
$y + \frac{7}{8} - \frac{7}{8} \le \frac{1}{2} - \frac{7}{8}$
$y \le \frac{4}{8} - \frac{7}{8}$
$y \le -\frac{3}{8}$
$\left\{y \middle| y \le -\frac{3}{8}\right\}$

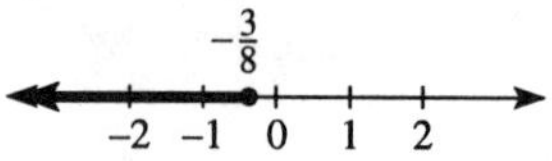

39. $-15y + 13 > 13 - 16y$
$-15y + 13 - 13 > 13 - 16y - 13$
$-15y > -16y$
$\frac{-15y}{y} > \frac{-16y}{y}$
$-15 > -16$
True for all y.
$\{y|-\infty < y < \infty\}$

41. $4x < 20$
$\frac{4x}{4} < \frac{20}{4}$
$x < 5$
$\{x|x < 5\}$

43. $3x \ge -15$
$x \ge -5$
$\{x|x \ge -5\}$

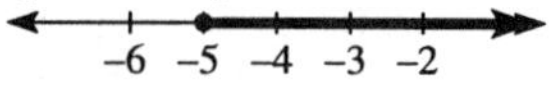

45. $-3x < 15$
$\frac{-3x}{-3} > \frac{15}{-3}$
$x > -5$
$\{x|x > -5\}$

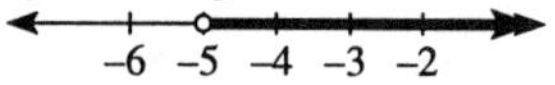

47. $-3x \ge -15$
$\frac{-3x}{-3} \le \frac{-15}{-3}$
$x \le 5$
$\{x|x \le 5\}$

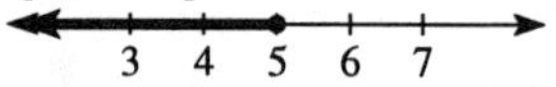

49. $-2y > \frac{1}{5}$

$\frac{-2y}{-2} < \frac{1}{5} \div (-2)$

$y < \frac{1}{5} \cdot \left(-\frac{1}{2}\right)$

$y < -\frac{1}{10}$

$\left\{y \middle| y < -\frac{1}{10}\right\}$

$-\frac{1}{10}$

−2 −1 0 1 2

51. $-\frac{1}{3}x < 7$

$-3\left(-\frac{1}{3}x\right) > -3(7)$

$x > -21$

$\{x|x > -21\}$

−23 −22 −21 −20 −19

53. $2y - 3 > 7$

$2y - 3 + 3 > 7 + 3$

$2y > 10$

$\frac{2y}{2} > \frac{10}{2}$

$y > 5$

$\{y|y > 5\}$

3 4 5 6 7

55. $3(x - 1) < 9$

$\frac{3(x-1)}{3} < \frac{9}{3}$

$x - 1 < 3$

$x < 4$

$\{x|x < 4\}$

3 4 5 6 7

57. $-2x - 3 < 3$

$-2x < 6$

$x > -3$

$\{x|x > -3\}$

−6 −5 −4 −3 −2

59. $3 - 7y \le 17$

$-7y \le 14$

$y \ge -2$

$\{y|y \ge -2\}$

−4 −3 −2 −1 0

61. $-x < 4$

$(-1)(-x) > (-1)(4)$

$x > -4$

$\{x|x > -4\}$

−6 −5 −4 −3 −2

63. $5 - y \le 1$

$-y \le -4$

$y \ge 4$

$\{y|y \ge 4\}$

3 4 5 6 7

65. $2y - 5 > -y + 6$

$2y > -y + 11$

$3y > 11$

$y > \frac{11}{3}$

$\left\{y \middle| y > \frac{11}{3}\right\}$

$\frac{11}{3}$

3 4 5 6 7

67. $2y - 5 < 5y - 11$

$2y < 5y - 6$

$-3y < -6$

$y > 2$

$\{y|y > 2\}$

0 1 2 3 4

69. $3(x + 1) - 5 < 2x + 1$

$3x + 3 - 5 < 2x + 1$

$3x - 2 < 2x + 1$

$3x < 2x + 3$

$x < 3$

$\{x|x < 3\}$

1 2 3 4 5

71. $8x+3>3(2x+1)-x+5$
$8x+3>6x+3-x+5$
$8x+3>5x+8$
$8x>5x+5$
$3x>5$
$x>\frac{5}{3}$
$\left\{x \middle| x>\frac{5}{3}\right\}$

$\frac{5}{3}$

0 1 2 3 4

73. $7(y+4)-13<12+13(3+y)$
$7y+28-13<12+39+13y$
$7y+15<51+13y$
$7y<36+13y$
$-6y<36$
$y>-6$
$\{y|y>-6\}$

−7 −6 −5 −4 −3

75. $\frac{x}{4}-\frac{3}{8}<2$
$8\left(\frac{x}{4}-\frac{3}{8}\right)<8(2)$
$2x-3<16$
$2x<19$
$x<\frac{19}{2}$
$\left\{x \middle| x<\frac{19}{2}\right\}$

$\frac{19}{2}$

7 8 9 10 11

77. $\frac{y}{3}+\frac{y}{4}\geq 1$
$12\left(\frac{y}{3}+\frac{y}{4}\right)\geq 12(1)$
$4y+3y\geq 12$
$7y\geq 12$
$y\geq\frac{12}{7}$
$\left\{y \middle| y\geq\frac{12}{7}\right\}$

$\frac{12}{7}$

0 1 2 3 4

79. $-0.4y+2>-1.2y-0.4$
$-0.4y>-1.2y-2.4$
$0.8y>-2.4$
$y>-3$
$\{y|y>-3\}$

−5 −4 −3 −2 −1

81. Let x = the number of births per 1000 population.
$x\leq 125$
$\{x|x\leq 125\}$

83. Let x = the number of births per 1000 population.
$x>65$
$\{x|x>65\}$

85. Let x = the number of births per 1000 population.
$x\geq 66$
$\{x|x\geq 66\}$

87. Let x equal the score on third test.
$\frac{44+72+x}{3}\geq 60$
$116+x\geq 180$
$x\geq 64$
The maximum score is 100: $x\leq 100$
$64\leq x\leq 100$
The score must be at least 64 or between and including 64 and 100.

89. Let x equal the number of bags of cement.
$245+95x\leq 3000$
$95x\leq 2755$
$x\leq 29$
29 or less bags of cement can be safely lifted on the elevator in one trip.

91. $N = 4t - 160$
$160 \ge 4t - 160$
$160 + 160 \ge 4t - 160 + 160$
$320 \ge 4t$
$\frac{320}{4} \ge \frac{4t}{4}$
$80 \ge t$
The temperature must be no more than 80°F.

93. Let x equal the number of customers.
Profit $= 40x - 200$
$40x - 200 > 12{,}000$
$40x > 12{,}200$
$x > 305$
The company must have more than 305 customers or it will be sold by the stockholders.

95. a. $12{,}000 + 700x < 5000 + 1100x$
$12{,}000 - 5000 < 1100x - 700x$
$7000 < 400x$
$17.5 < x$
After 17.5 years

b. The line representing gas heat is above (more expensive) the line representing electric up to 17.5 years. Then the line representing gas is below (less expensive) the line representing electric.

97. d is true.
a is false: zero is greater than both negative numbers.
b is false:

$3y - 2 \le y$ and	$y < 3y - 2$
$2y < 2$	$-2y < -2$
$y < 1$	$y > 1$

have different solutions.
c is false: $x > -3$ and $-4 < -3$.
d is true: $5 < x$ and $x > 5$ are the same.

99. b is true.
a is false: $-2x + 5 \ge 13$
The largest integer is -4.
$-2x \ge 8$
$x \le -4$
b is true:
$-\frac{x}{3} > -7$ is equivalent to $x < 21$
$-x > -21$
$x < 21$
c is false: 8,000% of $10 = 80(10) = 800$ not 80.
d is false: $8x > 4x$ is equivalent to $4x > 0$ or $x > 0$.

101. a. $3(x - 2) + 4 < 8(x + 1)$
$3x - 6 + 4 < 8x + 8$
$3x - 2 < 8x + 8$
$-2 - 8 < 8x - 3x$
$-10 < 5x$
$-2 < x$
$\{x | x > -2\}$

b. $x = -2$ is the x-coordinate where the two lines meet.

103. Answers may vary.

105. $4x - 4 < 4(x - 5)$
$4x - 4 < 4x - 20$
$4x - 4 - 4x < 4x - 20 - 4x$
$-4 < -20$
This is not true. There are no x-terms. There is no solution.

107. $y \le ax + b$
$y - b \le ax + b - b$
$y - b \le ax$
$\frac{y - b}{a} \ge \frac{ax}{a}$
$\frac{y - b}{a} \ge x$
$\left\{x \middle| x \le \frac{y - b}{a}\right\}$

109. $|x| > 2$

−2 −1 0 1 2

111. Let x equal monthly sales.
$0.30(x-1000) > 700$
$0.3x - 300 > 700$
$0.3x > 1{,}000$
$x > 3333.33$
Monthly sales greater than \$3333.33

113. b is true; Let x equal Mia's age. $2x + 3$ equals Eleanor's age.
$x + (2x + 3) \geq 24$
$3x + 3 \geq 24$
$3x \geq 21$
$x \geq 7$
$2x + 3 \geq 17$
Mia is at least 7.
Eleanor is at least 17.
a is false: Mia can be 8.
b is true: Eleanor can be 19.
c is false: Eleanor is at least 17 so she can be 17.
d is false: Mia cannot be 6 since she is at least 7.

Review Problems

114. $b^2 - 4ac$
$= (-2)^2 - 4(-1)(3)$
$a = -1,\ b = -2,\ c = 3$
$= 4 + 12 = 16$

115. $2[10 - (y-1)]$
$= 2(10 - y + 1)$
$= 2(11 - y)$
$= 22 - 2y$

116. $-\frac{1}{3}(6x - 9) = 23$
$-2x + 3 = 23$
$-2x = 20$
$x = -10$

Chapter 2 Review

1. $2y - 5 = 7$
$2y = 12$
$y = 6$
The solution is 6.

2. $5z + 20 = 3z$
$2z = -20$
$z = -10$
The solution is -10.

3. $7(y-4) = y + 2$
$7y - 28 = y + 2$
$6y = 30$
$y = 5$
The solution is 5.

4. $1 - 2(6 - y) = 3y + 2$
$1 - 12 + 2y = 3y + 2$
$-11 + 2y = 3y + 2$
$-13 = y$
The solution is -13.

5. $2(y-4) + 3(y+5) = 2y - 2$
$2y - 8 + 3y + 15 = 2y - 2$
$5y + 7 = 2y - 2$
$3y = -9$
$y = -3$
The solution is -3.

6. $2z - 4(5z + 1) = 3z + 17$
$2z - 20z - 4 = 3z + 17$
$-18z - 4 = 3z + 17$
$-21z = 21$
$z = -1$
The solution is -1.

7. $\frac{2}{3}x = \frac{1}{6}x + 1$
$4x = x + 6 \quad (\times 6)$
$3x = 6$
$x = 2$
The solution is 2.

8. $\frac{1}{2}y-\frac{1}{10}=\frac{1}{5}y+\frac{1}{2}$
$5y-1=2y+5 \quad (\times 10)$
$3y=6$
$y=2$
The solution is 2.

9. $0.2y-0.3=0.8y-0.3$
$-0.6y=0$
$y=0$
The solution is 0.

10. $17.4-3.6y=-16.08$
$17.4-3.6y-17.4=-16.08-17.4$
$-3.6y=-33.48$
$\frac{-3.6y}{-3.6}=\frac{-33.48}{-3.6}$
$y=9.3$
The solution is 9.3.

11. $-2(y-4)-(3y-2)=-2-(6y-2)$
$-2y+8-3y+2=-2-6y+2$
$-5y+10=0-6y$
$y=-10$
The solution is –10.

12. $\frac{x}{4}=2+\frac{x-3}{3}$
$3x=24+4(x-3) \quad (\times 12)$
$3x=24+4x-12$
$-x=12$
$x=-12$
The solution is –12.

13. $R=110+3A-4(A-27.5)$
$190=110+3A-4(A-27.5)$
$190=110+3A-4A+110$
$190=220-A$
$A=220-190$
$A=30$
30 years old

14. a. $H=2.2F+69.1$
$179.1=2.2F+69.1$
$110=2.2F$
$50=F$
50 cm
There is a point on the graph at coordinates (50, 179.1).

b. $H=2.2F+69.1$
$H-69.1=2.2F$
$\frac{H-69.1}{2.2}=F$

c. $F=\frac{H-69.1}{2.2}$
$F=\frac{157.1-69.1}{2.2}$
$F=40$
40 cm
There is a point on the graph at coordinates (40, 157.1).

15. $A=PB$
$33.6=0.70B$
$\frac{33.6}{0.70}=B$
$B=48$

16. $A=PB$
$A=0.28(26)$
$A=7.28$

17. $A=PB$
$1.2=P(60)$
$\frac{1.2}{60}=P$
$0.02=P$
2%

18. $A=PB$
$\frac{A}{B}=\frac{PB}{B}$
$P=\frac{A}{B}$

19. $P=\frac{A}{B}$
$P=\frac{14}{40}$
$P=0.35$
35%

20. a. $F = \frac{9}{5}C + 32$

$104 = \frac{9}{5}C + 32$

$104 - 32 = \frac{9}{5}C$

$72 = \frac{9}{5}C$

$40 = C$

40°C

b. $F = \frac{9}{5}C + 32$

$F - 32 = \frac{9}{5}C$

$\frac{5}{9}(F - 32) = C$

$C = \frac{5}{9}(F - 32)$

21. $P = 2L + 2W$ for W

$P - 2L = 2W$

$\frac{P - 2L}{2} = W$

$W = \frac{P - 2L}{2}$

22. $I = Prt$ for P

$\frac{I}{rt} = \frac{Prt}{rt}$

$P = \frac{I}{rt}$

23. $A = \frac{B + C}{2}$ for B

$2A = B + C$

$2A - C = B$

$B = 2A - C$

24. $F = f(1 - M)$ for M

$F = f - fM$

$fM = f - F$

$M = \frac{f - F}{f}$ or $1 - \frac{F}{f}$

25. $P = \frac{RT}{V}$ for V

$PV = RT$

$V = \frac{RT}{P}$

26. $2x - y = 14$

$-y = -2x + 14$

$y = 2x - 14$

$y = 2(6) - 14$

$y = 12 - 14$

$y = -2$

27. $3x - 2y = -6$

$-2y = -6 - 3x$

$y = \frac{6 + 3x}{2}$

$y = \frac{6 + 3(-2)}{2}$

$y = \frac{6 - 6}{2}$

$y = \frac{0}{2}$

$y = 0$

28. $-3 = 3y - 4x$

$4x - 3 = 3y$

$y = \frac{4x - 3}{3}$

$y = \frac{4\left(-\frac{1}{2}\right) - 3}{3}$

$y = \frac{-2 - 3}{3}$

$y = -\frac{5}{3}$

29. $Ax + By = C$ for y

$By = C - Ax$

$y = \frac{C - Ax}{B}$

Replace each value in the right hand side of the equation (C, A, x, and B) with the number values and solve for y.

30. Let n = the number.

$6n - 20 = 4n$

$2n = 20$

$n = 10$

The number is 10.

31. Let x equal the number.

$0.60x + 8 = 332$

$0.60x = 324$

$x = 540$
The number is 540.

32. Let x equal the first integer. Then $x + 1$ equals the second integer.
$x + (x + 1) = 39$
$2x + 1 = 39$
$2x = 38$
$x = 19$
$x + 1 = 20$
The scores were 19 and 20.

33. Let x equal the first odd integer.
Then $x + 2$ equals the second odd integer.
$2x + 2(x + 2) = 56$
$2x + 2x + 4 = 56$
$4x + 4 = 56$
$4x = 52$
$x = 13$
$x + 2 = 15$
The gate's dimensions are 13 ft by 15 ft.

34. Let x = the shortest piece. Then $2x$ = the largest piece and $x + 3$ = the middle-sized piece.
$x + 2x + x + 3 = 51$
$4x + 3 = 51$
$4x = 48$
$x = 12$
$2x = 24$
$x + 3 = 15$
The lengths of the pieces are 12 cm, 15 cm, and 24 cm.

35. Let x = the number of atoms of oxygen. Then $2x + 1$ = the number of atoms of carbon and $2x$ = the number of atoms of hydrogen.
$x + 2x + 1 + 2x = 21$
$5x + 1 = 21$
$5x = 20$
$x = 4$
$2x + 1 = 9$
9 atoms of carbon

36. Let x = the days in New York. Then $5x + 29$ = the days in Los Angeles.
$x + 5x + 29 = 185$
$6x + 29 = 185$
$6x = 156$
$x = 26$
$5x + 29 = 159$
New York and Los Angeles have 26 and 159 unhealthy air days, respectively. Houston, 40; San Diego, 27; Philadelphia, 24

37. Let x = the population.
Black = 11%
$0.11x = 1.2$ million
$x = \frac{1.2 \text{ million}}{0.11}$
$x \approx 10.91$ million

38. Let x equal the price before reduction.
$x - 0.45x = 247.50$
$0.55x = 247.50$
$x = 450$
The price of the VCR before reduction was $450.

39. Let x equal the yearly income.
$x - 0.35x = 32{,}630$
$0.65x = 32{,}630$
$x = 50{,}200$
Yearly income for all recreational needs to be satisfied is $50,200.

40. Let x equal the number of days that the book was on loan.
$1.25 + 0.55(x - 1) = 10.05$
$1.25 + 0.55x - 0.55 = 10.05$
$0.55x + 0.70 = 10.05$
$0.55x = 9.35$
$x = 17$
The book was on loan for 17 days.

41. Let x equal the present age of math club president.
$x + 10 = 2x - 5$
$15 = x$
No, Dora does not date the math club president. The president is only 15 years old, younger than 18.

42. Let x = the number of concerts to ensure an audience of 2627 people.
$2627 - 2987 = -8(x - 50)$
$-360 = -8x + 400$
$8x = 400 + 360$
$8x = 760$
$x = 95$
95 concerts
The line crosses through the point represented by the coordinates (95, 2627).

43. Let x = the year that annual sales equal \$525,000.
$525{,}000 = 100{,}000 + 25{,}000x$
$525{,}000 - 100{,}000 = 25{,}000x$
$425{,}000 = 25{,}000x$
$17 = x$
17 years

44. $\{x|x > 4\}$

45. $\{x|x \le -3\}$

46. $2y - 5 < 3$
$2y < 8$
$y < 4$
$\{y|y < 4\}$
3 4 5 6 7

47. $3 - 5x \le 18$
$-5x \le 15$
$x \ge -3$
$\{x|x \ge -3\}$
−5 −4 −3 −2 −1

48. $4x + 6 < 5x$
$-x < -6$
$x > 6$
$\{x|x > 6\}$
3 4 5 6 7

49. $9(z - 1) \ge 10(z - 2)$
$9z - 9 \ge 10z - 20$
$-z \ge -11$
$z \le 11$
$\{z|z \le 11\}$
9 10 11 12 13

50. $-3(4 - x) < 4x + 3 + x$
$-12 + 3x < 5x + 3$
$-2x < 15$
$x > -\frac{15}{2}$
$\left\{x \middle| x > -\frac{15}{2}\right\}$
$-\frac{15}{2}$
−9 −8 −7 −6 −5

51. $4y - (y - 3) \le -3(2y - 7)$
$4y - y + 3 \le -6y + 21$
$3y + 3 \le -6y + 21$
$9y \le 18$
$y \le 2$
$\{y|y \le 2\}$
0 1 2 3 4

52. $\frac{5y}{4} - \frac{1}{4} \le \frac{6y}{5} + \frac{1}{5}$
$\frac{5y}{4} - \frac{6y}{5} \le \frac{1}{5} + \frac{1}{4}$
$25y - 24y \le 4 + 5$ $(\times 20)$
$y \le 9$
$\{y|y \le 9\}$
7 8 9 10 11

53. $1.1y - 0.2 \le 1.0 - 0.4y$
$1.5y \le 1.2$
$y \le 0.8$

$\{y|y \le 0.8\}$

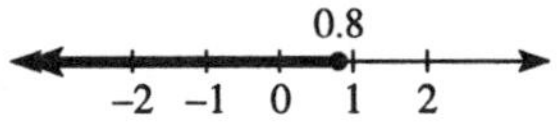

54. $-2x \ge 0$
$x \le 0$
$\{x|x \le 0\}$

55. $x \ge 24.3$

56. $x \le 33.9$

57. $24 < x < 35$

58. $x \le 33.9$

59. Let x equal the score on third test.
$\frac{42+74+x}{3} \ge 60$
$116 + x \ge 180$
$x \ge 64$
The student needs at least 64 on the third test.

60. Let x equal the number of customers.
profit $= 90n - 300$
$90n - 300 > 150000$
$90n > 150300$
$n > 1670$
The number of customers must be greater than 1670 for the company to be nationalized.

61. Let x equal the number of small dogs.
$240 + 160 + 25x \le 1000$
$400 + 25x \le 1000$
$25x \le 600$
$x \le 24$
The number of dogs is at most 24.

62. $1.1x + 24 > -0.5x + 42$
$1.1x + 0.5x > 42 - 24$
$1.6x > 18$
$x > 11.25$
$1983 + 11 = 1994$
Year 1994

Chapter 2 Test

1. $4x - 15 = 13$
$4x = 28$
$x = 7$

2. $12x + 4 = 7x - 21$
$12x - 7x = -21 - 4$
$5x = -25$
$x = -5$

3. $8x - 5(x-2) = x + 26$
$8x - 5x + 10 = x + 26$
$8x - 5x - x = 26 - 10$
$2x = 16$
$x = 8$

4. $-\frac{3}{4}x = -15$
$-3x = -60$
$x = 20$

5. $\frac{x}{10} + \frac{1}{3} = \frac{x}{5} + \frac{1}{2}$
$3x + 10 = 6x + 15$
$3x - 6x = 15 - 10$
$-3x = 5$
$x = -\frac{5}{3}$

6. $x = 0.97 + 0.03x$
$100x = 97 + 3x$
$100x - 3x = 97$
$97x = 97$
$x = 1$

7. $3x - 11 \le -23$
$3x \le -23 + 11$
$3x \le -12$
$x \le -4$

$\{x|x \le -4\}$

-5 -4 -3 -2 -1

8. $-5x > 30$
$x < -6$
$\{x|x < -6\}$

-7 -6 -5 -4 -3

9. $2x + 3 < 4x - 1$
$2x - 4x < -1 - 3$
$-2x < -4$
$x > 2$
$\{x|x > 2\}$

-2 -1 0 1 2

10. $3(x + 4) \ge 5x - 12$
$3x + 12 \ge 5x - 12$
$3x - 5x \ge -12 - 12$
$-2x \ge -24$
$x \le 12$
$\{x|x \le 12\}$

10 11 12 13 14

11. $\frac{x}{6} + \frac{1}{8} \le \frac{x}{2} - \frac{3}{4}$
$4x + 3 \le 12x - 18$
$4x - 12x \le -18 - 3$
$-8x \le -21$
$x \ge \frac{21}{8}$
$\left\{x \middle| x \ge \frac{21}{8}\right\}$

$\frac{21}{8}$

1 2 3 4 5

12. $4x + 3y = 8$ for y
$3y = 8 - 4x$
$y = \frac{8 - 4x}{3}$

13. $V = \pi r^2 h$ for h
$\frac{V}{\pi r^2} = h$
$h = \frac{V}{\pi r^2}$

14. $L = \frac{P - 2W}{2}$ for W
$2L = P - 2W$
$2L - P = -2W$
$\frac{2L - P}{-2} = W$
$W = \frac{P - 2L}{2}$

15. 13.2 is 60% of what number?
$13.2 = 0.60 \cdot x$
$\frac{13.2}{0.60} = x$
$x = \frac{132}{6} = 22$

16. What percent of 90 is 12.6?
$x \cdot 90 = 12.6$
$x = \frac{12.6}{90}$
$x = 0.14$
14%

17. Substitute $S = 48{,}196$ and solve for t.
$48{,}196 = 1472t + 21{,}700$
$\frac{48{,}196 - 21{,}700}{1472} = t$
$t = 18$
18 years after 1984, in 2002, the average annual salary for teachers will be $48,196.

18.

Six	more than	twice a number	is	34
6	+	$2x$	=	34

$2x = 34 - 6$
$2x = 28$
$x = 14$

19. The longest word in Spanish contains 22 letters. The longest word in Portuguese has 2 more letters than the longest word in French. Let x = the number of letters in the longest French word. Then $x + 2$ = the number of letters in the longest Portuguese word.
$x + (x + 2)$ exceeds the longest Spanish word by 30 letters.
$x + (x + 2) = 22 + 30$
$2x + 2 = 52$
$2x = 50$
$x = 25$
The longest French word has 25 letters and the longest Portuguese word has 27 letters.

20. Let Y = length (in miles) of the Yukon River.
Then Missouri is 560 miles longer than Y.
Missouri = $Y + 560$
And Mississippi is 1620 miles shorter than $2Y$.
Mississippi = $2Y - 1620$
Also,
Missouri + Mississippi + Yukon = 6860.
$(Y + 560) + (2Y - 1620) + Y = 6860$
$Y + 2Y + Y = 6860 - 560 + 1620$
$4Y = 7920$
$Y = 1980$
Yukon is 1980 miles long.
Mississippi is 2(1980) – 1620
= 2340 miles long.
Missouri is 1980 + 560 = 2540 miles long.

21. Let x be the original price.
$x - 0.35x = 1430$
$0.65x = 1430$
$x = 2200$
The original price was $2200.

22. Let x be the number of years of education.
$\$7200 + \$2600x = \$35,800$
$2600x = 35,800 - 7200$
$2600x = 28,600$
$x = 11$
11 years of education are needed.

23. 12% of what population is 24 million?
$0.12x = 24$
$x = 200$
The 1995 U.S. population over age 15 is 200 million.

24. $43t + 381 < 639$
$43t < 258$
$t < 6$
For years 1970 to 1975, the federal debt was less than $639 billion.

25. $\frac{76 + 80 + 72 + x}{4} \geq 80$
$76 + 80 + 72 + x \geq 320$
$x \geq 92$
The student must earn 92% or more on the fourth examination.

Chapter 3

Answers to Enrichment Essay: Critical Thinking: Is This a Trick Question?

1. Of course not. It's illegal to bury live people.
2. Yes; however it's not a holiday.
3. Moses didn't take any animals onto the ark, Noah did.
4. They all do. Some, of course, have more than that.
5. The match.
6. They weren't playing chess with each other.
7. Zero; since one of the factors is $(x - x)$, which equals 0, making the product 0.

Problem Set 3.1

1. Let n equal the number.
$7(n - 11) = 588$
$7n - 77 = 588$
$7n = 665$
$n = 95$

3. Let n equal the number.
$7n - 11 = 290$
$7n = 301$
$n = 43$

5. Let x equal the maximum price of dinner.
$x + 0.15x = 32.20$
$1.15x = 32.20$
$x = 28$
$28

7. Let x equal the selling price.
$x - 0.07x = 111{,}600$
$0.93x = 111{,}600$
$x = 120{,}000$
$120,000

9. Let x equal the number of hours to repair the car.
parts + labor = 448
$63 + 35x = 448$
$35x = 385$
$x = 11$
11 hours

11. Let x equal the number of years after 1939.
$280 + 1.44x = 559.36$
$1.44x = 279.36$
$x = 194$
The CO_2 level will reach 559.36 ppm in 194 years, that is, in the year 2133.

13. Let x equal the number of months after birth.
$7 + 1.5x = 16$
$1.5x = 9$
$x = 6$
after 6 months

15. Let x equal the amount invested at 9%.
$0.09x + 0.12(25{,}000 - x) = 2550$
$0.09x + 3000 - 0.12x = 2550$
$450 = 0.03x$
$15{,}000 = x$
$15,000 at 9% and $10,000 at 12%

17. Let x equal the amount at 14%. $2x$ equals the amount at 12%.
$0.14x + 0.12(2x) = 256.50$
$0.14x + 0.24x = 256.50$
$0.38x = 256.50$
$x = 675$
$2x = 1350$
$675 at 14%, $1350 at 12%

19. Amount of acid at 30% + amount of acid at 12% = amount of acid at 20%.
Let x equal the amount of acid at 30%.
$50 - x$ equals the amount of acid at 12%.
$0.30x + 0.12(50 - x) = 0.20(50)$
$0.30x + 6 - 0.12x = 10$
$0.18x = 4$
$x = 22\frac{2}{9}$
$50 - x = 27\frac{7}{9}$
$22\frac{2}{9}$ liters at 30%; $27\frac{7}{9}$ liters at 12%

21. 50 mi/hr → ← 55 mi/hr
315 miles

Let t equal time.
$50t + 55t = 315$
$105t = 315$
$t = 3$
The vehicles meet after 3 hours.

23.

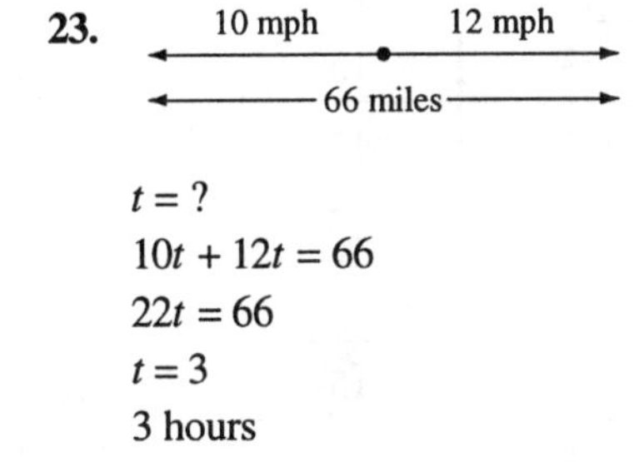

$t = ?$
$10t + 12t = 66$
$22t = 66$
$t = 3$
3 hours

25. 2, 8, 14, 20, ____, ____, ____
Add 6 repeatedly: 26, 32, 38

27. 15, 11, 7, 3, ____, ____, ____
Subtract 4 repeatedly: –1, –5, –9

29. Answers will vary.
3, 4, 7, 11, 18, 29, ____, ____, ____
Add the two previous numbers to obtain the next number in the sequence: 47, 76, 123
$18 + 29 = 47$
$29 + 47 = 76$
$47 + 76 = 123$

31. $1 + 3 = 4$
$1 + 3 + 5 = 9$
$1 + 3 + 5 + 7 = 16$
$1 + 3 + 5 + 7 + 9 = 25$
The sum of the first n odd numbers is n^2.
$1 + 3 + \cdots + 99 + 101 + \cdots + 197 + 199$
$= 100^2 = 10,000$

33.	1 point free throw		2 point field goal		3 point long shot		total
	no.	points	no.	points	no.	points	
1.					5	15	15
2.	1	1	1	2	4	12	15
3.	3	3			4	12	15
4.			3	6	3	9	15
5.	2	2	2	4	3	9	15
6.	4	4	1	2	3	9	15
7.	6	6			3	9	15
8.	1	1	4	8	2	6	15
9.	3	3	3	6	2	6	15
10.	5	5	2	4	2	6	15
11.	7	7	1	2	2	6	15
12.	9	9			2	6	15
13.			6	12	1	3	15
14.	2	2	5	10	1	3	15
15.	4	4	4	8	1	3	15
16.	6	6	3	6	1	3	15
17.	8	8	2	4	1	3	15
18.	10	10	1	2	1	3	15
19.	12	12			1	3	15
20.	1	1	7	14			15
21.	3	3	6	12			15
22.	5	5	5	10			15
23.	7	7	4	8			15
24.	9	9	3	6			15
25.	11	11	2	4			15
26.	13	13	1	2			15
27.	15	15					15

There are 27 ways.

35. We know the sum is $2 + 5 + 2 = 9$ since one of the area codes is 252. No area code has a first digit of 4.
One area code begins with 6:
$6 + a + b = 9$
Another area code ends with 1:
$c + d + 1 = 9$
No area code contains a digit that is in one of the other area codes.
$a + b = 3$ only 0 and 3 are possible
$c + d = 8$ only 7 and 1 are possible
Thus, the number can be 711 or 171.

37. pattern: $1, 9, 9 + 8 = 17, 17 + 8 = 25, \ldots$
$= 1, 1 + 8(1), 1 + 8(2), 1 + 8(3), \ldots$
$= 1, 1 + 8(2 - 1), 1 + 8(3 - 1), 1 + 8(4 - 1), \ldots$
number of dots in 10th term: $1 + 8(10 - 1) = 1 + 8(9) = 73$
number of dots in nth term: $1 + 8(n - 1) = 1 + 8n - 8 = 8n - 7$

39.

1st triangular Number	2nd	3rd	4th	5th	6th	12th	nth
1	3	6	10	15	21	78	$\frac{n(n+1)}{2}$

Note pattern: $\frac{1(1+1)}{2} = 1$
$\frac{2(2+1)}{2} = 3$
$\frac{3(3+1)}{2} = 6$
$\frac{4(4+1)}{2} = 10$
$\frac{5(5+1)}{2} = 15$
$\frac{6(6+1)}{2} = 21$
$\frac{12(12+1)}{2} = 78$

	1st term ($n = 1$)	2nd term ($n = 2$)	3rd term ($n = 3$)	4th term ($n = 4$)	5th term ($n = 5$)	Formula for the nth term
41.	1	3	5	7	9	$2n - 1$
43.	5	7	9	11	13	$2n + 3$
45.	2	8	18	32	50	$2n^2$

47. $9 + 8 + 7 + 65 + 4 + 3 + 2 + 1 = 99$

49. Given: $3x + 1 = 16$
$3x = 15$
$x = 5$
Conclusion: the number is 5.
True; argument is valid.

51. Given: Two numbers are both even.
(numbers could be 8 and 12 or 8 and 10)
Conclusion: The numbers differ by 2.
Not necessarily true; argument is invalid.

53. Given: My number is 8
Your number is 10
Fred's number is $8 + 3 = 11$
Conclusion: Fred's number is $10 - 1 = 9$.
False; argument is invalid.

55. Given: Tony's number is divisible by 5.
Maria's number is divisible by 3.
Conclusion: Tony's number and Maria's number cannot be the same.
False; 15 is one example of a number that is divisible by 3 and 5; argument is invalid.

57. Given: (winning team) – (losing team) = 4
Let x equal the points scored by losing team.
$x + 4$ equals points scored by winning team.
2(winning team) = 3(losing team) – 14
$2(x + 4) = 3x - 14$
$2x + 8 = 3x - 14$
$22 = x$
$x + 4 = 26$
Conclusion: The final score of the game was 26 to 22.
True; argument is valid.

59. Given: Let A = one number
$B - A$ = other number
Conclusion: The other number is $B - A$.
True; the argument is valid.

61. $\text{GP\%} = \dfrac{\text{sales} - \text{overhead}}{\text{sales}} \times 100$
Given: GP% = 25%
Sales = *not given*
Overhead = *not given*
Too little information

63. Let x equal Lorna's weight.
$x + 10$ equals Ada's weight.
$(x + 10) + 10 = x + 20$ equals Ana's weight.
More information needs to be given such as the sum of weights in order to solve the individual weight.
Too little information

65. Let x equal the original price.
$x - 0.35x = 0.65x$ equals the sale price.
$0.65x = 55$
Just the right amount of information given

67. Let x equal the length of the middle-sized piece.
$x - 16$ equals the length of the shortest piece.
$x + 12$ equals the length of the longest piece.
$x + (x - 16) + (x + 12) = 56$
Find $x - 16$.
Just the right amount of information given

69. 727

71. Answers may vary.

73. Answers may vary.

75. 12, a dozen stamps

77. The doctor was a female. She had a brother, but he had no brother.
Relationship: brother and sister

79. One word

81. Let x equal the number of tennis balls worth \$3.
$15 - x$ equals the number of tennis balls worth \$4.
$3x + 4(15 - x) = 49$
$3x + 60 - 4x = 49$

$-x = -11$
$x = 11$
$15 - x = 15 - 11 = 4$
eleven \$3 balls, four \$4 balls

83. Let x equal the number of ducks.
$20 - x$ equals the number of horses.
$2x + 4(20 - x) = 64$
$2x + 80 - 4x = 64$
$-2x = -16$
$x = 8$
$20 - x = 20 - 8 = 12$
8 ducks, 12 horses

Review Problems

85. $5(2 - y) + 3 = 3 + 4(3 - y)$
$10 - 5y + 3 = 3 + 12 - 4y$
$13 - 5y = 15 - 4y$
$-y = 2$
$y = -2$

86. $2y - (y + 7) < 3(y + 2) - 5$
$2y - y - 7 < 3y + 6 - 5$
$y - 7 < 3y + 1$
$-2y < 8$
$y > -4$
$\{y | y > -4\}$

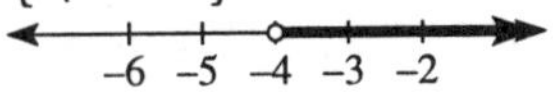

87. $P = 2s + b$ for s
$2s + b = P$
$2s = P - b$
$s = \frac{P - b}{2}$

Problem Set 3.2

1. $\frac{24 \text{ feet}}{36 \text{ feet}} = \frac{24}{36} = \frac{2}{3}$

3. $\frac{3\frac{1}{2} \text{ yards}}{5 \text{ yards}} = \frac{3\frac{1}{2}}{5} = \frac{\frac{7}{2}}{5} = \frac{7}{2} \cdot \frac{1}{5} = \frac{7}{10}$

5. $\frac{4 \text{ inches}}{3 \text{ feet}} = \frac{4 \text{ inches}}{3(12) \text{ inches}} = \frac{4}{36} = \frac{1}{9}$

7. $\frac{3 \text{ gallons}}{2 \text{ quarts}} = \frac{3(4 \text{ quarts})}{2 \text{ quarts}} = \frac{12}{2} = 6$

9. $\frac{30 \text{ centimeters}}{1 \text{ meter}} = \frac{30 \text{ centimeters}}{100 \text{ centimeters}}$
$= \frac{30}{100} = \frac{3}{10}$

11. $\frac{2000 \text{ pounds}}{6 \text{ tons}} = \frac{1 \text{ ton}}{6 \text{ tons}} = \frac{1}{6}$

13. $\frac{24}{x} = \frac{12}{7}$
$7 \cdot 24 = 12x$
$12x = 7(24)$
$x = \frac{7(24)}{12} = 7(2) = 14$
The solution is 14.

15. $\frac{y}{6} = \frac{18}{4}$
$4y = 6(18)$
$y = \frac{6(18)}{2 \cdot 2} = 3 \cdot 9 = 27$
The solution 27.

17. $\frac{y}{3} = -\frac{3}{4}$
$4y = -9$
$y = -\frac{9}{4}$
The solution is $-\frac{9}{4}$.

19. $\frac{-3}{8} = \frac{x}{40}$
$-120 = 8x$
$x = -15$
The solution is –15.

21. $\frac{x-2}{5}=\frac{3}{10}$
$10(x-2)=15$
$10x-20=15$
$10x=35$
$x=\frac{7}{2}$
The solution is $\frac{7}{2}$.

23. $\frac{y+10}{10}=\frac{y-2}{4}$
$4(y+10)=10(y-2)$
$4y+40=10y-20$
$60=6y$
$10=y$
The solution is 10.

25. $\frac{\text{actual height}}{\text{height in picture}}=\frac{4\text{ feet}}{0.6\text{ inch}}$
$=\frac{48\text{ inches}}{0.6\text{ inches}}=80$ or 80:1

27. $\frac{250}{390}=\frac{25}{39}$

29. $\frac{107+228}{587}=\frac{335}{587}$

31. $\frac{5.3}{10}$

33. $\frac{16\text{ hours}}{8\text{ hours}}=\frac{2}{1}$ or 2:1

35. $\frac{2\text{ hours}}{6\text{ hours}}=\frac{1}{3}$ or 1:3

37. $\frac{31}{18}$

39. The ratio increases.

41. 20-ounce size: $\frac{\$0.96}{20\text{ oz}}=\$0.048/\text{oz}$
50-ounce size: $\frac{\$2.20}{50\text{ oz}}=\$0.044/\text{oz}$
best buy: 50-ounce size

43. 5 years after installation:
$C_{\text{electric}}=5000+1100(5)=\$10,500$
$C_{\text{solar}}=30,000+150(5)=\$30,750$
40 years after installation:
$C_{\text{electric}}=5000+1100(40)=\$49,000$
$C_{\text{solar}}=30,000+150(40)=\$36,000$
Ratio after 5 years:
$\frac{\text{electric}}{\text{solar}}:\ \frac{10,500}{30,750}=\frac{14}{41}$
Ratio after 40 years:
$\frac{\text{electric}}{\text{solar}}:\ \frac{49,000}{36,000}=\frac{49}{36}$
$\frac{14}{41};\ \frac{49}{36}$

45. $\frac{11\text{ pounds}}{500\text{ sheets}}=\frac{x}{3200\text{ sheets}}$
$500x=11(3200)$
$x=\frac{11(3200)}{500}$
$x=70.4$ pounds

47. $\frac{55\text{ kg}}{8.8\text{ kg}}=\frac{90\text{ kg}}{x}$
$55x=90(8.8)$
$x=\frac{90(8.8)}{55}$
$x=14.4$ kg

49. $\frac{x}{50}=\frac{108}{27}$
$27x=50(108)$
$x=\frac{50(108)}{27}$
$x=200$ bass

51. $\frac{x}{82.4}=\frac{\$60}{148.2}$
$148.2x=4944$
$x=\$33.36$

53. batting average
$=\frac{\text{hits made}}{\text{number of times player came to bat}}$
Let x equal the hits made.
$\frac{x}{40}=0.325$
$x=13$
13 hits made

55. d is true;

a. $\frac{3 \text{ yards}}{4 \text{ feet}} = \frac{3(3 \text{ feet})}{4 \text{ feet}} = \frac{9}{4} \neq \frac{3}{4}$

b. $\frac{4}{3} = \frac{30}{\text{women}} \rightarrow \text{women} = \frac{90}{4} \approx 22 \neq 40$

c. $\frac{4}{y} = \frac{5}{7} \rightarrow 5y = 28$ *not* $7y = 20$

d. $\frac{y-4}{y} = \frac{3}{4} \rightarrow 4y - 16 = 3y$; true

57. $\frac{7.32}{2y-5} = \frac{-19.03}{28-5y}$

$7.32(28 - 5y) = -19.03(2y - 5)$

$204.96 - 36.6y = -38.06y + 95.15$

$1.46y = -109.81$

$y \approx -75.21$

59. Answers may vary.

61. Answers may vary.

63. Fran: $\frac{3 \text{ boats}}{2 \text{ days}}$

Martell: $\frac{1}{2}\left(\frac{3}{2} \text{ boats / day}\right) = \frac{3 \text{ boats}}{4 \text{ days}}$

$\frac{3 \text{ boats}}{4 \text{ days}} = \frac{9 \text{ boats}}{x}$

$3x = 36$

$x = 12$ days

65. Total invested by the three people:

$A + B + B = A + 2B$ dollars.

ratio of earnings:

first person: $\frac{A \text{ dollars}}{A + 2B \text{ dollars}}$

Amount of profit from \$1000 for the first person:

$\frac{A(\$1000)}{A+2B} = \frac{1000A}{A+2B}$ dollars

67. Group Activity Problem.

Review Problems

68. $\frac{1}{2}x + 7 = 13 - \frac{1}{4}x$

$\frac{3}{4}x = 6$

$x = \frac{4}{3}(6)$

$x = 8$

The solution is 8.

69. $2x - 3 \leq 5$

$2x \leq 8$

$x \leq 4$

$\{x | x \leq 4\}$

0 1 2 3 4

70. $\frac{35}{100} = \frac{x}{40}$

$100x = 1400$

$x = 14$ ml

Problem Set 3.3

1. Let x equal the measure of the angle.

$90 - x$ equals the measure of its complement.

$x = (90 - x) + 60$

$2x = 150$

$x = 75$

The angle measures 75°.

3. Let x equal the measure of angle.

$180 - x$ equals the measure of its supplement.

$x = 3(180 - x)$

$x = 540 - 3x$

$4x = 540$

$x = 135$

The angle measures 135°.

5. Let x equal the measure of the angle.

$90 - x$ equals the measure of its complement.

$180 - x$ equals the measure of its supplement.

$180 - x = 3(90 - x) + 10$

$180 - x = 270 - 3x + 10$

$2x = 100$

$x = 50$

The angle measures 50°.

7. $x + x + (x + 30) = 180$
$3x = 150$
$x = 50$
$50°, 50°, 80°$

9. $(4x) + (3x + 4) + (2x + 5) = 180$
$9x + 9 = 180$
$9x = 171$
$x = 19$
$(4x)° = (4 \cdot 19)° = 76°$
$(3x + 4)° = (3 \cdot 19 + 4)° = (57 + 4)° = 61°$
$(2x + 5)° = (2 \cdot 19 + 5)° = (38 + 5)° = 43°$

11. Let x equal the measure of the smallest angle.
$2x$ equals the measure of the second angle.
$x + 20$ equals the measure of the third angle.
$x + 2x + (x + 20) = 180$
$4x + 20 = 180$
$4x = 160$
$x = 40$
$2x = 80$
$x + 20 = 60$
The angles of the triangle measure 40°, 80°, and 60°.

13. $2(x) + 2(3x - 20) = 360$
$2x + 6x - 40 = 360$
$8x = 400$
$x = 50$
2 angles are 50°, 2 angles are 130°
Opposite interior angles are equal.

15. length = x, width = $x - 90$
$2x + 2(x - 90) = 1280$
$4x - 180 = 1280$
$4x = 1460$
$x = 365$
365 miles by 275 miles

17. width = x, length = $2x + 6$
$2x + 2(2x + 6) = 228$
$2x + 4x + 12 = 228$
$6x = 216$
$x = 36$
36 feet by 78 feet

19. Let x equal the length of the second side.
$2x - 1$ equals the length of the first side.
$2x + 1$ equals the length of the third side.
$x + (2x - 1) + (2x + 1) = 30$
$5x = 30$
$x = 6$
$2x - 1 = 11$
$2x + 1 = 13$
The length of the sides of the triangle are 6 inches, 11 inches, and 13 inches.

21. $L \cdot W = A$
$L \cdot (25) = 1250$
$L = 50$ meters

23. $\frac{1}{2}bh = A$
$\frac{1}{2}(21)h = 147$
$h = \frac{2}{21}(147)$
$h = 14$
The height is 14 m.

25. radius = $\frac{1}{2}$(diameter) = 3"
$A = \pi r^2 = \pi(3^2) = 9\pi \approx 9(3.14) = 28.26$
$C = 2\pi r = 2\pi(3) = 6\pi \approx 6(3.14) = 18.84$
$A = 9\pi$ sq. in. ≈ 28.26 sq in.;
$C = 6\pi" \approx 18.84"$

27. pool radius = $\frac{1}{2}(40) = 20$ m.
pool plus walkway radius = 20 + 2 = 22 m
$A_{\text{pool+walkway}} - A_{\text{pool}} = A_{\text{walkway}}$
$A_w = \pi(22)^2 - \pi(20)^2$
$= 484\pi - 400\pi = 84\pi$
$\approx 84(3.14) = 263.76$
$A_w = 84\pi \text{ m}^2 \approx 263.76 \text{ m}^2$

29. $V = \pi r^2 h$

$\frac{V_{\text{larger}}}{V_{\text{smaller}}} = \frac{\pi[3(3)]^2(4)}{\pi(3)^2(4)} = \frac{324\pi}{36\pi} = 9$

The larger cylinder has 9 times the volume of the smaller cylinder.

31. $A = \frac{1}{2}(d+b)h$

$A = \frac{1}{2}(10+6)5 = \frac{1}{2}(16)(5)$

$= 40$ square feet

33. $\frac{18}{9} = \frac{10}{x}$

$2 = \frac{10}{x}$

$2x = 10$

$x = 5$ in.

35. $\frac{20}{15} = \frac{x}{12}$

$15x = 20(12)$

$x = \frac{240}{15} = 16$ in.

37. $\frac{8}{6} = \frac{x}{12}$

$x = \frac{8(12)}{6} = 16$ feet

39. $2x + 2x + 40 = 180$

$4x = 140$

$x = 35$

The angle is $35°$.

41. **a.** $C = 2\pi r$

$446 \approx 2(3.14)r = 6.28r$

$r = 71.02 \approx 71$ feet

b. radius of floor $= 71 - 4 = 67$ feet

$A = \pi r^2 = \pi(67)^2 = 4489\pi$

$\approx 4489(3.14) = 14{,}095.46$

$A = 4489\pi$ sq ft $\approx 14{,}095.5$ sq ft

c. $V = \pi r^2 h$

$1{,}691{,}455 \approx (3.14)(67)^2 h$

$= 14{,}095.46h$

$h \approx 120$ feet

43. b is true.

a. A circle a radius $r = 2$ has $C = 2\pi r = 2\pi(2) = 4\pi$ and $A = \pi r^2 = \pi(2^2) = 4\pi$.

b. $x + 3(90 - x) = (180 - x) + (90 - x)$

$x + 270 - 3x = 270 - 2x$, true

c. $90 - x$ is an acute angle, not an obtuse angle.

d. They can be equal if the angles are equal.

45. Answers may vary.

47. length of box = 10 cm − 2(2 cm) = 6 cm

width of box = 10 cm − 2(2 cm) = 6 cm

height of box = 2 cm

$V = lwh$ = 6 cm(6 cm)(2 cm) = 72 cubic cm

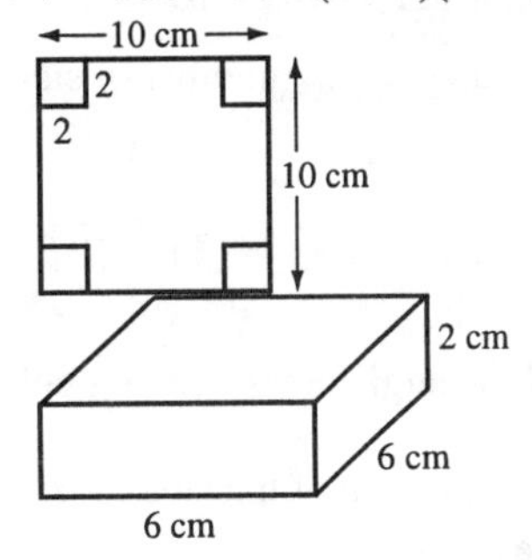

49. $l + w = \frac{40}{2} = 20$

$l = 20 - w$

Make a table.

w	l	Area
1	19	19
2	18	36
3	17	51
4	16	64
5	15	75
6	14	84
7	13	91
8	12	96
9	11	99
10	10	100
11	9	99

etc.

(The remaining areas are all less than 100)

The dimensions of 10 yd × 10 yd provide a region of greatest possible area.

51.

Polygon	number of sides	number of triangles	sum of the angles' measures
1.	4	2	$2 \times 180 = 360$
2.	5	3	$3 \times 180 = 540$
3.	6	4	$4 \times 180 = 720$

The sum of the angle measures of a polygon of n sides is $(n - 2)180$.

53. Let x equal the length of the side of the triangle. $x - 3$ equals the height of the rectangle. x equals the width of the rectangle.

width of rectangle = length of side of triangle

perimeter of figure = 2 length of side of triangle + 2 height of rectangle + width of rectangle

$34 = 2x + 2(x - 3) + x$

$34 = 2x + 2x - 6 + x$

$40 = 5x$

$8 = x$

The length of a side of the triangle is 8 meters.

55. area of square + area of triangle = 1020
$30(30)+\frac{1}{2}(30)h=1020$
$900 + 15h = 1020$
$15h = 120$
$h = 8$
The height of the triangle is 8 ft.

Review Problems

57. $2(x-7)-3(x+4)=4-(5x-2)$
$2x-14-3x-12=4-5x+2$
$-x-26=6-5x$
$4x=32$
$x=8$

58. $14=25\%(x)=0.25x$
$x=\frac{14}{0.25}=56$

59. $[3(12\div 2^2-3)^2]^2$
$=[3(12\div 4-3)^2]^2$
$=[3(3-3)^2]^2=0$

Chapter 3 Review Problems

1. $4(x-8)=24$
$4x-32=24$
$4x=56$
$x=14$

2. $4x-8=24$
$4x=32$
$x=8$

3. Let x equal the maximum price of dinner.
$x+0.25x=21.25$
$1.25x=21.25$
$x=17$
\$17

4. $x+0.05x=126$
$1.05x=126$
$x = 120$ pounds

5. US = 3 + 3G, Ch = 2 + 2G,
US + Ch + G = 41, so
US + Ch + G
= (3 + 3G) + (2 + 2G) + G = 41
5 + 6G = 41
6G = 36
G = 6
then US = 3 + 3(6) = 21
and Ch = 2 + 2(6) = 14
US: 21 million tons, China: 14 million tons, Germany: 6 million tons

6. Let x be the number of years after 1960.
$179.5+2.35x=320.5$
$2.35x=141$
$x=60$
The population reaches 320,500,000 in the year $1960 + x = 1960 + 60 = 2020$

7. Group A: \$30,000 + \$1500x
Group B: \$21,000 + \$2000x

a. After 10 years
Group A: 30,000 + 1500(10) = \$45,000
Group B: 21,000 + 2000(10) = \$41,000
Group A, \$45,000 after 10 years

b. $30{,}000 + 1500x = 21{,}000 + 2000x$
$9{,}000 = 500x$
$x = 18$
After 18 years, Group B's salary will be the same as Group A's salary.

8. Let x equal the total weight of the banana.
$\frac{1}{8}x$ equals the weight of the banana peel.
$\frac{7}{8}x$ equals the weight of the peeled banana.
$x=\frac{7}{8}x+\frac{7}{8}$
$\frac{1}{8}x=\frac{7}{8}$
$x=7$
7 ounces

9. Let x equal the amount at 8%.
$1000 - x$ equals the amount at 10%.
$0.08x + 0.10(1000 - x) = 94$
$0.08x + 100 - 0.10x = 94$
$-0.02x = -6$
$x = 300$
$1000 - x = 700$
\$300 at 8%, \$700 at 10%

10. Let x equal the amount invested at 8%.
$2x + 100$ equals the amount invested at 9%.
$0.08x + 0.09(2x + 100) = 1910$
$0.08x + 0.18x + 9 = 1910$
$0.26x = 1901$
$x \approx 7311.54$
$2x + 100 = 14{,}723.08$
\$7311.54 at 8%; \$14,723.08 at 9%

11. Let x equal the amount of 75% salt solution.
$10 - x$ equals the amount of 50% salt solution.
$0.75x + 0.50(10 - x) = 0.60(10)$
$0.75x + 5 - 0.5x = 6$
$0.25x = 1$
$x = 4$
$10 - x = 6$
4 gallons of 75% salt solution,
6 gallons of 50% salt solution

12. Let x equal the number of students in school with 10% African American. $1000 - x$ equals the number of students in school with 90% African American.
$0.10x + 0.90(1000 - x) = 0.42(1000)$
$0.10x + 900 - 0.90x = 420$
$-0.80x = -480$
$x = 600$
$1000 - x = 400$
600 students at school with 10% African American; 400 students at school with 90% African American

13. $60t + 80t = 400$
$140t = 400$
$t = 2\frac{6}{7}$
$2\frac{6}{7}$ hours

14.

pennies		nickels		dimes		total
no.	amount	no.	amount	no.	amount	
		1	5	1	10	15
5	5			1	10	15
		3	15			15
5	5	2	10			15
10	10	1	5			15
15	15					15

There are 6 ways of making change.

15. $\frac{-6}{9} = -\frac{2}{3}$; $\frac{4}{-6} = -\frac{2}{3}$;
To obtain the next term, multiply the previous term by $-\frac{2}{3}$.
$4\left(-\frac{2}{3}\right) = -\frac{8}{3}$
Check: $-\frac{8}{3}\left(-\frac{2}{3}\right) = \frac{16}{9}$

16. $\frac{A^2}{B} = C$
$\frac{7^2}{5} = \frac{49}{5} = \frac{49}{5}$
$\frac{8^2}{2} = \frac{64}{2} = 32$
$C = \frac{A^2}{B}$

17. Label boxes a, b, and c.
$a + b = 13$
$a + c = 16$
$b + c = 21$

$c - b = 3$
$\underline{c + b = 21}$
$2c = 24$
$c = 12$

$b = 21 - 12$
$b = 9$
$a = 13 - 9$
$a = 4$

4, 9, 12

18. Let $A = B = C = 0$
Then ABC since

$$\begin{array}{r} ABC \\ ABC \\ \underline{+ABC} \\ BBB \end{array} \qquad \begin{array}{r} 000 \\ 000 \\ \underline{+000} \\ 000 \end{array}$$

19. Row a: 3^2, 5^2, 3^3, 5^3
Row a contains 3^2 and 5^3
(or 3^3 and 5^2)

20. $(99-9)(99-19)(99-29) \cdots (99-199)$
$= 90(80)(70)\cdots(-100)$
$= 90(80)(70)\cdots(10)(0)(-10)(-20)\cdots(-100)$
$= 0$

21. from

to

where the 4 smaller squares combine to form a fifth, larger square

22. $\frac{\text{6 inches}}{\text{4 feet}} = \frac{\text{6 inches}}{\text{4(12 inches)}} = \frac{6}{48}$
$= \frac{1}{8}$ or 1:8

23. $\frac{\text{10 centimeters}}{\text{3 meters}} = \frac{\text{10 cm}}{\text{3(100 cm)}}$
$= \frac{10}{300} = \frac{1}{30}$ or 1:30

24. 40 people: 2(12) = 24 men;
40 − 24 = 16 women
$\frac{\text{women}}{\text{men}} = \frac{16}{24} = \frac{2}{3}$ or 2:3

25. a. $\frac{\text{Brazil}}{\text{US}} = \frac{30}{22} = \frac{15}{11}$

b. $\frac{\text{Australia and India}}{\text{Colombia}} = \frac{16+10}{18}$
$= \frac{26}{18} = \frac{13}{9}$

26. $\frac{\$3.96}{\text{18 oz}} = 0.22$
22¢ per ounce

27. $\frac{3}{x} = \frac{15}{25}$
$15x = 3(25)$
$x = \frac{75}{15}$
$x = 5$

28. $\frac{-3}{8} = \frac{x}{64}$
$-192 = 8x$
$x = -24$

29. $\frac{3 \text{ teachers}}{50 \text{ students}} = \frac{x}{5400 \text{ students}}$
$50x = 3(5400)$
$x = 324$ teachers

30. $\frac{32 \text{ tagged}}{82 \text{ caught}} = \frac{112 \text{ tagged}}{x}$
$32x = 82(112)$
$x = 287$
287 trout in lake

31. $\frac{\$20,000}{38.8} = \frac{x}{148.2}$
$2,964,000 = 38.8x$
$x \approx \$76,391.75$

32. Let x equal the measure of the angle.
$90 - x$ equals the measure of its complement.
$90 - x = 3x - 10$
$100 = 4x$
$25 = x$
$90 - x = 65$
angle, 25°; complement, 65°

33. Let x equal the measure of the angle.
$180 - x$ equals the measure of the supplement.
$180 - x = 4x - 45$
$225 = 5x$
$45 = x$
$180 - x = 135$
angle, 45°; supplement, 135°

34. $x + 2x + 3x = 180$
$6x = 180$
$x = 30$
$2x = 60$
$3x = 90$
30°, 60°, 90°

35. Let x = measure of angle A.
Then $7x + 11$ = measure of angle B, and
$5x$ = measure of angle C.
$x + 7x + 11 + 5x = 180$
$13x = 169$
$x = 13$
$7x + 11 = 102$
$5x = 65$
$m\angle A = 13^\circ, m\angle B = 102^\circ,$
$m\angle C = 65^\circ$

36. $5x - 3 = 2x + 6$
$3x = 9$
$x = 3$
$2x + 6 = 2 \cdot 3 + 6 = 6 + 6 = 12$
$5x - 3 = 5 \cdot 3 - 3 = 15 - 3 = 12$
$9x + 2 = 9 \cdot 3 + 2 = 27 + 2 = 29$

37. width = W, length = $14 + 2W$
$P = 2L + 2W$
$346 = 2(14 + 2W) + 2W = 28 + 6W$
$318 = 6W$
$W = 53$
so $L = 14 + 2(53) = 120$
53 m × 120 m

38. Let x, $x + 2$, $x + 4$ represent the sides of the triangle.
$x + (x + 2) + (x + 4) = 87$
$3x + 6 = 87$
$3x = 81$
$x = 27$
$x + 2 = 29$
$x + 4 = 31$
The lengths of the sides of the triangle are 27 yards, 29 yards, and 31 yards.

39. 4 shelves of length $3x$ plus 3 "sides"
$4(3x) + 3(x) = 60$
$15x = 60$
$x = 4$
4 ft high by 12 ft wide

40. $A = \frac{1}{2}bh$
$42 = \frac{1}{2}(14)h$
$h = 6$ feet

41. radius $= \frac{1}{2}(\text{diameter}) = \frac{1}{2}(10) = 5$ m
$A = \pi r^2 = \pi(5^2) = 25\pi \approx 25(3.14) = 78.5$
$C = 2\pi r = 2\pi(5) = 10\pi \approx 10(3.14) = 31.4$
$A = 25\pi \text{ m}^2 \approx 78.5 \text{ m}^2$
$C = 10\pi \text{ m} \approx 31.4 \text{ m}$

42. $A = \frac{1}{2}h(d+b)$
$A = 36$ sq yd, $d = 7$ yd, $h = 6$ yd, $b = ?$
$36 = \frac{1}{2}6(7+b)$
$36 = 3(7 + b)$
$12 = 7 + b$
$5 = b$
5 yards

43. $V = \pi r^2 h$
$$\frac{V_{\text{smaller}}}{V_{\text{larger}}} = \frac{\pi(3^2)5}{\pi(6^2)(10)} = \frac{45\pi}{360\pi} = \frac{1}{8}$$

44. $V = \left(\frac{4}{3}\right)\pi r^3$
$$\frac{V_{\text{larger}}}{V_{\text{smaller}}} = \frac{\left(\frac{4}{3}\right)\pi(6)^3}{\left(\frac{4}{3}\right)\pi(3)^3} = \frac{288\pi}{36\pi} = 8$$
8 times larger

45. $\frac{x}{10} = \frac{5}{8}$
$8x = 50$
$x = 6\frac{1}{4}$ feet

46. $\frac{x}{8} = \frac{15}{24}$
$24x = 120$
$p = 5$ feet

Chapter 3 Test

1. Let s be the previous salary.
$s + 0.05s = 33{,}600$
$1.05s = 33{,}600$
$s = 32{,}000$
The previous salary was $32,000.

2. Let H be the number of Hispanics.
Then the number of Blacks is $H + 63{,}770$
and the number of Whites is $3H - 2188$.
$H + (H + 63{,}770) + (3H - 2188) = 460{,}622$
$5H = 399{,}040$
$H = 79{,}808$
79,808 Hispanics,
$79{,}808 + 63{,}770$
$= 143{,}578$ Blacks,
$3(79{,}808) - 2188$
$= 237{,}236$ Whites

3. Let t be the number of years after 1993.
$462 + 15t = 807$
$15t = 345$
$t = 23$
The average weekly salary will reach $807
in $1993 + 23 = 2016$.

4. Let x be the amount invested at 9%
$0.09x + 0.06(6000 - x) = 480$
$0.03x = 120$
$x = 4000$
$4000 at 9% and $2000 at 6%.

5. Let x be the number of liters of the 50% acid solution.
$0.50x + 0.80(100 - x) = 0.68(100)$
$-0.30x = -12$
$x = 40$
40 liters of the 50% acid solution and 60 liters of the 80% acid solution.

6. Let x = time until the cars meet.
$45x + 35x = 400$
$80x = 400$
$x = 5$
The cars will meet in 5 hours.

7. $9^2 - 6^2 = 45$

8. Find the pattern.

Base (n)	Interior triangles
3	9
4	16
12	144

Number of interior triangles = n^2

9. $1 \times 2 = 2$
$2 \times 3 = 6$
$6 \times 4 = 24$
$24 \times 5 = 120$
$120 \times 6 = 720$

10. $\frac{3 \text{ inches}}{5 \text{ feet}} = \frac{3 \text{ inches}}{60 \text{ inches}} = \frac{1}{20}$

11. $\frac{\$6.66}{18 \text{ ounces}} = \$0.37$ per ounce

12. $\frac{-7}{5} = \frac{91}{x}$
$-7x = 5 \cdot 91$
$-7x = 455$
$x = -65$

13. Let x be the number of deer in the park.
$\frac{200}{x} = \frac{5}{150}$
$200 \cdot 150 = 5x$
$30{,}000 = 5x$
$6000 = x$
6000 deer

14. $\frac{\$1.87}{1000} = \frac{x}{20{,}000}$
$1.87 \cdot 20{,}000 = 1000x$
$37{,}400 = 1000x$
$37.4 = x$
$37.40

15. Let x be the measure of the angle.
Then $90 - x$ = complement
$x = (90 - x) + 16$
$2x = 106$
$x = 53°$

16. $x + 3x + (3x - 30) = 180$
$7x = 210$
$x = 30$
The angles are 30°, $3(30)° = 90°$,
$(3 \cdot 30 - 30)° = 60°$.

17. Area of circle: πr^2
Difference: $\pi(12)^2 - \pi(9)^2$.
$= 144\pi - 81\pi = 63\pi$
The larger dartboard is 63π square inches larger.

18. $4x + 2(x + 3) = 18$
$4x + 2x + 6 = 18$
$6x = 12$
$x = 2$
The length is 2 feet and the height is 5 feet.

19. Let h be the height of the sail.
$\left(\frac{1}{2}\right)(8)(h) = 56$
$4h = 56$
$h = 14$
14 feet

20. $\frac{x}{8} = \frac{4}{10}$
$10x = 32$
$x = 3.2$
3.2 inches

Cumulative Review Problems (Chapters 1–3)

1. $330 + 1700E = 1700E + 330$

2. $A_c = \frac{1}{2}A$
$= \frac{1}{2}[W(600 - 1.5W)]$
$= \frac{1}{2}[200(600 - 1.5(200))]$
$= \frac{1}{2}(200)(600 - 300)$
$= 100(300) = 30{,}000$ sq ft

3. $\frac{-9(3-6)}{(-12)(3)+(-3-5)(8-4)} = \frac{-9(-3)}{-36+(-8)(4)}$
$= \frac{27}{-36-32} = \frac{27}{-68} = -\frac{27}{68}$

4. temperature at 10 P.M. + change in temperature at 3 A.M. + change in temperature at 12 A.M.
$= -4°\text{F} - 11°\text{F} + 21°\text{F} = 6°\text{F}$

5. Possible answers:
1992: 37 to 24
1993: 52 to 17
1994: 25 to 13

6. **a.** Natural numbers: $\{8, \sqrt{25}\}$

b. Whole numbers: $\{0, 8, \sqrt{25}\}$

c. Integers: $\{-3, 0, 8, \sqrt{25}\}$

d. Rational numbers:
$\{-3, -\frac{1}{2}, \frac{1}{7}, 0, 8, 9.\overline{3}, \sqrt{25}\}$

e. Irrational numbers: $\{\sqrt{29}\}$

f. Real numbers:
$\{-3, -\frac{1}{2}, \frac{1}{7}, 0, 8, 9.\overline{3}, \sqrt{25}, \sqrt{29}\}$

7. Possible answers:

a. 1929, 1971

b. 550,000 in 1956

8. $\frac{1}{5}y + \frac{2}{3}y = y + \frac{1}{15}$
$15\left(\frac{1}{5}y + \frac{2}{3}y\right) = 15\left(y + \frac{1}{15}\right)$
$3y + 10y = 15y + 1$
$13y = 15y + 1$
$-2y = 1$
$y = -\frac{1}{2}$
The solution is $-\frac{1}{2}$.

9. Pages that face each other have consecutive page numbers
$x + (x + 1) = 385$
$2x + 1 = 385$
$2x = 384$
$x = 192$
pages 192 and 193

10. Let x be the number of trips across the toll bridge.
$0.50x = 10 + 0.10x$
$0.4x = 10$
$x = 25$ trips

11. $x - 0.25x = 135$
$0.75x = 135$
$x = 180$ pounds

12. $10(2x - 1) = 8(2x + 1) + 14$
$20x - 10 = 16x + 8 + 14$
$4x = 32$
$x = 8$

13. $-4y + 7 \le 15$
$-4y \le 8$
$y \ge -2$

−3 −2 −1 0 1

14. Let x = number in Belgium.
Colombia = $2x - 20$
Brazil = $3x$
Colombia + Brazil + Belgium = 520
so $(2x - 20) + (3x) + x = 520$
$6x - 20 = 520$
$6x = 540$
$x = 90$
so Columbia = $2(90) - 20 = 180 - 20 = 160$
Brazil = $3(90) = 270$
Colombia: 160, Belgium: 90, Brazil: 270 vehicles per km
France: 40
Norway: 50
Canada, UK: 70
Germany, Thailand, USA: 80
Spain, Turkey: 110 vehicles/km

15. $4(2x - 1) - 3(x - 11) - 2(-4x - 5)$
$= 8x - 4 - 3x + 33 + 8x + 10 = 13x + 39$

16. a. (7, 600)

b. Answers may vary.

17. $\frac{5}{x} = \frac{6}{6+9}$
$\frac{5}{x} = \frac{6}{15}$
$75 = 6x$
$x = 12.5$ feet

18. $\frac{x}{21} = \frac{5}{20}$
$20x = 105$
$x = \frac{105}{20}$
$x = \frac{21}{4}$

19. $\frac{9}{135} = \frac{5}{x}$
$9x = 5(135)$
$x = 75$
\$75

20. W = width, $L = 2W - 10$
$P = 2L + 2W$
$400 = 2(2W - 10) + 2W$
$400 = 4W - 20 + 2W$
$400 = 6W - 20$
$420 = 6W$
$W = 70$
so $L = 2(70) - 10 = 130$
70 yds × 130 yds

21. $170 = 0.34x$
$x = 500$ hate crimes overall
$0.23(500) = 115$ hate crimes targeted at others
$0.16(500) = 80$ hate crimes targeted at Jews
$0.27(500) = 135$ hate crimes targeted at Gays and Lesbians

22. $3^2 + 4^2 = 5^2$
$10^2 + 11^2 + 12^2 = 13^2 + 14^2$
$21^2 + 22^2 + 23^2 + 24^2 = 25^2 + 26^2 + 27^2$

23. d is not possible:
$2x > 23$
$x > \frac{23}{2}$
$x > 11.5$

$9x < 865$
$x < 96\frac{1}{9}$

$11.5 < x < 96\frac{1}{9}$
a. $x = 96$; possible
b. $x < 50$; possible
c. $x > 23$; possible
d. $x > 100$ *not* possible

24. $3x - 17 = 8443$
$3x = 8460$
$x = 2820$
(Gunyashev) + (gorilla) = $x - 5$
Let G equal the amount of Gunyashev lifted.
$G + (G + 775) = 2820 - 5$
$2G = 2815 - 775$
$2G = 2040$

$G = 1020$
1020 pounds

25. Let x equal the measure of the angle.
$90 - x$ equals the measure of its complement.
$x - (90 - x) = 16$
$2x - 90 = 16$
$2x = 106$
$x = 53$
$53°$

26. Let x equal the number of sheets of paper.
$2x + 4 \leq 29$
$2x \leq 25$
$x \leq 12.5$
12 sheets or less

27. Let x be the amount invested at 8% and
$15{,}000 - x$ be the amount invested at 6%.
$x(0.08) + (15{,}000 - x)(0.06) = 1100$
$(0.08 - 0.06)x + 900 = 1100$
$0.02x = 200$
$x = 10{,}000$
\$10,000 at 8% and \$5,000 at 6%

28. $d = rt$
$6t + 8t = 21$
$14t = 21$
$t = 1.5$ hours

29. $12 \div 2 \div 3 = 2$
$x = 12,\ y = 2,\ z = 3$

30. $A = \frac{m+n}{2}$ for m
$2A = m + n$
$2A - n = m$
$m = 2A - n$

Chapter 4

Section 4.1

1. $y = 3x$

   (2, 3):
   $3 = 3(2)$
   $3 = 6$ False
   Not a solution

   (3, 2):
   $2 = 3(3)$
   $2 = 9$ False
   Not a solution

   (–4, –12):
   $-12 = 3(-4)$
   $-12 = -12$ True
   A solution

3. $y = -4x$

   (–5, –20):
   $-20 = -4(-5)$
   $-20 = 20$ False
   Not a solution

   (0, 0):
   $0 = -4(0)$
   $0 = 0$ True
   A solution

   (9, –36):
   $-36 = -4(9)$
   $-36 = -36$ True
   A solution

5. $y = 2x + 6$

   (0, 6):
   $6 = 2(0) + 6$
   $6 = 6$ True
   A solution

   (–3, 0):
   $0 = 2(-3) + 6$
   $0 = 0$ True
   A solution

   (2, –2)
   $-2 = 2(2) + 6$
   $-2 = 10$ False
   Not a solution

7. $3x + 5y = 15$

   (–5, 6):
   $3(-5) + 5(6) = 15$
   $-15 + 30 = 15$
   $15 = 15$ True
   A solution

   (0, 5):
   $3(0) + 5(5) = 15$
   $0 + 25 = 15$
   $25 = 15$ False
   Not a solution

   (10, –3):
   $3(10) + 5(-3) = 15$
   $30 - 15 = 15$
   $15 = 15$ True
   A solution

9. $x + 3y = 0$

   (0, 0):
   $0 + 3(0) = 0$
   $0 = 0$ True
   A solution

   $\left(1, \frac{1}{3}\right)$:
   $1 + 3\left(\frac{1}{3}\right) = 0$
   $1 + 1 = 0$
   $2 = 0$ False
   Not a solution

$\left(2, -\frac{2}{3}\right)$:

$2+3\left(-\frac{2}{3}\right)=0$

$2-2=0$

$0=0$ True

A solution

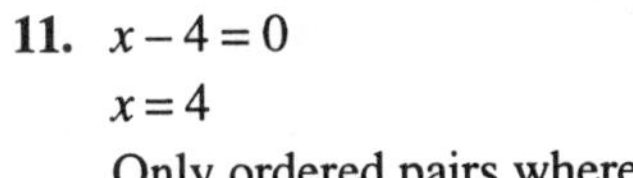

11. $x-4=0$

$x=4$

Only ordered pairs where $x = 4$ are solutions.

(4, 7) is a solution.

(3, 4) not a solution $x \neq 4$

(0, –4) not a solution $x \neq 4$

13. $y=x$

$f(x)=x$

x	–1	0	1	2
y	–1	0	1	2

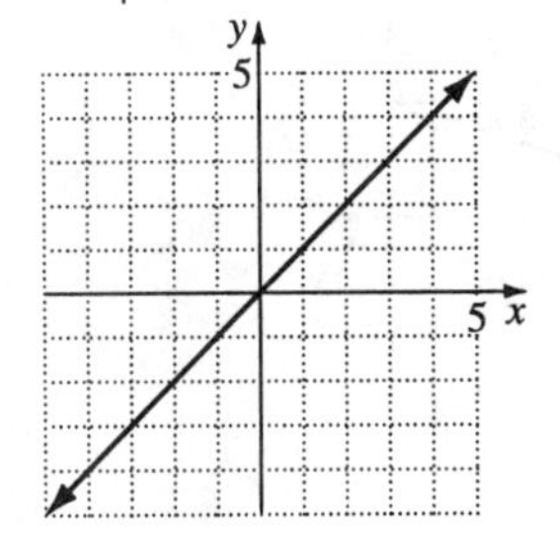

15. $y=2x$

$f(x)=2x$

x	–1	0	1	2
y	–2	0	2	4

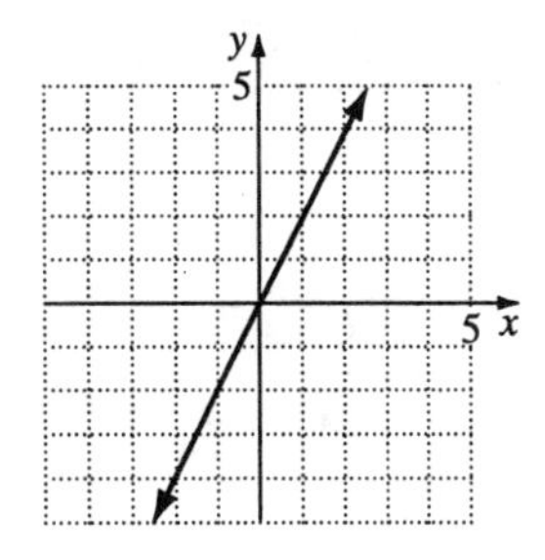

17. $y=-2x$

$f(x)=-2x$

x	–1	0	1	2
y	2	0	–2	–4

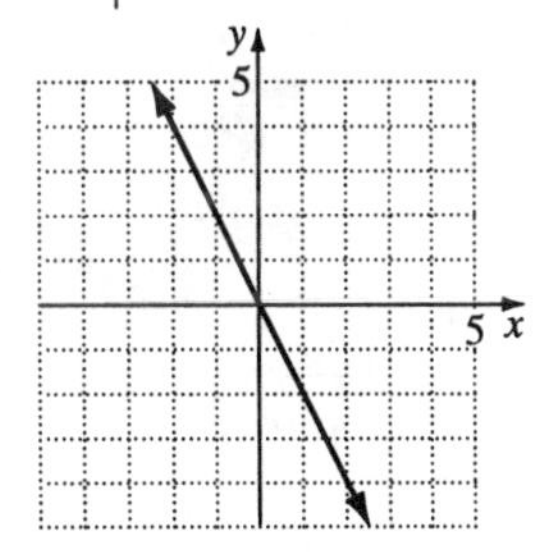

19. $y=\frac{1}{2}x$

$f(x)=\frac{1}{2}x$

x	–2	0	2	4
y	–1	0	1	2

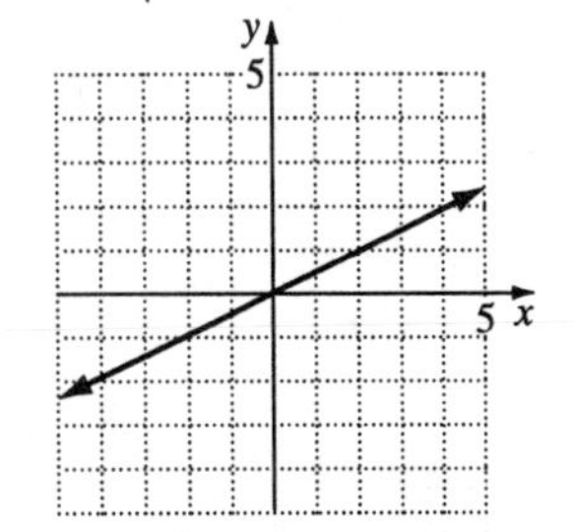

21. $y = -\frac{2}{3}x$

$f(x) = -\frac{2}{3}x$

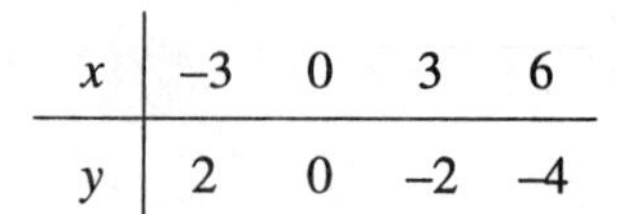

x	−3	0	3	6
y	2	0	−2	−4

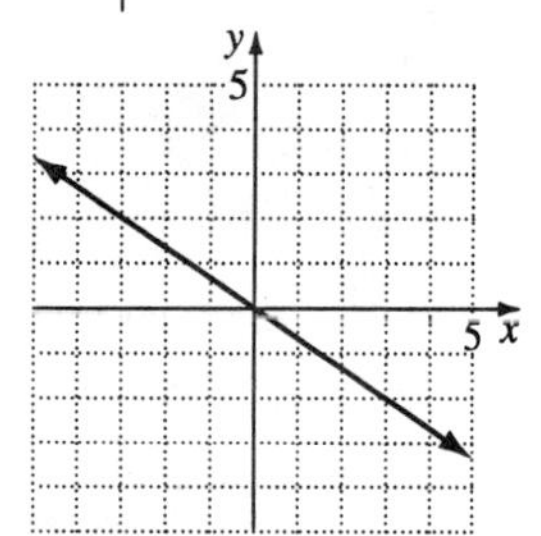

23. $y = x + 2$

$f(x) = x + 2$

x	−2	0	1	2
y	0	2	3	4

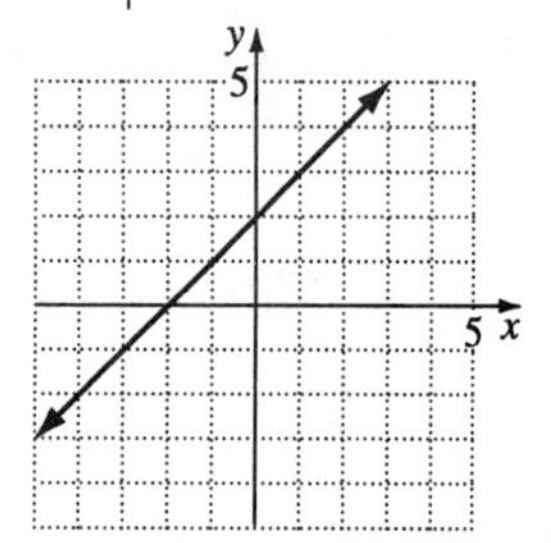

25. $y = x - 3$

$f(x) = x - 3$

x	−3	0	3	6
y	−6	−3	0	3

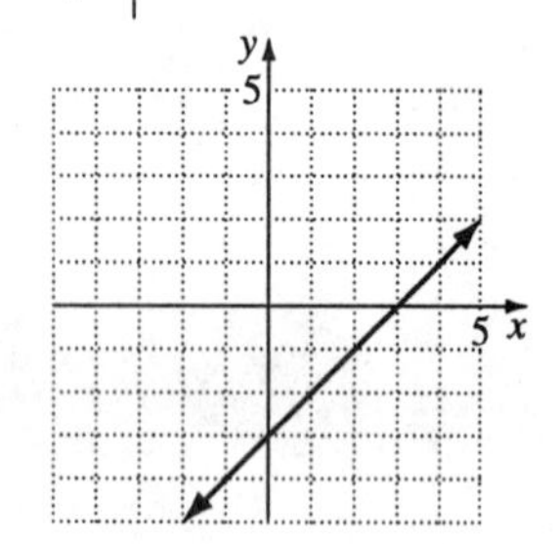

27. $y = 2x + 1$

$f(x) = 2x + 1$

x	−1	0	1	2
y	−1	1	3	5

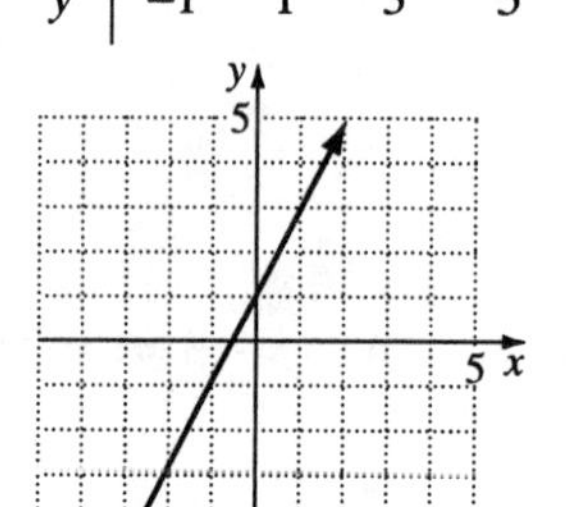

29. $y = \frac{1}{3}x + 1$

$f(x) = \frac{1}{3}x + 1$

x	−3	0	3	6
y	0	1	2	3

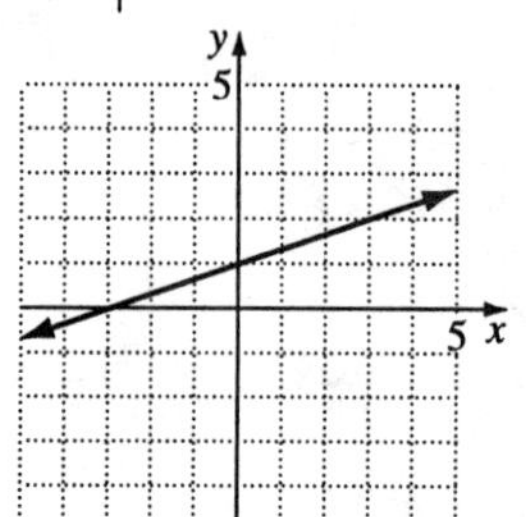

31. $x + y = -1$

$y = -x - 1$

$f(x) = -x - 1$

x	−2	−1	0	1
y	1	0	−1	−2

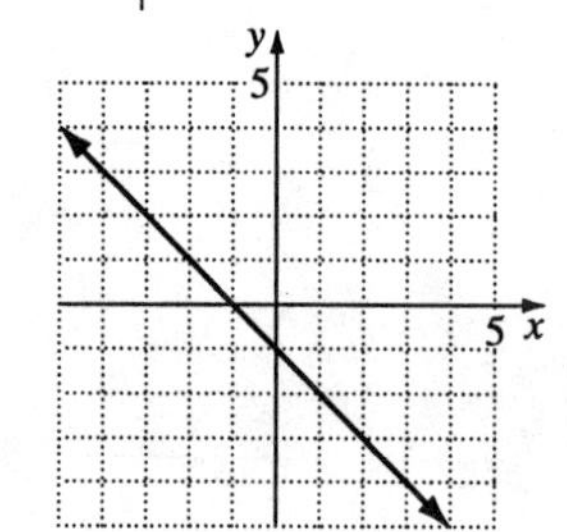

33. $y = \frac{3}{2}x - 1$

$f(x) = \frac{3}{2}x - 1$

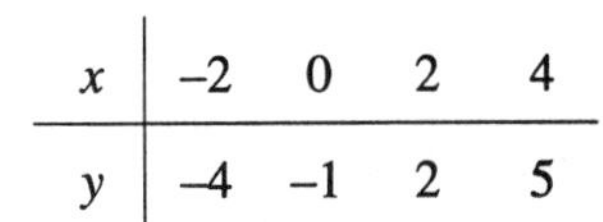

x	–2	0	2	4
y	–4	–1	2	5

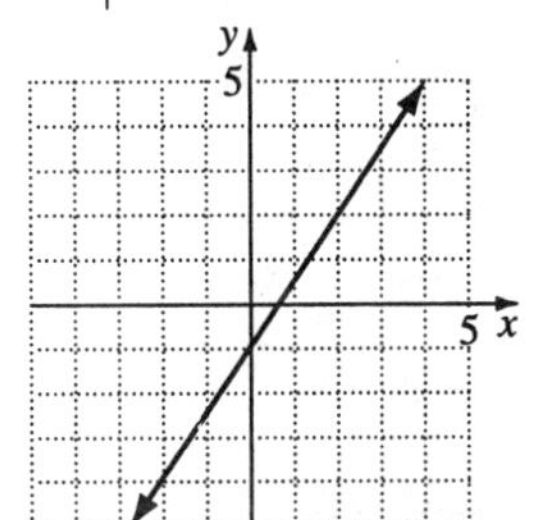

35. $y = -\frac{5}{2}x - 1$

$f(x) = -\frac{5}{2}x - 1$

x	–4	–2	0	2
y	9	4	–1	–6

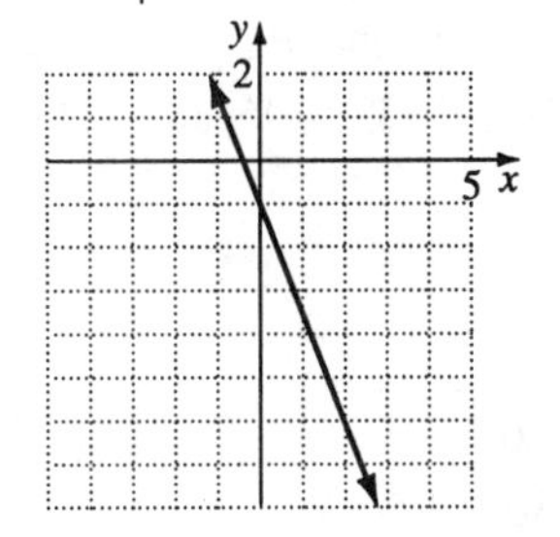

37. $y = \frac{1}{2}x - 3$

$f(x) = \frac{1}{2}x - 3$

x	–2	0	4	8
y	–4	–3	–1	1

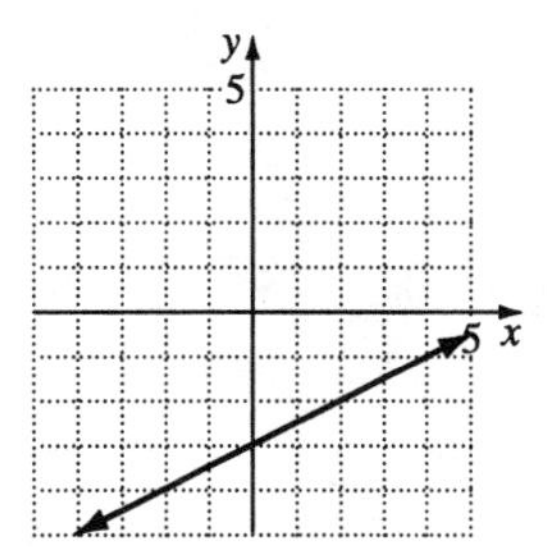

39. $2x + y = 1$

$y = -2x + 1$

$f(x) = -2x + 1$

x	–1	0	1	2
y	3	1	–1	–3

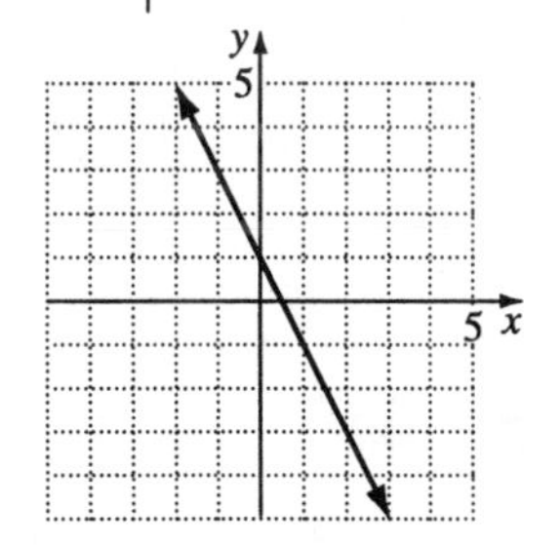

41. $y = x + \frac{1}{2}$

$f(x) = x + \frac{1}{2}$

x	–1	0	1	2
y	$-\frac{1}{2}$	$\frac{1}{2}$	$1\frac{1}{2}$	$2\frac{1}{2}$

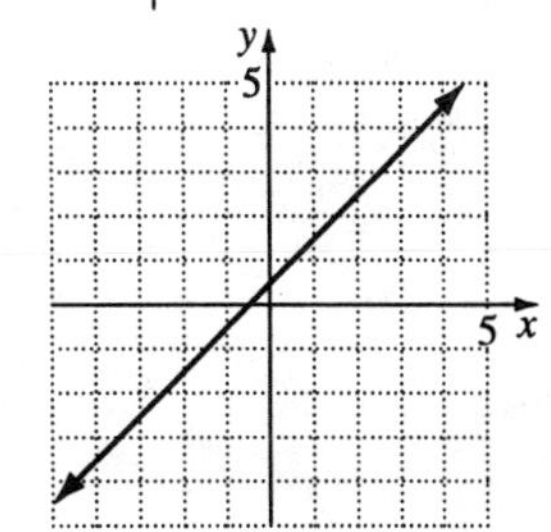

43. $x + 2y = -2$
$2y = -x - 2$
$y = -\frac{1}{2}x - 1$
$f(x) = -\frac{1}{2}x - 1$

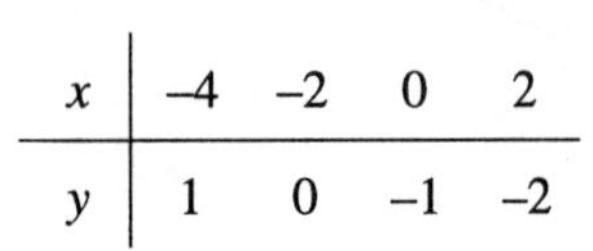

x	–4	–2	0	2
y	1	0	–1	–2

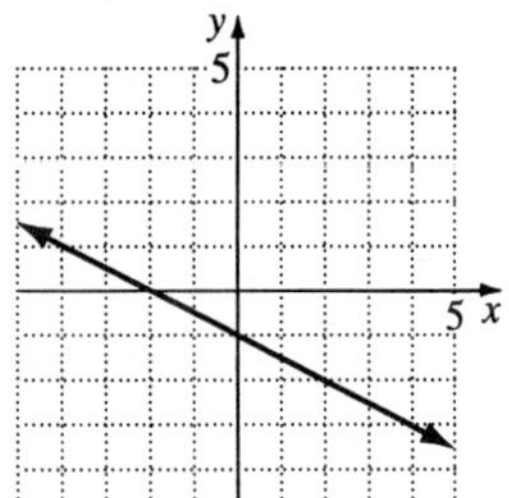

45. $6x - 3y = 6$
$2x - y = 2$
$-y = -2x + 2$
$y = 2x - 2$
$f(x) = 2x - 2$

x	0	1	2	3
y	–2	0	2	4

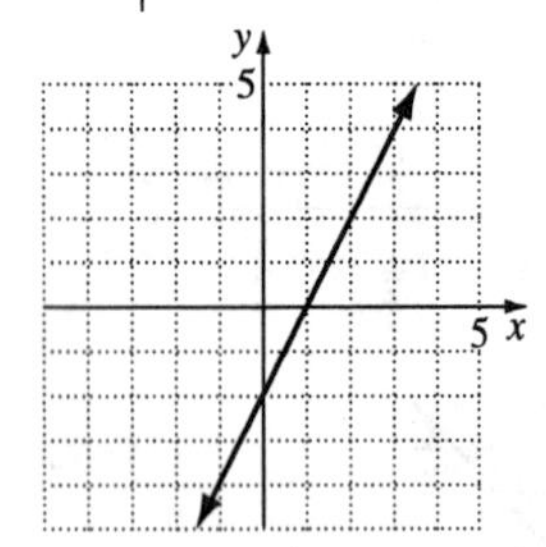

47. $6y + 2x = -6$
$3y + x = -3$
$3y = -x - 3$
$y = -\frac{1}{3}x - 1$
$f(x) = -\frac{1}{3}x - 1$

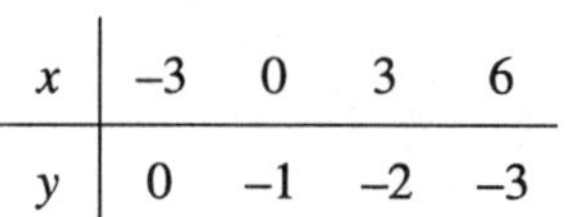

x	–3	0	3	6
y	0	–1	–2	–3

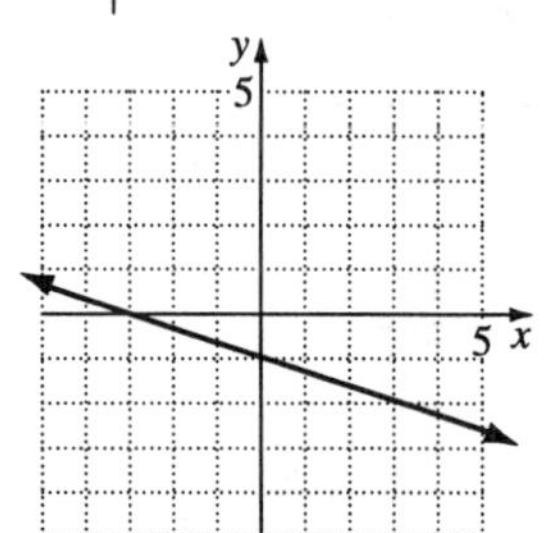

49. $f(x) = 1496x + 16{,}116$
$f(0) = 1496(0) + 16{,}116 = 16{,}116$
$f(5) = 1496(5) + 16{,}116 = 23{,}596$
$f(10) = 1496(10) + 16{,}116 = 31{,}076$
In 1980, 1985, and 1990, the average yearly salary was $16,116, $23,596, and $31,076, respectively.

51. $f(x) = 4x$
$f(2) = 4(2) = 8$
$f(5) = 4(5) = 20$
$f(100) = 4(100) = 400$
A square measuring 2 m, 5 m, and 100 m on a side has a perimeter of 8 m, 20 m, and 400 m, respectively.
$x \le 0$ are meaningless because length is a positive measure.

53. a.

t	0	1	5	10	15	20
M	0	$\frac{1}{5}$	1	2	3	4

b. 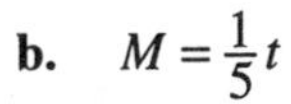
$M = \frac{1}{5}t$

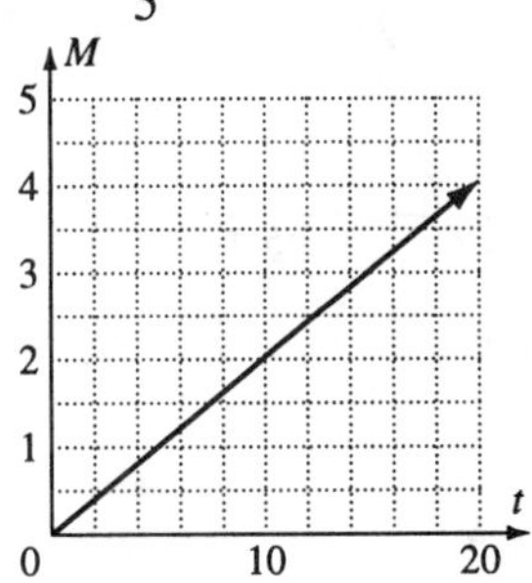

55. $f(x) = 30{,}000 + 50x$

a.

x	$f(x) = 30{,}000 + 50x$	(x, y)
0	$30{,}000 + 50(0) = 30{,}000$	(0, 30,000)
10	$30{,}000 + 50(10) = 30{,}500$	(10, 30,500)
100	$30{,}000 + 50(100) = 35{,}000$	(100, 35,000)
1000	$30{,}000 + 50(1000) = 80{,}000$	(1000, 80,000)

b.

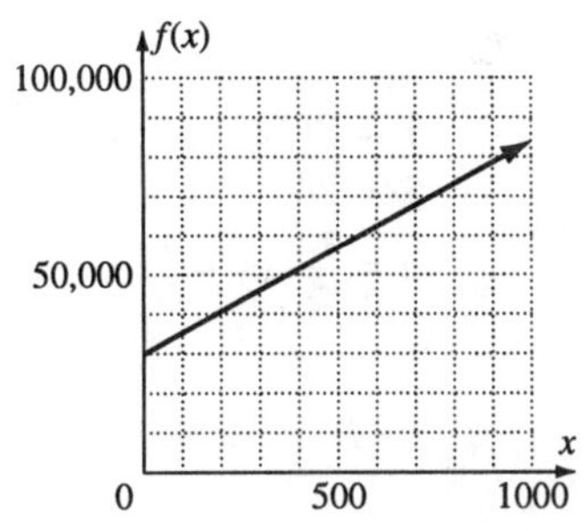

57. $y = 60{,}000 - 5000x$
$0 \le x \le 12$

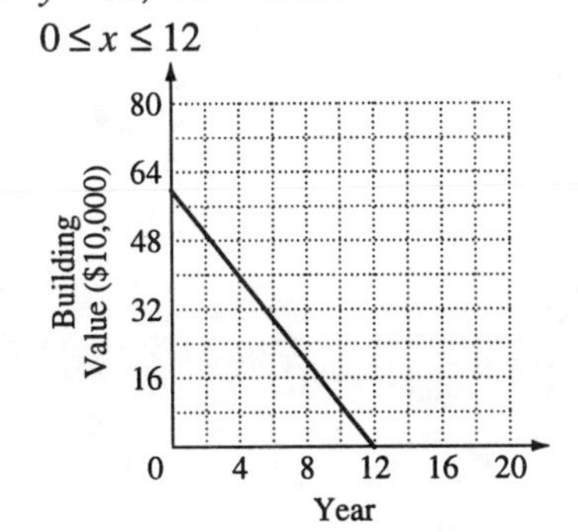

59. b is true.
Given $y = mx + b$, $y = b$ when $x = 0$.

61.

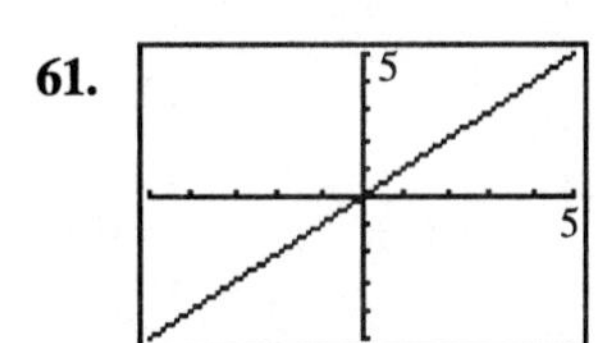

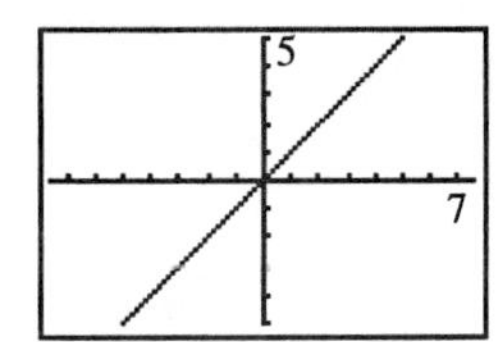

63. $y = 2x - 1$
$y = -1 + 2x$
Commutative property of addition

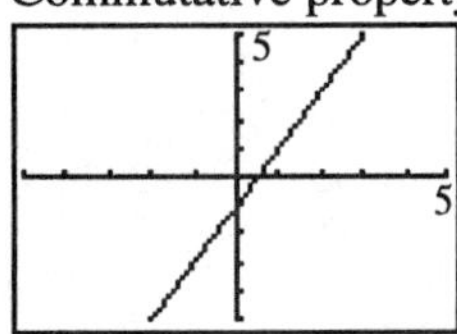

65. $y = 2 + (x + 3)$
$y = (2 + x) + 3$
Associative property of addition

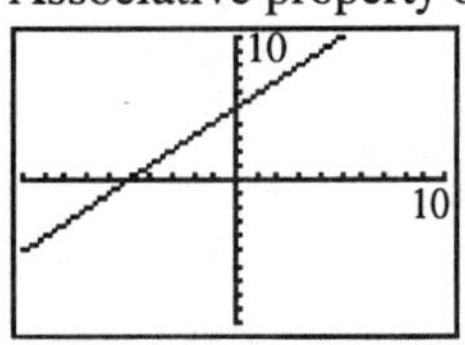

67. Answers may vary.

69.

x	−3	−2	−1	0	1	2	3
$y = x^2 - 1$	8	3	0	−1	0	3	8

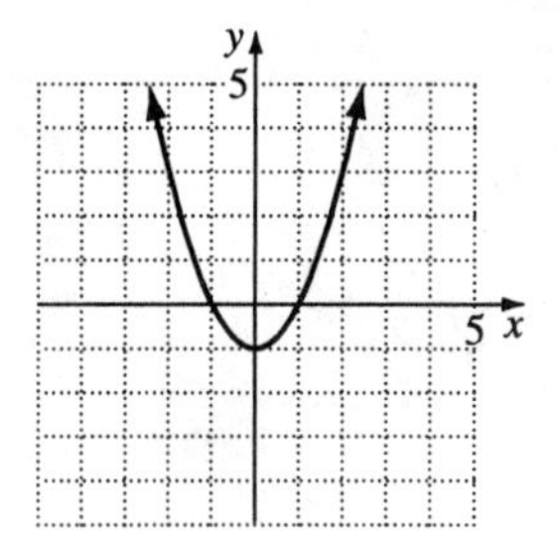

Review Problems

70. $2(x - 8) = 3(x - 4) - 5x$
$2x - 16 = 3x - 12 - 5x$
$2x - 16 = -2x - 12$
$4x = 4$
$x = 1$
The solution is 1.

71. $2(x + 6) \leq 4x - 2$
$2x + 12 \leq 4x - 2$
$-2x \leq -14$
$x \geq 7$

72. $\frac{12}{x} = \frac{5}{2}$
$5x = 2(12)$
$5x = 24$
$x = \frac{24}{5}$ or $4\frac{4}{5}$

Problem Set 4.2

1. $x - y = 3$

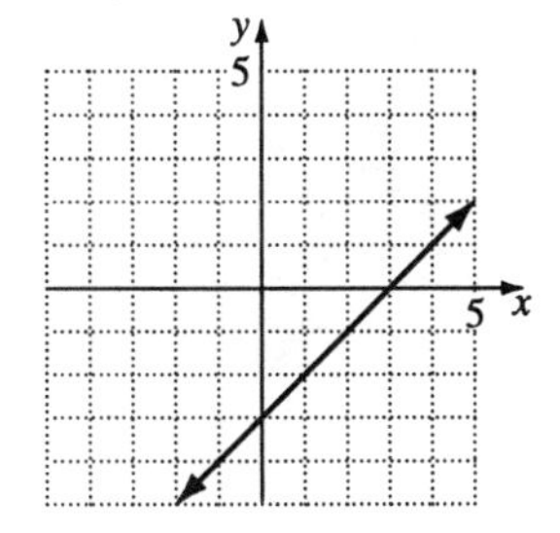

3. $3x = 4y - 12$

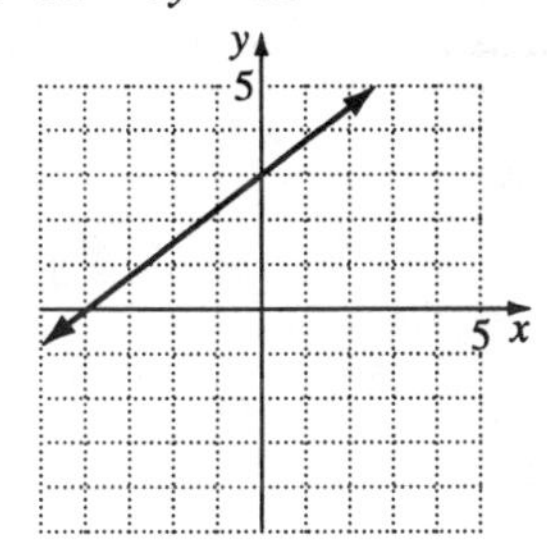

5. $7x - 2y = 14$

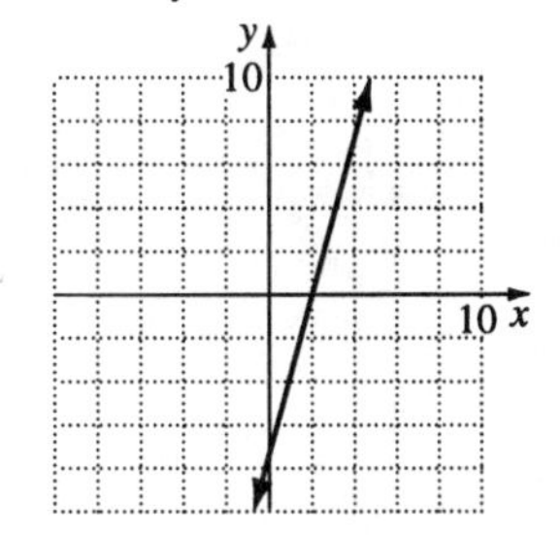

7. $2x - y = 0$

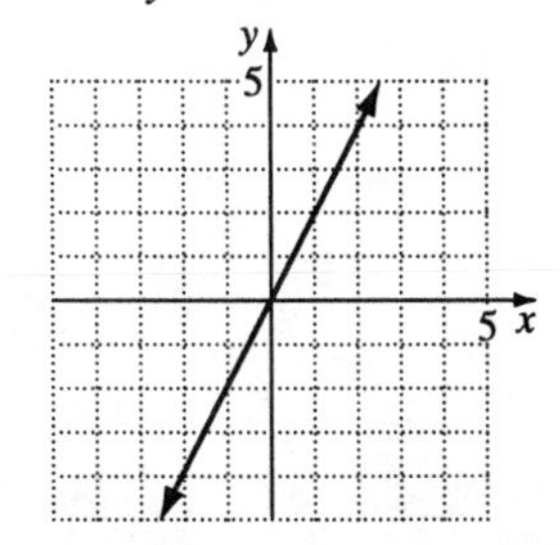

9. $y = -3x$

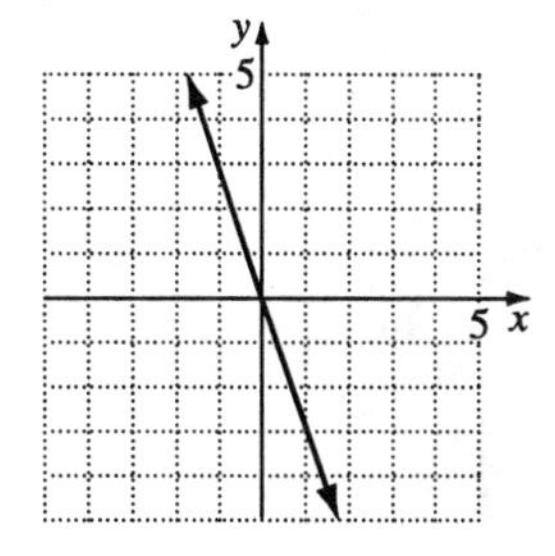

11. $y = 3x + 1$

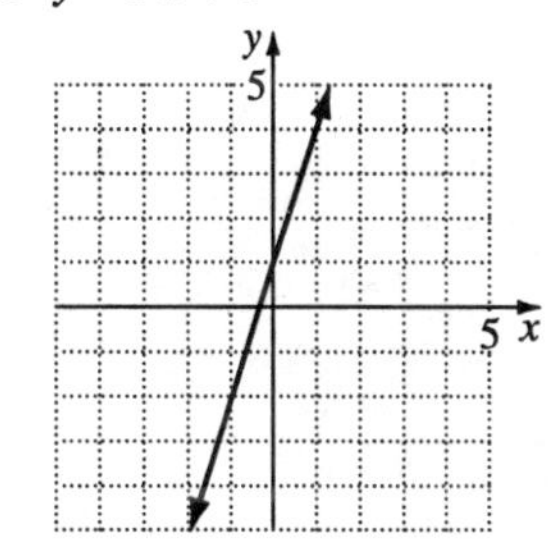

13. $x = 4$

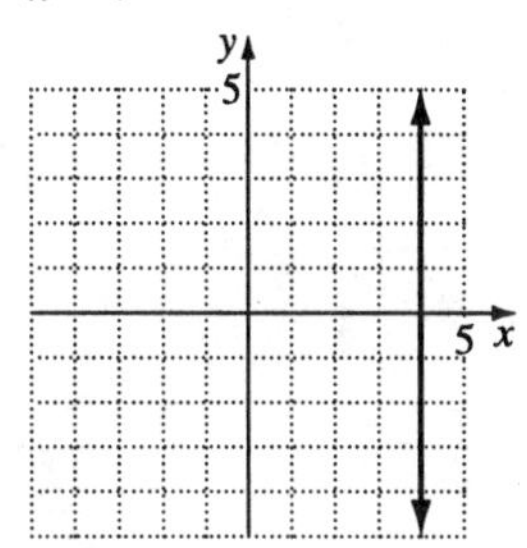

15. $x = -2$

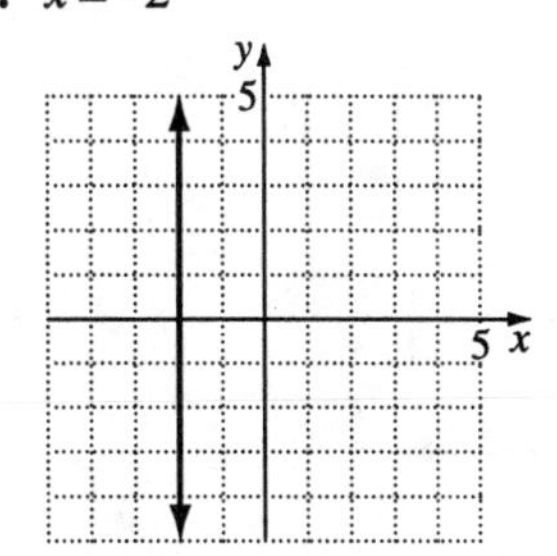

17. $x-6=0$

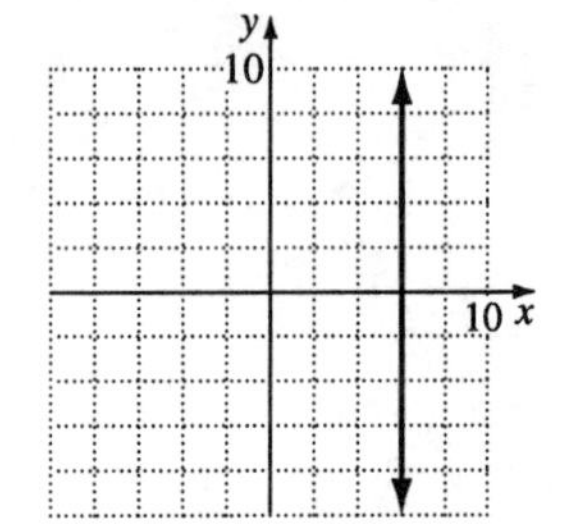

19. $y=5$

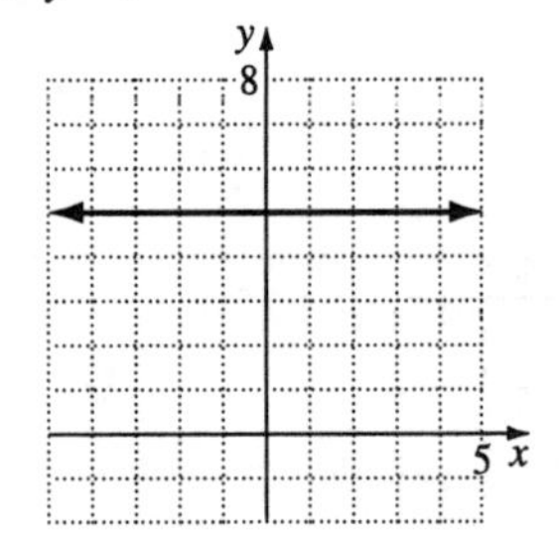

21. $y=-3$

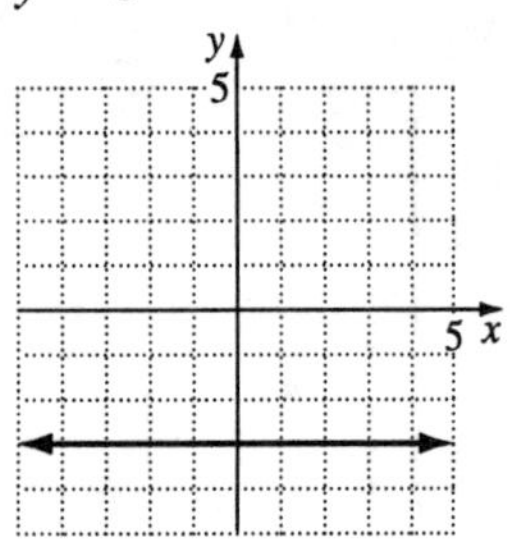

23. $y+6=0$

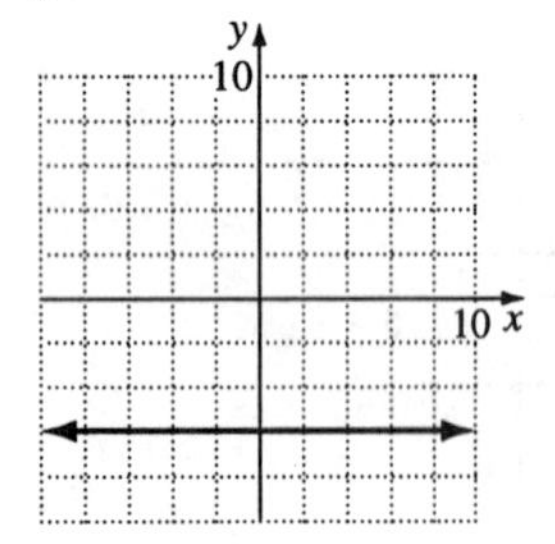

25. $x=0$

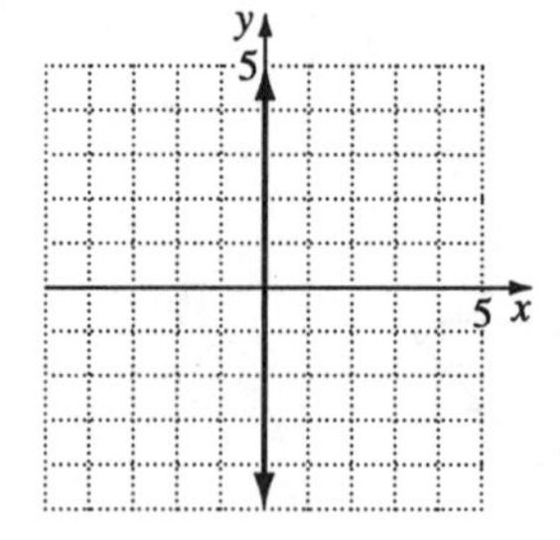

27. $3y=9$

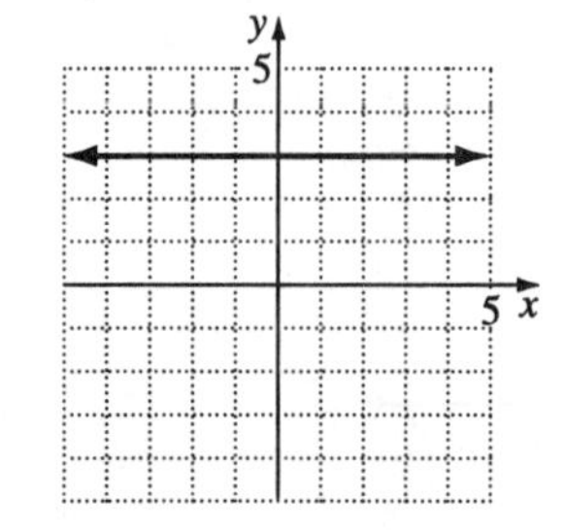

29. $-3x-2y=6$

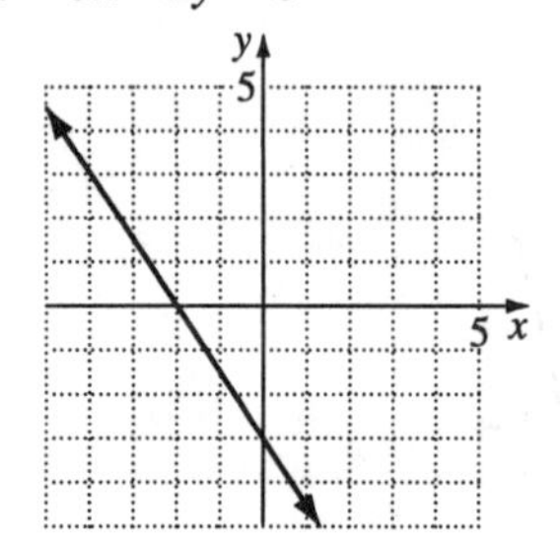

31. $20x-240=-60y$

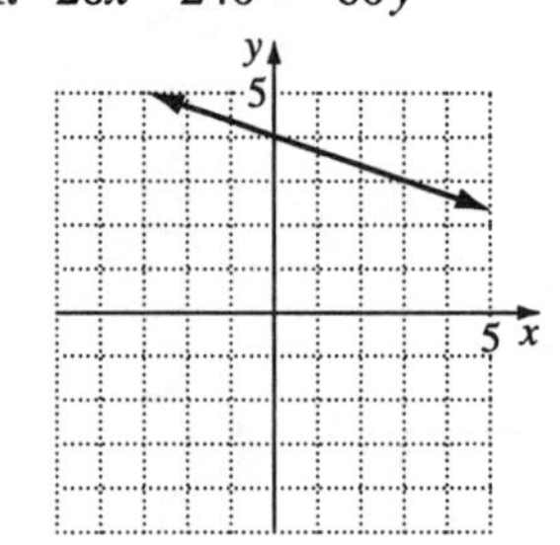

33. $\frac{1}{3}x + \frac{1}{4}y = 12$

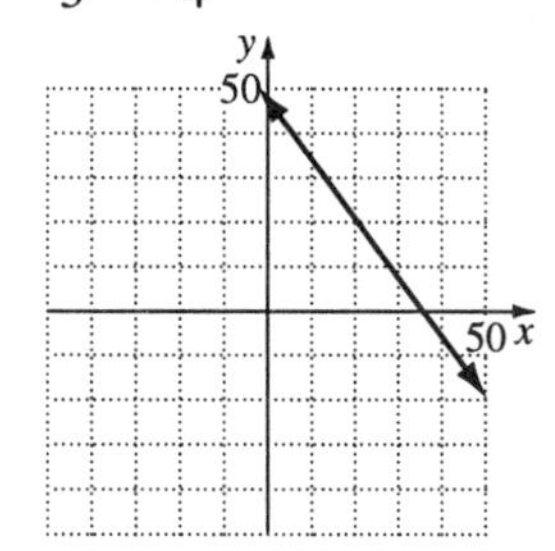

35. a. $C = 300t$
$C = 300(1) = 300$ cal
$C = 300(2) = 600$ cal
$C = 300(2.5) = 750$ cal
$C = 300(4) = 1200$ cal

b.

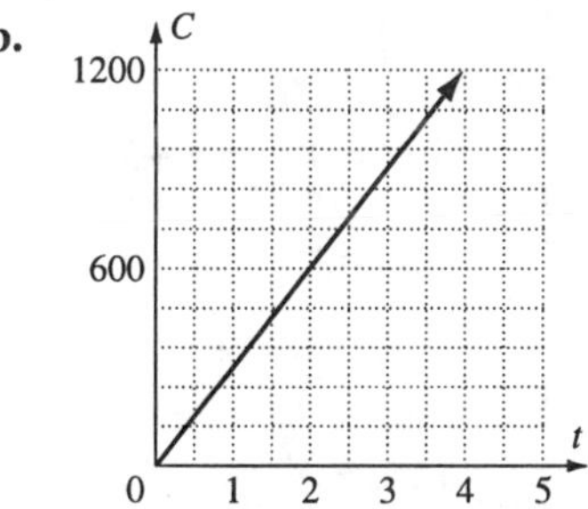

37. a. $C = 50 + 2x$

b.

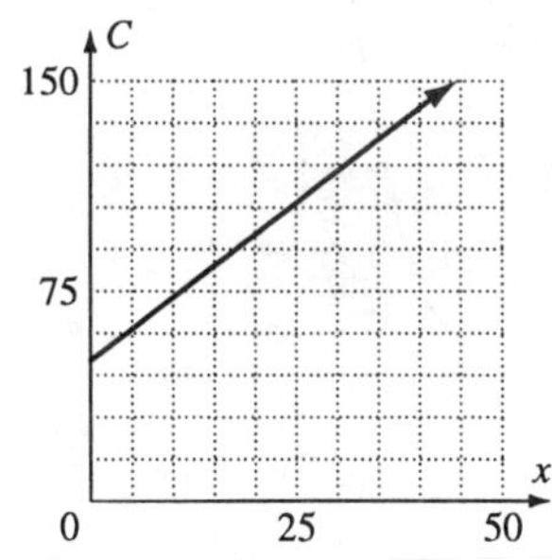

c. $C = \$74$

39. a. $y = 577 + \frac{69}{5}x$

b. 1990: $y = 577 + \frac{69}{5}(5) = 646$
1995: $y = 577 + \frac{69}{5}(10) = 715$
646 thousand and 715 thousand

c.

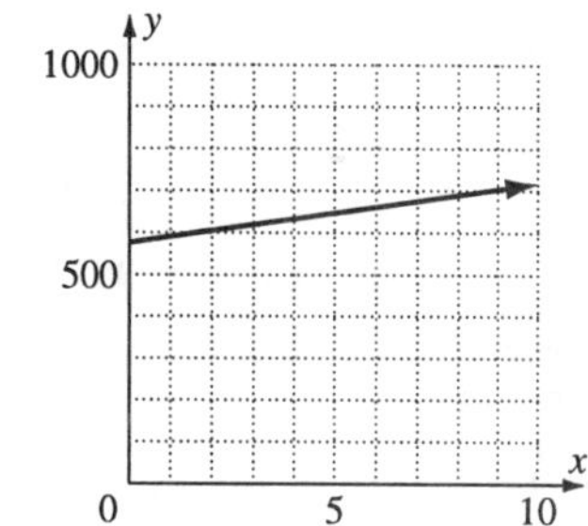

d. 604.6 thousand

41. d is true.

43. a. $y = 300 + 0.04x$

b.

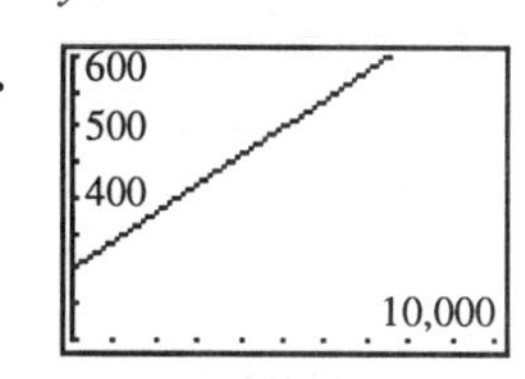

c. \$480

45. Answers may vary.

47. a. $2(10 + x) + 2(8 + y) = 58$
$20 + 2x + 16 + 2y = 58$
$2x + 2y = 22$
$x + y = 11$

b.

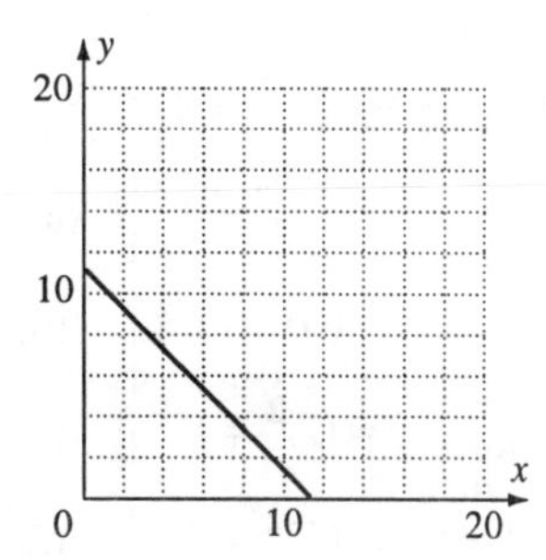

$y = 6.5$; 14.5 m by 14.5 m

Review Problems

49. $3(y-2)+y=y-7$
$3y-6+y=y-7$
$3y=-1$
$y=-\frac{1}{3}$
The solution is $-\frac{1}{3}$.

50. $x+2(x+12)+3(x+10)=180$
$x+2x+24+3x+30=180$
$6x+54=180$
$6x=126$
$x=21$
$2(x+12)=2(21+12)=2(33)=66$
$3(x+10)=3(21+10)=3(31)=93$
The measures of the angles are 21°, 66°, and 93°.

51. Let x equal the value of the lot.
$5x$ equals the value of the house.
$x+5x=112{,}200$
$6x=112{,}200$
$x=18{,}700$
The lot is worth $18,700.

Problem Set 4.3

1. $y=x^2$

x	$y=x^2$	(x, y)
–3	$y=(-3)^2=9$	(–3, 9)
–2	$y=(-2)^2=4$	(–2, 4)
–1	$y=(-1)^2=1$	(–1, 1)
0	$y=(0)^2=0$	(0, 0)
1	$y=(1)^2=1$	(1, 1)
2	$y=(2)^2=4$	(2, 4)
3	$y=(3)^2=9$	(3, 9)

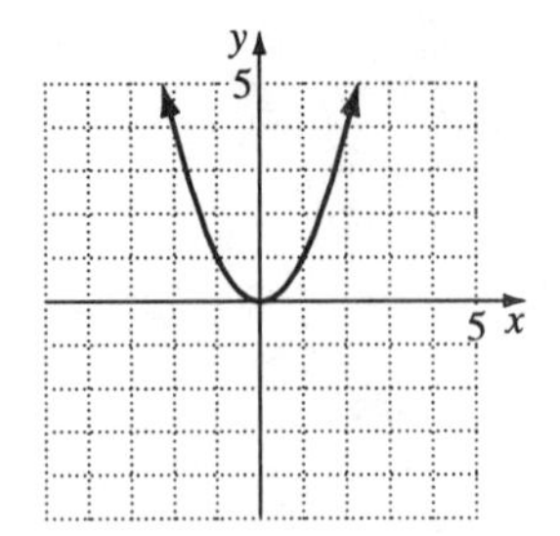

3. $f(x)=x^2-5$

x	$f(x)=x^2-5$	(x, y)
–3	$f(x)=(-3)^2-5=4$	(–3, 4)
–2	$f(x)=(-2)^2-5=-1$	(–2, –1)
–1	$f(x)=(-1)^2-5=-4$	(–1, –4)
0	$f(x)=0^2-5=-5$	(0, –5)
1	$f(x)=1^2-5=-4$	(1, –4)
2	$f(x)=2^2-5=-1$	(2, –1)
3	$f(x)=3^2-5=4$	(3, 4)

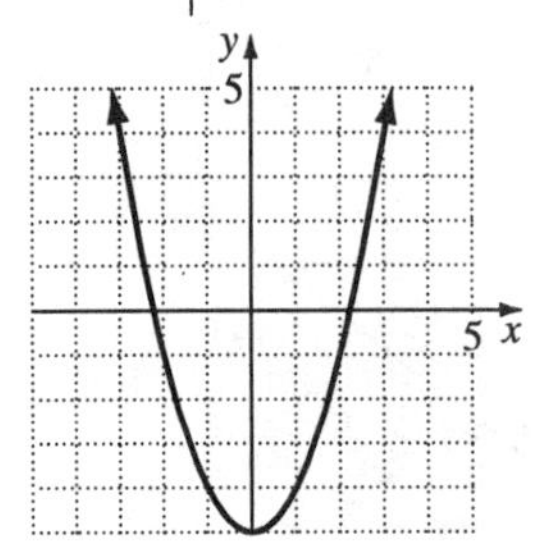

5. $y = -x^2$

x	$y = -x^2$	(x, y)
–3	$y = -(-3)^2 = -9$	(–3, –9)
–2	$y = -(-2)^2 = -4$	(–2, –4)
–1	$y = -(-1)^2 = -1$	(–1, –1)
0	$y = -(0)^2 = 0$	(0, 0)
1	$y = -1^2 = -1$	(1, –1)
2	$y = -2^2 = -4$	(2, –4)
3	$y = -3^2 = -9$	(3, –9)

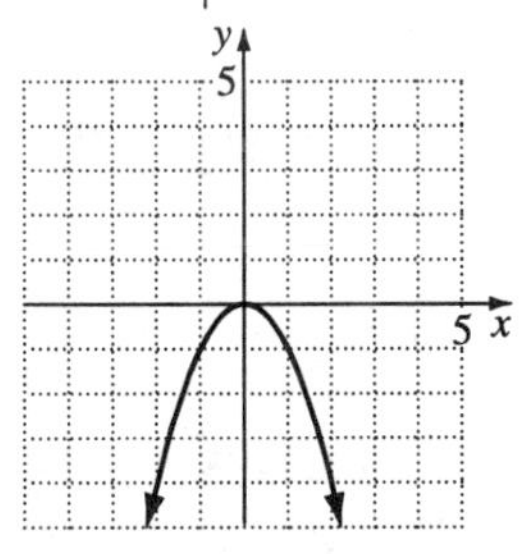

7. $f(x) = x^2 + x - 6$

x	$f(x) = x^2 + x - 6$	(x, y)
–2	$f(x) = (-2)^2 + (-2) - 6 = -4$	(–2, –4)
–1	$f(x) = (-1)^2 + (-1) - 6 = -6$	(–1, –6)
0	$f(x) = 0^2 + 0 - 6 = -6$	(0, –6)
1	$f(x) = 1^2 + 1 - 6 = -4$	(1, –4)
2	$f(x) = 2^2 + 2 - 6 = 0$	(2, 0)
3	$f(x) = 3^2 + 3 - 6 = 6$	(3, 6)

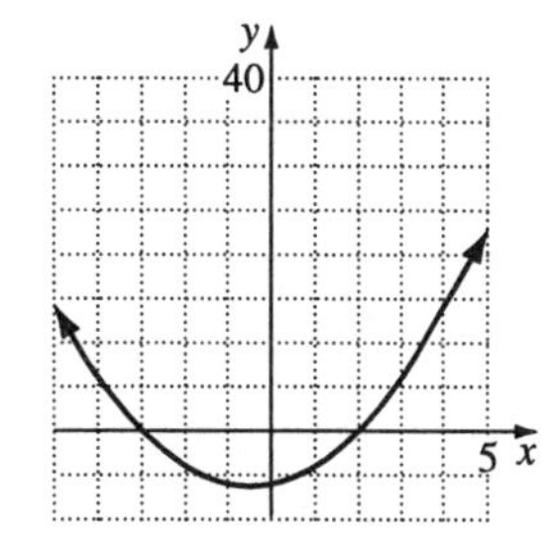

9. $y = x^2 - x - 2$

x	$y = x^2 - x - 2$	(x, y)
–2	$y = (-2)^2 - (-2) - 2 = 4$	(–2, 4)
–1	$y = (-1)^2 - (-1) - 2 = 0$	(–1, 0)
0	$y = 0^2 - 0 - 2 = -2$	(0, –2)
1	$y = 1^2 - 1 - 2 = -2$	(1, –2)
2	$y = 2^2 - 2 - 2 = 0$	(2, 0)
3	$y = 3^2 - 3 - 2 = 4$	(3, 4)

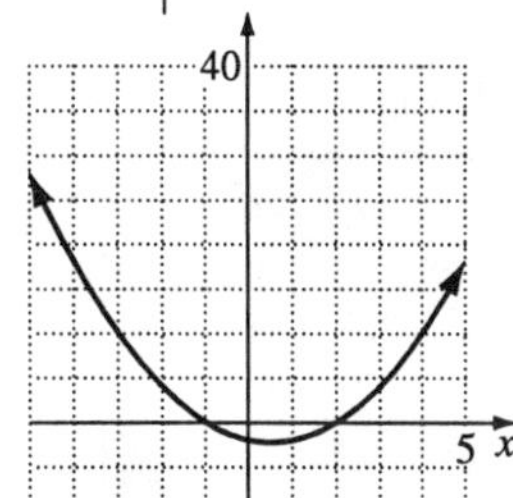

11. $f(x) = x^3$

x	$f(x) = x^3$	(x, y)
–2	$f(x) = (-2)^3 = -8$	(–2, –8)
–1	$f(x) = (-1)^3 = -1$	(–1, –1)
0	$f(x) = 0^3 = 0$	(0, 0)
1	$f(x) = 1^3 = 1$	(1, 1)
2	$f(x) = 2^3 = 8$	(2, 8)

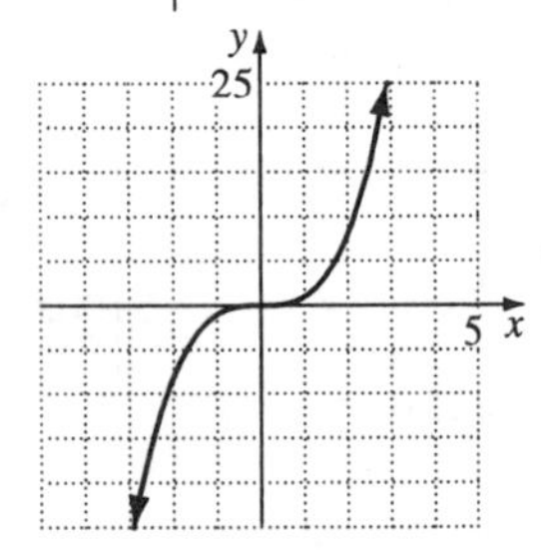

13. $f(x) = \sqrt{x}$

x	$f(x) = \sqrt{x}$	(x, y)
0	$f(x) = \sqrt{0} = 0$	(0, 0)
1	$f(x) = \sqrt{1} = 1$	(1, 1)
4	$f(x) = \sqrt{4} = 2$	(4, 2)
9	$f(x) = \sqrt{9} = 3$	(9, 3)
16	$f(x) = \sqrt{16} = 4$	(16,4)

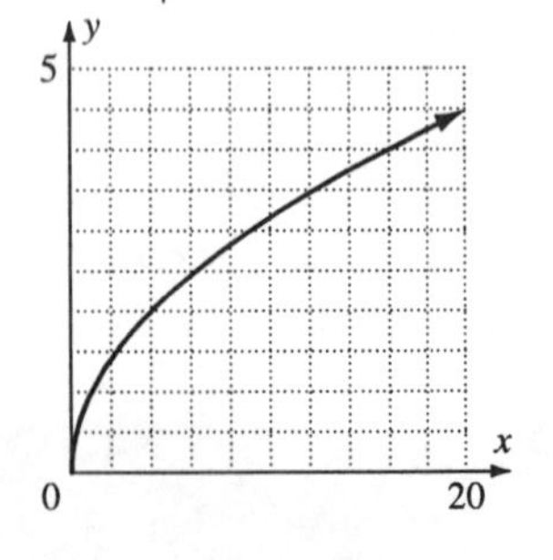

15. $f(x) = \frac{11}{3}x^2 + \frac{94}{3}x + 23$

$f(10) = \frac{11}{3}(10)^2 + \frac{94}{3}(10) + 23$

$f(10) = 703$

In 1995, 703 million CDs sold.

Jazz CDs sold = 0.033(703)

= 23.199 million

17. **a.** $f(0) = 0$

After 0 seconds, the ball is at 0 ft.

b. $f(1) = 48$

After 1 second, the ball is at 48 ft.

c. $f(2) = 64$

After 2 seconds, the ball is at 64 ft.

d. $f(3) = 48$

After 3 seconds, the ball is at 48 ft.

e. $f(4) = 0$

After 4 seconds, the ball is at 0 feet.

19. $f(x) = 0.67x^2 - 27.74x + 387$

$f(12) = 0.67(12)^2 - 27.74(12) + 387$

= 150.6

$f(20) = 0.67(20)^2 - 27.74(20) + 387$

= 100.2

$f(30) = 0.67(30)^2 - 27.74(30) + 387$

= 157.8

(12, 150.6), (20, 100.2), (30, 157.8)

When flying 12 mph, 20 mph, and 30 mph, the power expenditure is 150.6 calories, 100.2 calories, and 157.8 calories, respectively. 20 mph results in the least amount of power expenditure.

21. c is true

23. $f(x) = 0.0075x^2 - 0.2676x + 14.8$

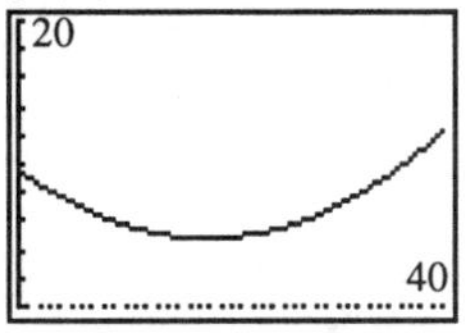

17.9 years after 1940, or in the later part of 1957, fuel efficiency was at its worst, which was approximately 12.4 miles per gallon.

25. Students should verify graphs.

27. Answers may vary.

Review Problems

29. Let x = the fourth grade.

$\frac{x+96+82+91}{4} \geq 90$

$\frac{x+269}{4} \geq 90$

$x + 269 \geq 360$

$x \geq 91$

Scores of 91 or greater

30. Let x = the length of the shorter piece.
Then $2x$ = length of longer piece.

$x + 2x = 36$

$3x = 36$

$x = 12$

$2x = 24$

12 inches and 24 inches

31. $150t + 250t = 800$

$400t = 800$

$t = 2$

2 hours

Problem Set 4.4

1. $(x_1, y_1) = (2, 6)$, $(x_2, y_2) = (3, 5)$

$m = \frac{5-6}{3-2} = \frac{-1}{1} = -1$; line falls

3. (4, 7), (8, 10)

$m = \frac{10-7}{8-4} = \frac{3}{4}$; line rises

5. (–2, 1), (2, 2)

$m = \frac{2-1}{2-(-2)} = \frac{1}{4}$; line rises

7. (4, –2), (3, –2)

$m = \frac{-2-(-2)}{3-4} = \frac{0}{-1} = 0$; line is horizontal

9. (–2, 4), (–1, –1)

$m = \frac{-1-4}{-1-(-2)} = \frac{-5}{1} = -5$; line falls

11. (5, 3), (5, –2)

$m = \frac{-2-3}{5-5} = \frac{-5}{0}$ = undefined;
line is vertical

13. (5, –2), (1, 0)

$m = \frac{0-(-2)}{1-5} = \frac{2}{-4} = -\frac{1}{2}$; line falls

15. (2, 0), (0, 8)

$m = \frac{8-0}{0-2} = \frac{8}{-2} = -4$; line falls

17. (5, 1), (–2, 1)

$m = \frac{1-1}{-2-5} = \frac{0}{-7} = 0$; line is horizontal

19. (–1, 2), (–1, 3)

$m = \frac{3-2}{-1-(-1)} = \frac{1}{0}$ = undefined;
line is vertical

21. (1, 2), (3, 6)

$m = \frac{6-2}{3-1} = \frac{4}{2} = 2$

23. (3, 1), (6, –2)

$m = \frac{-2-1}{6-3} = \frac{-3}{3} = -1$

25. (–2, 2), (–2, –5)

$m = \frac{-5-2}{-2-(-2)} = \frac{-7}{0}$ = undefined

27. $(-4,-1), (-4, 5)$

$m = \dfrac{5-(-1)}{-4-(-4)} = \dfrac{6}{0} = \text{undefined}$

29. x-intercept: 6 $(6, 0)$

y-intercept: -2 $(0, -2)$

$m = \dfrac{-2-0}{0-6} = \dfrac{-2}{-6} = \dfrac{1}{3}$

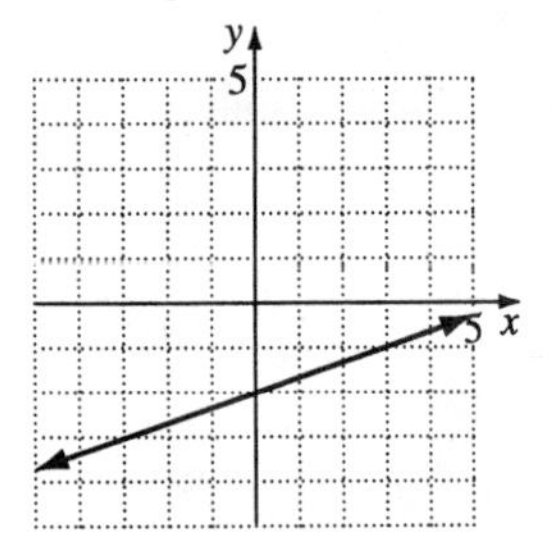

31. **a.** L_2: $m = 0$

b. L_1: (0, 2), (3, 3)

$m = \dfrac{1}{3}$

c. L_3: (–2, 0), (3, 1) $m = \dfrac{1}{5}$

33. y-intercept = 4; $m = 3$

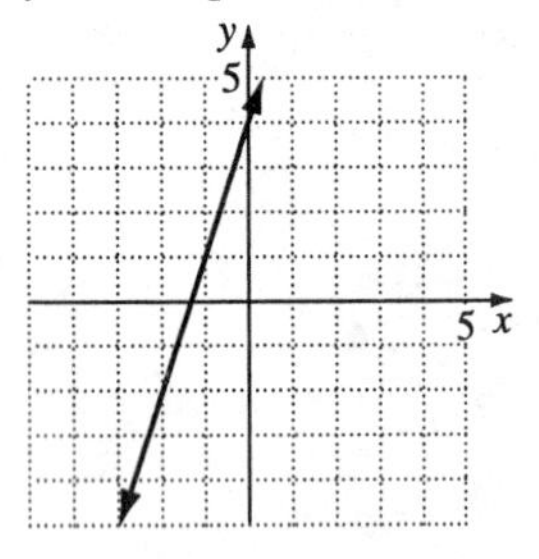

35. y-intercept = –1, $m = \dfrac{1}{2}$

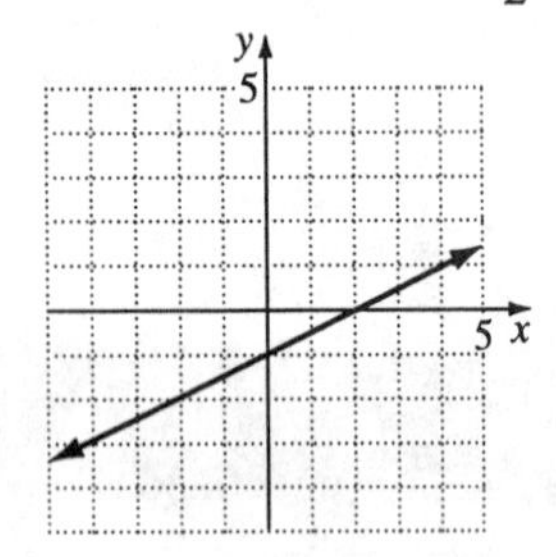

37. y-intercept = 1, $m = -\dfrac{1}{2}$

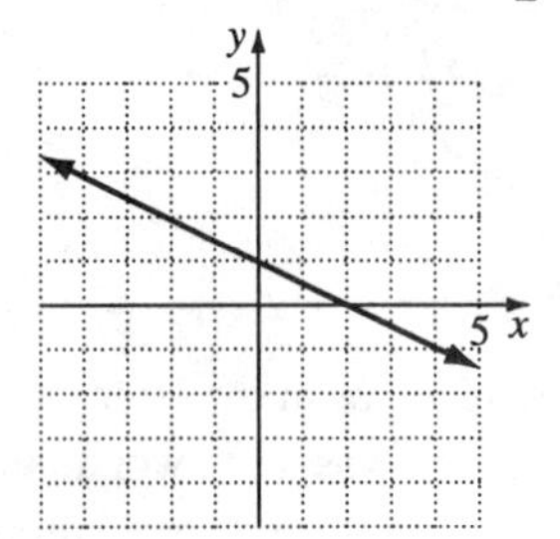

39. y-intercept = –3, $m = -\dfrac{2}{3}$

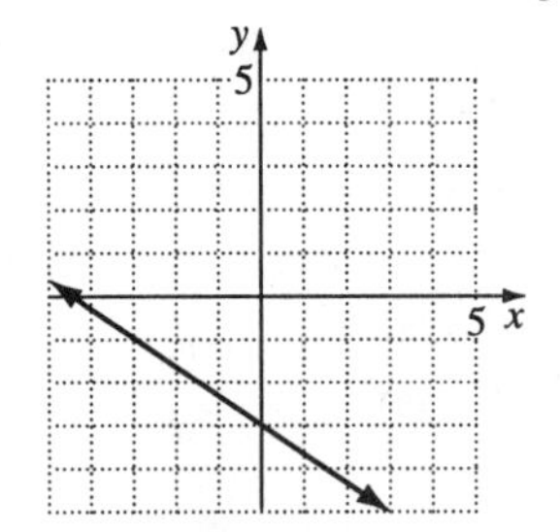

41. y-intercept = 0, $m = \dfrac{5}{3}$

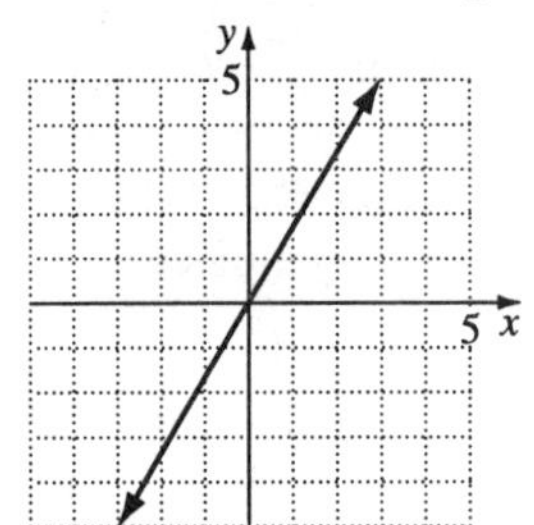

43. y-intercept = 0, $m = -4$

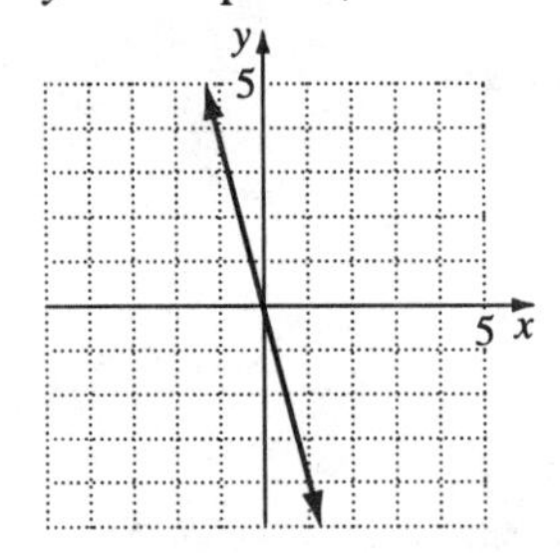

45. x-intercept = 2, $m = \frac{2}{3}$

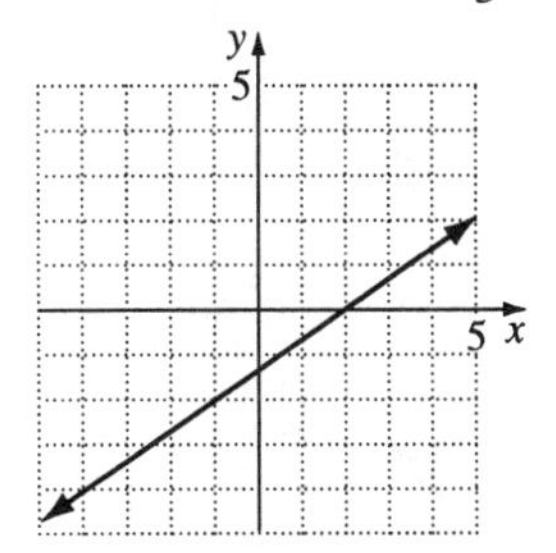

47. x-intercept = 1, $m = -\frac{3}{4}$

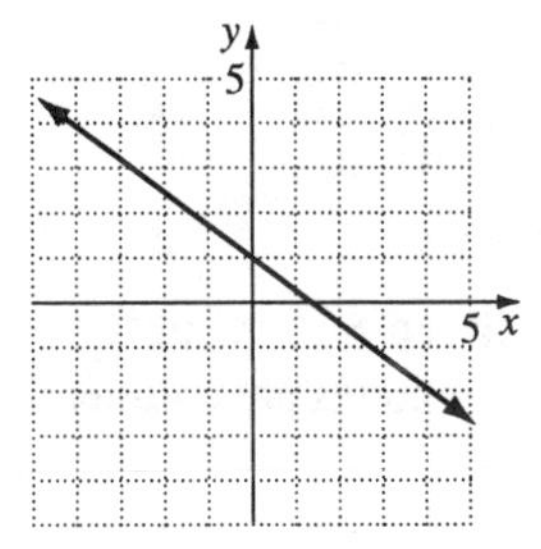

49. y-intercept = −3, $m = 0$

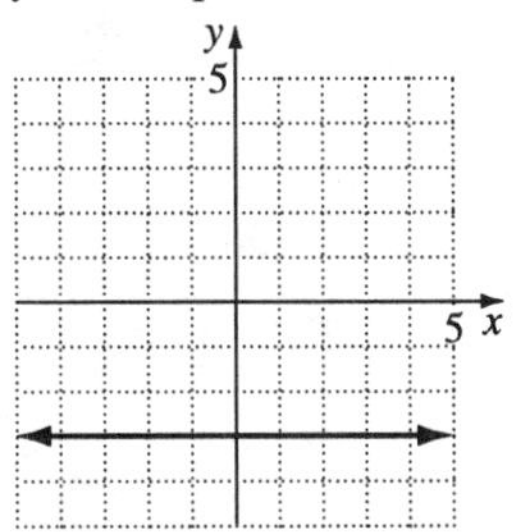

51. x-intercept = 3, slope is undefined

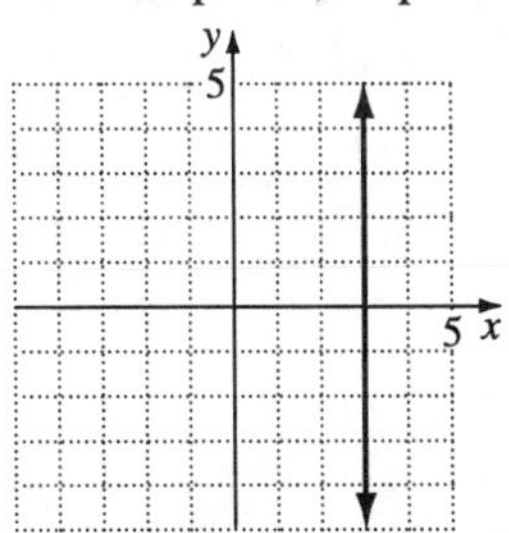

53. (0, −3), (−1, −5):

$$m = \frac{-5-(-3)}{-1-0} = \frac{-5+3}{-1} = 2$$

(1, 3), (−2, −3):

$$m = \frac{-3-3}{-2-1} = \frac{-6}{-3} = 2$$

The slopes are the same.
Therefore, the lines are parallel.

55. The slopes of the opposite sides are
(−3, −3), (2, −5)

$$m = \frac{-5+3}{2+3} = -\frac{2}{5}$$

(2, −5), (5, −1):

$$m = \frac{-1+5}{5-2} = \frac{4}{3}$$

(5, −1), (0, 1)

$$m = \frac{1+1}{0-5} = -\frac{2}{5}$$

(0, 1), (−3, −3)

$$m = \frac{-3-1}{-3-0} = \frac{4}{3}$$

The slopes of the opposite sides are equal. Since the opposite sides are parallel, the figure is a parallelogram.

57. $\text{pitch} = \frac{\text{rise}}{\text{run}} = \frac{3/2}{5/2} = \frac{3}{5}$

59. $\frac{50}{625} = 0.08$
8% grade

61. points (1950, 70), (2000, 80)

$$m = \frac{80-70}{2000-1950} = \frac{10}{50} = \frac{1}{5}$$

For each year increase between 1950 and 2000, the average life expectancy increases by $\frac{1}{5}$ year.

63. points (1990, 19.2), (1992, 17)

$$m = \frac{17-19.2}{1992-1990} = \frac{-2.2}{2} = -1.1\%$$

The slope is negative because the average percent is decreasing between these years. The average percent decreases by 1.1% each year from 1990 to 1992.

65. points:
$A(62, 7{,}000)$
$B(64, 8{,}000)$
$C(67, 10{,}000)$
$D(70, 12{,}400)$
$m_{AB} = \frac{8000-7000}{64-62} = \$500$
$m_{BC} = \frac{10{,}000-8{,}000}{67-64} = \$666.67$
$m_{CD} = \frac{12{,}400-10{,}000}{70-67} = \$800$
Between ages 62 and 64 yrs, the average annual benefit increases by \$500 per year. Between ages 64 and 67 years, the average annual benefit increases by \$666.67 per year. Between ages 67 and 70 years, the average annual benefit increases by \$800 per year.

67. points (1987, 420,000), (1994, 1,200,000)
$m = \frac{1{,}200{,}000-420{,}000}{1994-1987} \approx \$111{,}429$
The salaries have increased substantially in the last seven years.

69. Statement b is true.
(2, 2) and (0, 0); $m = \frac{2-0}{2-0} = 1$ true

71. Answers may vary.

73.

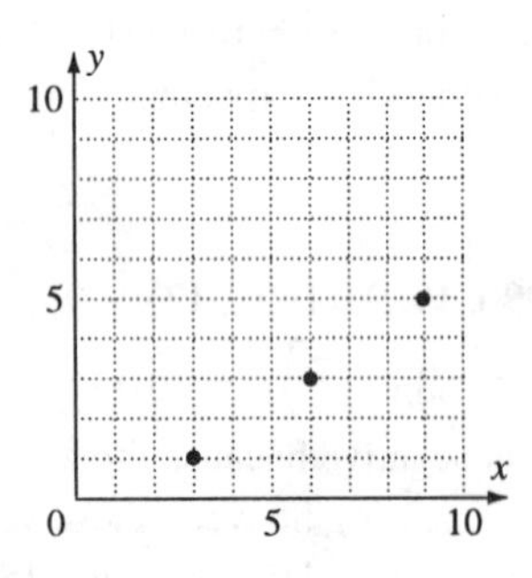

Yes; they are collinear.
Check three slopes.
$m_1 = \frac{3-1}{6-3} = \frac{2}{3}$
$m_2 = \frac{5-3}{9-6} = \frac{2}{3}$
$m_3 = \frac{5-1}{9-3} = \frac{2}{3}$

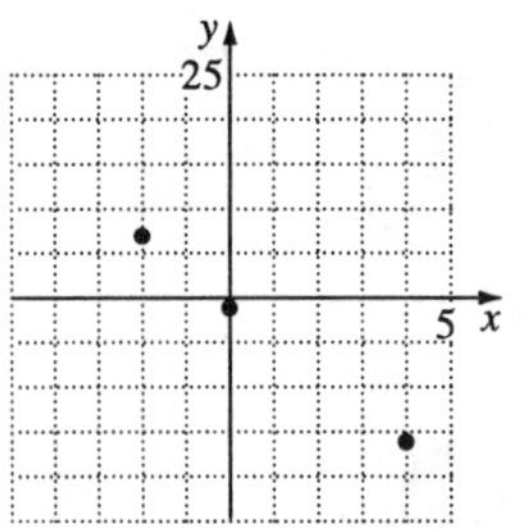

No; they are not collinear.
Check three slopes.
$m_1 = \frac{-16+1}{4-0} = -\frac{15}{4}$
$m_2 = \frac{7+16}{-2-4} = -\frac{23}{6}$
$m_3 = \frac{7+1}{-2-0} = -4$
If the slopes between each set of two points are equal, then the points are collinear.

Review Problems

75. $\frac{x}{5} = \frac{7}{3}$
$x = \frac{5(7)}{3}$
$x = 11\frac{2}{3}$; $11\frac{2}{3}$ in.

76. $0.05x = 36$
$x = 720$

77. $y = 2x + 7.98$
where y = number of personal computers (in millions) x years after 1996.
$35.98 = 2x + 7.98$
$28 = 2x$
$14 = x$
$1996 + 14 = 2010$

Problem Set 4.5

1. $y = 3x - 4$
slope = 3
y-intercept = –4

3. $y = -\frac{1}{2}x + 5$
slope = $-\frac{1}{2}$
y-intercept = 5

5. $y = \frac{3}{4}x$
slope = $\frac{3}{4}$
y-intercept = 0

7. $y = -5 - 7x$
$y = -7x - 5$
slope = −7
y-intercept = −5

9. $-5x + y = 7$
$y = 5x + 7$
slope = 5
y-intercept = 7

11. $x + y = 6$
$y = -x + 6$
slope = −1
y-intercept = 6

13. $y = 2$
slope = 0
y-intercept = 2

15. $8x + 4y = 8$
$2x + y = 2$
$y = -2x + 2$
slope = −2
y-intercept = 2

17. $3x - 2y = 6$
$-2y = -3x + 6$
$y = \frac{3}{2}x - 3$
slope = $\frac{3}{2}$
y-intercept = −3

19. $x - y = 0$
$-y = -x$
$y = x$
slope = 1
y-intercept = 0

21. slope = 6
y-intercept = 5
$y = 6x + 5$

23. slope = −4
y-intercept = −2
$y = -4x - 2$

25. slope = $\frac{1}{2}$
y-intercept = −3
$y = \frac{1}{2}x - 3$

27. slope = $-\frac{3}{5}$
y-intercept = −4
$y = -\frac{3}{5}x - 4$

29. $y = 2x + 3$
$m = 2$, (0, 3)

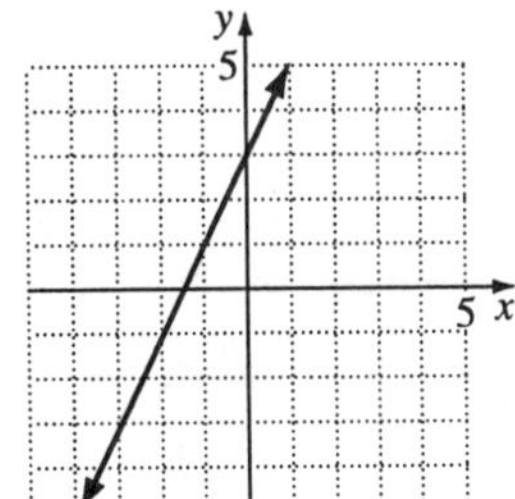

31. $y = -2x + 4$
$m = -2$, $(0, 4)$

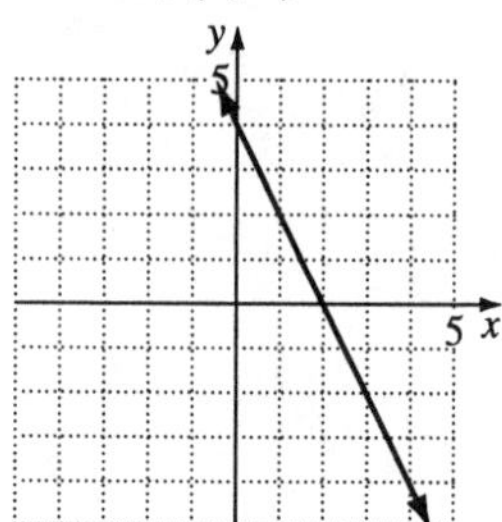

33. $y = \frac{1}{2}x + 3$
$m = \frac{1}{2}$, $(0, 3)$

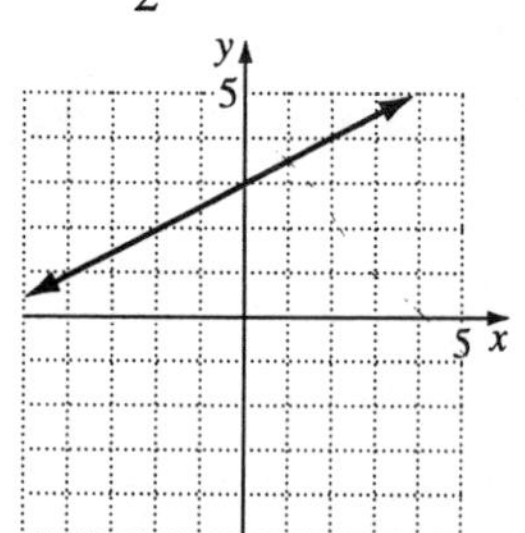

35. $y = \frac{2}{3}x - 4$
$m = \frac{2}{3}$, $(0, -4)$

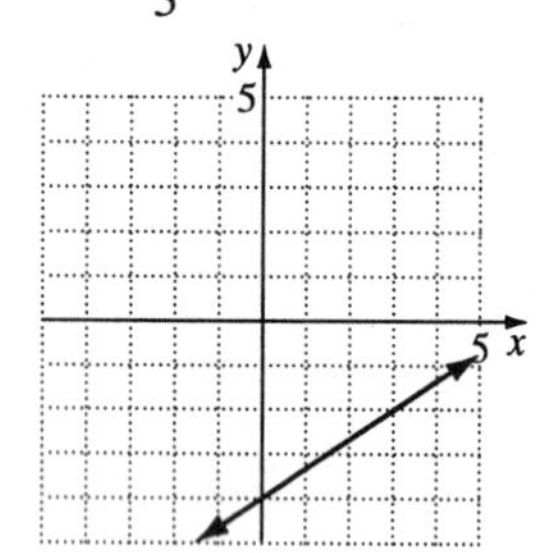

37. $y = -\frac{3}{4}x + 4$
$m = -\frac{3}{4}$, $(0, 4)$

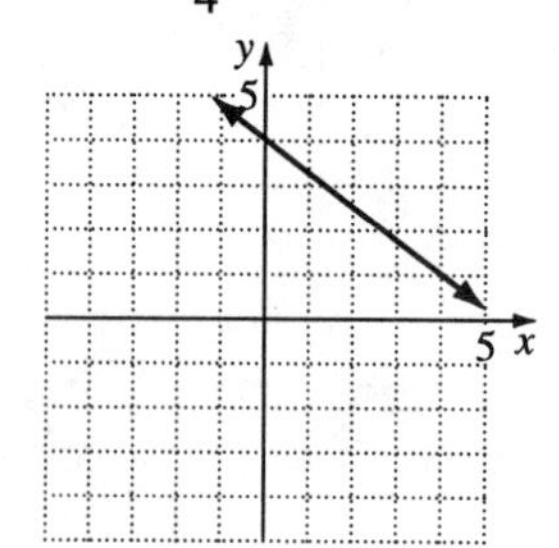

39. $y = -\frac{3}{2}x - 1$
$m = -\frac{3}{2}$, $(0, -1)$

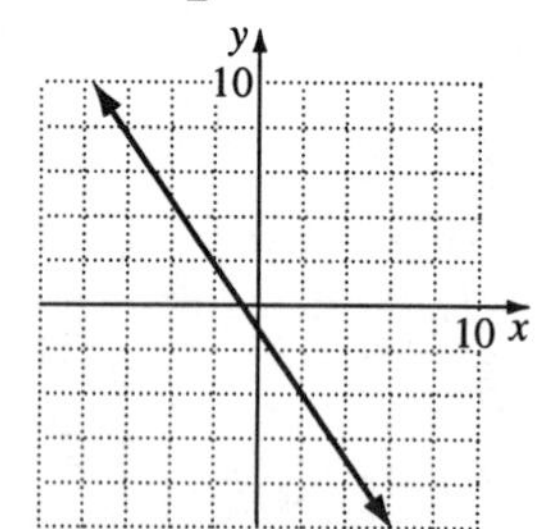

41. $y = 3x$
$m = 3$, $(0, 0)$

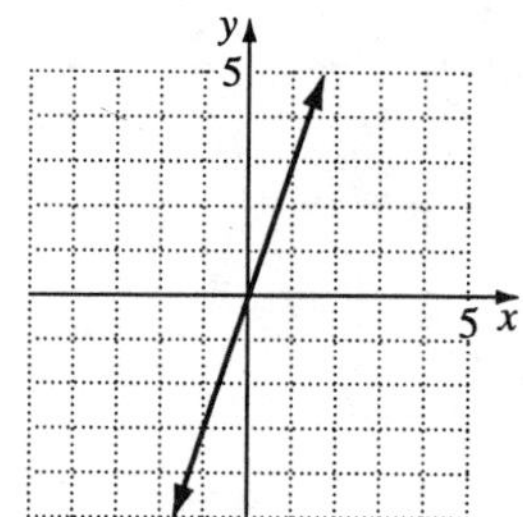

43. $y = -\frac{5}{3}x$

$m = -\frac{5}{3}$, (0, 0)

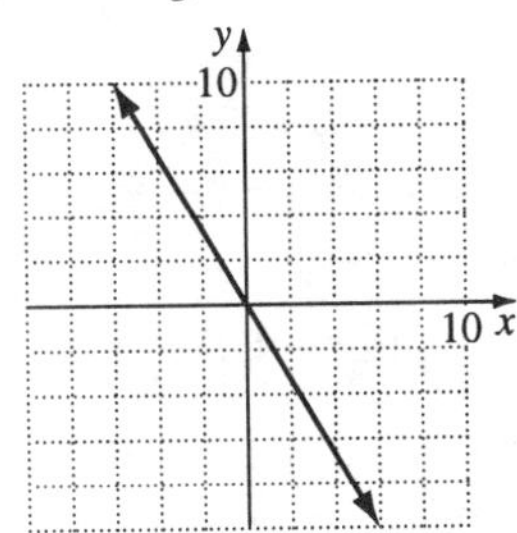

45. $y = 3x + 1$

$y = 3x - 3$

Same slope

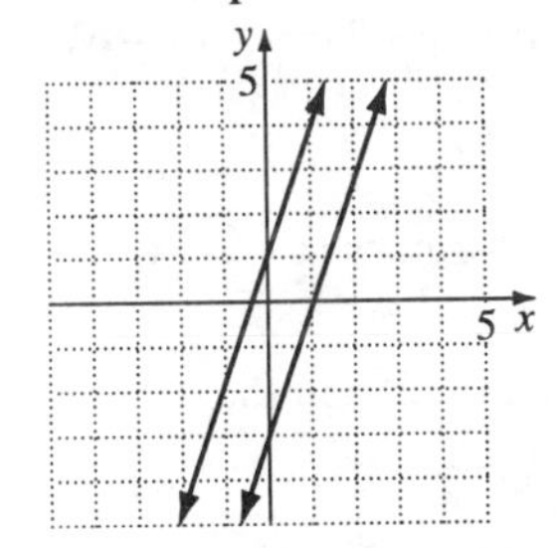

47. $4x - y = 2$ or $y = 4x - 2$

$y = 4x + 2$

Same slope

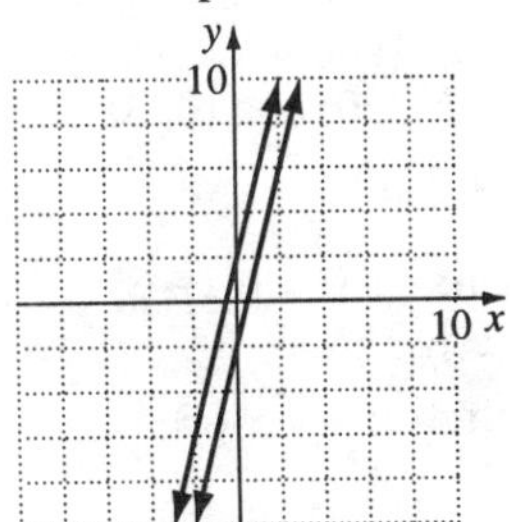

49. $S = -82t + 1972$

y-intercept = 1972

In 1980, 1972 thousand turntables were sold.

Slope = −82

The average number of turntables sold decreases by 82 thousand each year.

51. **a.** $y = \frac{1}{33}x + 1$

$p = \frac{1}{33}d + 1$

b. $p = \frac{1}{33}(99) + 1$

$p = 4$ atmospheres

53. **a.** $y = -185x + 20{,}151$

b. $y = -185(15) + 20{,}151$

$y = \$17{,}376$

55. c is true

$3x + 2y = 5$

$2y = -3x + 5$

$y = -\frac{3}{2}x + \frac{5}{2}$

$m = -\frac{3}{2}$

57. Verify graphs

59. $T = 10d + 20$

700

60

61. Positive slope;

Number of turntables sold is decreasing because the number of CD players sold is increasing.

63. $y = 5x + 2$

$y = \frac{1}{5}x$

$5\left(\frac{1}{5}\right) = 1$

No

65. $y = 2x - 3$

$x + 2y = 1$ or $y = -\frac{1}{2}x + \frac{1}{2}$

$(2)\left(-\frac{1}{2}\right) = -1$

Yes

67. Group activity

Review Problems

68. Let x = width
Then $2x - 33$ = length
$2(x + 2x - 33) = 378$
$6x - 66 = 378$
$6x = 444$
$x = 74$
$2x - 33 = 115$
74 yards by 115 yards

69. $-3x + 7 \le -38$
$-3x \le -45$
$x \ge 15$

70. $\frac{x}{600} = \frac{3}{1.2}$
$x = 1500$ yards

Problem Set 4.6

1. slope = 2, passing through (3, 5)
point-slope form: $y - 5 = 2(x - 3)$
$y - 5 = 2x - 6$
slope-intercept form: $y = 2x - 1$

3. slope = 6, passing through (–2, 5)
$y - 5 = 6[x - (-2)]$
point-slope form: $y - 5 = 6(x + 2)$
$y - 5 = 6x + 12$
slope-intercept form: $y = 6x + 17$

5. slope = –3, passing through (–2, –3)
point slope form: $y + 3 = -3(x + 2)$
$y + 3 = -3x - 6$
slope-intercept form: $y = -3x - 9$

7. slope = –4, passing through (–4, 0)
$y - 0 = -4(x + 4)$
point-slope form: $y = -4(x + 4)$
slope-intercept form: $y = -4x - 16$

9. slope = –1, passing through $\left(-\frac{1}{2}, -2\right)$
point-slope form: $y + 2 = -1\left(x + \frac{1}{2}\right)$
$y + 2 = -x - \frac{1}{2}$
slope-intercept form: $y = -x - \frac{5}{2}$

11. slope = $\frac{1}{2}$, passing through the origin: (0, 0)
point-slope form: $y - 0 = \frac{1}{2}(x - 0)$
slope-intercept form: $y = \frac{1}{2}x$

13. slope = $-\frac{2}{3}$, passing through (6, –2)
point-slope form: $y + 2 = -\frac{2}{3}(x - 6)$
$y + 2 = -\frac{2}{3}x + 4$
slope-intercept form: $y = -\frac{2}{3}x + 2$

15. passing through (1, 2) and (5, 10)
slope = $\frac{10-2}{5-1} = \frac{8}{4} = 2$
point-slope form: $y - 2 = 2(x - 1)$
or $y - 10 = 2(x - 5)$
$y - 2 = 2x - 2$
slope-intercept form: $y = 2x$

17. passing through (–3, 0) and (0, 3)
slope = $\frac{3-0}{0+3} = \frac{3}{3} = 1$
point-slope form: $y - 0 = 1(x + 3)$
or $y - 3 = 1(x - 0)$
slope-intercept form: $y = x + 3$

19. passing through (–3, –1) and (2, 4)
slope = $\frac{4+1}{2+3} = \frac{5}{5} = 1$
point-slope form: $y + 1 = 1(x + 3)$ or
$y - 4 = 1(x - 2)$
slope-intercept form: $y = x + 2$

21. passing through (–3, –2) and (3, 6)

slope = $\frac{6+2}{3+3} = \frac{8}{6} = \frac{4}{3}$

point-slope form: $y+2 = \frac{4}{3}(x+3)$ or

$y-6 = \frac{4}{3}(x-3)$

$y+2 = \frac{4}{3}x+4$

slope-intercept form: $y = \frac{4}{3}x+2$

23. passing through (–3, –1) and (4, –1)

slope = $\frac{-1+1}{4+3} = \frac{0}{7} = 0$

point-slope form: $y + 1 = 0(x + 3)$ or

$y + 1 = 0(x-4)$

slope-intercept form: $y = -1$

25. passing through (2, 4) with x-intercept = –2; (–2, 0)

slope = $\frac{0-4}{-2-2} = \frac{-4}{-4} = 1$

point-slope form: $y - 4 = 1(x-2)$

slope-intercept form: $y = x+2$

27. x-intercept = $-\frac{1}{2}$ and y-intercept = 4

$\left(-\frac{1}{2}, 0\right)$, (0, 4)

slope = $\frac{4-0}{0+\frac{1}{2}} = \frac{4}{\frac{1}{2}} = 8$

point-slope form: $y-0 = 8\left(x+\frac{1}{2}\right)$ or

$y-4 = 8(x-0)$

slope-intercept form: $y = 8x+4$

29. a. slope = $\frac{125-115}{30-10} = \frac{1}{2}$

b. $y-115 = \frac{1}{2}(x-10)$ or

$y-125 = \frac{1}{2}(x-30)$

c. $y-115 = \frac{1}{2}x-5$

$y = \frac{1}{2}x+110$

d. $y = \frac{1}{2}(80)+110$

$y = 150$

Blood pressure = 150

31. a. slope = $\frac{178-114}{6.4-1.3} = \frac{64}{5.1}$

b. $y-114 = \frac{64}{5.1}(x-1.3)$ or

$y-178 = \frac{64}{5.1}(x-6.4)$

c. $y-114 = \frac{64}{5.1}x - \frac{83.2}{5.1}$

$y = \frac{64}{5.1}x + \frac{498.2}{5.1}$

d. $y = \frac{64}{5.1}(11.6) + \frac{498.2}{5.1}$

$y \approx 243.3$

About 243 per 100,000

33. Statement c is true.

a. A line with no slope is a vertical line.

b. (–3, 2)

$y-3 = 7(x+2)$

$2-3 = 7(-3+2)$

$-1 \neq -7$

c. true; slope = $\frac{6+5}{2-2}$ is undefined

d. $y = -3x + 7$; $m = -3$ not 3

35. Statement b is true.

a. (1, 4) $m = 2$

$y-4 = 2(x-1)$

$y = 2x + 2$ *not* $2x+4$

b. true; A vertical line (except $x = 0$) has no y-intercept

c. $y = \frac{5}{2}x + \frac{3}{2}$; $m = \frac{5}{2}$ not 5 and $b = \frac{3}{2}$ not 3

d. only one line can be drawn with both conditions y-intercept = 3, slope = $\frac{1}{2}$

37. $y = 1.75x - 2$

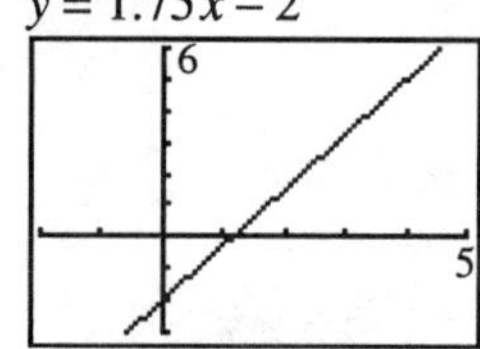

39. Students should verify graphs.

41. Answers may vary.

43. If the line is parallel it has the same slope as
$y = 2x + 1$ or $m = 2$.
point-slope: $y - 2 = 2(x + 3)$
$y - 2 = 2x + 6$
slope-intercept form: $y = 2x + 8$

45. Group activity

Review Problems

46. $4 - 3(x - 5) < -2x$
$4 - 3x + 15 < -2x$
$-3x + 19 < -2x$
$-x < -19$
$x > 19$

17 18 19 20 21

47. Let x = height
$\frac{1}{2}bh = A$
$\frac{1}{2}(12)x = 54$
$6x = 54$
$x = 9$
9 cm

48. Let x = smallest angle measure.
Then $2x$ = 2nd largest angle measure and
$3x$ = largest angle measure.
$x + 2x + 3x = 180$
$6x = 180$
$x = 30$
$2x = 60$
$3x = 90$
The angles are 30°, 60°, and 90°.

Problem Set 4.7

1. $x + y > 4$
(2, 2): $2 + 2 > 4;\ 4 > 4$ false
(3, 2): $3 + 2 > 4;\ 5 > 4$ true
(3, 2) satisfies the inequality
(–3, 8): $-3 + 8 > 4;\ 5 > 4$ true
(–3, 8) satisfies the inequality

3. $2x + y \ge 5$
(4, 0): $8 + 0 \ge 5$ true
(4, 0) satisfies the inequality
(1, 3): $2 + 3 \ge 5$ true
(1, 3) satisfies the inequality
(0, 0): $0 + 0 \ge 5$ false

5. $y \ge -2x + 4$
(4, 0): $0 \ge -8 + 4 = -4$ true
(4, 0) satisfies the inequality
(1, 3): $3 \ge -2 + 4 = 2$ true
(1, 3) satisfies the inequality
(–2, –4): $-4 \ge 4 + 4 = 8$ false

7. $y > -2x + 1$
(2, 3): $3 > -4 + 1 = -3$ true
(2, 3) satisfies the inequality
(0, 0): $0 > 0 + 1 = 1$ false
(0, 5): $5 > 0 + 1 = 1$ true
(0, 5) satisfies the inequality

9. $x + y \geq 4$

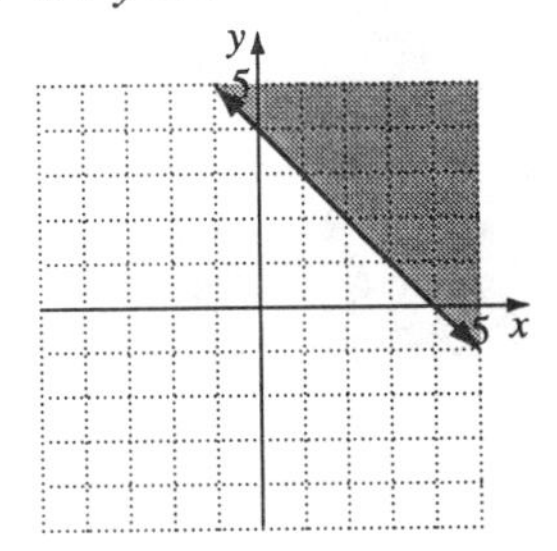

11. $x - y < 3$

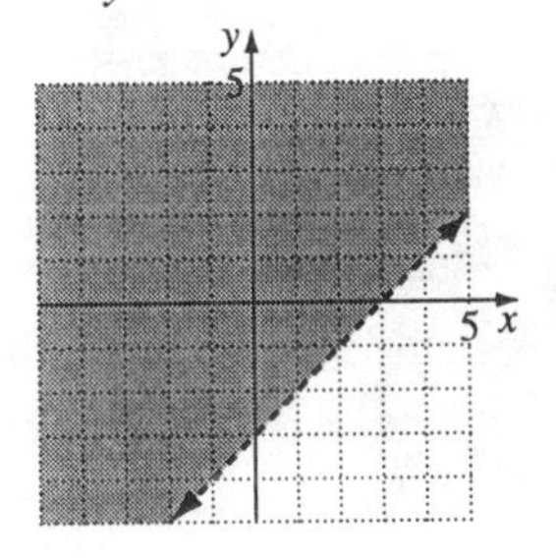

13. $2x + y > 4$

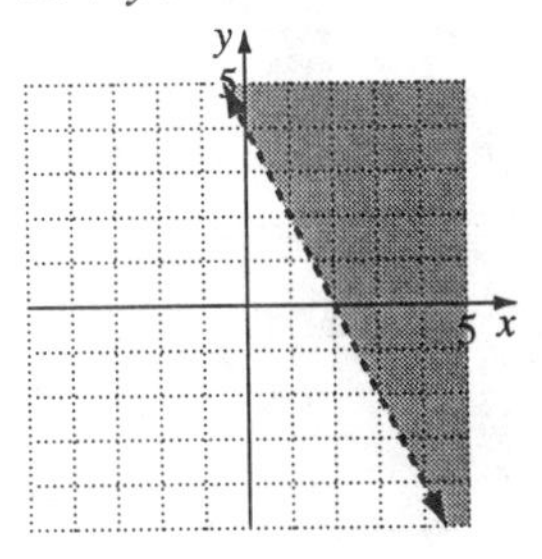

15. $x - 3y \leq 6$

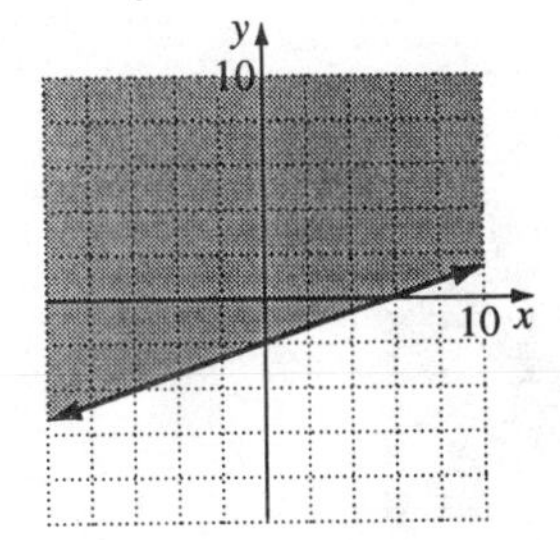

17. $3x - 2y \leq 6$

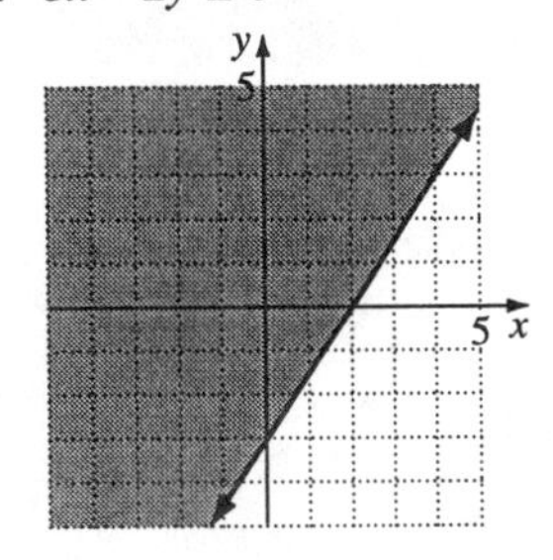

19. $4x + 3y > 12$

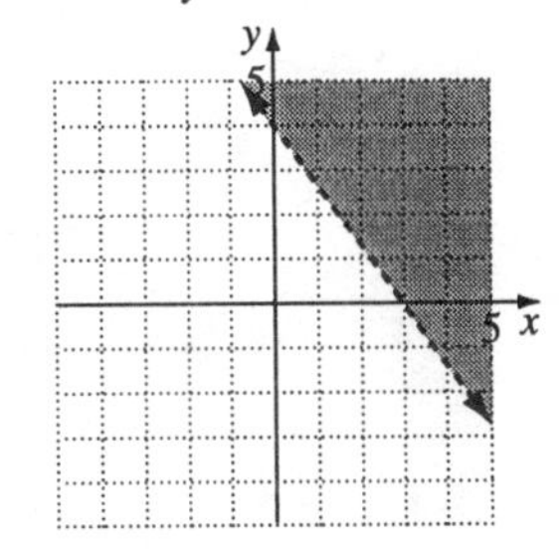

21. $5x - y < -10$

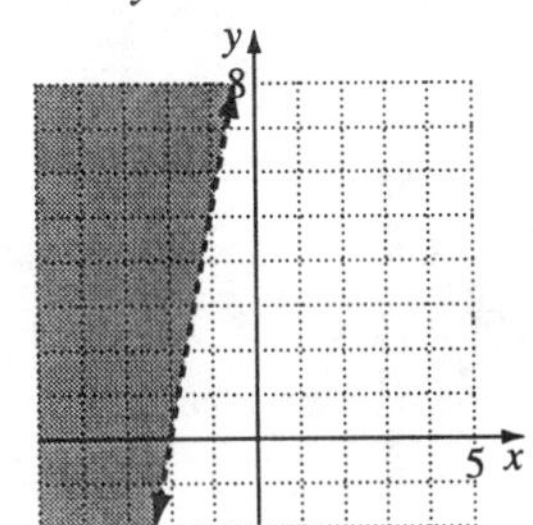

23. $2x - \frac{1}{2}y \geq 2$

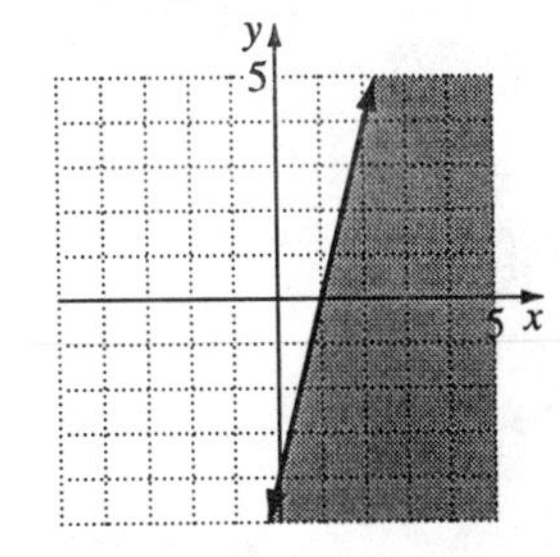

25. $x + y \leq 0$

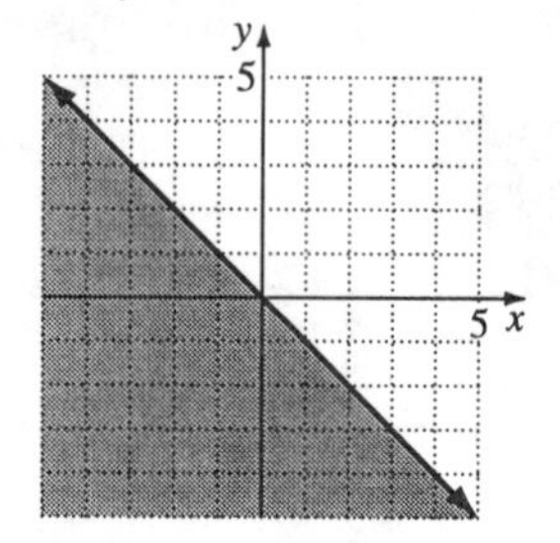

27. $x \geq 3$

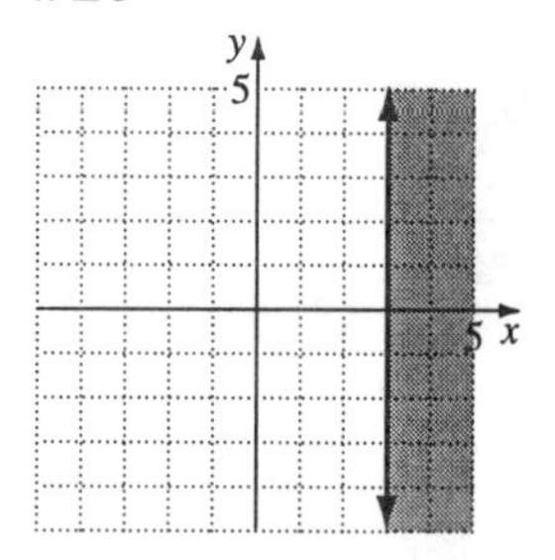

29. $x > -4$

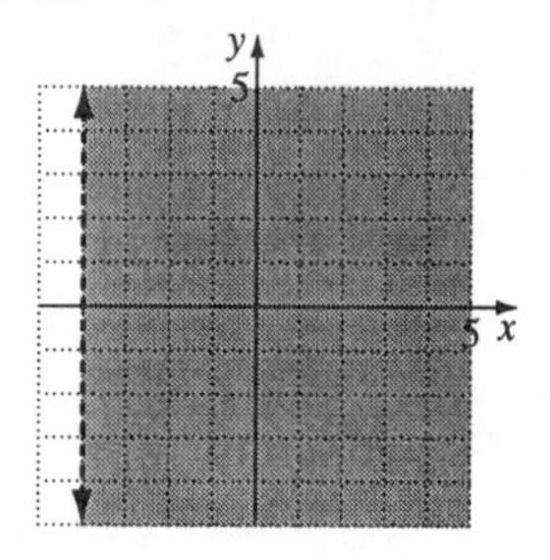

31. $y \leq 2$

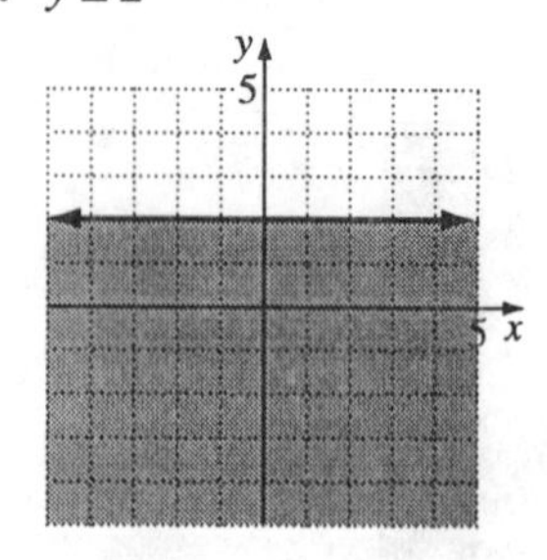

33. $y > -1$

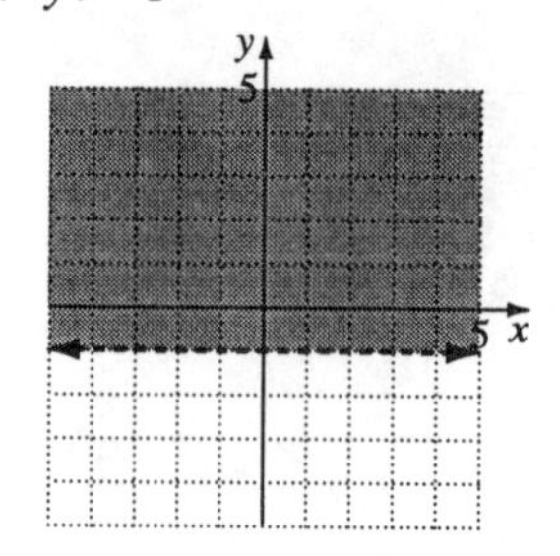

35. $x \geq 0$

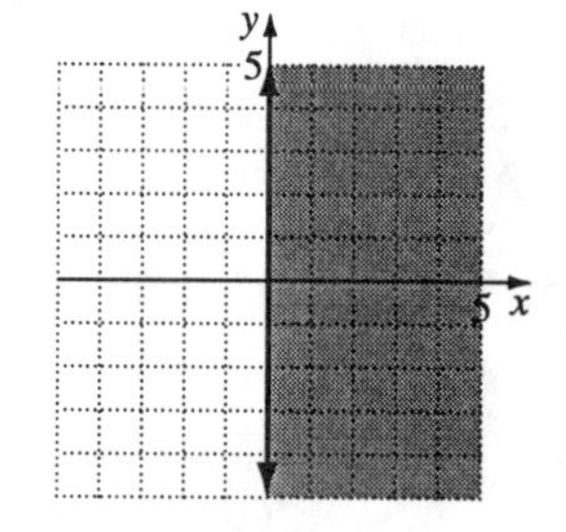

37. $y \geq x + 1$

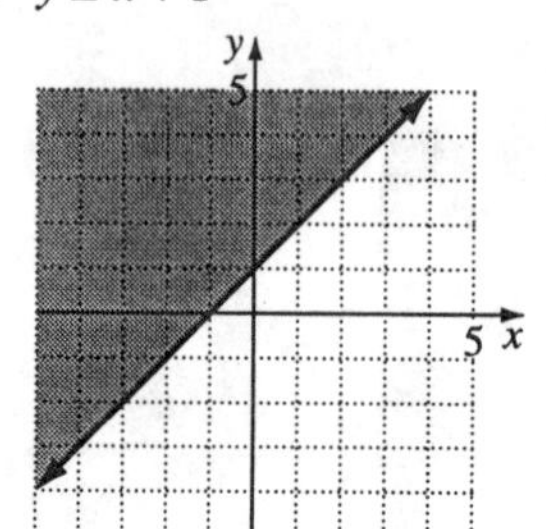

39. $y < -x + 4$

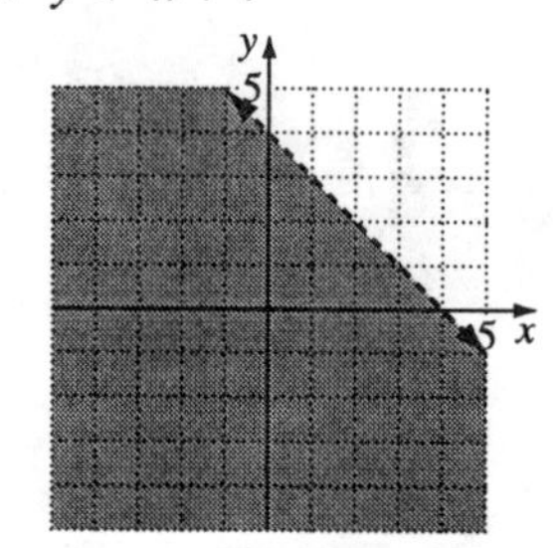

41. $y < 2x + 3$

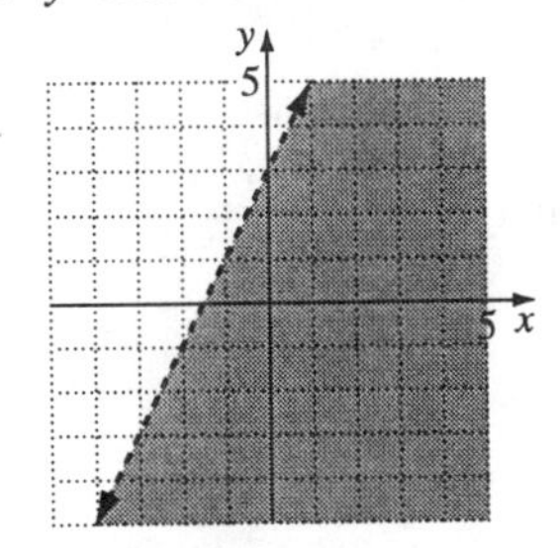

43. $y \geq 3x - 2$

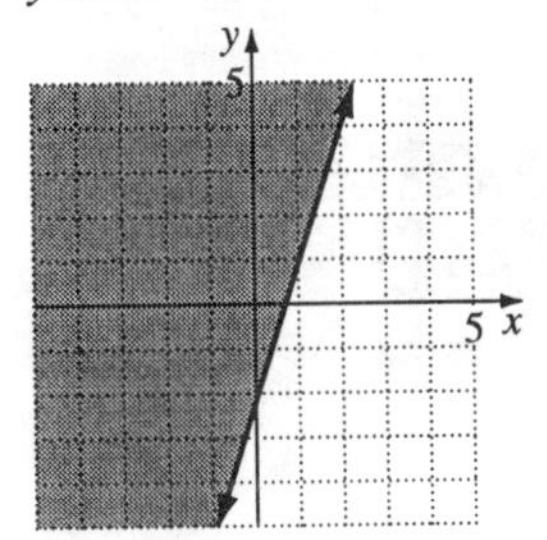

45. $y > \frac{1}{2}x + 2$

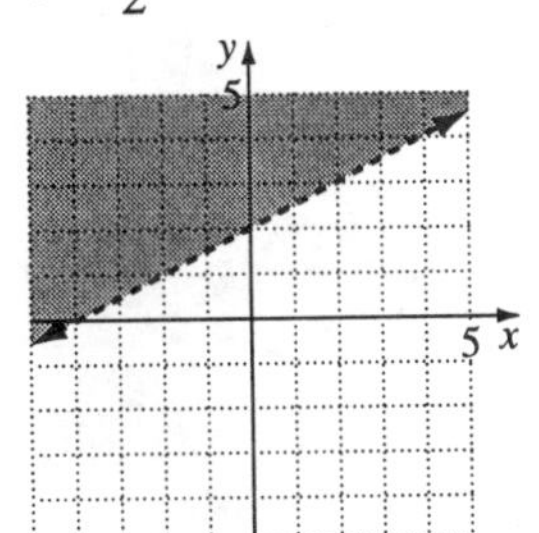

47. $y < \frac{3}{4}x - 3$

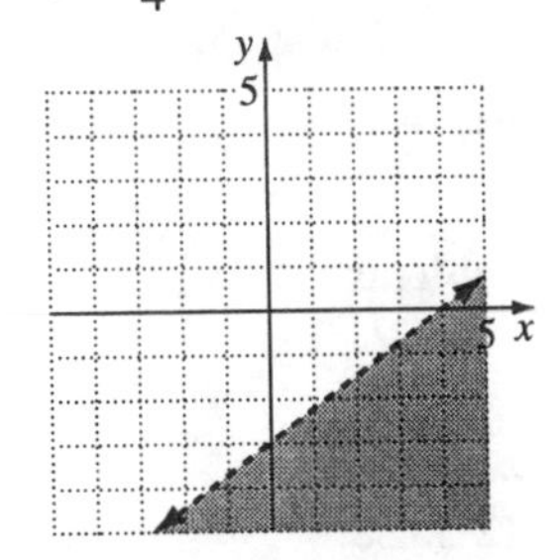

49. $y > 2x$

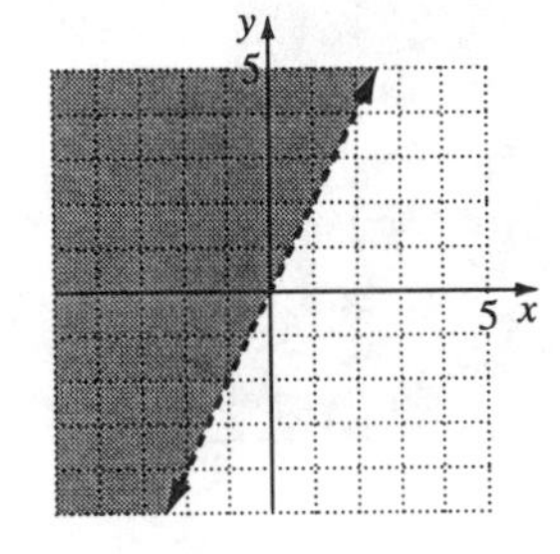

51. $y \leq \frac{5}{4}x$

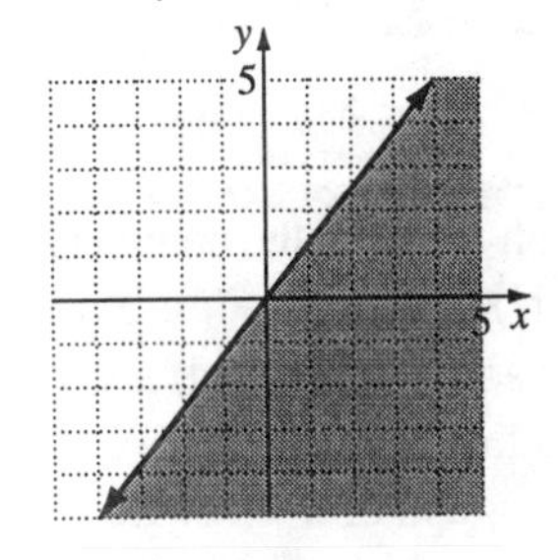

53. $y > -\frac{2}{3}x + 1$

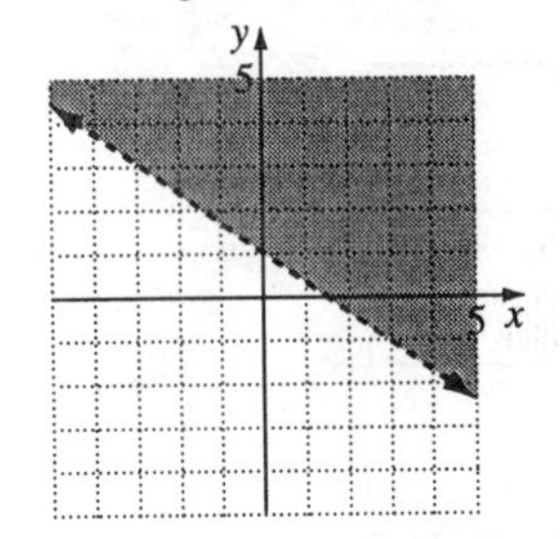

55. $2x + y \leq 4$; f

57. $y < 1$; b

59. $y \geq 3x$; e

61. a. $75x$

b. $50y$

c. $75x + 50y > 300$

d.

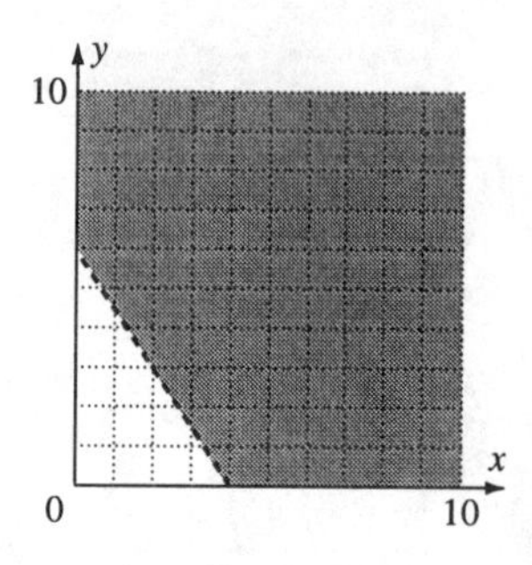

e. Answers may vary.

63. d is true;

a. (0, –3): $y > 2x - 3$
$-3 > 0 - 3$ false

b. The graph is above the boundary line $x = y + 1$, *not* below.

c. $y \ge 4x$, a solid line is used, *not* a dashed line.

d. true; The graph of $x < 4$ is the half-plane to the left of $x = 4$.

65. $y \le -3x + 4$

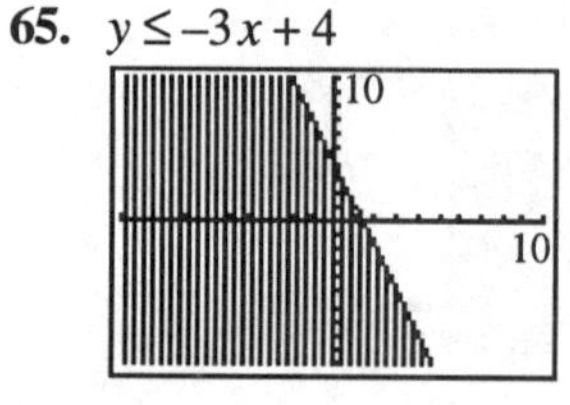

67. $y \ge \frac{1}{2}x + 4$

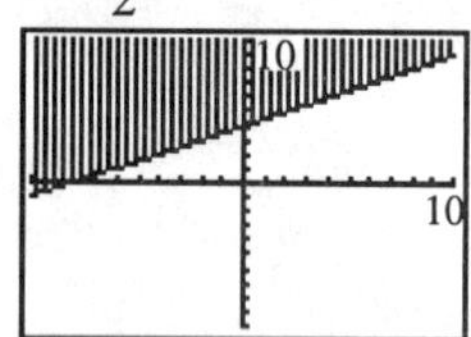

69. $y = 2(2x + 1) - 3x$

a. $x = -2$

b. $2(2x + 1) - 3x = 0$
$4x + 2 - 3x = 0$
$x + 2 = 0$
$x = -2$
They are equal.

c. $x > -2$

d. $x < -2$

71–75. Answers may vary.

77. $y > 2x + 3$

79. $xy \le 0$

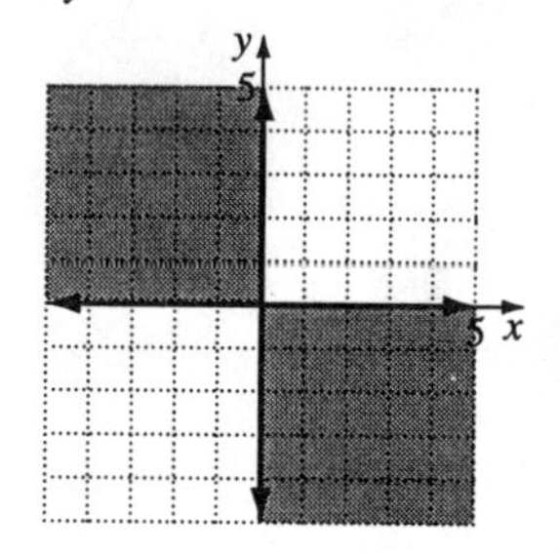

Review Problems

80. $-\frac{7}{8}\left(-\frac{4}{15}\right)$
$= \frac{(-7)(-4)}{(8)(15)}$
$= \frac{(-7)(-1)}{(2)(15)}$
$= \frac{7}{30}$

81. $-10 + 16 \div 2(-4)$
$= -10 + 8(-4)$
$= -10 - 32$
$= -42$

82. $-2 \le x < 4$

–2 0 2 4 6

Chapter 4 Review Problems

1. $3x - y = 12$
(0, –12): $0 + 12 = 12$ true
(0, 4): $0 - 4 = 12$ false
(–1, 15): $-3 - 15 = 12$ false
(–2, –18): $-6 + 18 = 12$ true

2. $y = -\frac{1}{2}x + 1$

x	$y = -\frac{1}{2}x + 1$	(x, y)
−4	$y = -\frac{1}{2}(-4) + 1 = 3$	(−4, 3)
−2	$y = -\frac{1}{2}(-2) + 1 = 2$	(−2, 2)
0	$y = -\frac{1}{2}(0) + 1 = 1$	(0, 1)
2	$y = -\frac{1}{2}(2) + 1 = 0$	(2, 0)
4	$y = -\frac{1}{2}(4) + 1 = -1$	(4, −1)

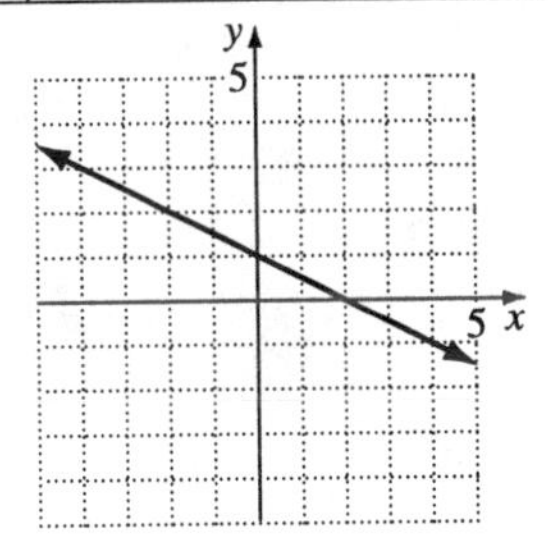

3. a. $x - 2y = 4$
$-2y = -x + 4$
$y = \frac{1}{2}x - 2$

b. $f(x) = \frac{1}{2}x - 2$

c.

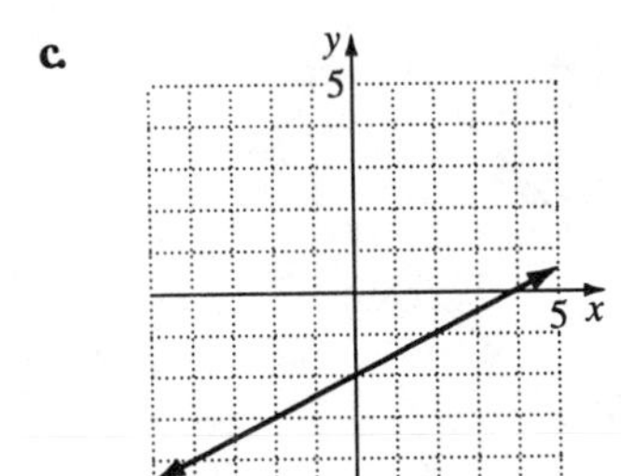

4. $f(x) = 2.35x + 179.5$
$f(20) = 2.35(20) + 179.5$
$f(20) = 226.5$
In 1980, the world population was 226.5 million.

5. $f(x) = -2.26x + 70.62$

a. $f(10) = -2.26(10) + 70.62 = 48.02$
$f(20) = -2.26(20) + 70.62 = 25.42$
$f(50) = -2.26(50) + 70.62 = -42.38$
$f(100) = -2.26(100) + 70.62 = -155.38$
If a bag of sugar is priced at 10¢, 20¢, 50¢, and $1.00, the quantity of bags purchased yearly is 48.02 million, 25.42 million, 0, and 0, respectively.

b. As price goes up, demand goes down.

c. $f(x) = -2.26x + 70.62$

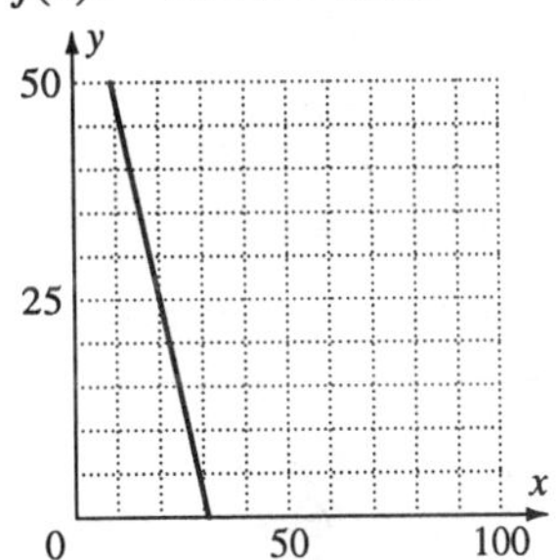

The line has a negative slope.

d. Verify graph.

6. $d = 30t$

a. $d = 30(1) = 30$ miles
$d = 30(2) = 60$ miles
$d = 30(2.5) = 75$ miles
$d = 30(4) = 120$ miles

b.

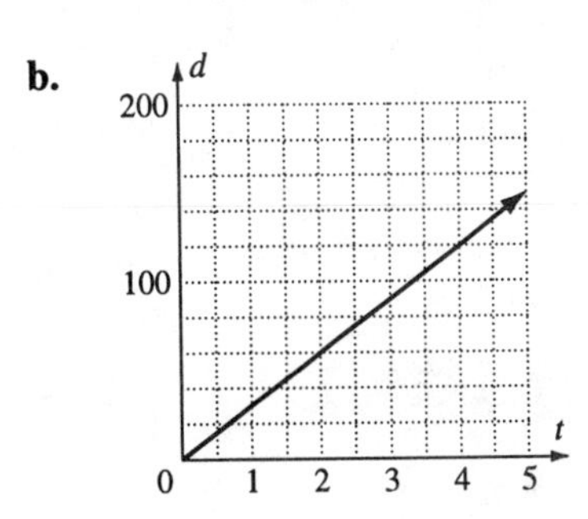

7. $2x + y = 4$

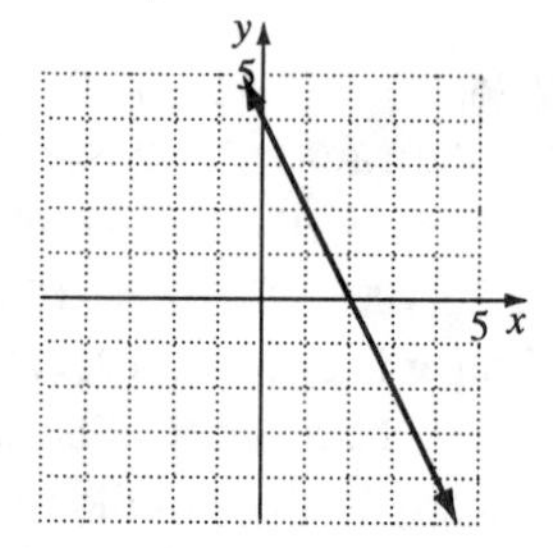

8. $3x - 2y = 12$

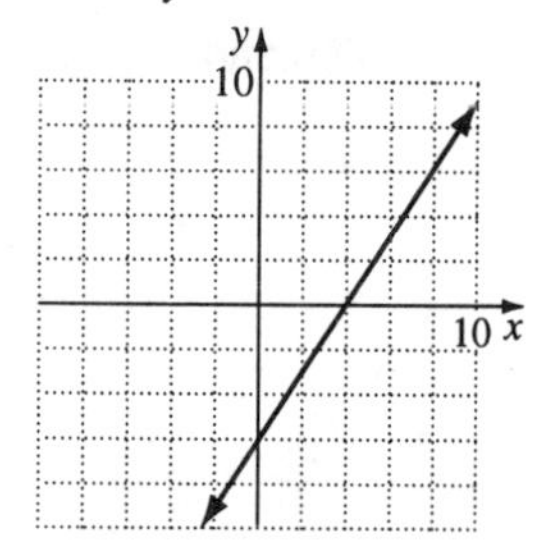

9. $3x = 6 - 2y$

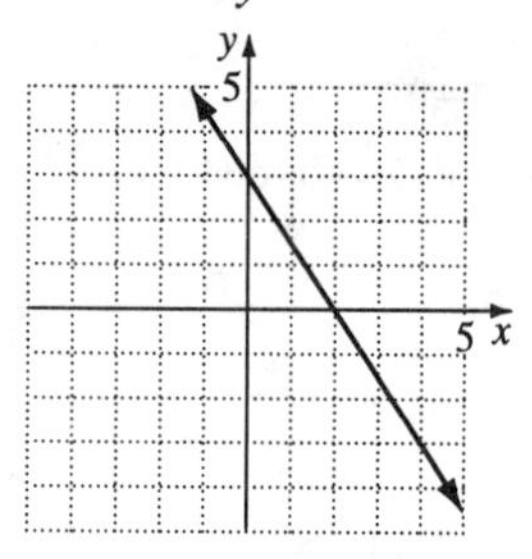

10. $3x - y = 0$

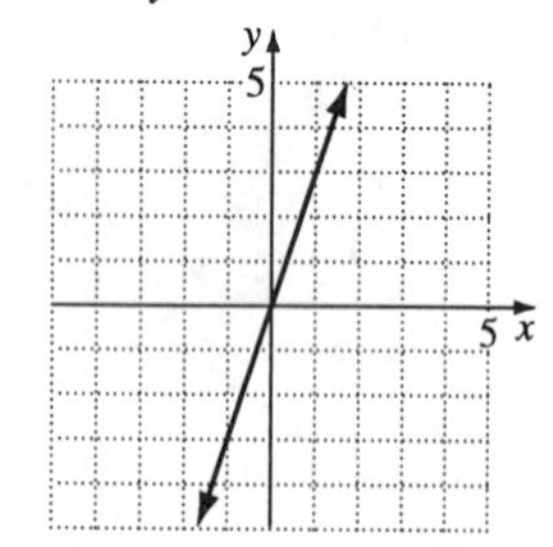

11. $x = 3$

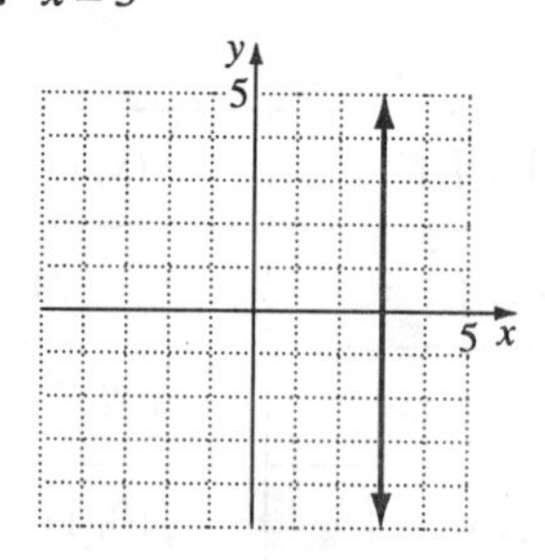

12. $2y = -10$

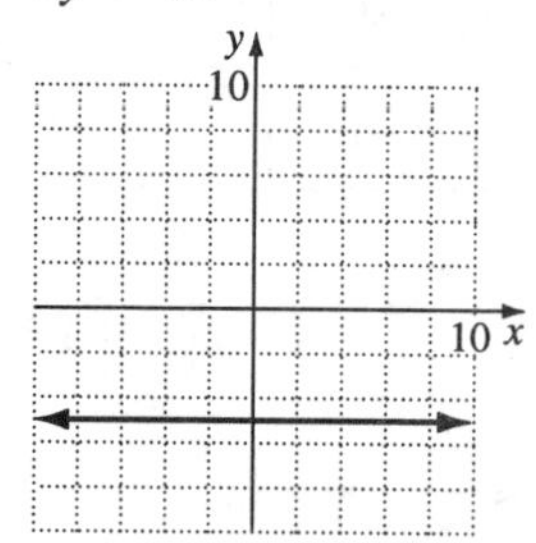

13. **a.** $S = 0.10x + 200$

b. $S = 0.10(0) + 200 = \$200$
$S = 0.10(10{,}000) + 200 = \$1200$
$S = 0.10(20{,}000) + 200 = \$2200$
$S = 0.10(30{,}000) + 200 = \$3200$

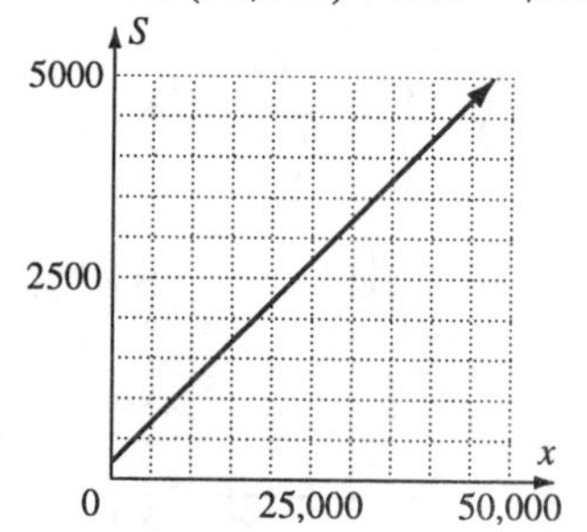

c. \$800

d. \$15,000

14. **a.** $C = 0.25x + 30$

b. $C = 0.25(100) + 30 = \$55$
$C = 0.25(200) + 30 = \$80$
$C = 0.25(300) + 30 = \$105$
$C = 0.25(400) + 30 = \$130$

$C = 0.25(500) + 30 = \$155$

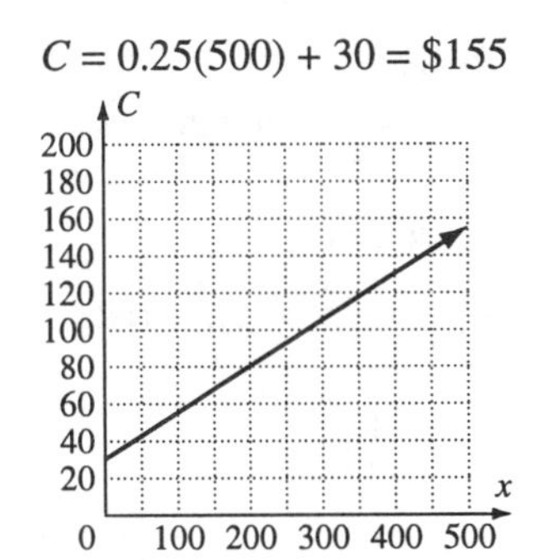

c. $117.50

d. 280 miles

15. $y = x^2 - 2$

x	$y = x^2 - 2$	(x, y)
–3	$y = (-3)^2 - 2 = 7$	(–3, 7)
–2	$y = (-2)^2 - 2 = 2$	(–2, 2)
–1	$y = (-1)^2 - 2 = -1$	(–1, –1)
0	$y = 0^2 - 2 = -2$	(0, –2)
1	$y = 1^2 - 2 = -1$	(1, –1)
2	$y = 2^2 - 2 = 2$	(2, 2)
3	$y = 3^2 - 2 = 7$	(3, 7)

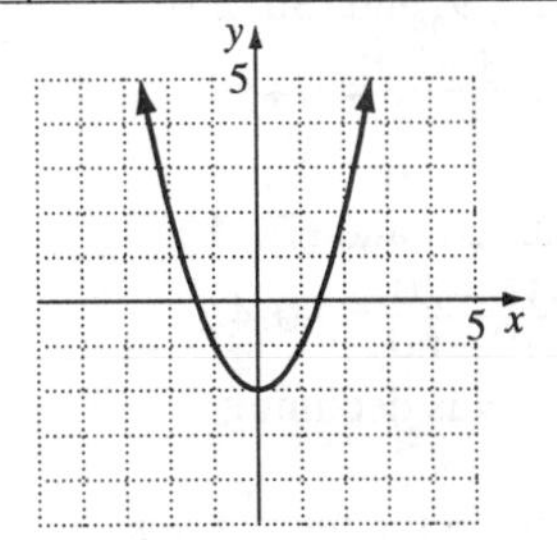

16. $f(x) = x^2 + 2x + 1$

x	$f(x) = x^2 + 2x + 1$	(x, y)
–3	$f(x) = (-3)^2 + 2(-3) + 1 = 4$	(–3, 4)
–2	$f(x) = (-2)^2 + 2(-2) + 1 = 1$	(–2, 1)
–1	$f(x) = (-1)^2 + 2(-1) + 1 = 0$	(–1, 0)
0	$f(x) = 0^2 + 2(0) + 1 = 1$	(0, 1)
1	$f(x) = 1^2 + 2(1) + 1 = 4$	(1, 4)
2	$f(x) = 2^2 + 2(2) + 1 = 9$	(2, 9)

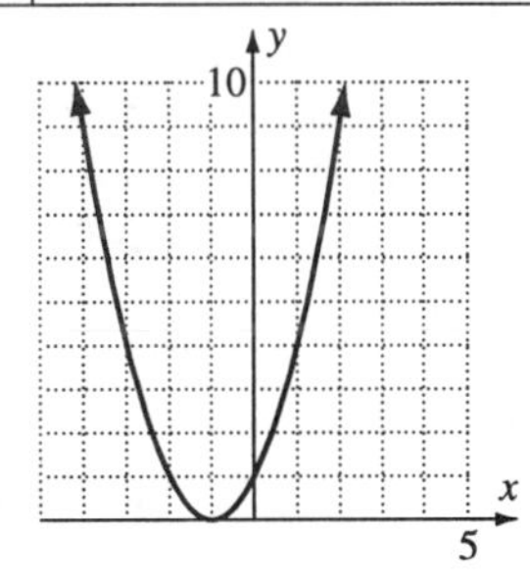

17. $f(x) = 2x^2 + 22x + 320$

a. $f(5) = 2(5)^2 + 22(5) + 320$
$f(5) = 480$
In 1985, there were 480 thousand inmates.

b. $x = 2010 - 1980 = 30$
$f(30) = 2(30)^2 + 22(30) + 320$
$f(30) = 2780$
2780 thousand

18. $f(x) = -0.0013x^3 + 0.078x^2 - 1.43x + 18.1$
$f(10) = -0.0013(10)^3 + 0.078(10)^2 - 1.43(10) + 18.1$
$f(10) = 10.3$
In 1970, 10.3% of families were below the poverty level.

19. $f(x) = 4.98x - 41.34$
$f(10) = 4.98(10) - 41.34 = 8.46\%$
$f(12) = 4.98(12) - 41.34 = 18.42\%$
$f(14) = 4.98(14) - 41.34 = 28.38\%$
$f(16) = 4.98(16) - 41.34 = 38.34\%$
The percentage of people with 10 years, 12 years, 14 years, and 16 years of education that do volunteer work is 8.46, 18.42, 28.38, and 38.34, respectively. It models the real-world data very well.

20. a. time: 5 P.M.;
minimum temperature: –4°F

b. time: 8 P.M.;
maximum temperature: 16°F

c. x-intercepts: 4 and 6
The Fahrenheit temperature is 0° at 4 P.M. and 6 P.M.

d. y-intercept: 12
At 12 noon , the Fahrenheit temperature is 12°.

e. For each value of time, there is only one value of temperature.
7 P.M.: 4°F, 8 P.M.: 16°F;
Between 7 P.M. and 8 P.M. the temperature rose 16° – 4° = 12°.

21. a. $f(10) = 30$
In 1990, the average age is 30 years.

b. $f(50) = 42$
In 2030, the average age will be 42 years.

c. $f(80) = 45$
In 2060, the average age will be 45 years.

d. As time passes, the average age increases.

22. (3, 2), (5, 1)
$m = \frac{1-2}{5-3} = -\frac{1}{2}$, line falls

23. (–1, –2), (–3, –4)
$m = \frac{-4+2}{-3+1} = \frac{-2}{-2} = 1$; line rises

24. $\left(-3, \frac{1}{4}\right), \left(6, \frac{1}{4}\right)$
$m = \frac{\frac{1}{4}-\frac{1}{4}}{6+3} = \frac{0}{9} = 0$; line is horizontal

25. (–2, 5), (–2, 10)
$m = \frac{10-5}{-2+2} = \frac{5}{0}$ is undefined;
line is vertical

26. Points (–1, 3) and (1, –1)
$\text{slope} = \frac{-1-3}{1-(-1)} = \frac{-4}{2} = -2$

27. $\text{pitch} = \frac{\text{rise}}{\text{run}} = \frac{1}{6} = \frac{3}{x}$
$x = 18$ feet

28. points (1988, 29,625) and (1995, 33,496)
$\text{slope} = \frac{33,496 - 29,625}{1995 - 1988} = \frac{3871}{7} = 553$
Each year, the average salary increases by $553.

29. a. 40 years

b. points: (20, 9) and (30, 23)
$\text{slope} = \frac{23-9}{30-20} = 1.4$

c. points (60, 20) and (70, 16)
$\text{slope} = \frac{16-20}{70-60} = -0.4$
productivity is declining

d. Age 40–50 in the humanities
Between ages 40 and 50, productivity in the humanities is approximately constant.

e. Humanities

f. Age 60–70 in the arts
points: (60, 15) and (70, 6)
slope = $\frac{6-15}{70-60} = -0.9$
Between ages 60 and 70, productivity in the arts declines more rapidly than that of any of the other disciplines.

30. $y = 5x - 7$
slope = 5
y-intercept = –7

31. $y = -8 - 9x$
$y = -9x - 8$
slope = –9
y-intercept = –8

32. $2x + 3y = -6$
$3y = -2x - 6$
$y = -\frac{2}{3}x - 2$
slope = $-\frac{2}{3}$
y-intercept = –2

33. slope = –5
y-intercept = 3
$y = -5x + 3$

34. slope = $-\frac{1}{2}$
y-intercept = –2
$y = -\frac{1}{2}x - 2$

35. points (0, 2) and (2, –2)
slope = $\frac{-2-2}{2-0} = -\frac{4}{2} = -2$
$y - 2 = -2(x - 0)$
$y = -2x + 2$

36. points (0, –1) and (4, 1)
slope = $\frac{1-(-1)}{4-0} = \frac{2}{4} = \frac{1}{2}$
$y - (-1) = \frac{1}{2}(x - 0)$
$y + 1 = \frac{1}{2}x$
$y = \frac{1}{2}x - 1$

37. $y = 2x - 4$

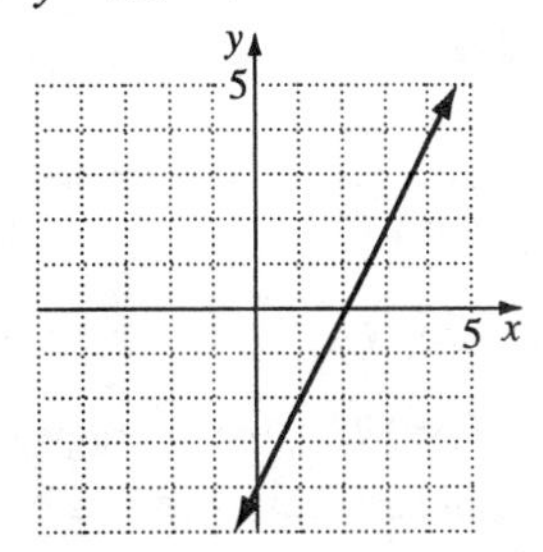

38. $y = -\frac{2}{3}x + 5$

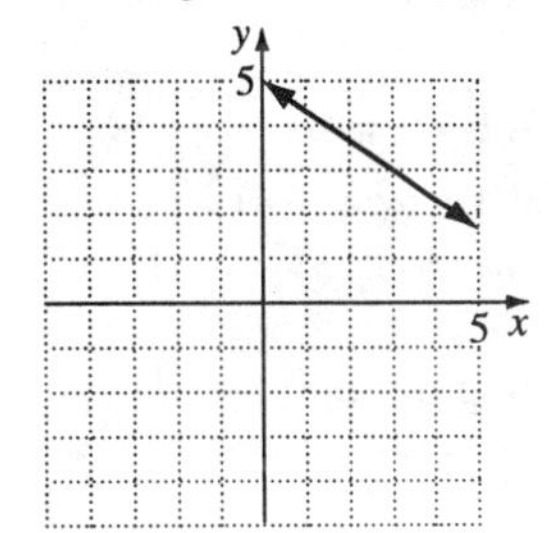

39. $y = \frac{3}{4}x - 2$

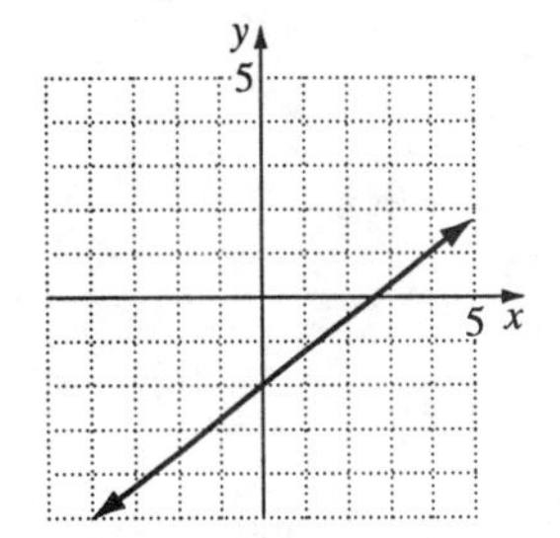

40. $y = -\frac{1}{3}x + 4$
$y = -\frac{1}{3}x - 1$
same slope

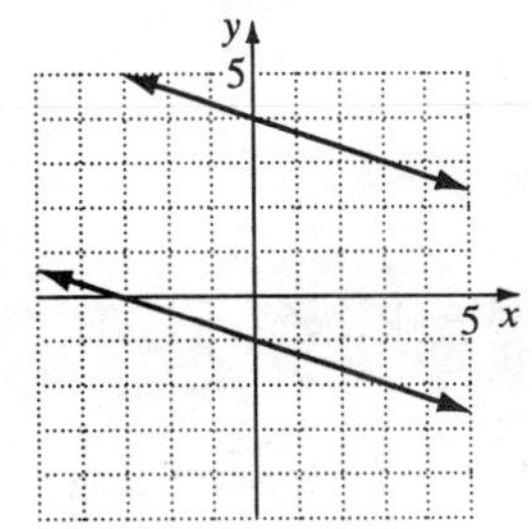

41. $y = 3.657x + 14.784$

a. 14.784
In 1980, there were 14.784 million subscribers.

b. 3.657
Each year after 1980, the average number of subscribers increases by 3.657 million.

42. a. $y = 3.14x + 87.1$

b. $x = 2000 - 1960 = 40$
$y = 3.14(40) + 87.1 = 212.7$
212.7 million tons

43. slope = 6, passing through (–4, 7)
point-slope: $y - 7 = 6(x + 4)$
$y - 7 = 6x + 24$
slope-intercept: $y = 6x + 31$

44. passing through (3, 4) and (2, 1)
$m = \frac{1-4}{2-3} = \frac{-3}{-1} = 3$
point-slope: $y - 4 = 3(x - 3)$ or
$y - 1 = 3(x - 2)$
$y - 4 = 3x - 9$
slope-intercept: $y = 3x - 5$

45. passing through (–2, –3) and (4, –1)
$m = \frac{-1+3}{4+2} = \frac{2}{6} = \frac{1}{3}$
point-slope: $y + 3 = \frac{1}{3}(x + 2)$ or
$y + 1 = \frac{1}{3}(x - 4)$
$y + 3 = \frac{1}{3}x + \frac{2}{3}$
slope-intercept: $y = \frac{1}{3}x - \frac{7}{3}$

46. a. slope = $\frac{75-61}{38-31} = \frac{14}{7} = 2$

b. $y - 61 = 2(x - 31)$ or $y - 75 = 2(x - 38)$

c. $y - 61 = 2x - 62$
$y = 2x - 1$

d. $y = 2(36) - 1$
$y = 71$
71 in. or 5 ft 11 in.

47. $3x - 4y > 7$
(0, 0): $0 - 0 > 7$ false
(–2, –1): $-6 + 4 > 7$ false
(–2, –5): $-6 + 20 > 7$ true
(–2, –5) satisfies the inequality
(–3, 4): $-9 - 16 > 7$ false
(3, –6): $9 + 24 > 7$ true
(3, –6) satisfies the inequality

48. $x - 2y > 6$

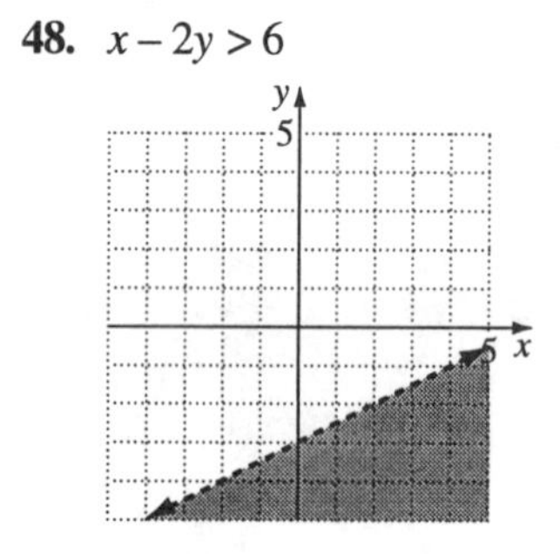

49. $4x - 6y \le 12$

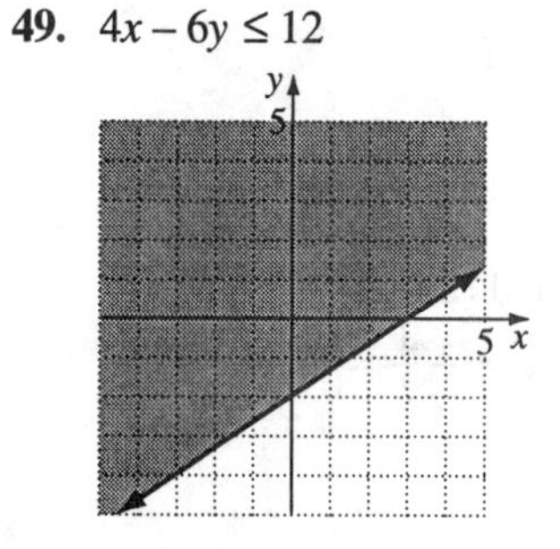

50. $x + 2y \le 0$

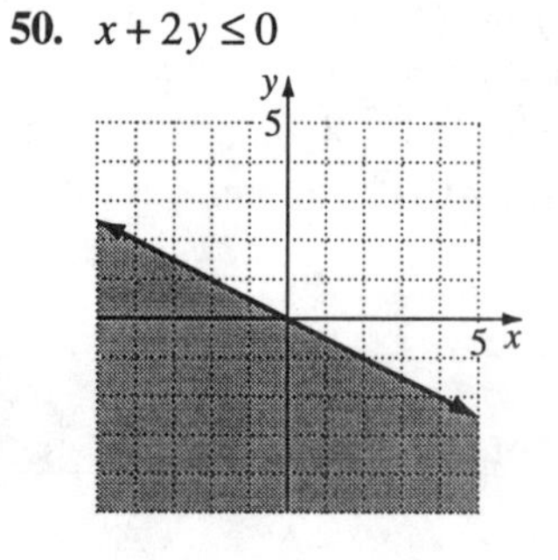

51. $y > 3x + 2$

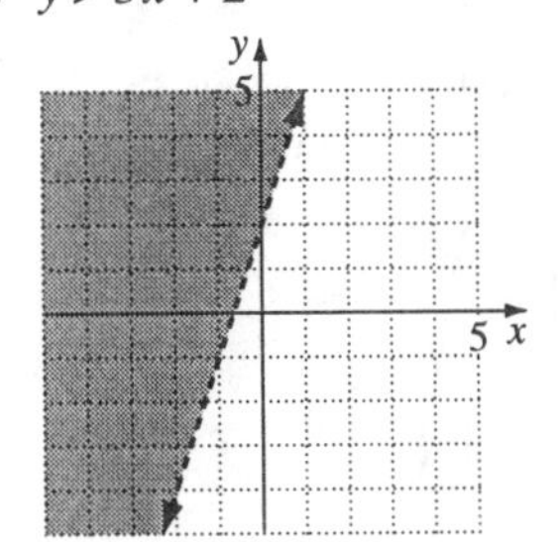

52. $y \le \frac{1}{3}x + 2$

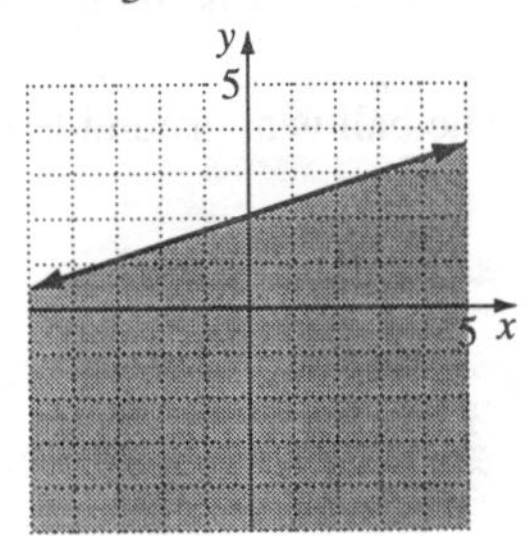

53. $y < -\frac{1}{2}x$

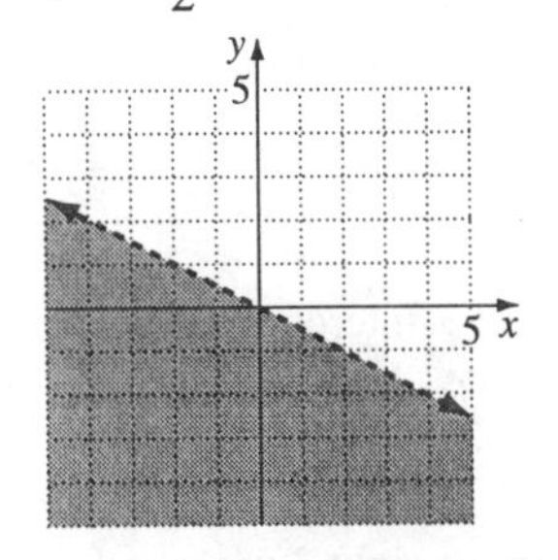

54. $x < 4$

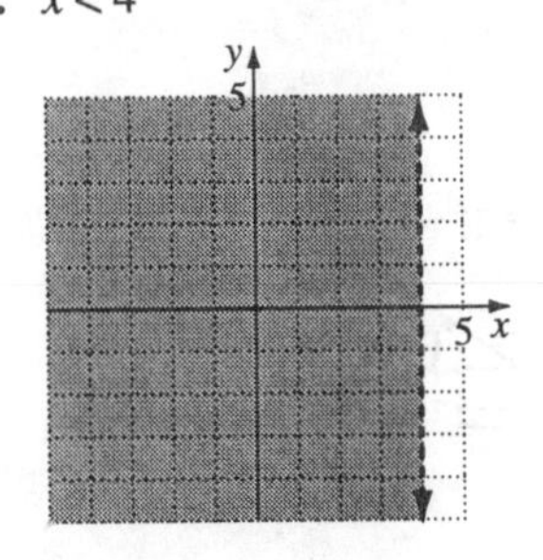

55. $y \ge -2$

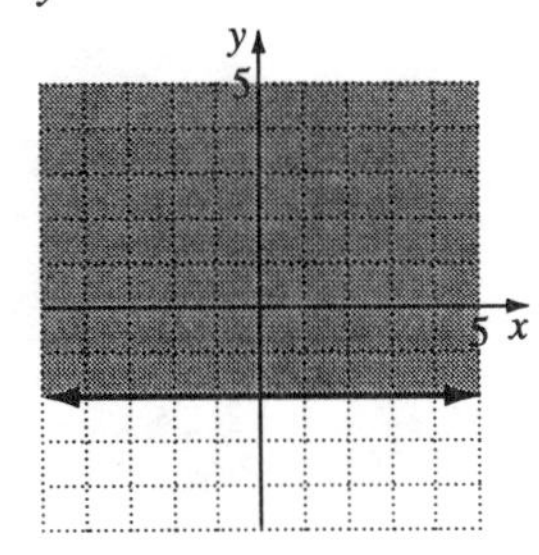

Chapter 4 Test

1. $x + 2y = 6$
$2y = -x + 6$
$y = -\frac{1}{2}x + 3$
$f(x) = -\frac{1}{2}x + 3$

2. $f(7) = 0.43(7) + 30.86$
$= 3.01 + 30.86$
$f(7) = 33.87$
There were 33.87 million married women in the U.S. in the civilian work force in 1997.

3. $f(30) = 0.002(30)^2 + 0.41(30) + 7.34$
$= 1.80 + 12.30 + 7.34$
$f(30) = 21.44$
21.44% of the U.S. population graduated from college in 1990.

4. $4x - 2y = -8$
x-intercept:
$4x - 2(0) = -8$
$4x = -8$
$x = -2$
y-intercept: $4(0) - 2y = -8$
$-2y = -8$
$-y = 4$

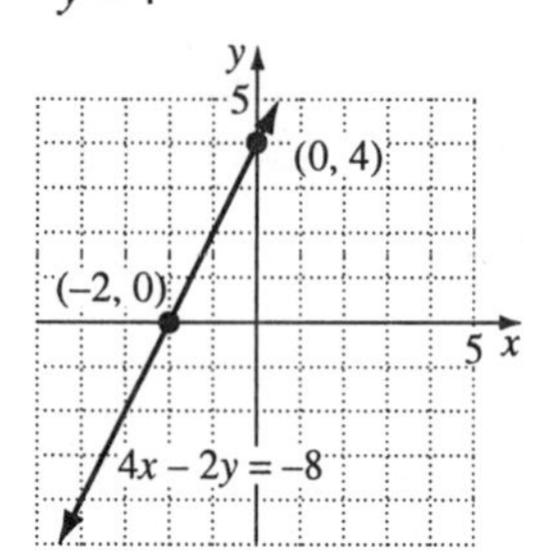

5. $2y = -6$
$y = -3$

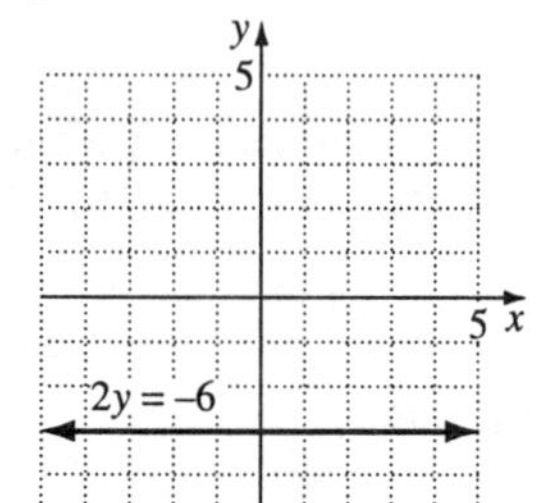

6. $V = 12{,}000 - 1250x$

7. $V = 7000$ when $t = 4$.
The car's value is $7000.

8.

x	$f(x) = 2 - x^2$	(x, y)
–3	$2-(-3)^2 = -7$	(–3, –7)
–2	$2-(-2)^2 = -2$	(–2, –2)
–1	$2-(-1)^2 = 1$	(–1, 1)
0	$2-(0)^2 = 2$	(0, 2)
1	$2-(1)^2 = 1$	(1, 1)
2	$2-(2)^2 = -2$	(2, –2)
3	$2-(3)^2 = -7$	(3, –7)

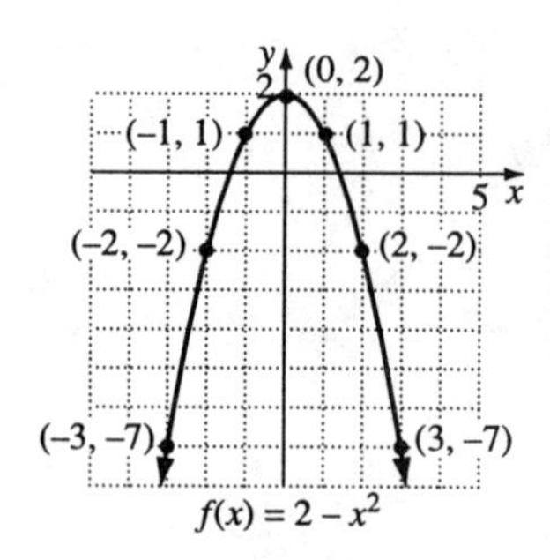

9. a. The maximum height of 30 meters was reached 2 seconds after being thrown.

b. $f(4.5) = 0$. The ball is at ground level.

10. $m = \frac{y_2 - y_1}{x_2 - x_1} = \frac{-2-4}{-5-(-3)} = \frac{-6}{-2} = 3$

11. $m = \frac{1-(-2)}{1-(-1)} = \frac{3}{2}$

12. Find the slope-intercept form of the equation: $y = mx + b$
$3x + 2y = 8$
$2y = -3x + 8$
$y = -\frac{3}{2}x + 4$
Slope is $-\frac{3}{2}$ and y-intercept is 4.

13. y-intercept: $y = \frac{2}{3}(0) - 1 = -1$
From (0, –1) move right 3 units and up 2 units $\left(m = \frac{2}{3}\right)$.

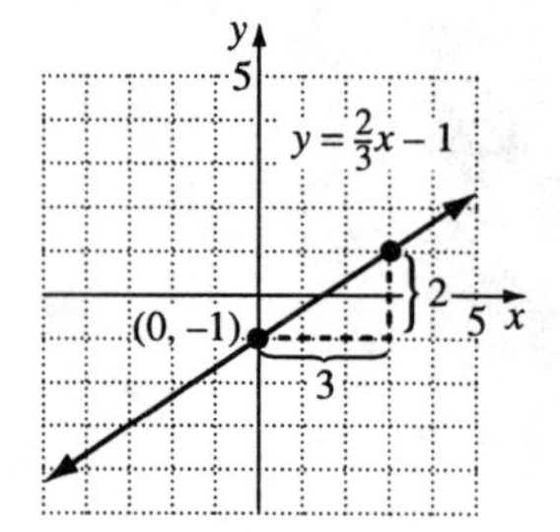

14. y-intercept: $y = -2(0) + 3 = 3$
From (0, 3) move right 1 unit and right down 2 units ($m = -2$).

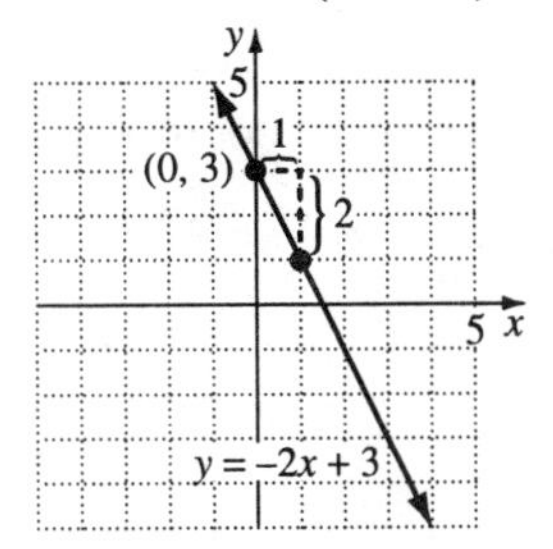

15. $y = mx + b$
$y = -6x + 4$

16. a. y-intercept is 3231; the population of Arizona, in thousands, in 1985

b. Slope is 89; the population increase, in thousands, each year after 1985 of Arizona

17. Point-slope form:
$y - y_1 = m(x - x_1)$
$y - 3 = \frac{1}{2}(x - (-2))$
$y - 3 = \frac{1}{2}(x + 2)$
Slope intercept-form: $y = \frac{1}{2}x + 4$

18. Point-slope form:
$y - y_1 = m(x - x_1)$
$m = \frac{-8+2}{3-1} = -3$
$y - (-2) = -3(x - 1)$
$y + 2 = -3(x - 1)$
Slope-intercept form:
$y + 2 = -3(x - 1)$ or $y + 8 = -3(x - 3)$
$y + 2 = -3x + 3$
$y = -3x + 1$

19. $2x - y \geq 4$
$-y \geq -2x + 4$
$y \leq 2x - 4$
Draw the solid line $y = 2x - 4$.
Test the point (0, 0).
$2(0) - 0 \geq 4$
$0 \geq 4$
False
Shade the half-plane that does not contain (0, 0).

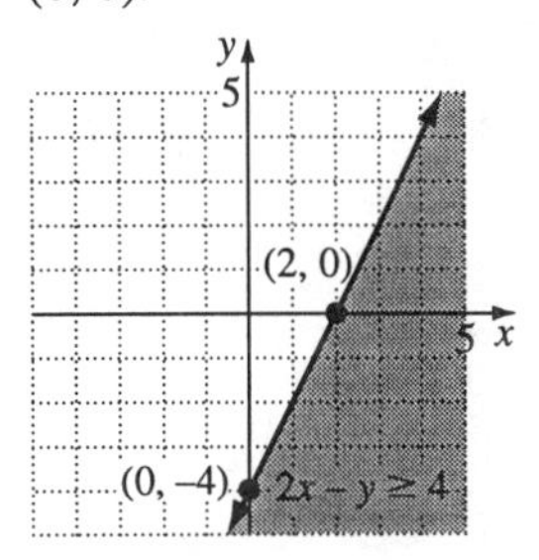

20. Draw a dashed line $y = 2x - 2$.
Test the point (0, 0).
$0 \leq 2(0) - 2$
$0 < -2$
False
Shade the half-plane that does not contain (0, 0).

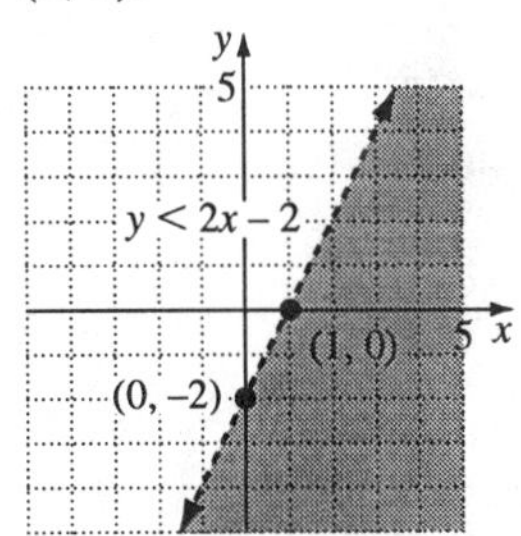

Cumulative Review Problems (Chapters 1–4)

1. $\frac{5(-3) - 3(-4)}{5(-10) + 2}$
$= \frac{-15+12}{-50+2} = \frac{-3}{-48} = \frac{1}{16}$

2. $3(y + 1) + 11 = 16 + 5y$
$3y + 3 + 11 = 16 + 5y$
$3y + 14 = 16 + 5y$
$-2 = 2y$
$-1 = y$
The solution is -1.

3. $\frac{1}{4}y+\frac{2}{3}y=\frac{1}{6}$
$12\left(\frac{1}{4}y+\frac{2}{3}y\right)=12\left(\frac{1}{6}\right)$
$3y+8y=2$
$11y=2$
$y=\frac{2}{11}$
The solution is $\frac{2}{11}$.

4. $-7(2y+1)>4(3-y)+1$
$-14y-7>12-4y+1$
$-10y>20$
$y<-2$
$\{y|y<-2\}$

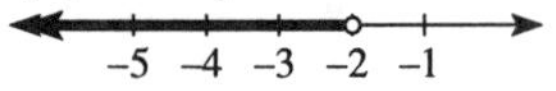

5. Let x equal the price before reduction.
$x-0.35x=185.25$
$0.65x=185.25$
$x=285$
$285

6. $y=4.5x-46.7$
$133.3=4.5x-46.7$
$180=4.5x$
$40=x$
40 mph

7. $0.04(6.6)=\$0.264$ billion

8. Let x = number in Europe
Then $x+4$ = number in Africa
and $2x$ = number in Latin America
$x+x+4+2x=12$
$4x=8$
$x=2$
2 in Europe
6 in Africa
4 in Latin America
Answers may vary.

9. $\frac{x}{400}=\frac{2.5}{10}$
$10x=400(2.5)$
$x=100$ feet

10. Let x equal the number of hours plumber worked.
$18+35x=228$
$35x=210$
$x=6$
6 hours

11. Let x equal the running speed of a cheetah.
$x-29$ equals the running speed of Robert Hayes.
$3x+5=242$
$3x=237$
Solution: $x=79$
sum of average running speeds exceeds solution by 4:
$x+(x-29)=79+4$
$2x=112$
$x=56$
$x-29=27$
Robert Hayes's running speed was 27 miles per hour.

12. $\frac{135}{6}=\frac{360}{x}$
$135x=360(6)$
$x=16$

13. Let x equal the measure of the angle.
$90-x$ equals the measure of its complement.
$180-x$ equals the measure of its supplement.
$(90-x)+(180-x)=114$
$270-2x=114$
$156=2x$
$78=x$
The angle measures 78°.

14. $f(x) = 9.2x^2 - 46.7x + 480$
$f(10) = 9.2(10)^2 - 46.7(10) + 480$
$f(10) = 933$
In 1980, the national debt was $933 billion.

15. $y = -3x + 2$
$m = -3,\ b = 2$

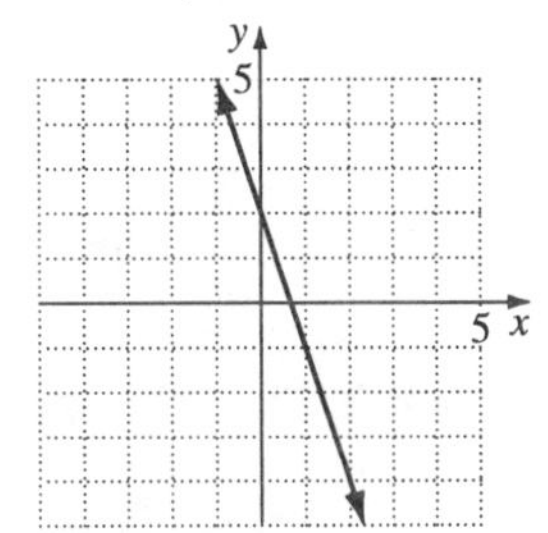

16. Possible answer:

Year	Ratio
1989	1000/2.15 million
1994	600/1.60 million

The policy seems to have decreased the ratio.

17. Let x equal the length of each side of the square.
$x + 6$ equals the length of each side of the equilateral triangle.
perimeter of square = perimeter of triangle
$4x = 3(x + 6)$
$4x = 3x + 18$
$x = 18$
$x + 6 = 24$
length of each side of the triangle:
24 decimeters

18. $T = \frac{1}{4}C + 37$

a.

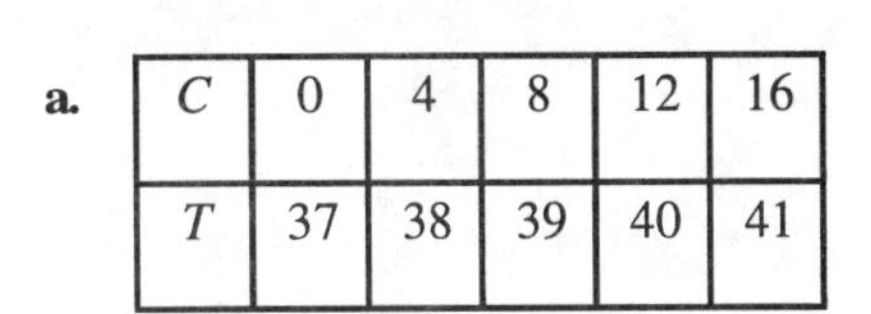

C	0	4	8	12	16
T	37	38	39	40	41

b.

19. $P = 2L + 2W$
$2L + 2W = P$
$2L = P - 2W$
$L = \frac{P - 2W}{2}$ or $\frac{P}{2} - W$

20. (1, 3), (3, 5)
$m = \frac{5-3}{3-1} = \frac{2}{2} = 1$
point-slope: $y - 3 = 1(x - 1)$ or
$y - 5 = 1(x - 3)$
slope-intercept: $y - 3 = x - 1$
$y = x + 2$

21. $3x - 4y > 12$

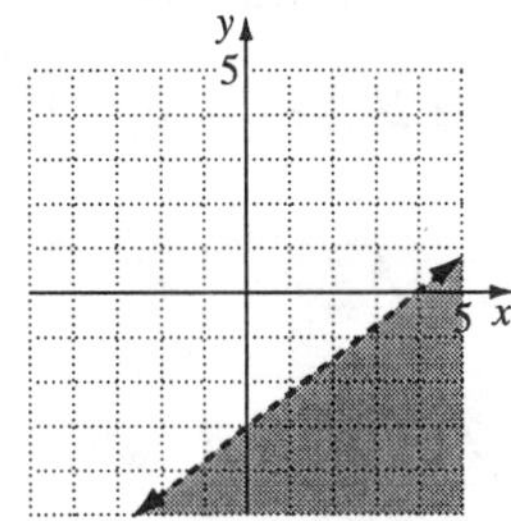

22. a. $3000

b. 1976

c. They are not evenly spaced. From 1960–1985, the space between the marks represent 5 years, whereas from 1986–1993, each space represents one year.

23. $\frac{4}{10} = \frac{2}{5}$

24. $\frac{12 \cdot 2}{6} = 4$
$x = 12,\ y = 2,\ z = 6$

25. $A = \frac{1}{2}bh$
$h = 20$ yards, $A = 150$ square yards
$150 = \frac{1}{2}b(20)$
$150 = 10b$
$15 = b$
base, 15 yards

26. $y = x^2 + 2x + 2$

x	$y = x^2 + 2x + 2$	(x, y)
–3	$y = (-3)^2 + 2(-3) + 2 = 5$	(–3, 5)
–2	$y = (-2)^2 + 2(-2) + 2 = 2$	(–2, 2)
–1	$y = (-1)^2 + 2(-1) + 2 = 1$	(–1, 1)
0	$y = 0^2 + 2(0) + 2 = 2$	(0, 2)
1	$y = 1^2 + 2(1) + 2 = 5$	(1, 5)
2	$y = 2^2 + 2(2) + 2 = 10$	(2, 10)

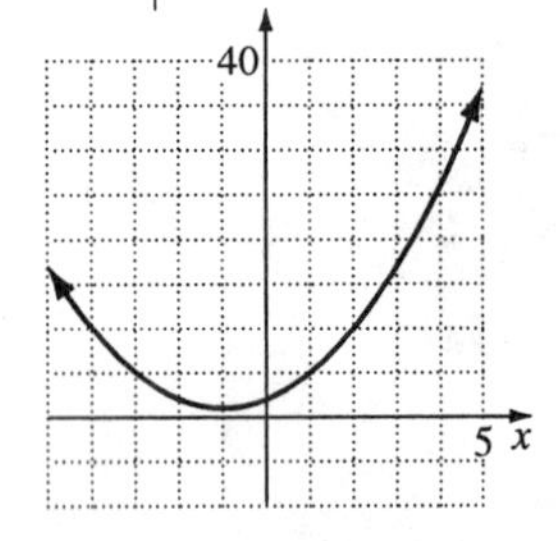

27. slope $= \frac{1006 - 1080}{1992 - 1980} = -\frac{74}{12} \approx -6.2$
For each year from 1980 to 1992, the average number of hectares of the tropical forests is decreasing by 6.2 hectares.

28. $40t + 60t = 350$
$100t = 350$
$t = 3.5$
3.5 hours

29. Possible answer: 18%

30. **a.** 6, 10, 15, 21, 28, 36

b. 1, 4, 9, 16, 25, 36

Chapter 5

Problem Set 5.1

1. (2, 3)
$x + 3y = 11$
$2 + 3(3) = 11$
$2 + 9 = 11$
True

$x - 5y = -13$
$2 - 5(3) = -13$
$2 - 15 = -13$
$-13 = -13$
True

Since the ordered pair (2, 3) satisfies both equations, it is a solution of the given system of equations.

3. (–3, –1)
$5x - 11y = -4$
$5(-3) - 11(-1) = -4$
$-15 + 11 = -4$
$-4 = -4$
True

$6x - 8y = -10$
$6(-3) - 8(-1) = -10$
$-18 + 8 = -10$
$-10 = -10$
True

(–3, –1) is the solution of the given system.

5. (2, 5)
$2x + 3y = 17$
$2(2) + 3(5) = 17$
$4 + 15 = 17$
$19 = 17$
False

$x + 4y = 16$
$2 + 4(5) = 16$
$2 + 20 = 16$
$22 = 16$
False

Since (2, 5) fails to satisfy both equations, it is not a solution of the given system.

7. $\left(\frac{1}{3}, 1\right)$
$6x - 9y = -7$
$6\left(\frac{1}{3}\right) - 9(1) = -7$
$2 - 9 = -7$
$-7 = -7$
True

$9x + 5y = 8$
$9\left(\frac{1}{3}\right) + 5(1) = 8$
$3 + 5 = 8$
$8 = 8$
True

$\left(\frac{1}{3}, 1\right)$ is a solution of the given system.

9. (8, 5)
$5x - 4y = 20$
$5(8) - 4(5) = 20$
$40 - 20 = 20$
$20 = 20$
True

$3y = 2x + 1$
$3(5) = 2(8) + 1$
$15 = 16 + 1$
$15 = 17$
False
(8, 5) fails to satisfy both equations; it is not a solution of the given system.

11. (0, 5)

$\frac{3}{5}x + \frac{2}{5}y = 2$

$\frac{3}{5}(0) + \frac{2}{5}(5) = 2$

$0 + 2 = 2$

$2 = 2$

True

$y = 5$

$5 = 5$

True

(0, 5) is the solution of the given system.

13. $x + y = 6$

$x - y = 2$

solution: (4, 2)

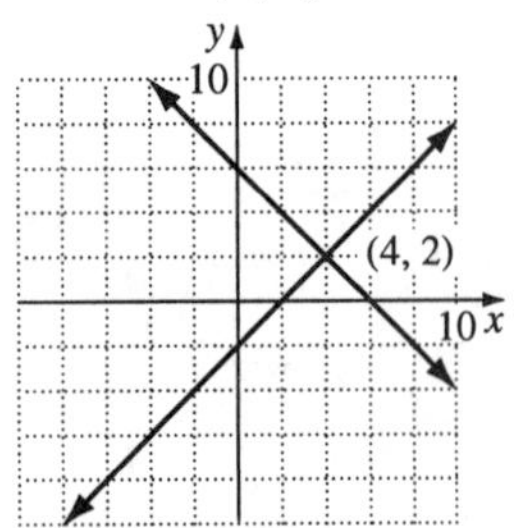

15. $x + y = 1$

$y - x = 3$

solution: (–1, 2)

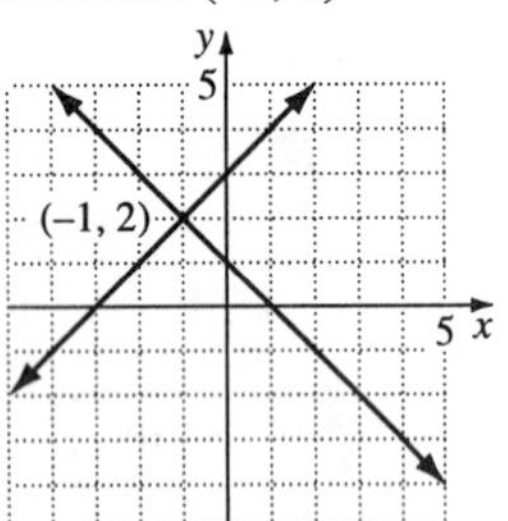

17. $3x + y = 3$

$6x + 2y = 12$

solution: ∅; inconsistent

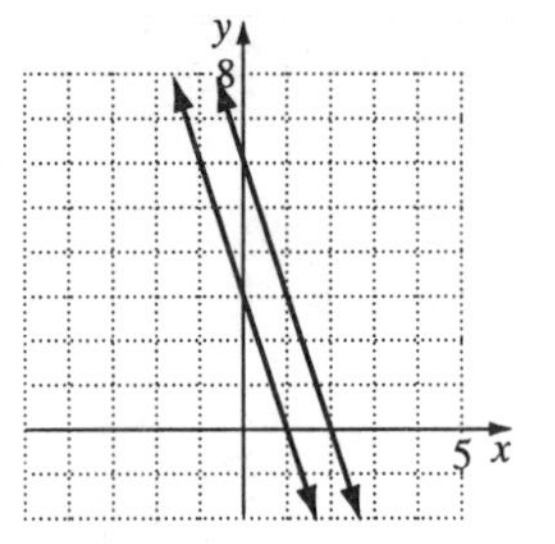

19. $2x - 3y = 6$

$4x + 3y = 12$

solution: (3, 0)

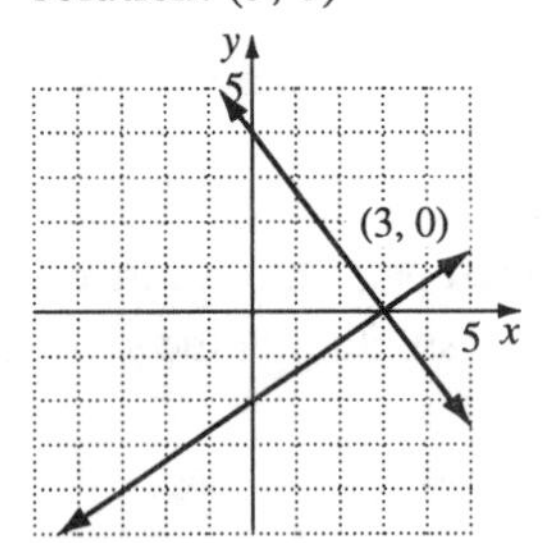

21. $x + y = 5$

$-x - y = -6$

solution: ∅; inconsistent

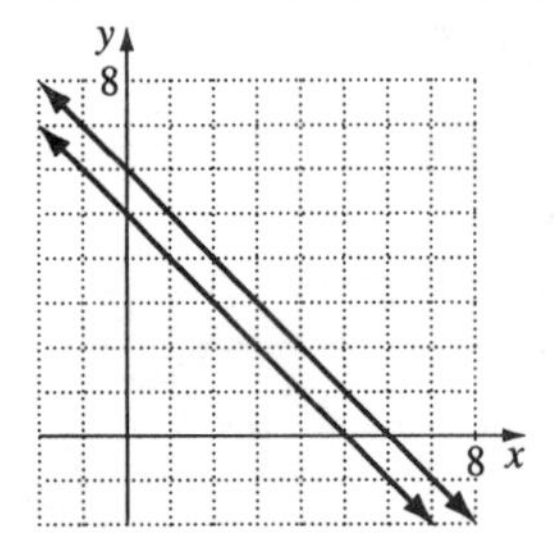

23. $x - y = 2$

$3x - 3y = -6$

solution: ∅; inconsistent

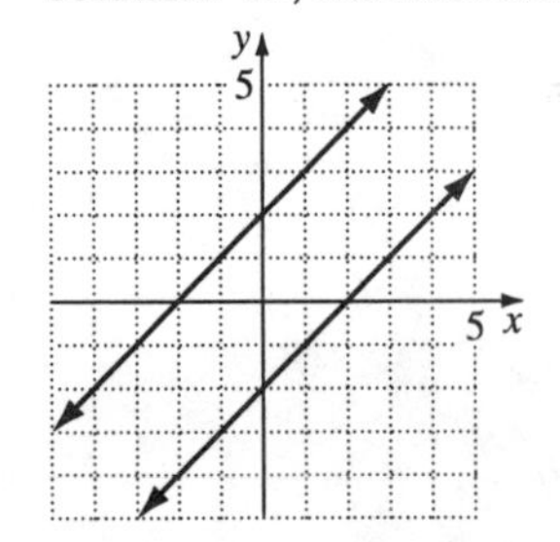

25. $4x + y = 4$
$3x - y = 3$
solution: (1, 0)

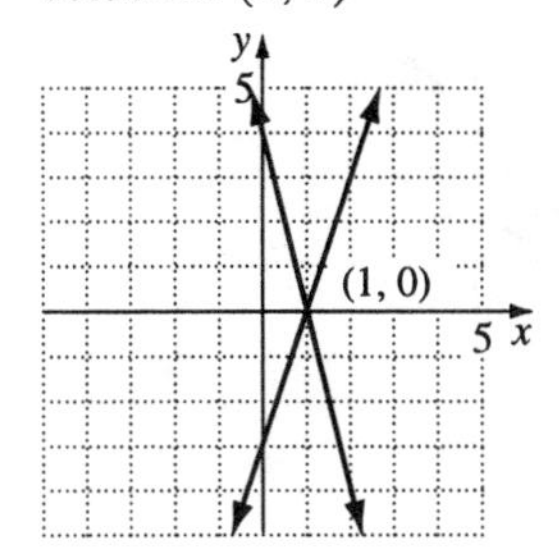

27. $x + y = 4$
$x = -2$
solution: (–2, 6)

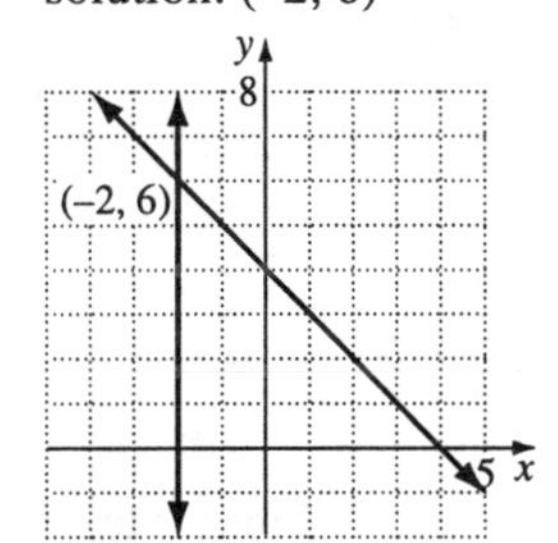

29. $x = -3$
$y = 5$
solution: (–3, 5)

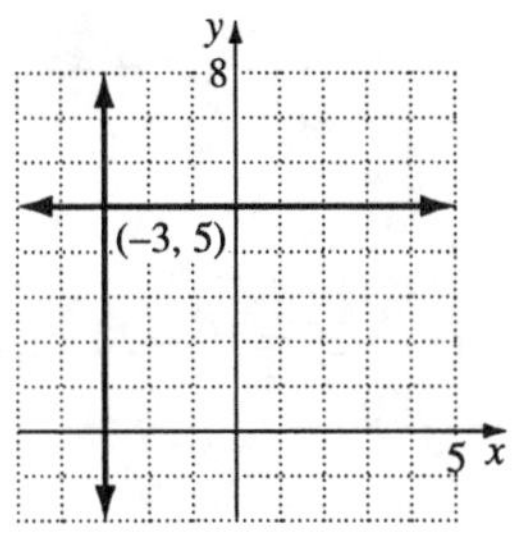

31. $y = x + 5$
$y = -x + 3$
solution: (–1, 4)

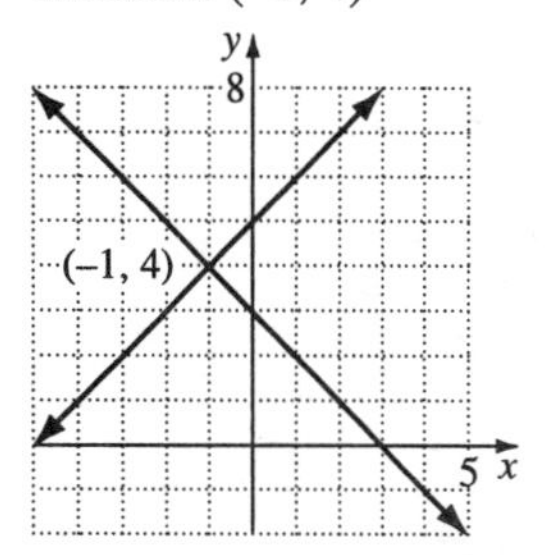

33. $y = 2x$
$y = -x + 6$
solution: (2, 4)

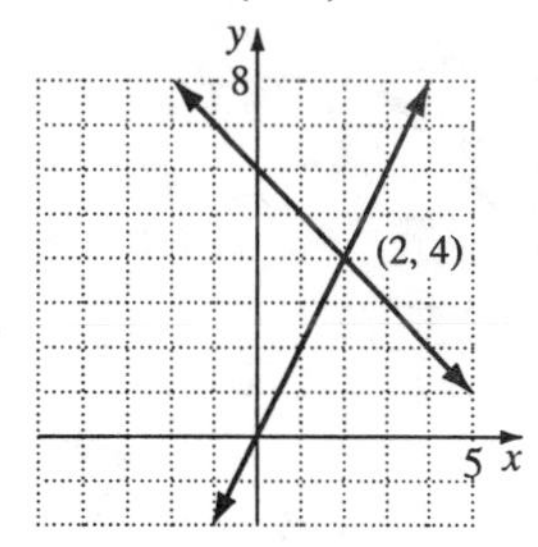

35. $y = 3x - 4$
$y = -2x + 1$
solution: (1, –1)

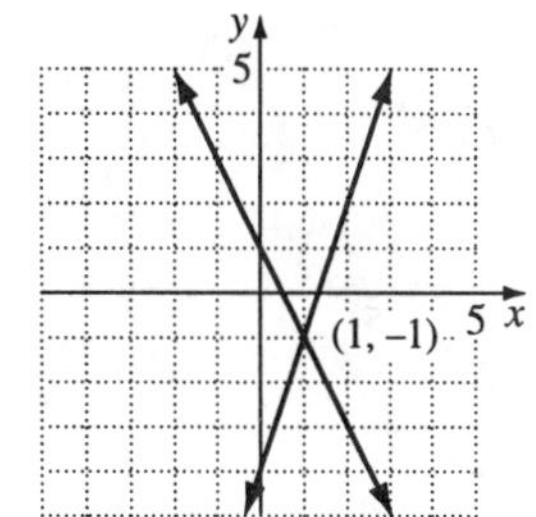

37. $y = 2x - 1$
$y = 2x + 1$
solution: $\varnothing$; inconsistent

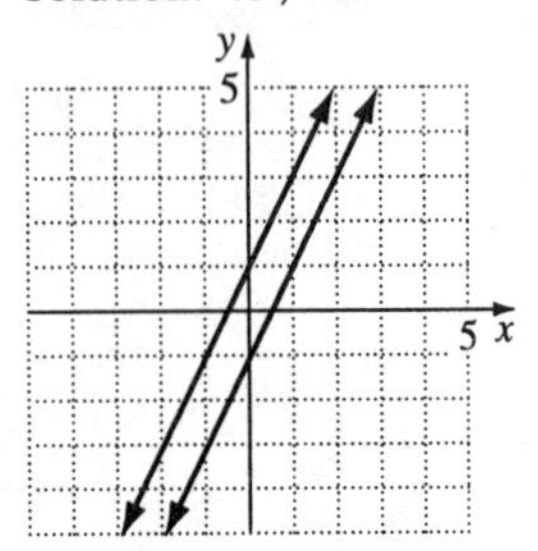

39. $x - y = 0$
$2x = 2y$
solution: infinite solutions; dependent

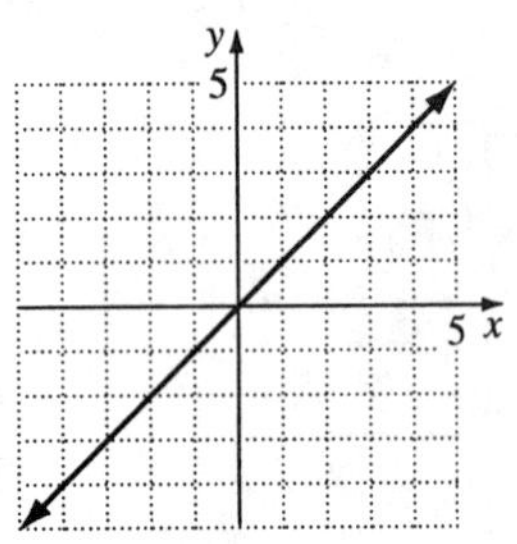

41. $y = 2x - 1$
$x - 2y = -4$
solution: (2, 3)

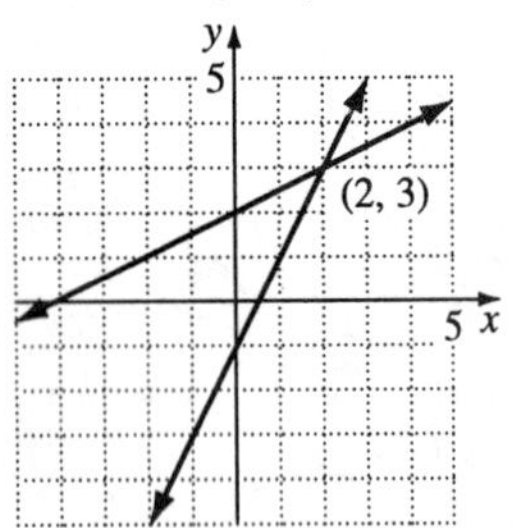

43. $y = \frac{1}{2}x - 1$
$x - y = -1$

solution: (–4, –3)

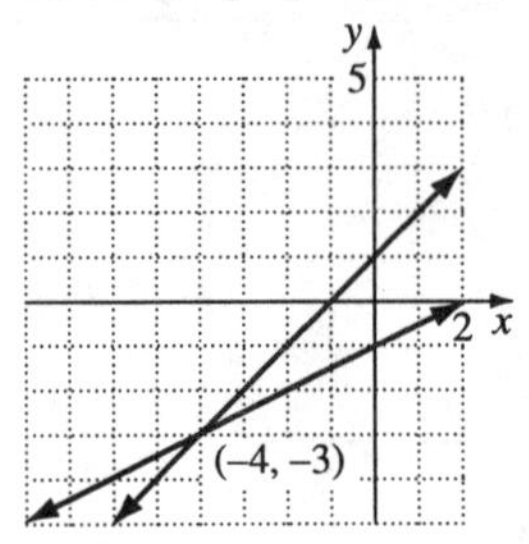

45. $x = 2$
$x = -1$
solution: $\varnothing$; inconsistent

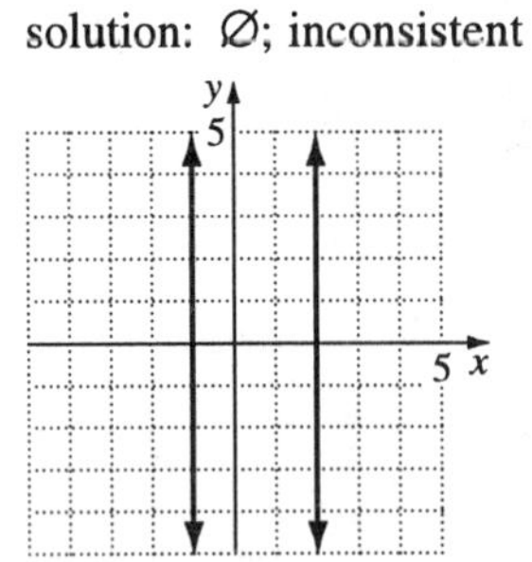

47. **a.** The artist must sell 4 pieces to break even. The graphs intersect at $x = 4$.

b. $x > 4$

c. Revenue – Cost $= 9x - (20 + 4x)$
Profit $= 9x - 20 - 4x$
$= 5x - 20$
$x = 2$:
$5x - 20 = 5(2) - 20 = 10 - 20 = -10$
artist's loss: \$10

d. Revenue – Cost
$= 9x - (20 + 4x)$
$= 5x - 20$
$x = 10$;
$5x - 20 = 5(10) - 20 = 30$
artist's profit: \$30

49. b is true

51. c is true

53. $y = -x + 5$
$y = x - 7$
Solution: (6, –1)

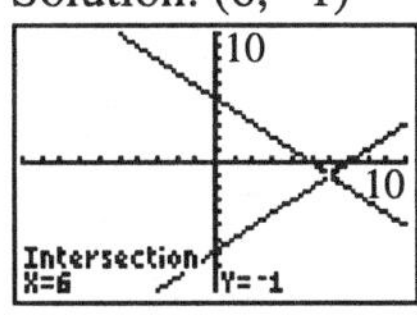

55. $2x - 3y = 6$
$4x + 3y = 12$
Solution: (3, 0)

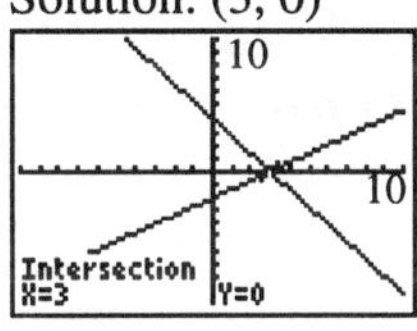

57. $2x - 3 - 5x = 13 + 4x - 2$
Solution: (–2, 3)

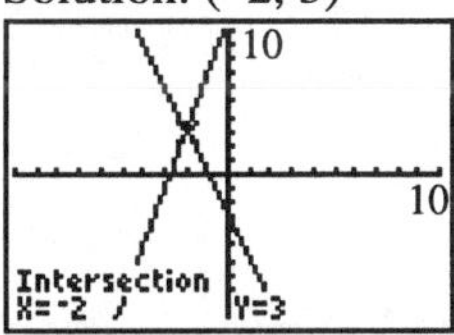

Check: $2(-2) - 3 - 5(-2) = 13 + 4(-2) - 2$
$-4 - 3 + 10 = 13 - 8 - 2$
$3 = 3$

59. $2x + 5 = 12 - 6x + 3(2x + 3)$
Solution: (8, 21)

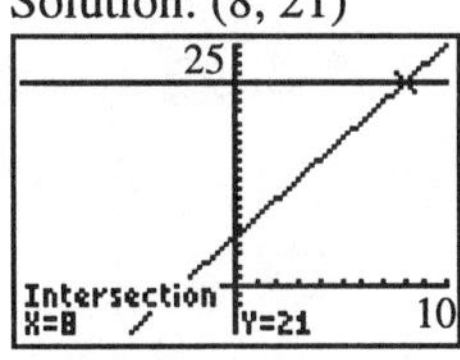

Check:
$2(8) + 5 = 12 - 6(8) + 3[2(8) + 3]$
$16 + 5 = 12 - 48 + 3(16 + 3)$
$21 = 12 - 48 + 3(19)$
$21 = 12 - 48 + 57$
$21 = 21$

61–63. Answers may vary.

65.

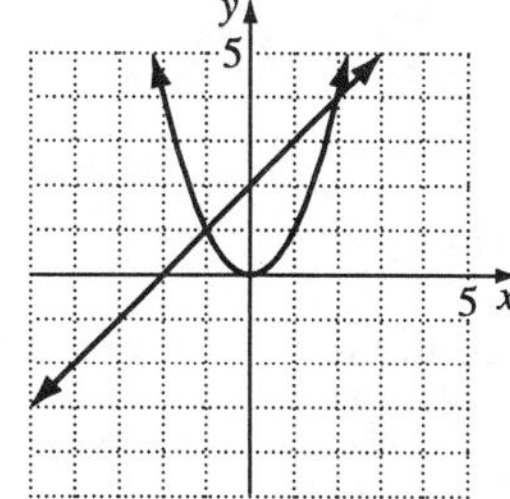

Solution {(–1, 1), (2, 4)}
Check: $y = (-1)^2 = 1$
$y = (2)^2 = 4$
$y = -1 + 2 = 1$
$y = 2 + 2 = 4$

67. Possible answer: $y = x - 4$, $y = 1$
There are infinitely many systems.
Answers may vary.

69. Answers may vary.

Review Problems

71. $3(y - 4) - (y + 7) - 2(3y - 6)$
$= 3y - 12 - y - 7 - 6y + 12$
$= -4y - 7$

72. $6y - 2(y + 4) - 2y = -4(y - 1)$
$6y - 2y - 8 - 2y = -4y + 4$
$2y - 8 = -4y + 4$
$6y = 12$
$y = 2$
2

73. $\frac{5 \text{ people}}{13 \text{ kilograms}} = \frac{700,000 \text{ people}}{x}$
$5x = 13(700,000)$ kilograms
$x = 1,820,000$ kilograms

Problem Set 5.2

1. $x+y=1$
$x-y=3$
$2x=4$ add to eliminate y
$x=2$
$2+y=1$ substitute $x=2$ into either equation and solve for y
$y=-1$
(2, –1)

3. $2x+3y=6$
$2x-3y=6$
$4x=12$
$x=3$
$2(3)+3y=6$
$3y=0$
$y=0$
(3, 0)

5. $x+2y=7$
$-x+3y=18$
$5y=25$
$y=5$
$x+2(5)=7$
$x+10=7$
$x=-3$
(–3, 5)

7. $5x-y=9$
$-5x+2y=-8$
$y=1$
$5x-1=9$
$5x=10$
$x=2$
(2, 1)

9. $x+2y=2$
$-4x+3y=25$
Multiply equation 1 by 4
Don't change equation 2
$4x+8y=8$
$-4x+3y=25$
$11y=33$
$y=3$
$x+2(3)=2$
$x+6=2$
$x=-4$
(–4, 3)

11. $2x-7y=2$
$3x+y=-20$
No change to equation 1.
Multiply equation 2 by 7.
$2x-7y=2$
$21x+7y=-140$
$23x=-138$
$x=-6$
$3(-6)+y=-20$
$-18+y=-20$
$y=-2$
(–6, –2)

13. $x+5y=-1$
$2x+7y=1$
Multiply equation 1 by –2.
No change to equation 2.
$-2x-10y=2$
$2x+7y=1$
$-3y=3$
$y=-1$
$x+5(-1)=-1$
$x-5=-1$
$x=4$
(4, –1)

15. $4x+3y=15$
$2x-5y=1$
No change to equation 1.
Multiply equation 2 by –2.
$4x+3y=15$
$-4x+10y=-2$
$13y=13$
$y=1$
$2x-5(1)=1$
$2x=6$
$x=3$
(3, 1)

17. $3x - y = 1$
$3x - y = 2$
Multiply equation 1 by –1.
No change to equation 2.
$-3x + y = -1$
$3x - y = 2$
$0 = 1$
False
No solution to this inconsistent system.
$\varnothing$

19. $3x - 4y = 11$
$2x + 3y = -4$
Multiply equation 1 by 3.
Multiply equation 2 by 4.
$9x - 12y = 33$
$8x + 12y = -16$
$17x = 17$
$x = 1$
$2(1) + 3y = -4$
$3y = -6$
$y = -2$
(1, –2)

21. $3x + 2y = -1$
$-2x + 7y = 9$
Multiply equation 1 by 2.
Multiply equation 2 by 3.
$6x + 4y = -2$
$-6x + 21y = 27$
$25y = 25$
$y = 1$
$3x + 2(1) = -1$
$3x = -3$
$x = -1$
(–1, 1)

23. $x + 3y = 2$
$3x + 9y = 6$
Multiply equation 1 by –3.
No change to equation 2.
$-3x - 9y = -6$
$3x + 9y = 6$
$0 = 0$
True
The system has infinitely many solutions. The equations are dependent. The solution set consists of all ordered pairs that satisfy either equation.
(x, y) where $x + 3y = 2$

25. $3x = 2y + 7$
$5x = 2y + 13$
Rewrite:
$3x - 2y = 7$
$5x - 2y = 13$
Multiply equation 1 by –1.
No change to equation 2.
$-3x + 2y = -7$
$5x - 2y = 13$
$2x = 6$
$x = 3$
$3(3) = 2y + 7$
$2 = 2y$
$1 = y$
(3, 1)

27. $2x = 3y - 4$
$-6x + 12y = 6$
Rewrite equation 1.
Divide equation 2 by 3.
$2x - 3y = -4$
$-2x + 4y = 2$
$y = -2$
$2x = 3(-2) - 4$
$2x = -10$
$x = -5$
(–5, –2)

29. $7x - 3y = 4$
$-14x + 6y = -7$
Multiply equation 1 by 2.
No change to equation 2.
$14x - 6y = 8$
$-14x + 6y = -7$
$0 = 1$
False
No solution; the system is inconsistent.
$\varnothing$

31. $2x - y = 3$
$4x + 4y = -1$
Multiply equation 1 by 4.

No change to equation 2.

$$\begin{array}{r} 8x-4y=12 \\ 4x+4y=-1 \\ \hline 12x=11 \end{array}$$

$$x=\frac{11}{12}$$

$$2\left(\frac{11}{12}\right)-y=3$$

$$y=\frac{22}{12}-\frac{36}{12}$$

$$y=-\frac{14}{12}=-\frac{7}{6}$$

$$\left(\frac{11}{12}, -\frac{7}{6}\right)$$

33. $4x=5+2y$

$2x+3y=4$

Rewrite equation 1.

Multiply equation 2 by –2.

$$\begin{array}{r} 4x-2y=5 \\ -4x-6y=-8 \\ \hline -8x=-3 \end{array}$$

$$y=\frac{3}{8}$$

$$4x=5+2\left(\frac{3}{8}\right)$$

$$4x=5+\frac{3}{4}=\frac{23}{4}$$

$$x=\frac{23}{16}$$

$$\left(\frac{23}{16}, \frac{3}{8}\right)$$

35. $4x-8y=36$

$3x-6y=27$

Divide equation 1 by –4.

Divide equation 2 by 3.

$$\begin{array}{r} -x+2y=-9 \\ x-2y=9 \\ \hline 0=0 \end{array}$$

True

The system has infinitely many solutions.

The equations are dependent.

(x, y) where $x-2y=9$

37. $2x+4y=5$

$3x+6y=6$

No change to equation 1.

Multiply equation 2 by $-\frac{2}{3}$.

$$\begin{array}{r} 2x+4y=5 \\ -2x-4y=-4 \\ \hline 0=1 \end{array}$$

False

No solution; the system is inconsistent.

$\varnothing$

39. $5x+y=2$

$3x+y=1$

No change to equation 1.

Multiply equation 2 by –1.

$$\begin{array}{r} 5x+y=2 \\ -3x-y=-1 \\ \hline 2x=1 \end{array}$$

$$x=\frac{1}{2}$$

$$3\left(\frac{1}{2}\right)+y=1$$

$$y=-\frac{1}{2}$$

$$\left(\frac{1}{2}, -\frac{1}{2}\right)$$

41. $2x+2y=-2-4y$

$3x+y=7y+27$

Rewrite:

$$\begin{array}{r} 2x+6y=-2 \\ 3x-6y=27 \\ \hline 5x=25 \end{array}$$

$$x=5$$

$$2(5)+2y=-2-4y$$

$$2(5)+6y=-2$$

$$6y=-12$$

$$y=-2$$

$(5, -2)$

43. $x+y=11$

$\frac{1}{5}x+\frac{1}{7}y=1$

Multiply equation 1 by –5.

Multiply equation 2 by 35.

$$\begin{array}{r} -5x-5y=-55 \\ 7x+5y=35 \\ \hline 2x=-20 \end{array}$$

$$x=-10$$

$$-10+y=11$$

$$y=21$$

$(-10, 21)$

45. $\frac{3}{5}x+\frac{4}{5}y=1$

$\frac{1}{4}x-\frac{3}{8}y=-1$

Multiply equation 1 by 5.
Multiply equation 2 by 8.

$3x+4y=5$
$2x-3y=-8$

Now multiply equation 1 by 3.
Multiply equation 2 by 4.

$$\begin{aligned} 9x+12y&=15 \\ 8x-12y&=-32 \\ \hline 17x&=-17 \\ x&=-1 \end{aligned}$$

$$\begin{aligned} \frac{3}{5}(-1)+\frac{4}{5}y&=1 \\ \frac{4}{5}y&=\frac{8}{5} \\ y&=2 \end{aligned}$$

(−1, 2)

47. $\frac{4}{5}x-y=-1$

$\frac{2}{5}x+y=1$

Multiply equation 1 by 5.
Multiply equation 2 by 5.

$$\begin{aligned} 4x-5y&=-5 \\ 2x+5y&=5 \\ \hline 6x&=0 \\ x&=0 \end{aligned}$$

$$\begin{aligned} \frac{2}{5}(0)+y&=1 \\ y&=1 \end{aligned}$$

(0, 1)

49. $2L+2W=1040$
$L=W+200$

$$\begin{aligned} 2L+2W&=1040 \\ 2L-2W&=400 \\ \hline 4L&=1440 \\ L&=360 \end{aligned}$$

$W=L-200=360-200=160$
Length = 360 feet
Width = 160 feet

51. c is true

53. Check for students.

55–59. Answers may vary.

Review Problems

61. Answers may vary. Samples given.

a. $12+3-6=9$

b. $12\div6+3=5$

c. $(2)(12)\div6=4$

62. Let x equal the measure of the angle.
$90-x$ equals the measure of its complement.
$180-x$ equals the measure of its supplement.
$(90-x)+(180-x)=196$
$270-2x=196$
$-2x=-74$
$x=37$
The angle measures 37°.

63. $6(y-5)-9y<-4y-5(2y-5)$
$6y-30-9y<-4y-10y+25$
$-3y-30<-14y+25$
$11y<55$
$y<5$
$\{y|y<5\}$

Problem Set 5.3

1. $x+y=4$
$y=3x$
Substitute $3x$ for y.
$x+(3x)=4$
$4x=4$
$x=1$
$y=3(1)=3$
(1, 3)

3. $x+3y=8$
$y=2x-9$
Substitute for y in equation 1.

$x+3(2x-9)=8$
$x+6x-27=8$
$7x=35$
$x=5$
$y=2(5)-9=10-9=1$
$(5, 1)$

5. $x=9-2y$
$x+2y=13$
Substitute for x into equation 2.
$(9-2y)+2y=13$
$9=13$ False
No solution;
The system is inconsistent.
$\varnothing$

7. $2(x-1)-y=-3$
$y=2x+3$
Substitute for y into equation 1.
$2(x-1)-(2x+3)=-3$
$2x-2-2x-3=-3$
$-5=-3$ False
No solution;
The system is inconsistent.
$\varnothing$

9. $x+3y=5$
$4x+5y=13$
Solve equation 1 for x and substitute into equation 2.
$x=5-3y$
$4(5-3y)+5y=13$
$20-12y+5y=13$
$-7y=-7$
$y=1$
$x=5-3(1)=5-3=2$
$(2, 1)$

11. $2x-y=-5$
$x+5y=14$
Solve equation 1 for y and substitute into equation 2.
$y=2x+5$
$x+5(2x+5)=14$
$x+10x+25=14$
$11x=-11$
$x=-1$
$y=2(-1)+5=3$
$(-1, 3)$

13. $21x-35=7y$
$y=3x-5$
Substitute for y into equation 1.
$21x-35=7(3x-5)$
$21x-35=21x-35$
$-35=-35$
$0=0$ True
The system has infinitely many solutions.
The equations are dependent.
(x, y) where $y=3x-5$

15. $x-y=11$
$x-6y=-9$
Solve equation 1 for x and substitute into equation 2.
$x=y+11$
$(y+11)-6y=-9$
$-5y=-20$
$y=4$
$x=4+11=15$
$(15, 4)$

17. $2x-y=3$
$5x-2y=10$
Solve equation 1 for y and substitute into equation 2.
$y=2x-3$
$5x-2(2x-3)=10$
$5x-4x+6=10$
$x=4$
$y=2(4)-3=8-3=5$
$(4, 5)$

19. $x+8y=6$
$2x+4y=-3$
Solve equation 1 for x and substitute into equation 2.

$x=-8y+6$
$2(-8y+6)+4y=-3$
$-16y+12+4y=-3$
$-12y=-15$
$y=\frac{5}{4}$
$x=-8\left(\frac{5}{4}\right)+6=-10+6=-4$
$\left(-4, \frac{5}{4}\right)$

21. $x=4y-2$
$x=6y+8$
Substitute for y into equation 2.
$4y-2=6y+8$
$-10=2y$
$-5=y$
$x=4(-5)-2=-20-2=-22$
$(-22, -5)$

23. $y=2x-8$
$y=3x-13$
Substitute for y into equation 2.
$2x-8=3x-13$
$5=x$
$y=2(5)-8$
$y=2$
$(5, 2)$

25. $5x+2y=0$
$x-3y=0$
Solve equation 2 for x and substitute into equation 1.
$x=3y$
$5(3y)+2y=0$
$17y=0$
$y=0$
$x=3(0)=0$
$(0, 0)$

27. $6x+2y=7$
$y=2-3x$
Substitute for y into equation 1.
$6x+2(2-3x)=7$
$6x+4-6x=7$
$4=7$ False
No solution; the system is inconsistent.
$\varnothing$

29. $2x+5y=-4$
$3x-y=11$
Solve equation 2 for y and substitute into equation 1.
$y=3x-11$
$2x+5(3x-11)=-4$
$2x+15x-55=-4$
$17x=51$
$x=3$
$y=3(3)-11=-2$
$(3, -2)$

31.
$$\begin{array}{r} 2x+3y=2 \\ x-3y=-6 \\ \hline 3x=-4 \\ x=-\frac{4}{3} \end{array}$$
$-\frac{4}{3}-3y=-6$
$-3y=-6+\frac{4}{3}=-\frac{14}{3}$
$y=\frac{14}{9}$
$\left(-\frac{4}{3}, \frac{14}{9}\right)$

33.
$$\begin{array}{r} x+y=1 \\ 3x-y=3 \\ \hline 4x=4 \\ x=1 \end{array}$$
$1+y=1$
$y=0$
$(1, 0)$

35. $3x+2y=-3$
$2x-5y=17$
Multiply equation 1 by 5.
Multiply equation 2 by 2.

$$\begin{array}{r} 15x+10y=-15 \\ 4x-10y=34 \\ \hline 19x=19 \\ x=1 \\ 2(1)-5y=17 \\ -5y=15 \\ y=-3 \end{array}$$

$(1, -3)$

37. $3x-2y=6$
$y=3$
Substitute for y in equation 1.
$3x-2(3)=6$
$3x=12$
$x=4$
$(4, 3)$

39. $3x+7y=-10$
$x+2=0$
Solve equation 2 for x and substitute into equation 1.
$x=-2$
$3(-2)+7y=-10$
$-6+7y=-10$
$7y=-4$
$y=-\frac{4}{7}$
$\left(-2, -\frac{4}{7}\right)$

41. $3x-2y=8$
$x=-2y$
Substitute for x in equation 1.
$3(-2y)-2y=8$
$-6y-2y=8$
$-8y=8$
$y=-1$
$x=-2(-1)=2$
$(2, -1)$

43.
$$\begin{array}{r} 4x+y=-12 \\ -3x-y=10 \\ \hline x=-2 \\ 4(-2)+y=-12 \\ y=-4 \end{array}$$
$(-2, -4)$

45. $3(1-2x)-2(3y+4)=1 \rightarrow$
$3-6x-6y-8=1$
$3(x-1)-2y=-5 \rightarrow$
$3x-3-2y=-5$
Simplify:
$-6x-6y=6$
$3x-2y=-2$
Divide equation 1 by -3.
No change to equation 2.
$$\begin{array}{r} 2x+2y=-2 \\ 3x-2y=-2 \\ \hline 5x=-4 \\ x=-\frac{4}{5} \end{array}$$
$y=-1-x=-1-\left(-\frac{4}{5}\right)=-\frac{1}{5}$
$\left(-\frac{4}{5}, -\frac{1}{5}\right)$

47. $y=3x-1$
$-12x+4y=-3$
Substitute for y into equation 2.
$-12x+4(3x-1)=-3$
$-12x+12x-4=-3$
$-4=-3$ False
No solution; the system is inconsistent.
$\varnothing$

49. $3x-4y=19$
$7x+18y=17$
Multiply equation 1 by -7.
Multiply equation 2 by 3.
$$\begin{array}{r} -21x+28y=-133 \\ 21x+54y=51 \\ \hline 82y=-82 \\ y=-1 \\ 3x-4(-1)=19 \\ 3x=15 \\ x=5 \end{array}$$
$(5, -1)$

51. $2L+2W=1294$
$L=W+253$
$2(W+253)+2W=1294$
$2W+506+2W=1294$
$4W+506=1294$

$4W = 788$
$W = 197$
$L = W + 253 = 197 + 253 = 450$
Length = 450 feet, Width = 197 feet

53. a. $7x + 8y = 14{,}066$
$x + 10y = 2120$
Solve 2nd equation for y.
$y = -\frac{1}{10}x + 212$
Substitute for y in 1st equation.
$7x + 8\left(-\frac{1}{10}x + 212\right) = 14{,}066$
$7x - \frac{4}{5}x + 1696 = 14{,}066$
$\frac{31}{5}x = 12{,}370$
$x \approx 1995$
In 1995 the deaths from motor vehicle accidents and from gunfire were equal.

b. $x + 10y = 2120$
$1995 + 10y = 2120$
$10y = 125$
$y = 12.5$
In 1995, the deaths from motor vehicle accidents and from gunfire were both 12.5 per hundred thousand.

c. $7x + 8y = 14{,}066;\ y = -\frac{7}{8}x + \frac{7033}{4}$
$x + 10y = 2120;\ y = -\frac{1}{10}x + 212$
Slope $= -\frac{7}{8},\ -\frac{1}{10}$
The deaths from motor vehicle accidents are declining faster than the deaths from gunfire.

55. Statement a is true.

a. $x + y = 4$
$\underline{x - y = 0}$
$2x = 4$
$x = 2$
$y = 2$
substitution: (2, 2)
$y - 2 = 3(x - 2)$ True

57. Answers may vary.

59. $x = 3 - y - z$
$2x + y - z = -6$
$3x - y + z = 11$
Substitute for x in equation 2.
$2(3 - y - z) + y - z = -6$
Substitute for x in equation 3.
$3(3 - y - z) - y + z = 11$
Simplify:
$6 - 2y - 2z + y - z = -6$, so $-y - 3z = -12$
$9 - 3y - 3z - y + z = 11$, so $-4y - 2z = 2$
Multiply $-y - 3z = -12$ by -2, and divide $-4y - 2z = 2$ by 2:
$2y + 6z = 24$
$\underline{-2y - z = 1}$
$5z = 25$
$z = 5$
$y = -3z + 12 = -15 + 12 = -3$
$x = 3 - y - z = 3 + 3 - 5 = 1$
$x = 1,\ y = -3,\ z = 5$

Review Problems

61. $2x - 3y < 6$

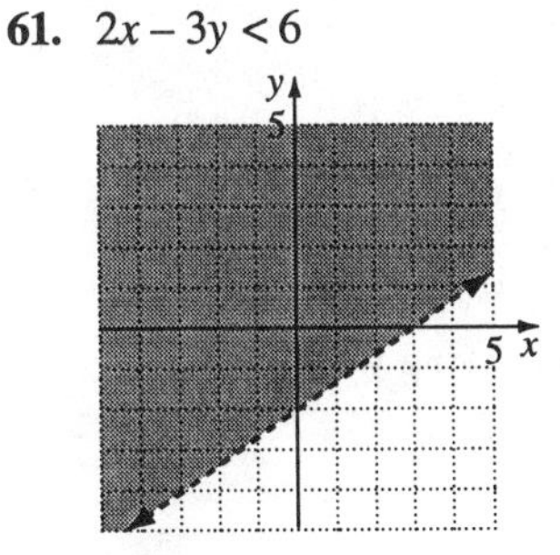

62. passing through (–1, 6)
$m = -4$
point-slope: $y - 6 = -4(x + 1)$
$y - 6 = -4x - 4$
slope-intercept: $y = -4x + 2$

63. Let x equal the amount invested at 7.5%.
$x + 800$ equals the amount invested at 9%.
$0.075x + 0.09(x + 800) = 270$
$0.075x + 0.09x + 72 = 270$
$0.165x = 198$
$x = 1200$
$x + 800 = 2000$
$1200 at 7.5%; $2000 at 9%

Problem Set 5.4

1. a. Let x = the number in Sweden.
Let y = the number in Norway.
$x + y = 206$
$2x + 3y = 477$
Multiply first equation by –2 and add to 2nd equation.
$$\begin{array}{r} -2x - 2y = -412 \\ 2x + 3y = \ \ 477 \\ \hline y = \ \ \ 65 \end{array}$$
Substitute back.
$x + 65 = 206$
$x = 141$
There are 141 in Sweden and 65 in Norway.

b. $\frac{141}{349} \approx 0.40$
$\frac{65}{165} \approx 0.39$
Sweden has 40% and Norway has 39%.

3. Let x = percent living alone
Let y = percent living with relatives
$x + y = 23$
$2x + 3y = 40 + 13$
$x + y = 23$
$2x + 3y = 53$
Multiply first equation by –2 and add to second equation.
$$\begin{array}{r} -2x - 2y = -46 \\ 2x + 3y = \ \ 53 \\ \hline y = \ \ \ 7 \end{array}$$
Substitute back.
$x + y = 23$
$x + 7 = 23$
$x = 16$
16% live alone and 7% live with relatives.

5. Let x = mg of cholesterol in eggs
Let y = mg of cholesterol in Whopper
$x + y = 300 + 241$
$2x + 3y = 1257$

$x + y = 541$
$2x + 3y = 1257$

Multiply first equation by –2 and add to second equation.

$$\begin{array}{r} -2x - 2y = -1082 \\ 2x + 3y = \ \ 1257 \\ \hline y = \ \ \ 175 \end{array}$$
Substitute back.
$x + 175 = 541$
$x = 366$
The eggs have 366 mg of cholesterol.
The Whopper has 175 mg of cholesterol.

7. Let x = servings of macaroni
Let y = servings of broccoli
$3x + 2y = 14$
$16x + 4y = 48$
Multiply first equation by –2 and add to second equation.

$$\begin{array}{r} -6x - 4y = -28 \\ 16x + 4y = \ \ 48 \\ \hline 10x \ \ \ \ \ \ \ = \ \ 20 \\ x \ \ \ \ \ \ \ = \ \ \ 2 \end{array}$$
Substitute back.
$3(2) + 2y = 14$
$2y = 8$
$y = 4$
2 servings of macaroni and 4 servings of broccoli.

9. Let x equal the cost of each sweater and y equal the cost of each shirt.
$x+3y=42$
$3x+2y=56$

$$\begin{array}{r} -3x-9y=-126 \\ 3x+2y=56 \\ \hline -7y=-70 \\ y=10 \end{array}$$

$x=-3y+42=-30+42=12$
cost of one sweater, $12; cost of one shirt, $10

11. $9y-2=10x+10$
$4x+2y+10x+10+9y-2=180$

Simplifying, we get
$10x-9y=-12$
$14x+11y=172$

$$\begin{array}{r} -70x+63y=84 \\ 70x+55y=860 \\ \hline 118y=944 \\ y=8 \\ x=6 \end{array}$$

$(4x+2y)° = (4\cdot 6+2\cdot 8)°$
$=(24+16)° = 40°$
$(9y-2)° = (9\cdot 8-2)° = (72-2)° = 70°$
$(10x+10)° = (10\cdot 6+10)°$
$=(60+10)° = 70°$
Measures of angles are 40°, 70°, and 70°.

13. $5x-(2y-80)=3x$
$5x-(2y-80)+2y=180$

Simplifying, we get
$2x-2y=-80$
$5x=100$, or

$x-y=-40$
$x=20$
$y=x+40=60$

$(3x)° = 60°$
$(2y)° = 120°$
$[5x-(2y-80)]° = [5\cdot 20-(2\cdot 60-80)]°$
$=(100-40)° = 60°$

15. Let l = the length
Let w = the width

$2l+2w=228$
$7l+4w=690$

Multiply first equation by –2 and add to second equation.

$$\begin{array}{r} -4l-4w=-456 \\ 7l+4w=\ \ 690 \\ \hline 3l=\ \ 234 \\ l=\ \ \ 78 \end{array}$$

$2(78)+2w=228$
$156+2w=228$
$2w=72$
$w=36$
length = 78 feet
width = 36 feet

17. Let l equal the length of rectangle, w equals the width
$l+4w=31$
$2l+2w=32$

$$\begin{array}{r} l+4w=31 \\ -4l-4w=-64 \\ \hline -3l=-33 \\ l=11 \\ 11+4w=31 \\ 4w=20 \\ w=5 \end{array}$$

$l=11$ m, $w=5$ m
area $= l\cdot w = 55\text{ m}^2$

19. Let x equal the speed of boat in still water, y equal the speed of current.

	R	$\times$	T	$= D$
Trip with current	$x+y$		1 hour	12 km
Trip against current	$x-y$		2 hours	12 km

$2(x-y) = 12$
$1(x+y) = 12$

$2x - 2y = 12$
$x + y = 12$

$$\begin{aligned} x - y &= 6 \\ x + y &= 12 \\ \hline 2x &= 18 \\ x &= 9 \\ 9 + y &= 12 \\ y &= 3 \end{aligned}$$

$x = 9$ km/hour, $y = 3$ km/hour
speed of the boat in still water, 9 km/hr; speed of current, 3 km/hr

21. Let x = speed of hawk in still air, y = speed of wind.

	R	$\times$	T	$= D$
Trip with wind	$x+y$		8 hours	300 miles
Trip against wind	$x-y$		7 hours	100 miles

$8(x+y) = 300$
$7(x-y) = 100$

$8x + 8y = 300$
$7x - 7y = 100$

$$\begin{aligned} 28x + 28y &= 1050 \\ 28x - 28y &= 400 \\ \hline 56x &= 1450 \\ x &\approx 25.893 \\ 8(x+y) &= 300 \\ x + y &= 37.5 \\ 25.893 + y &= 37.5 \\ y &\approx 11.607 \end{aligned}$$

$x \approx 25.893$ miles/hour, $y \approx 11.607$ miles/hour
speed of wind, about 11.6 miles per hour

23. $y = 125x$

25. 10 tables, $1250 spent and taken in

27. The company loses money if $x < 10$ and makes a profit if $x > 10$

29. Student should check solution.

31. Answers may vary.

33. length of large rectangle = $y + 8$
width of the large rectangle = $x + 10$
$2(x + 10) + 2(y + 8) = 58$
$y + x + y = 17.5$
Simplify.
$2x + 2y = 22$
$x + 2y = 17.5$

$$\begin{aligned} 2x + 2y &= 22 \\ -x - 2y &= -17.5 \\ \hline x &= 4.5 \end{aligned}$$

$y = 11 - x = 11 - 4.5 = 6.5$
$x = 4.5,\ y = 6.5$

35.	beginning	*A*	*B*
Number of people in upstairs apt.	x	$x-1$	$x+1$
Number of people in downstairs apt.	y	$y+1$	$y-1$
equation		$x-1=y+1$	$x+1$ $=2(y-1)$

A: one upstairs moves downstairs
B: one downstairs moves upstairs

$x-1=y+1$
$x+1=2y-2$

$-x+y=-2$
$x-2y=-3$
$-y=-5$
$y=5$
$x=y+2=7$
7 people upstairs; 5 people downstairs

37. $(-1, 0)$:
$m=\frac{2}{3}$
$y-0=\frac{2}{3}(x+1)$
$y=\frac{2}{3}x+\frac{2}{3}$
$(14, -2)$: $m=-\frac{2}{3}$
$y+2=-\frac{2}{3}(x-14)$
$y+2=-\frac{2}{3}x+\frac{28}{3}$
$y=-\frac{2}{3}x+\frac{22}{3}$
The intersection of the two equations gives the coordinates of the boat in distress.
$\frac{2}{3}x+\frac{2}{3}=-\frac{2}{3}x+\frac{22}{3}$
$\frac{4x}{3}=\frac{20}{3}$
$x=5$

$y=\frac{2}{3}x+\frac{2}{3}=\frac{10}{3}+\frac{2}{3}=\frac{12}{3}=4$
$(5, 4)$

Review Problems

38. $4x-2y>8$

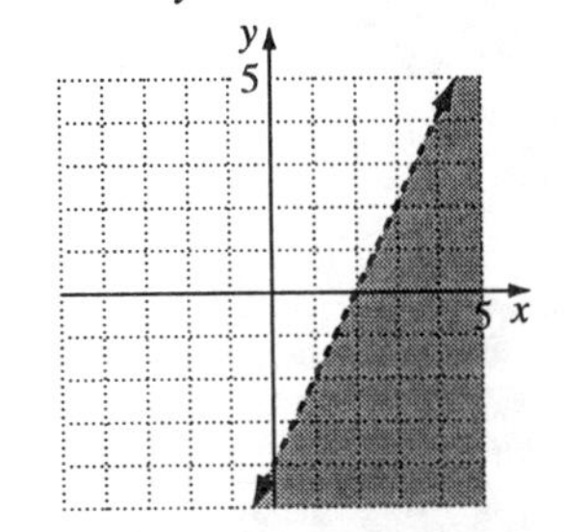

39. Let x equal the number of twenties.
$32+x$ equals the number of tens.
$20x+10(32+x)=800$
$20x+320+10x=800$
$30x=480$
$x=16$
$32+x=32+16=48$ tens
16 \$20 bills

40. $f(x)=x^2-x-2$
$f(-1)=(-1)^2+1-2$
$=1+1-2=0$

Problem Set 5.5

1. $x+y\le 4$
$x-y\le 1$

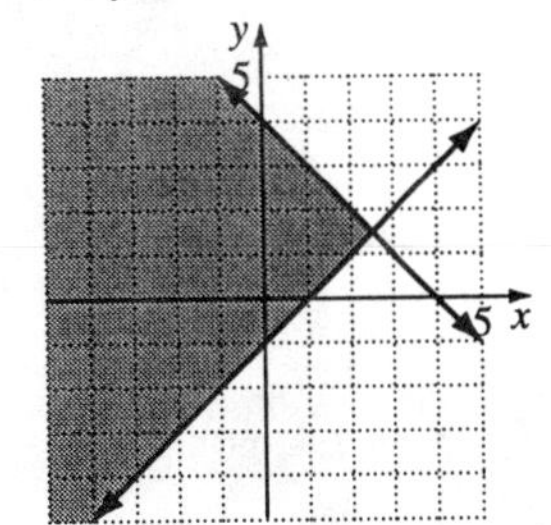

3. $2x - 4y \le 8$
$x + y \ge -1$

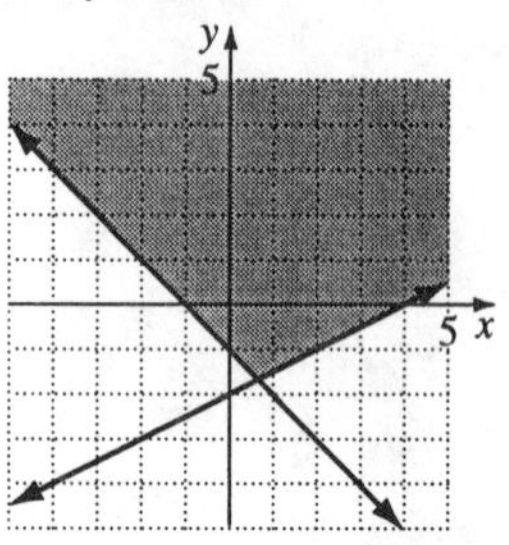

5. $x + 3y \le 6$
$x - 2y \le 4$

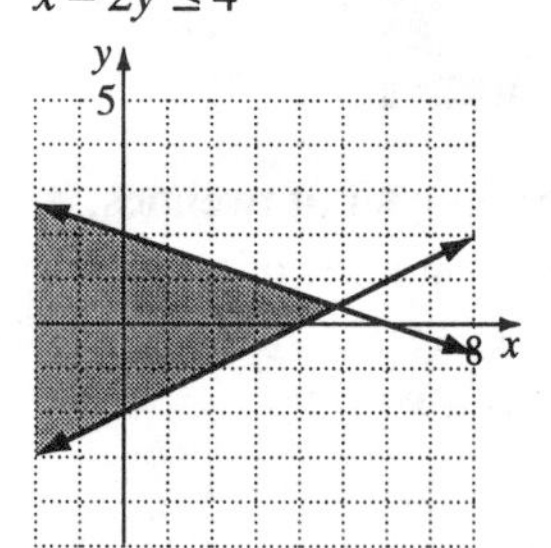

7. $x - 4y \le 4$
$x \ge 2y$

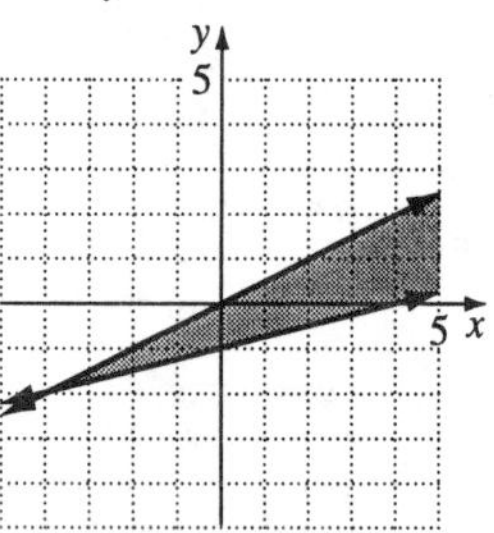

9. $2x + y \le 4$
$x + 2 \ge y$

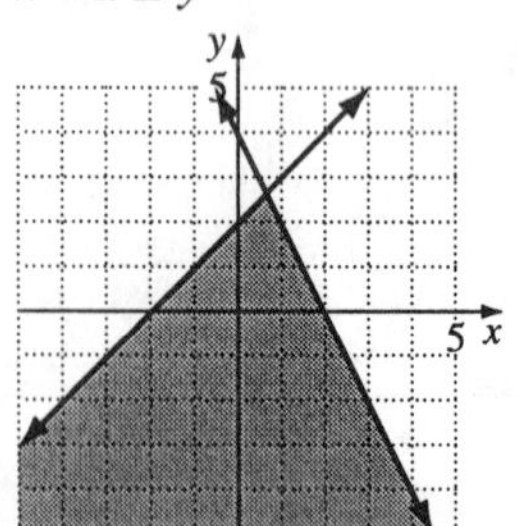

11. $y \le 2x + 2$
$y \ge 2x + 1$

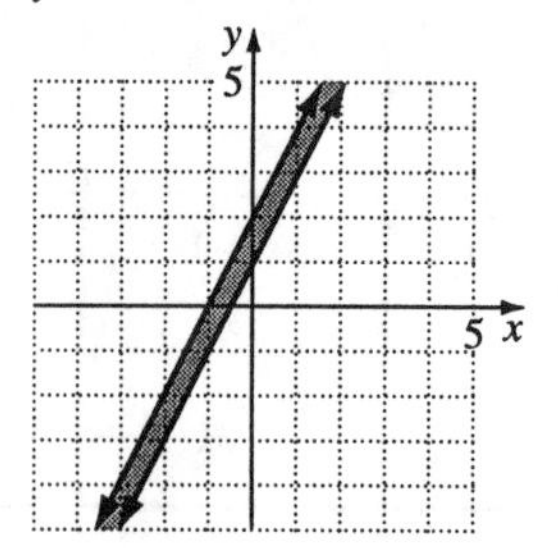

13. $y > 2x - 3$
$y < 2x + 1$

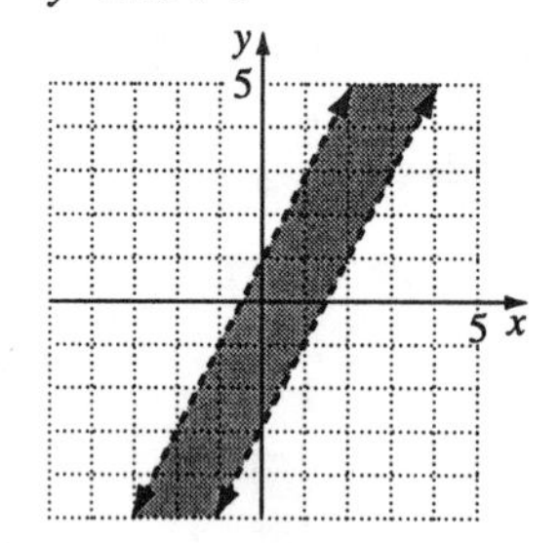

15. $x - 2y > 4$
$2x + y \ge 6$

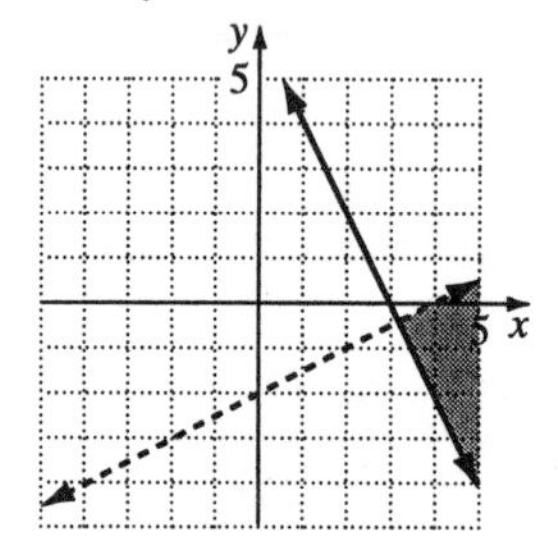

17. $x \ge 3$
$y \ge 3$

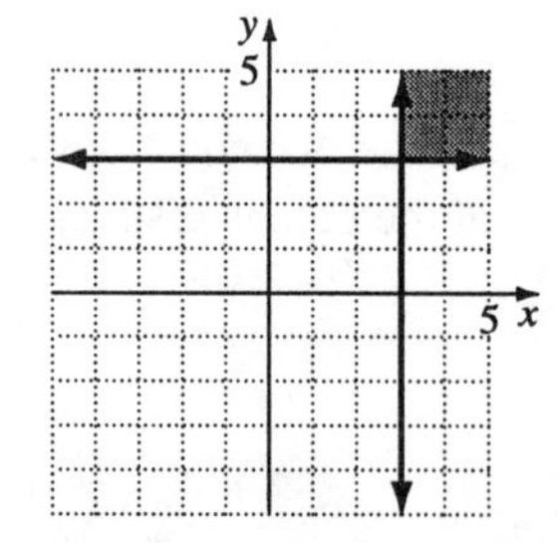

19. $x \geq 2$
$y < 3$

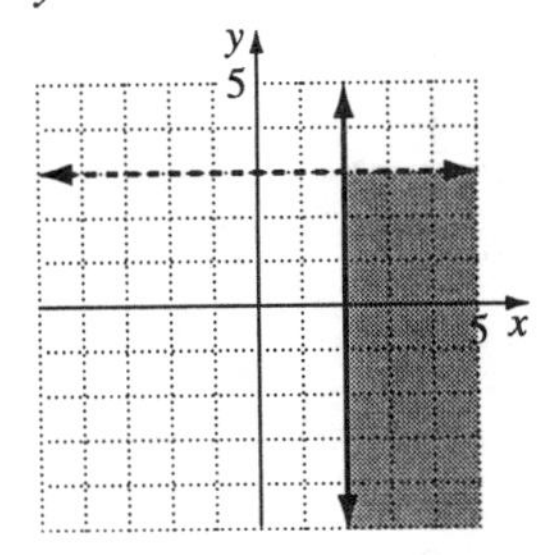

21. $x + y < 1$
$x + y > 4$
No solution

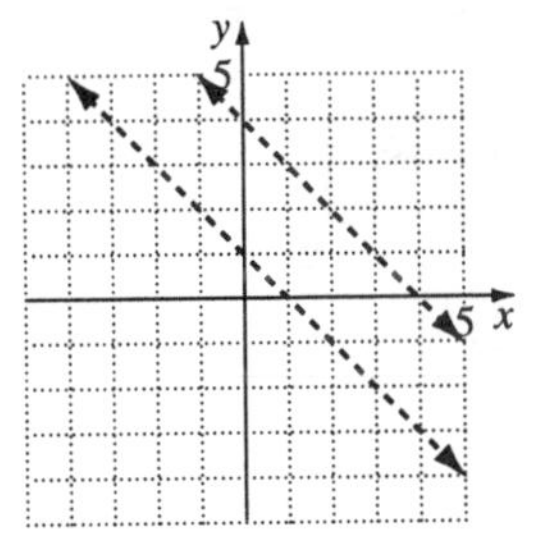

23. $x > 0$
$y \leq 0$

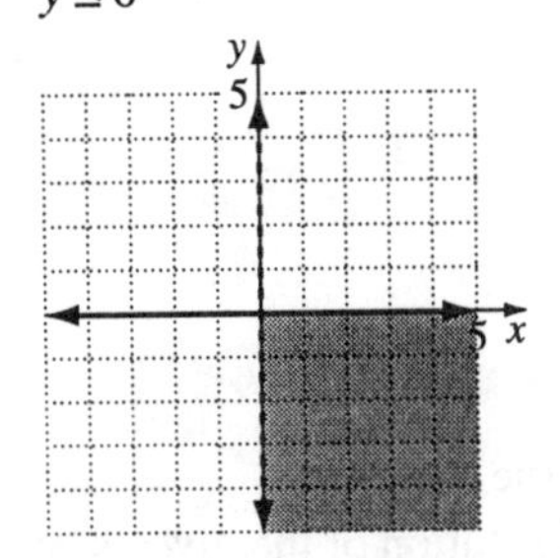

25. $2x + y \geq 6$
$y \leq -2x - 4$
No solution

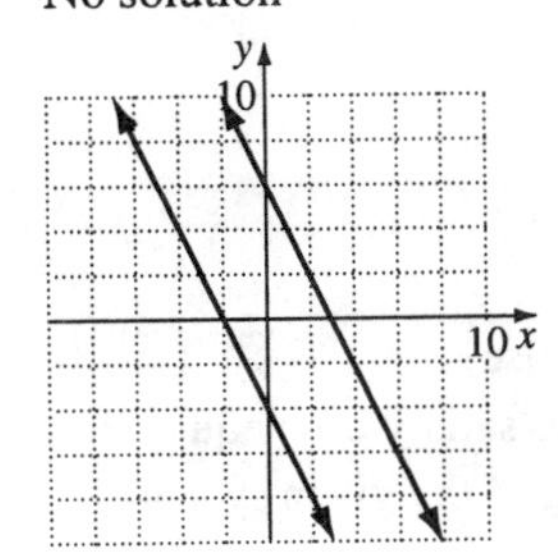

27. $y \geq 2x + 1$
$y \leq 5$

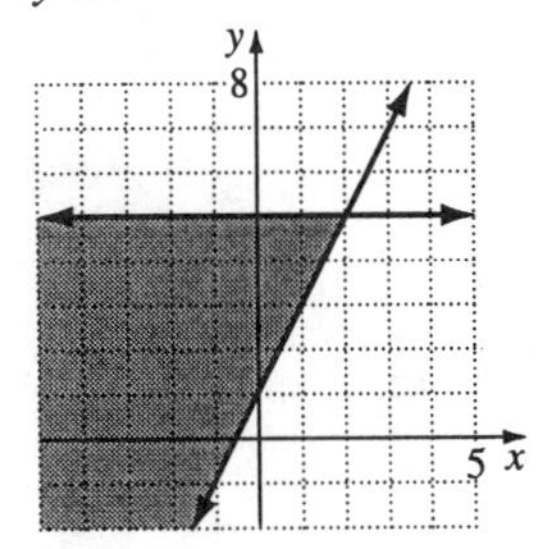

29. $x + y \leq 5$
$x \geq 0$
$y \geq 0$

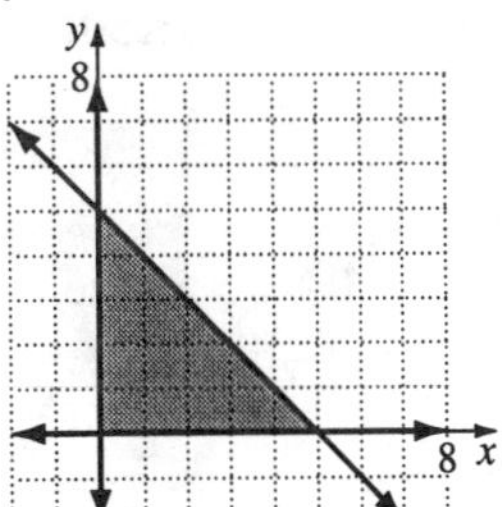

31. $4x - 3y > 12$
$x \geq 0$
$y \leq 0$

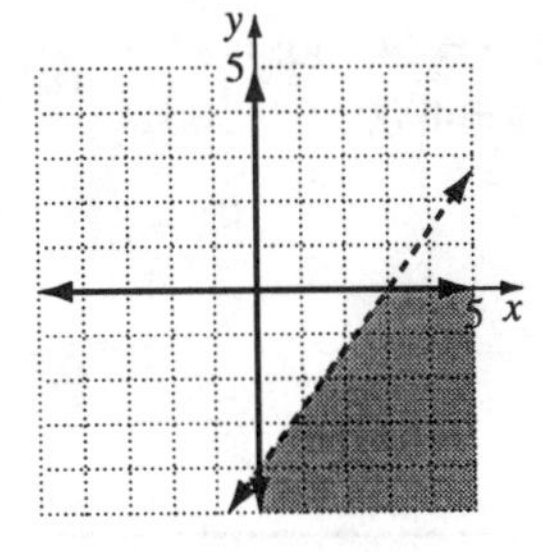

33. $0 \leq x \leq 3$
$0 \leq y \leq 3$

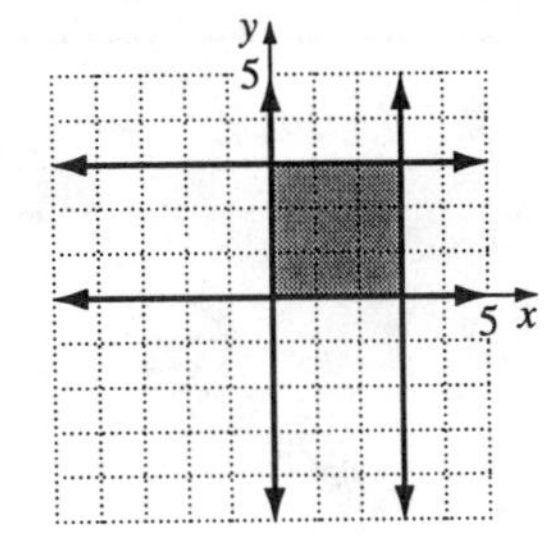

35. $x - y \le 4$
$x + 2y \le 4$
$x \ge 0$

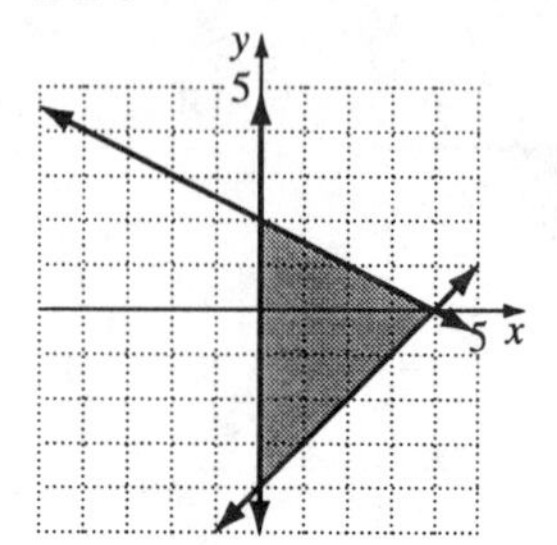

37. a. $y \ge 60$ and $y \le 80$ $(60 \le y \le 80)$

b. $y \ge 30$ and $y \le 50$ $(30 \le y \le 50)$

39. b is true.
(0, 6):
$x + y > 5$ $\quad 0 + 6 > 5$ true
$x - y < 0$ $\quad 0 - 6 < 0$ true

41–43. Answers may vary.

45. $20x + 10y \le 80{,}000$

47. Students should verify graph.

49. maximum value = 44,000;
$x = 2000$ and $y = 4000$;
44,000, 2000, 4000

Review Problems

50. $y = x^2 - 1$

x	−2	−1	0	1	2
y	3	0	−1	0	3

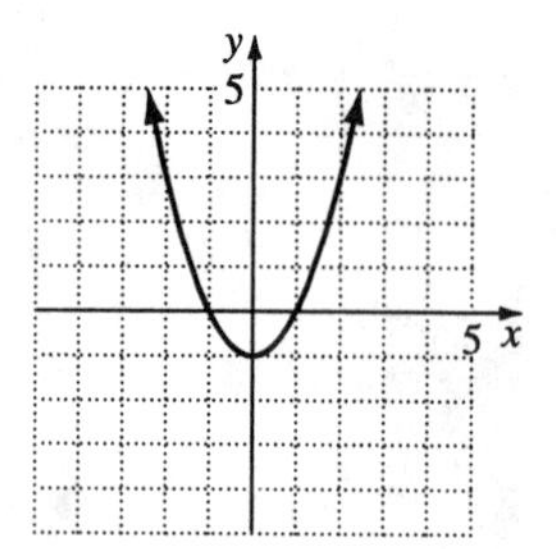

51. (−5, −2), (−1, 6)
$m = \frac{6+2}{-1+5} = \frac{8}{4} = 2$
point-slope:
$y + 2 = 2(x + 5)$ or $y - 6 = 2(x + 1)$
slope-intercept: $y + 2 = 2x + 10$
$y = 2x + 8$
standard: $2x - y = -8$

52. $-5 + [(-11 + 3) - (-1 - 9)]$
$= -5 + (-8 + 10)$
$= -5 + 2$
$= -3$

Chapter 5 Review Problems

1. (1, −5):
$4x - y = 9$
$4 + 5 = 9$
$9 = 9$ True

$2x + 3y = -13$
$2 - 15 = -13$
$-13 = -13$ True
(1, −5) is the solution of the given system.

2. (−5, 2):
$2x + 3y = -4$
$-10 + 6 = -4$
$-4 = -4$ True

$x - 4y = -10$
$-5 - 8 = -10$
$-13 = -10$ False
(−5, 2) fails to satisfy both equations. It is not a solution of the given system.

3. $x + y = 6$
$x - y = 6$
solution: (6, 0)

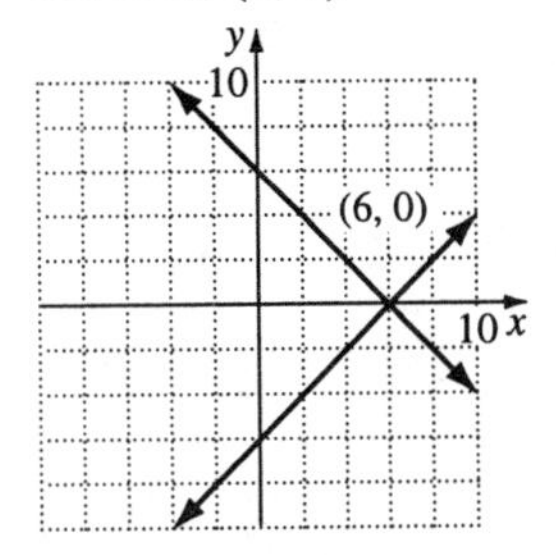

4. $2x - 3y = 12$
$-2x + y = -8$
solution: (3, –2)

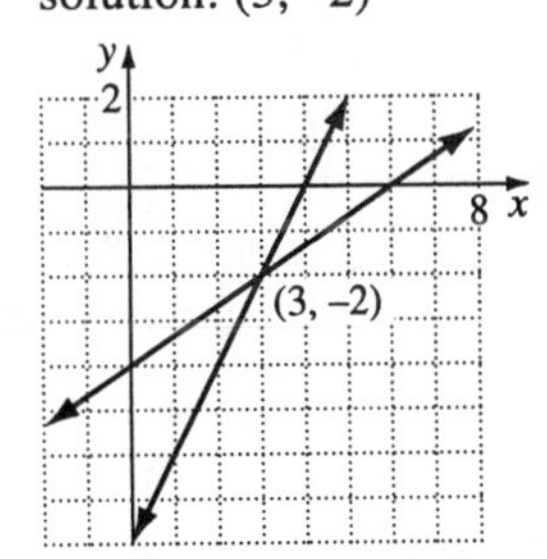

5. $y = \frac{1}{2}x$
$y = 2x - 3$
solution: (2, 1)

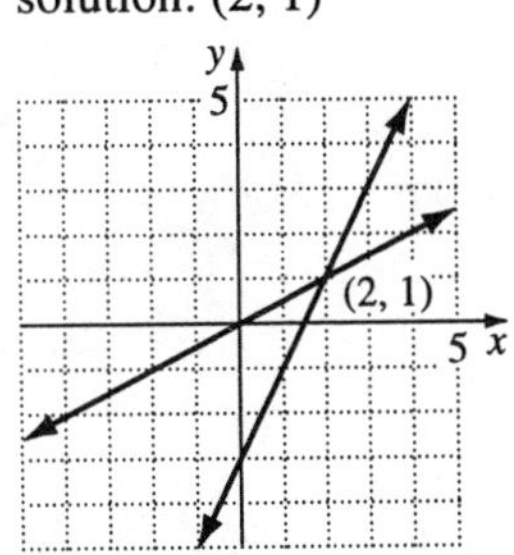

6. $3x + 2y = 6$
$3x - 2y = 6$
solution: (2, 0)

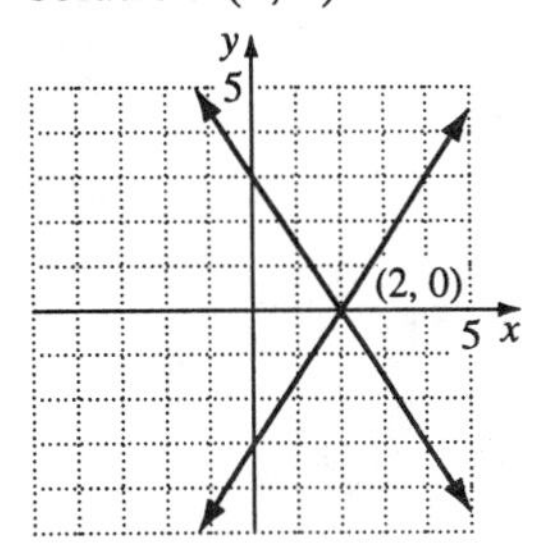

7. $y = 4x$
$y = 4x - 2$
solution: $\varnothing$; inconsistent

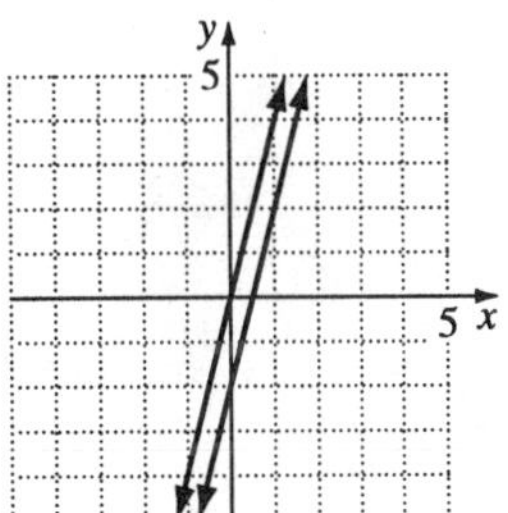

8. $2x - 4y = 8$
$x = 2y + 4$
solution: infinite solutions; dependent

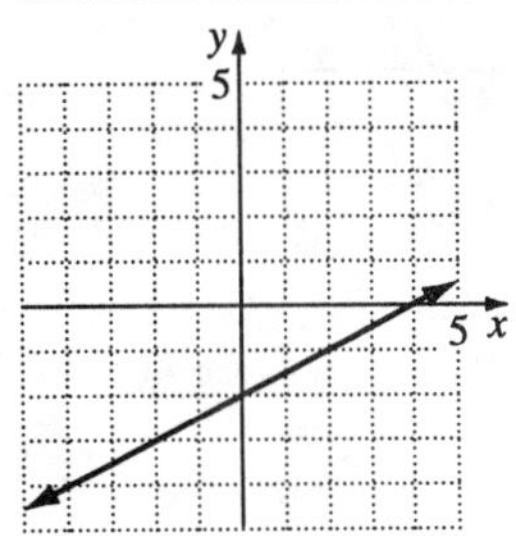

9. $x - y = 4$
$x = -2$
solution: (–2, –6)

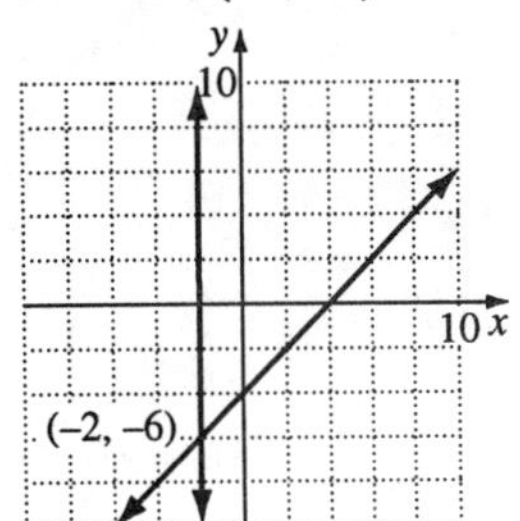

10. $x = -3$
$y = 6$
solution: $(-3, 6)$

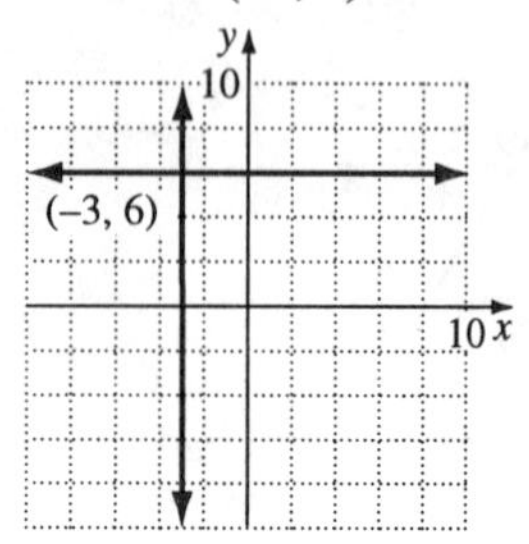

11. $x + y = 6$
$2x + y = 8$

$$\begin{array}{r} x + y = 6 \\ \underline{-2x - y = -8} \\ -x = -2 \\ x = 2 \\ y = 4 \end{array}$$

$(2, 4)$

12. $3x - 4y = 1$
$12x - y = -11$

$$\begin{array}{l} -3x + 4y = -1 \\ \underline{48x - 4y = -44} \\ \quad 45x = -45 \\ \qquad x = -1 \\ \qquad y = 12x + 11 = -1 \end{array}$$

$(-1, -1)$

13. $3x - 7y = 13$
$6x + 5y = 7$

$$\begin{array}{r} -6x + 14y = -26 \\ \underline{6x + 5y = 7} \\ 19y = -19 \\ y = -1 \\ 3x + 7 = 13 \\ 3x = 6 \\ x = 2 \end{array}$$

$(2, -1)$

14. $8x - 4y = 16$
$4x + 5y = 22$

$$\begin{array}{r} -8x + 4y = -16 \\ \underline{8x + 10y = 44} \\ 14y = 28 \\ y = 2 \\ 4x + 10 = 22 \\ 4x = 12 \\ x = 3 \end{array}$$

$(3, 2)$

15. $5x - 2y = 8$
$3x - 5y = 1$

$$\begin{array}{r} 25x - 10y = 40 \\ \underline{-6x + 10y = -2} \\ 19x = 38 \\ x = 2 \\ 10 - 2y = 8 \\ -2y = -2 \\ y = 1 \end{array}$$

$(2, 1)$

16. $x = 2y$
$2x + 6y = 5$

$$\begin{array}{r} -2x + 4y = 0 \\ \underline{2x + 6y = 5} \\ 10y = 5 \\ y = \frac{1}{2} \\ x = 2\left(\frac{1}{2}\right) = 1 \end{array}$$

$\left(1, \frac{1}{2}\right)$

17. $4(x + 3) = 3y + 7$
$2(y - 5) = x + 5$

$$\begin{array}{r} 4x + 12 = 3y + 7 \\ 2y - 10 = x + 5 \end{array}$$

$$\begin{array}{r} 4x - 3y = -5 \\ -x + 2y = 15 \\ 4x - 3y = -5 \\ \underline{-4x + 8y = 60} \\ 5y = 55 \\ y = 11 \end{array}$$

$4x+12=3y+7$
$2y-10=x+5$
$x=2y-15=22-15=7$
$(7, 11)$

18. $2x+y=5$
$2x+y=7$

$$\begin{array}{r} 2x+y=5 \\ -2x-y=-7 \\ \hline 0=-2 \end{array} \text{ false}$$

No solution; the system is inconsistent.
$\varnothing$

19. $3x-4y=-1$
$-6x+8y=2$

$$\begin{array}{r} 3x-4y=-1 \\ -3x+4y=1 \\ \hline 0=0 \end{array} \text{ true}$$

The system has infinitely many solutions.
The equations are dependent.
(x, y) where $3x-4y=-1$

20. $2x+7y=0$
$7x+2y=0$

$$\begin{array}{r} -4x-14y=0 \\ 49x+14y=0 \\ \hline 45x=0 \\ x=0 \\ y=0 \end{array}$$

$(0, 0)$

21. $x=-3y$
$3y+x=-1$
Substituting for x in equation 2,
$3y+(-3y)=-1$
$0=-1$ False
No solution; the system is inconsistent.
$\varnothing$

22. $x+y=3$
$3x+2y=9$
Solving for y in equation 1 and substituting in equation 2,
$y=3-x$
$3x+2(3-x)=9$
$3x+6-2x=9$
$x=3$
$y=3-3=0$
$(3, 0)$

23. $x+3y=-4$
$3x+2y=3$
Solving equation 1 for x and substituting into equation 2,
$x=-3y-4$
$3(-3y-4)+2y=3$
$-9y-12+2y=3$
$-7y=15$
$y=-\frac{15}{7}$
$x=-3\left(-\frac{15}{7}\right)-4=\frac{45}{7}-\frac{28}{7}=\frac{17}{7}$
$\left(\frac{17}{7}, -\frac{15}{7}\right)$

24. $y+1=3x$
$8x-1=4y$
Solving equation 1 for y and substituting into equation 2,
$y=3x-1$
$8x-1=4(3x-1)$
$8x-1=12x-4$
$3=4x$
$\frac{3}{4}=x$
$y=3\left(\frac{3}{4}\right)-1=\frac{5}{4}$
$\left(\frac{3}{4}, \frac{5}{4}\right)$

25. $3x - 2y = -4$
$x = -2$
Substituting for x in equation 1,
$-6 - 2y = -4$
$-2y = 2$
$y = -1$, $x = -2$
$(-2, -1)$

26. $y = 39 - 3x$
$y = 2x - 61$
Substituting for y in equation 2,
$39 - 3x = 2x - 61$
$-5x = -100$
$x = 20$
$y = 39 - 60 = -21$
$(20, -21)$

27. $3x + 4y = 6$
$y - 6x = 6$
Solving equation 2 for y and substituting into equation 1,
$y = 6x + 6$
$3x + 4(6x + 6) = 6$
$3x + 24x + 24 = 6$
$27x = -18$
$x = -\frac{2}{3}$
$y = 6\left(-\frac{2}{3}\right) + 6 = -4 + 6 = 2$
$\left(-\frac{2}{3}, 2\right)$

28. $2x - y = 4$
$x = y + 1$
Substituting for x in equation 1,
$2(y + 1) - y = 4$
$y = 2$
$x = 2 + 1 = 3$
$(3, 2)$

29. $4x + y = 5$
$12x = 15 - 3y$
Solving equation 1 for y and substituting into equation 2,
$y = 5 - 4x$
$12x = 15 - 3(5 - 4x)$
$12x = 15 - 15 + 12x$
$0 = 0$ True
The system has infinitely many solutions.
The equations are dependent.
(x, y) where $4x + y = 5$

30. $4x - y = -3$
$y = 4x$
Substituting for y in equation 1,
$4x - 4x = -3$
$0 = -3$ False
No solution; the system is inconsistent.
$\varnothing$

31. $3x + 4y = -8$
$2x + 3y = -5$

$$\begin{array}{r} -6x - 8y = 16 \\ 6x + 9y = -15 \\ \hline y = 1 \end{array}$$

$3x + 4(1) = -8$
$3x + 4 = -8$
$3x = -12$
$x = -4$
$(-4, 1)$

32. $6x + 8y = 39$
$y = 2x - 2$

$6x + 8(2x - 2) = 39$
$6x + 16x - 16 = 39$
$22x = 55$
$x = \frac{5}{2}$

$y = 2\left(\frac{5}{2}\right) - 2$
$y = 5 - 2$
$y = 3$
$\left(\frac{5}{2}, 3\right)$

33. $x+2y=7$
$2x+y=8$

$$\begin{array}{r} -2x-4y=-14 \\ 2x+y=8 \\ \hline -3y=-6 \\ y=2 \end{array}$$

$x+2(2)=7$
$x+4=7$
$x=3$
$(3, 2)$

34. $y=2x-3$
$y=-2x-1$

$2x-3=-2x-1$
$2x+2x=3-1$
$4x=2$
$x=\frac{1}{2}$

$y=2\left(\frac{1}{2}\right)-3$
$y=1-3$
$y=-2$
$\left(\frac{1}{2}, -2\right)$

35. Let x = weight of gorilla
Let y = weight of orangutan

$2x+3y=1465$
$x+2y=815$

$$\begin{array}{r} 2x+3y=1465 \\ -2x-4y=-1630 \\ \hline -y=-165 \\ y=165 \end{array}$$

$2x+3(165)=1465$
$2x+495=1465$
$2x=970$
$x=485$

Gorilla weighs 485 lbs
Orangutan weighs 165 lb

36. Let x = percent of women
Let y = percent of men

$$\begin{array}{r} y+x=32.5 \\ y-x=9.9 \\ \hline 2y=42.4 \\ y=21.2 \end{array}$$

$21.2+x=32.5$
$x=11.3$

11.3% of women and 21.2% of men have extramarital affairs.

37. Let x = glasses of grape juice
Let y = glasses of apple juice

$165x+120y=735$
$42x+30y=186$

$$\begin{array}{r} 165x+120y=735 \\ -168x-120y=-744 \\ \hline -3x=-9 \\ x=3 \end{array}$$

$165(3)+120y=735$
$495+120y=735$
$120y=240$
$y=2$
3 glasses of grape juice
2 glasses of apple juice

38. Let x equal the cost of one pen.
y equals the cost of one pad.
$8x+6y=3.90$
$3x+2y=1.40$

$$\begin{array}{r} -8x-6y=-3.90 \\ 9x+6y=4.20 \\ \hline x=0.30 \\ 2y=1.40-0.90=0.50 \\ y=0.25 \end{array}$$

cost of one pen, \$0.30

39. Let x = number of full-size cars
Let y = number of compact cars
$3x + 2y = 108{,}000$
$4x + 3y = 149{,}000$

$$\begin{array}{r} 9x + 6y = 324{,}000 \\ -8x - 6y = -298{,}000 \\ \hline x \quad = 26{,}000 \end{array}$$

$3(26{,}000) + 2y = 108{,}000$
$78{,}000 + 2y = 108{,}000$
$2y = 30{,}000$
$y = 15{,}000$
Full-size car costs $26,000
Compact car costs $15,000

40. $m\angle A = m\angle B \rightarrow 3y + 20 = 4x - 30$
$(3y + 20) + (4x - 30) + (x + 5y + 10) = 180$
$-4x + 3y = -50$
$5x + 8y = 180$

$$\begin{array}{r} -20x + 15y = -250 \\ 20x + 32y = 720 \\ \hline 47y = 470 \\ y = 10 \\ -4x + 30 = -50 \\ -4x = -80 \\ x = 20 \end{array}$$

$m\angle A = (3y + 20)° = 50°$
$m\angle B = (4x - 30)° = 50°$
$m\angle C = (x + 5y + 10)° = 80°$

41. $10y + 5 = 3x + 10$
$8x + 5 + 10y + 5 = 180$

$$\begin{array}{r} 3x - 10y = -5 \\ 8x + 10y = 170 \\ \hline 11x = 165 \\ x = 15 \\ 10y = 3x + 5 = 50 \\ y = 5 \end{array}$$

$(8x + 5)° = (8 \cdot 15 + 5)° = 125°$
$(10y + 5)° = (10 \cdot 5 + 5)° = 55°$
$(3x + 10)° = (3 \cdot 5 + 10)° = 55°$

42. Let l = length and w = width
$2l + 2w = 28$
$4l - 3w = 21$

$$\begin{array}{r} -4l - 4w = -56 \\ 4l - 3w = 21 \\ \hline -7w = -35 \\ w = 5 \end{array}$$

$2l + 2(5) = 28$
$2l + 10 = 28$
$2l = 18$
$l = 9$
width = 5 feet
length = 9 feet

43. Let w equal the width.
l equals the length.
$2w + 2l = 24$
$2(2w) + 3(2l) = 62$

$$\begin{array}{r} -2w - 2l = -24 \\ 2w + 3l = 31 \\ \hline l = 7 \\ w = 12 - 7 = 5 \end{array}$$

length, 7 yards; width, 5 yards

44. Let x equal the speed of the plane in still air.
y equals the speed of the wind.
$6(x + y) = 1080$
$3(x - y) = 360$

$$\begin{array}{r} x + y = 180 \\ x - y = 120 \\ \hline 2x = 300 \\ x = 150 \\ y = 30 \end{array}$$

speed of plane in still air, 150 mph; speed of wind, 30 mph

45. a. 1000 copies must be produced and sold to break even. That is the x-coordinate of the intersection on the graph.

b. There is a profit if $x > 1000$.

c. Loss = 400 + 0.85(400) – 1.25(400) = $240

d. Profit = 1.25(2000) – 400 – 0.85(2000) = $400

46. $2x + y < 6$
$y - 2x < 6$

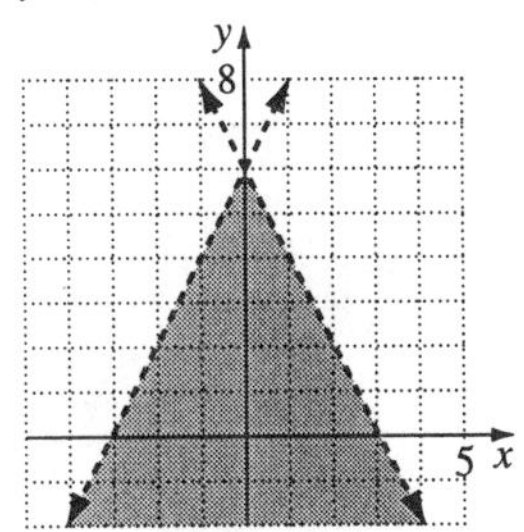

47. $2x + 3y \leq 6$
$y > 3x$

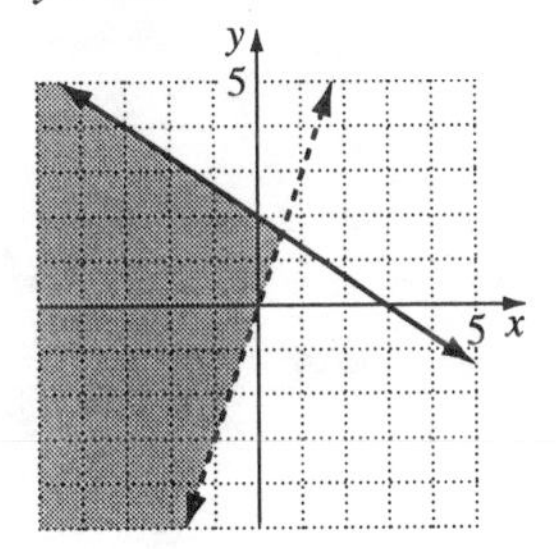

48. $y < 2x - 2$
$x > 3$

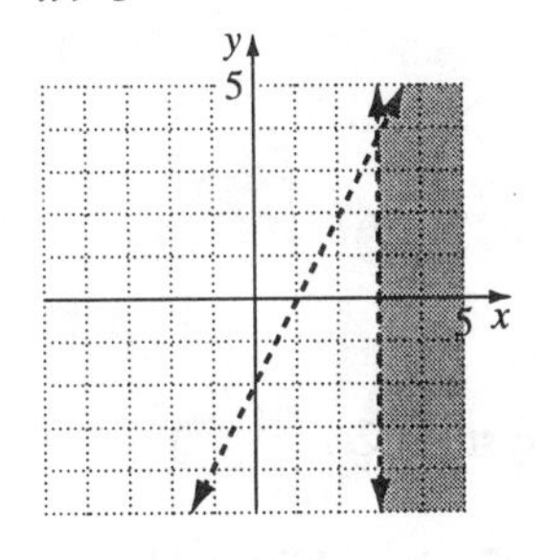

49. $y \geq 5x - 4$
$y \leq 5x + 1$

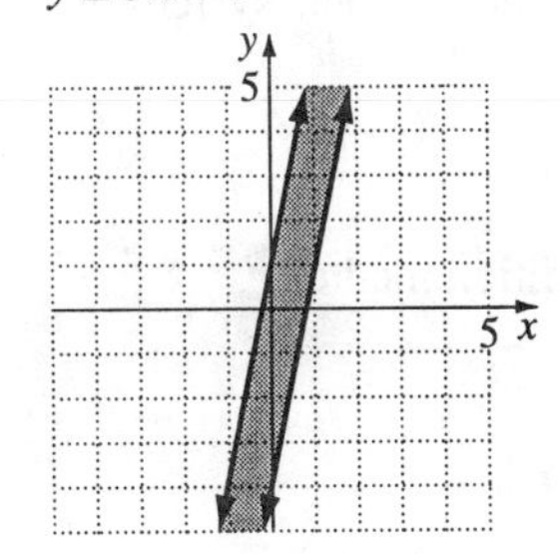

50. $x < 6$
$y \geq -1$

51. $2x + 3y \geq 6$
$3x - y \leq 3$

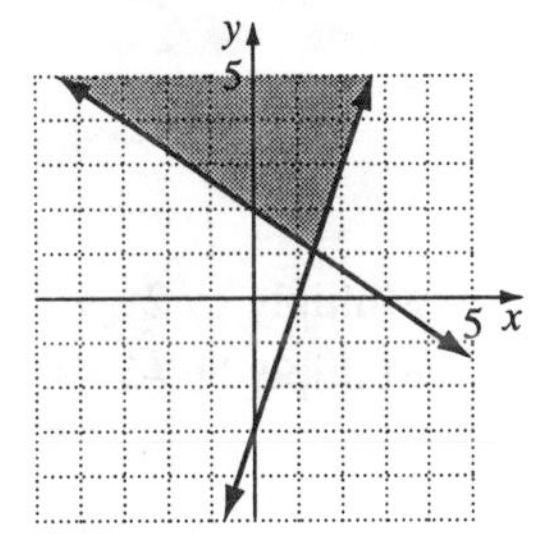

Chapter 5 Test

1. $3 = 2\left(-\frac{3}{2}\right) + 6$　　$3(3) - 2\left(-\frac{3}{2}\right) = 12$
$3 = 3$　　$12 = 12$
True　　True
Yes, $\left(3, -\frac{3}{2}\right)$ is a solution of the system.

2. $2x + y = 6$　　$x - 2y = 8$
$y = -2x + 6$　　$-2y = -x + 8$
$y = \frac{1}{2}x - 4$

Graph $y = -2x + 6$ and $y = \frac{1}{2}x - 4$.

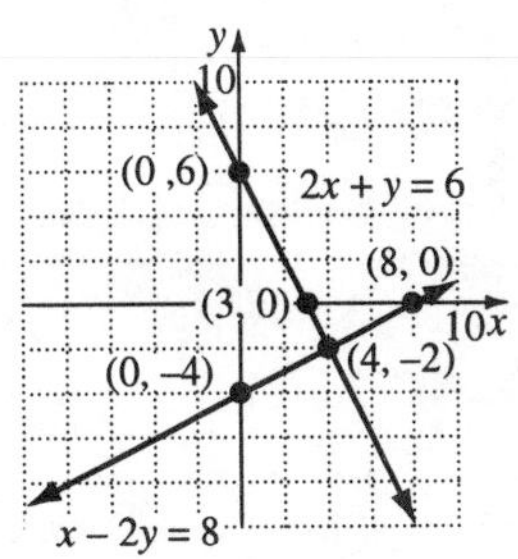

The point of intersection is (4, –2).

3.
$$\begin{array}{r} 2x+y=2 \\ +\ 4x-y=-8 \\ \hline 6x \quad =-6 \\ x \quad =-1 \end{array}$$
$2(-1)+y=2$
$y=4$
$(-1, 4)$

4. $2x+3y=1$ Multiply by 3 $\rightarrow$ $6x+9y=3$
$3x+2y=-6$ Multiply by -2 $\rightarrow$ $+-6x-4y=12$
$$\begin{array}{r} 5y=15 \\ y=3 \end{array}$$
$2x+3(3)=1$
$2x=-8$
$x=-4$
$(-4, 3)$

5. $4x-5y=9$ Multiply by 2 $\rightarrow$ $8x-10y=18$
$5x-2y=24$ Multiply by -5 $\rightarrow$ $+-25x+10y=-120$
$$\begin{array}{r} -17x \quad =-102 \\ x \quad =6 \end{array}$$
$4(6)-5y=9$
$-5y=-15$
$y=3$
$(6, 3)$

6. $3x-5y=2$
$3x-5(32-3x)=2$
$3x-160+15x=2$
$18x=162$
$x=9$
$3(9)-5y=2$
$-5y=-25$
$y=5$
$(9, 5)$

7. $2x-7y=-3$
$2(3y)-7y=-3$
$-y=-3$
$y=3$
$2x-7(3)=-3$
$2x=18$
$x=9$
$(9, 3)$

8. $y=3x-9$
$3x+8=3x-9$
$0=-17$
False
Inconsistent system $\varnothing$.

9. Let x be the cost, in billions of current dollars, of World War II and y be the cost of the Vietnam Conflict. The system of equations is:
$x+y=500$
$x-y=120$
Using the addition method:
$2x=620$
$x=310$
$310+y=500$
$y=190$

World War II cost \$310 billion and the Vietnam Conflict cost \$190 billion.

10. Let x be the price of a sweater and y be the price of a shirt. The system of equations is:
$x+3y=32$
$2x+4y=52$
Solve the first equation for x and substitute into the second equation:
$2(-3y+32)+4y=52$
$-2y=-12$
$y=6$
$x+3(6)=32$
$x=14$
A sweater cost \$14 and a shirt cost \$6.

11. Let x be the number of servings of macaroni and y be the number of servings of broccoli. The system of equations is:
$3x+2y=13$
$16x+4y=56$
Multiply the first equation by –2 and add to the second equation:
$10x=30$
$x=3$
$3(3)+2y=13$
$2y=4$
$y=2$
Three servings of macaroni and two servings of broccoli.

12. The system of equations is:
$2x=y$
$2x+y+(3y-x)=180$
Substitute $y=2x$ into the second equation:
$2x+2x+[3(2x)-x]=180$
$9x=180$
$x=20$
Angle $A=[2x]^\circ=[2(20)]^\circ=40^\circ$
Angle $B=$ Angle $A=40^\circ$
Angle $C=(3y-x)^\circ=[3(2\cdot 20)-20]^\circ$
$=100^\circ$

13. Let x be the speed of the motorboat and y be the speed of the current. The system of equations is:
$\begin{Bmatrix}(x+y)2=48\\(x-y)3=48\end{Bmatrix}$
or
$\begin{Bmatrix}x+y=24\\x-y=16\end{Bmatrix}$
Using the addition method:
$2x=40$
$x=20$
$20+y=24$
$y=4$
The speed of the motorboat is 20 MPH and the speed of the current is 4 MPH.

14. Graph $y\geq 2x-4$

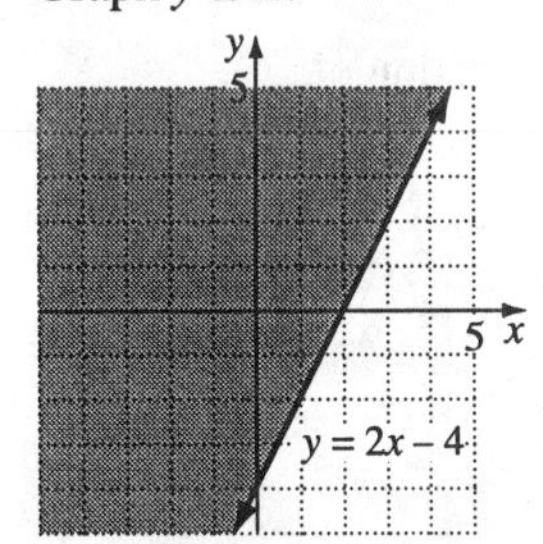

Include $y<2x+1$ in graph.

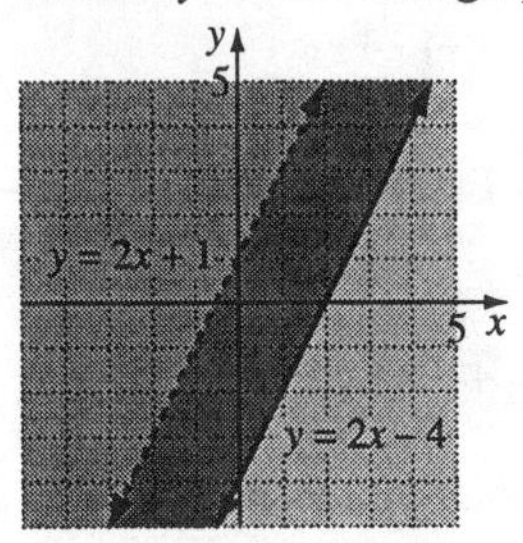

Graph the intersection of
$y\geq 2x-4$ and $y<2x+1$

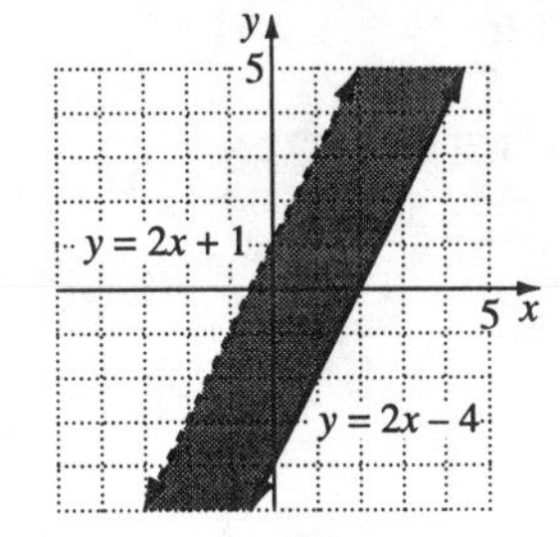

15. Graph $2x - 3y \le 6$.

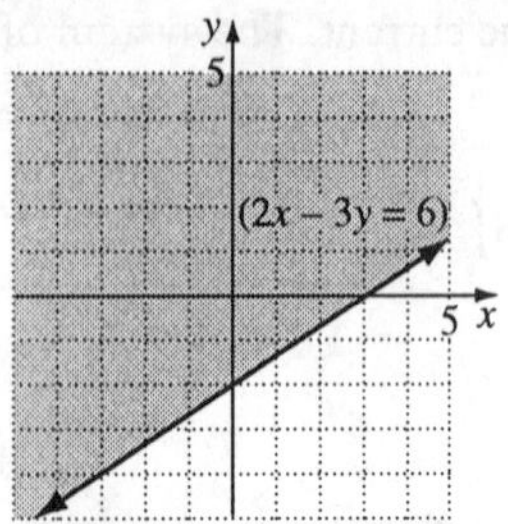

Include $x \ge 3$ in graph.

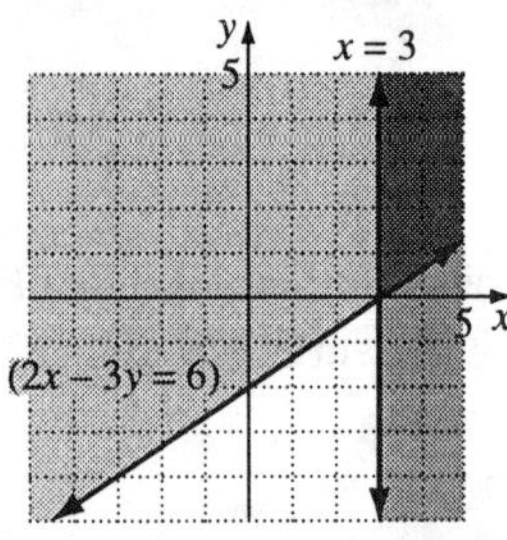

Graph the intersection of $2x - 3 \le 6$ and $x \ge 3$.

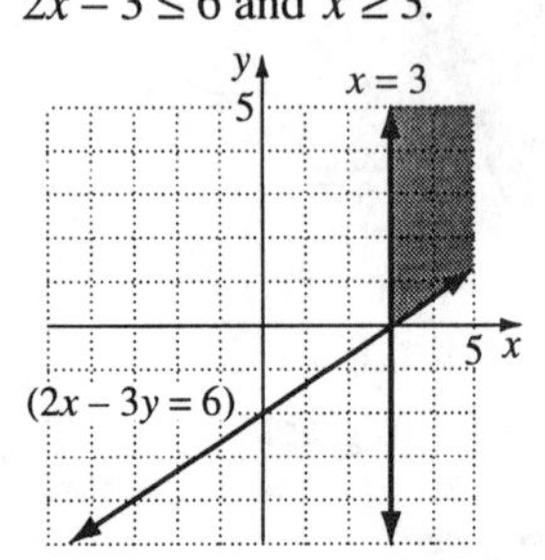

Cumulative Review Problems (Chapters 1–5)

1. $6(3y - 2) - (y - 14) - 2(8y + 7)$
$= 18y - 12 - y + 14 - 16y - 14$
$= y - 12$

2. $-14 - [18 - (6 - 10)]$
$= -14 - (18 + 4)$
$= -14 - 22 = -36$

3. point on line: (3, 1), (0, −2)
slope $= \dfrac{-2-1}{0-3} = \dfrac{-3}{-3} = 1$
y-intercept at −2
$y = 1x - 2$
$y = x - 2$

4. $3y + 2y - 7(y + 1) = -3(y + 4)$
$-2y - 7 = -3y - 12$
$y = -5$
-5

5. Let A equal the number of ounces of health bar A.
Let B equal the number of ounces of health bar B.
Ascorbic acid:
$15A + 10B = 45$
Niacin:
$2A + 4B = 14$
Divide equation 1 by 5.
Divide equation 2 by −2.

$$\begin{array}{r} 3A + 2B = 9 \\ -A - 2B = -7 \\ \hline 2A = 2 \\ A = 1 \\ 2B = 7 - A = 6 \\ B = 3 \end{array}$$

1 ounce of A; 3 ounces of B

6. $A = p + prt$
$A - p = prt$
$\dfrac{A-p}{pr} = t$ or $t = \dfrac{A}{pr} - \dfrac{1}{r}$

7. 5 years

8. $a > 2$: 3, 5, 7, 9
$b < 8$: 1, 3, 5, 7
c, d even: 2, 4, 6, 8
$a < c < d$
$c < d < b$
$a < c < d < b$
$3 < 4 < 6 < 7$
$d = 6$

9. Let x equal the length of each side of the square.
$2x - 10$ equals the length of each side of the equilateral triangle.
perimeter of triangle = perimeter of square
$3(2x - 10) = 4x$

$6x - 30 = 4x$
$2x = 30$
$x = 15$
$2x - 10 = 30 - 10 = 20$
length of each side of triangle:
20 centimeters

10. $t5 + 37 + 51 + 4u = 161$
$t5 + 4u = 161 - 88$
$t5 + 4u = 73$
$t + 4 = 7 - 1 \qquad t = 2$
$5 + u = 13 \qquad u = 8$

11. $6x - 3y = 12$

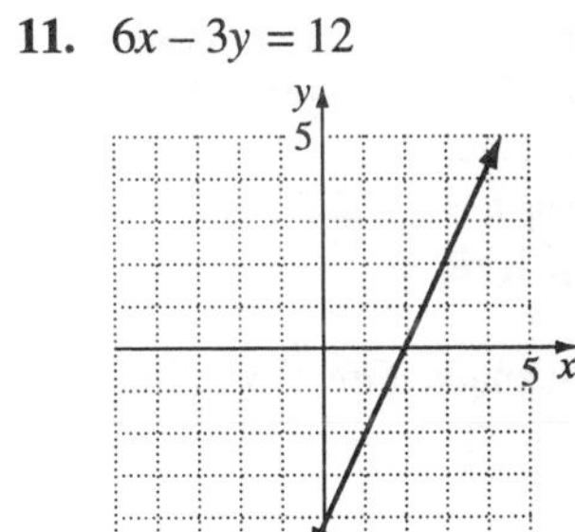

12. $y = \frac{1}{2}x - 2$

$m = \frac{1}{2},\ b = -2$

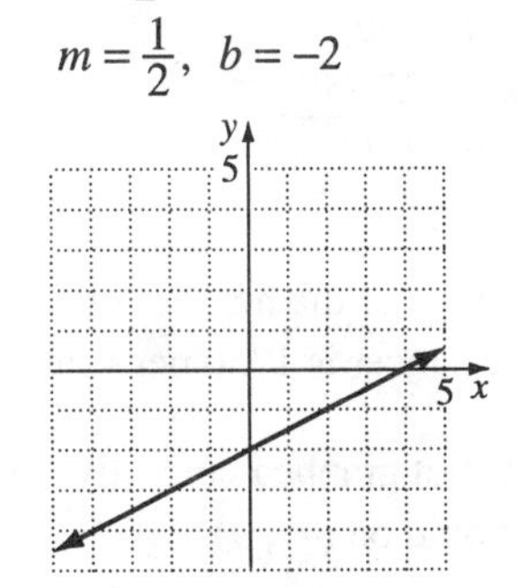

13. $y \geq 3x - 1$

14. $3x - 4y = 8$
$4x + 5y = -10$

$$\begin{aligned} 15x - 20y &= 40 \\ \underline{16x + 20y} &\underline{= -40} \\ 31x &= 0 \\ x &= 0 \\ -4y &= 8 - 0 \\ y &= -2 \end{aligned}$$

(0, –2)

15. Let x = 1995 population of U.S.
$0.02x = 5.1$ million
$x = 255$ million
1995 population of U.S. is 255 million
Protestant = 0.56(255) = 142.8 million
Roman Catholic = 0.28(255) = 71.4 million
Other = 0.14(255) = 35.7 million

16. $f(x) = -0.05x^2 + 2x + 1$
$f(2) = -0.05(2)^2 + 2(2) + 1$
$= -0.2 + 4 + 1 = 4.8$
After 2 hours there is 4.8 parts per million of the drug in the bloodstream.

17. $2x - y < 0$

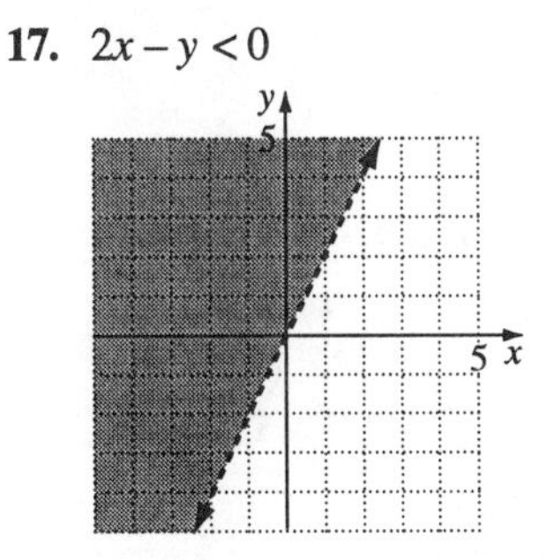

18. $\frac{3}{9} = \frac{1}{3}$

19. $3(y + 1) \leq 5(2y - 4) + 2$
$3y + 3 \leq 10y - 20 + 2$
$-7y \leq -21$
$y \geq 3$
$\{y | y \geq 3\}$

1 2 3 4 5

20. Let x = life span for black men
Then $x + 8.2$ = life span for white men
$\frac{x+x+8.2}{2} = 68.6$
$x + 4.1 = 68.6$
$x = 64.5$
$x + 8.2 = 72.7$
Black men = 64.5 yrs
White men = 72.7 yrs
Answers may vary.

21. Approximately $2.3 billion

22. $\frac{x}{5} = \frac{10}{8}$
$8x = 50$
$x = 6.25$
6.25 in.

23. $2x - 3y = 9$
$y - 4x = -8$

$y = 4x - 8$
$2x - 3(4x - 8) = 9$
$2x - 12x + 24 = 9$
$-10x = -15$
$x = \frac{3}{2}$
$y = 4\left(\frac{3}{2}\right) - 8$
$y = 6 - 8$
$y = -2$
$\left(\frac{3}{2}, -2\right)$

24. Let x equal the number.
$2x + 15 = 9 + x$
$x = -6$

25. $y = -x^2 + 4x - 3$

x	−1	0	1	2	3	4
y	−8	−3	0	1	0	−3

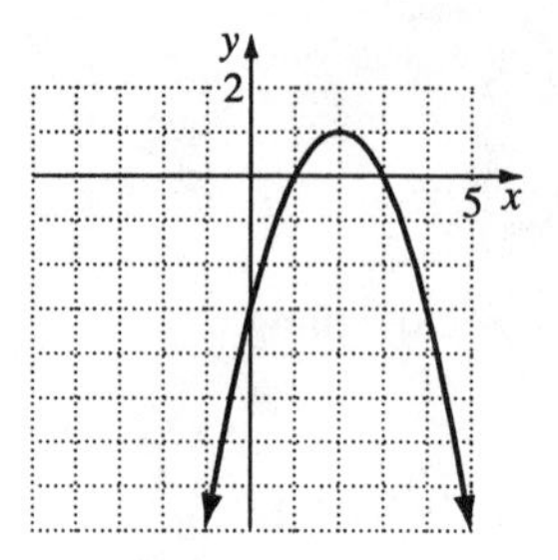

26. $y < -3$

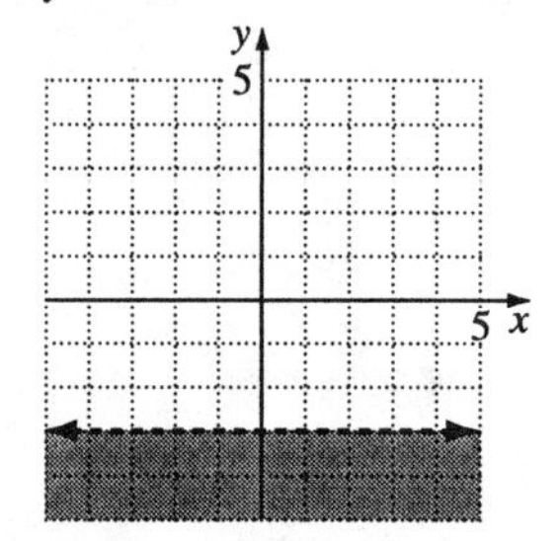

27. Sweden, U.S., Canada, Great Britain, France, Australia, Germany

28. Answers may vary.

29. 1988 ≈ (1988, 27,000)
1996 ≈ (1996, 41,000)
$\text{slope} = \frac{41,000 - 27,000}{1996 - 1988}$
$= \frac{14,000}{8} = 1750$
The rate of change of the number of deaths due to prostate cancer is 1750 per year.

30. No; one line is not graphed correctly. The intersection should be (−7, 9).

Chapter 6

Problem Set 6.1

1. $3x + 7$ is a binomial of degree 1.

3. -9 is a monomial of degree zero.

5. $x^3 - 2x$ is a binomial of degree 3.

7. $x^2 - 3x + 4$ is a trinomial of degree 2.

9. $3y^{17}$ is a monomial of degree 17.

11. $7y^2 - 9y^4 + 5$ is a trinomial of degree 4.

13. $4x - 10x$ is a binomial of degree 1.

15. $5x - 10x^2 = -10x^2 + 5x$; degree 2

17. $3x + 4x^5 - 3x^2 - 2$
$= 4x^5 - 3x^2 + 3x - 2$; degree 5

19. $3 - 3y^4 = -3y^4 + 3$; degree 4

21. $13 = 13$; degree 0

23. $(5x + 7) + (-8x + 3)$
$= (5x - 8x) + (7 + 3)$
$= -3x + 10$

25. $(3x^2 + 7x - 9) + (7x^2 + 8x - 2)$
$= (3x^2 + 7x^2) + (7x + 8x) + (-9 - 2)$
$= 10x^2 + 15x - 11$

27. $(5x^2 - 3x) + (2x^2 - x)$
$= (5x^2 + 2x^2) + (-3x - x)$
$= 7x^2 - 4x$

29. $(3x^2 - 7x + 10) + (x^2 + 6x + 8)$
$= (3x^2 + x^2) + (-7x + 6x) + (10 + 8)$
$= 4x^2 - x + 18$

31. $(4y^3 + 7y - 5) + (10y^2 - 6y + 3)$
$= 4y^3 + 10y^2 + (7y - 6y) + (-5 + 3)$
$= 4y^3 + 10y^2 + y - 2$

33. $(2x^2 - 6x + 7) + (3x^3 - 3x)$
$= 3x^3 + 2x^2 + (-6x - 3x) + 7$
$= 3x^3 + 2x^2 - 9x + 7$

35. $(4y^2 + 8y + 11) + (-2y^3 + 5y + 2)$
$= -2y^3 + 4y^2 + (8y + 5y) + (11 + 2)$
$= -2y^3 + 4y^2 + 13y + 13$

37. $(-2y^6 + 3y^4 - y^2) + (-y^6 + 5y^4 + 2y^2)$
$= (-2y^6 - y^6) + (3y^4 + 5y^4) + (-y^2 + 2y^2)$
$= -3y^6 + 8y^4 + y^2$

39. $\left(\frac{1}{2}x^3 + \frac{2}{3}x^2 - \frac{5}{8}x + 3\right) + \left(-\frac{3}{4}x^3 - \frac{3}{8}x - 11\right)$
$= \left(\frac{1}{2}x^3 - \frac{3}{4}x^3\right) + \frac{2}{3}x^2 + \left(-\frac{5}{8}x - \frac{3}{8}x\right) + (3 - 11)$
$= -\frac{1}{4}x^3 + \frac{2}{3}x^2 - x - 8$

41. $(0.03x^5 - 0.1x^3 + x + 0.03)$
$+ (-0.02x^5 + x^4 - 0.7x + 0.3)$
$= (0.03x^5 - 0.02x^5) + x^4 - 0.1x^3$
$+ (x - 0.7x) + (0.03 + 0.3)$
$= 0.01x^5 + x^4 - 0.1x^3 + 0.3x + 0.33$

43. $(x - 8) - (3x + 2)$
$= x - 8 - 3x - 2$
$= -2x - 10$

45. $(x^2 - 5x - 3) - (6x^2 + 4x + 9)$
$= x^2 - 5x - 3 - 6x^2 - 4x - 9$
$= -5x^2 - 9x - 11$

47. $(x^2 - 5x) - (6x^2 - 4x)$
$= -5x^2 - x$

49. $(x^2 - 8x - 9) - (5x^2 - 4x - 3)$
$= -4x^2 - 4x - 6$

51. $(y - 8) - (3y - 2)$
$= -2y - 6$

53. $(6y^3 + 2y^2 - y - 11) - (y^2 - 8y + 9)$
$= 6y^3 + y^2 + 7y - 20$

55. $(7n^3 - n^7 - 8) - (6n^3 - n^2 - 10)$
$= n^3 - n^7 + n^2 + 2$
$= -n^7 + n^3 + n^2 + 2$

57. $(y^6 - y^3) - (y^2 - y)$
$= y^6 - y^3 - y^2 + y$

59. $(7x^4 + 4x^2 + 5x) - (-19x^4 - 5x^2 - x)$
$= 26x^4 + 9x^2 + 6x$

61. Add:
$$\begin{array}{r} 5y^3 - 7y^2 \\ \underline{6y^3 + 4y^2} \\ 11y^3 - 3y^2 \end{array}$$

63. Add:
$$\begin{array}{r} 3x^2 - 7x + 4 \\ \underline{-5x^2 + 6x - 3} \\ -2x^2 - x + 1 \end{array}$$

65. Add:
$$\begin{array}{r} \frac{1}{4}x^4 - \frac{2}{3}x^3 - 5 \\ \underline{-\frac{1}{2}x^4 + \frac{1}{5}x^3 + 4.7} \end{array} \rightarrow$$
$$\begin{array}{r} \frac{1}{4}x^4 - \frac{10}{15}x^3 - 5 \\ \underline{-\frac{2}{4}x^4 + \frac{3}{15}x^3 + 4.7} \\ -\frac{1}{4}x^4 - \frac{7}{15}x^3 - 0.3 \end{array}$$

67. Add:
$$\begin{array}{r} y^3 + 5y^2 - 7y - 3 \\ \underline{-2y^3 + 3y^2 + 4y - 11} \\ -y^3 + 8y^2 - 3y - 14 \end{array}$$

69. Add:
$$\begin{array}{r} 4x^3 - 6x^2 + 5x - 7 \\ \underline{-9x^3 - 4x + 3} \\ -5x^3 - 6x^2 + x - 4 \end{array}$$

71. Add:
$$\begin{array}{r} 7x^4 - 3x^3 + x^2 \\ \underline{x^3 - x^2 + 4x - 2} \\ 7x^4 - 2x^3 + 4x - 2 \end{array}$$

73. Add:
$$\begin{array}{r} 7x^2 - 9x + 3 \\ 4x^2 + 11x - 2 \\ \underline{-3x^2 + 5x - 6} \\ 8x^2 + 7x - 5 \end{array}$$

75. Subtract:
$$\begin{array}{r} 7x + 1 \\ \underline{-(3x - 5)} \end{array} \rightarrow \begin{array}{r} 7x + 1 \\ \underline{-3x + 5} \\ 4x + 6 \end{array}$$

77. Subtract:
$$\begin{array}{r} 7x^2 - 3 \\ \underline{-(-3x^2 + 4)} \end{array} \rightarrow \begin{array}{r} 7x^2 - 3 \\ \underline{3x^2 - 4} \\ 10x^2 - 7 \end{array}$$

79. Subtract:
$$\begin{array}{r} 7y^2 - 5y + 2 \\ \underline{-(11y^2 + 2y - 3)} \end{array} \rightarrow \begin{array}{r} 7y^2 - 5y + 2 \\ \underline{-11y^2 - 2y + 3} \\ -4y^2 - 7y + 5 \end{array}$$

81. Subtract:
$$\begin{array}{r} 7x^3 + 5x^2 - 3 \\ \underline{-(-2x^3 - 6x^2 + 5)} \end{array} \rightarrow \begin{array}{r} 7x^3 + 5x^2 - 3 \\ \underline{+2x^3 + 6x^2 - 5} \\ 9x^3 + 11x^2 - 8 \end{array}$$

83. Subtract:
$$\begin{array}{r} 5y^3 + 6y^2 - 3y + 10 \\ \underline{-(6y^3 - 2y^2 - 4y - 4)} \end{array} \rightarrow$$
$$\begin{array}{r} 5y^3 + 6y^2 - 3y + 10 \\ \underline{-6y^3 + 2y^2 + 4y + 4} \\ -y^3 + 8y^2 + y + 14 \end{array}$$

85. Subtract:

$$\begin{array}{r} 7x^4 - 3x^3 + 2x^2 \\ -(-x^3 - x^2 + x - 2) \rightarrow \\ \hline \end{array}$$

$$\begin{array}{r} 7x^4 - 3x^3 + 2x^2 \\ x^3 + x^2 - x + 2 \\ \hline 7x^4 - 2x^3 + 3x^2 - x + 2 \end{array}$$

87. Subtract:

$$\begin{array}{r} 4y^3 - \frac{1}{2}y^2 + \frac{3}{8}y + 1 \\ -\left(\frac{9}{2}y^3 + \frac{1}{4}y^2 - y + \frac{3}{4}\right) \rightarrow \\ \hline \end{array}$$

$$\begin{array}{r} 4y^3 - \frac{1}{2}y^2 + \frac{3}{8}y + 1 \\ -\frac{9}{2}y^3 - \frac{1}{4}y^2 + y - \frac{3}{4} \\ \hline -\frac{1}{2}y^3 - \frac{3}{4}y^2 + \frac{11}{8}y + \frac{1}{4} \end{array}$$

89. $f(x) = 14x^3 - 17x^2 - 16x + 34$

$f(2) = 14(2)^3 - 17(2)^2 - 16(2) + 34$

$= 112 - 68 - 32 + 34 = 46$

A moth with an abdominal width of 2 mm has 46 eggs.

91. $f(n) = \frac{1}{3}n^3 + \frac{1}{2}n^2 + \frac{1}{6}n$

$f(1) = \frac{1}{3}(1)^3 + \frac{1}{2}(1)^2 + \frac{1}{6}(1) = 1$

$f(2) = \frac{1}{3}(2)^3 + \frac{1}{2}(2)^2 + \frac{1}{6}(2) = 5$

$f(3) = \frac{1}{3}(3)^3 + \frac{1}{2}(3)^2 + \frac{1}{6}(3) = 14$

$f(4) = \frac{1}{3}(4)^3 + \frac{1}{2}(4)^2 + \frac{1}{6}(4) = 30$

$f(5) = \frac{1}{3}(5)^3 + \frac{1}{2}(5)^2 + \frac{1}{6}(5) = 55$

$f(6) = \frac{1}{3}(6)^3 + \frac{1}{2}(6)^2 + \frac{1}{6}(6) = 91$

$f(10) = \frac{1}{3}(10)^3 + \frac{1}{2}(10)^2 + \frac{1}{6}(10) = 385$

$f(10) = 385$

93. a.

t	$f(t) = -16t^2 + 128t$	Ordered Pair
0	$f(0) = -16(0)^2 + 128(0) = 0$	(0, 0)
2	$f(2) = -16(2)^2 + 128(2) = 192$	(2, 192)
4	$f(4) = -16(4)^2 + 128(4) = 256$	(4, 256)
6	$f(6) = -16(6)^2 + 128(6) = 192$	(6, 192)
8	$f(8) = -16(8)^2 + 128(8) = 0$	(8, 0)

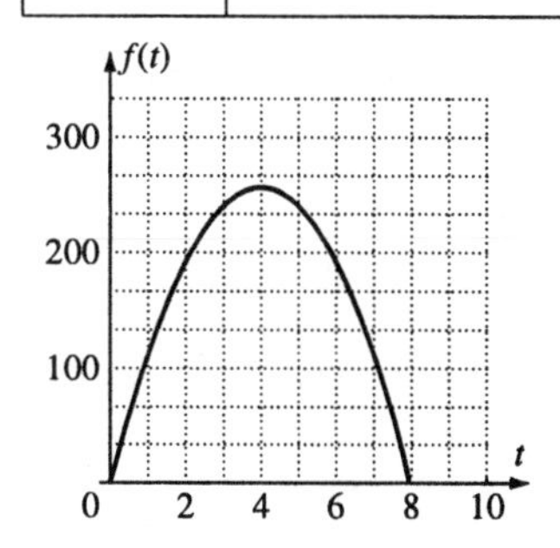

b. 8 seconds

c. 4 seconds; 256 feet

95. triangle: $\frac{1}{2}(3y)(2y) = 3y^2$
rectangle: $2y(y) = 2y^2$
sum: $3y^2 + 2y^2 = 5y^2$

97. Statement c is true. $3x^2 - 7x + \sqrt{5}$ has degree 2.

99. Answers may vary.

101. Answers may vary.

103. Surface area
$= 2(x)(x) + 4(x)(5) = 2x^2 + 20x$

Review Problems

105. $(-3)^4 = (-3)(-3)(-3)(-3) = 81$

106. $3(x-2) \le 9(x+2)$
$3x - 6 \le 9x + 18$
$3x - 9x \le 18 + 6$
$-6x \le 24$
$x \ge -4$

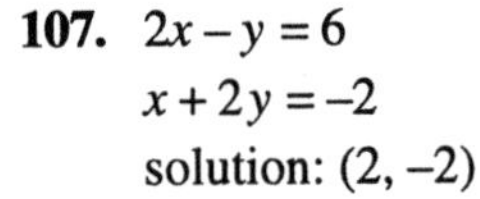

107. $2x - y = 6$
$x + 2y = -2$
solution: (2, –2)

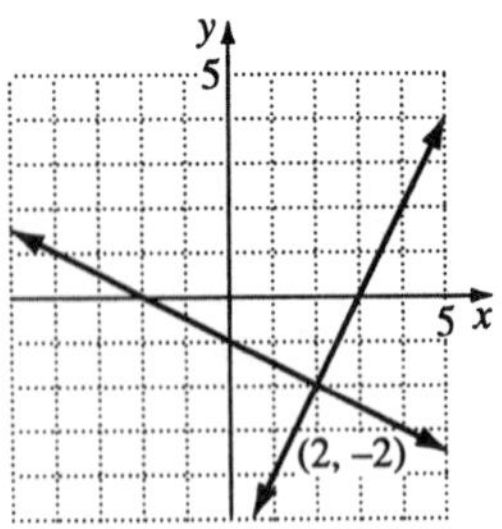

Problem Set 6.2

1. $2^2 \cdot 2^3 = 2^{2+3} = 2^5 = 32$

3. $(-3)(-3)^3 = (-3)^{1+3} = (-3)^4 = 81$

5. $x^3 \cdot x^7 = x^{3+7} = x^{10}$

7. $r \cdot r^8 = r^{1+8} = r^9$

9. $(2x^2)(4x^3) = 8x^{2+3} = 8x^5$

11. $(2y)(y^{13}) = 2y^{1+13} = 2y^{14}$

13. $(-7y)(3y^7) = -21y^{1+7} = -21y^8$

15. $(-2x^3)(-3x^2) = 6x^{3+2} = 6x^5$

17. $x^3 \cdot x^2 \cdot x = x^{3+2+1} = x^6$

19. $(2x^2)(-3x)(8x^4)$
$= (2)(-3)(8)x^{2+1+4}$
$= -48x^7$

21. $(2^2)^3 = 2^{2\cdot 3} = 2^6 = 64$

23. $(x^3)^4 = x^{3\cdot 4} = x^{12}$

25. $(r^8)^{12} = r^{8\cdot 12} = r^{96}$

27. $(5x)^2 = 5^2x^2 = 25x^2$

29. $(-2y)^3 = (-2)^3y^3 = -8y^3$

31. $(-4x)^2 = (-4)^2x^2 = 16x^2$

33. $(2x^2)^2 = 2^2(x^2)^2 = 4x^{2\cdot 2} = 4x^4$

35. $(4y^2)^3 = 4^3(y^2)^3 = 64y^{2\cdot 3} = 64y^6$

37. $(-3y^4)^3 = (-3)^3y^{4\cdot 3} = -27y^{12}$

39. $(-2x^7)^5 = (-2)^5x^{7\cdot 5} = -32x^{35}$

41. $(4x)(2x^2) + (4x^2)(3x)$
$= 8x^3 + 12x^3 = 20x^3$

43. $x(x-3) = x \cdot x - 3x = x^2 - 3x$

45. $-x(x+4) = -x^2 - 4x$

47. $2x(x-6) = 2x^2 - 12x$

49. $-4y(3y+5) = -12y^2 - 20y$

51. $4x^2(x-2) = 4x^2(x) - 4x^2(2) = 4x^3 - 8x^2$

53. $2x^2(x^2+3x)$
$= 2x^2(x^2) + 2x^2(3x)$
$= 2x^4 + 6x^3$

55. $-5x^2(x^2-x)$
$= -5x^2(x^2) - 5x^2(-x)$
$= -5x^4 + 5x^3$

57. $-y^3(3y^2-5)$
$= -y^3(3y^2) - y^3(-5)$
$= -3y^5 + 5y^3$

59. $3x(6x^2-5x)$
$= 3x(6x^2) - 3x(5x)$
$= 18x^3 - 15x^2$

61. $(4x-3)5x = 4x(5x) - 3(5x) = 20x^2 - 15x$

63. $(3x^3-4x^2)(-2x)$
$= 3x^3(-2x) - 4x^2(-2x)$
$= -6x^4 + 8x^3$

65. $x(3x^3-2x+5)$
$= x(3x^3) - x(2x) + x(5)$
$= 3x^4 - 2x^2 + 5x$

67. $-y(-3y^2-2y-4)$
$= -y(-3y^2) - y(-2y) - y(-4)$
$= 3y^3 + 2y^2 + 4y$

69. $x^2(3x^4-5x-3)$
$= x^2(3x^4) + x^2(-5x) + x^2(-3)$
$= 3x^6 - 5x^3 - 3x^2$

71. $2x^2(3x^2-4x+7)$
$= 2x^2(3x^2) + 2x^2(-4x) + 2x^2(7)$
$= 6x^4 - 8x^3 + 14x^2$

73. $(x^2+5x-3)(-2x)$
$= x^2(-2x) + 5x(-2x) - 3(-2x)$
$= -2x^3 - 10x^2 + 6x$

75. $-3x^2(-4x^2+x-5)$
$= -3x^2(-4x^2) - 3x^2(x) - 3x^2(-5)$
$= 12x^4 - 3x^3 + 15x^2$

77. $(x+3)(x+5)$
$= x(x+5) + 3(x+5)$
$= x^2 + 5x + 3x + 15$
$= x^2 + 8x + 15$

79. $(x+11)(x+9)$
$= x(x+9) + 11(x+9)$
$= x^2 + 9x + 11x + 99$
$= x^2 + 20x + 99$

81. $(2x+1)(x+4)$
$= 2x(x+4) + 1(x+4)$
$= 2x^2 + 8x + x + 4$
$= 2x^2 + 9x + 4$

83. $(x+7)(9x+10)$
$= x(9x+10) + 7(9x+10)$
$= 9x^2 + 10x + 63x + 70$
$= 9x^2 + 73x + 70$

85. $(x+3)(x-5)$
$= x(x-5) + 3(x-5)$
$= x^2 - 5x + 3x - 15$
$= x^2 - 2x - 15$

87. $(x-11)(x+9)$
$= x(x+9) - 11(x+9)$
$= x^2 + 9x - 11x - 99$
$= x^2 - 2x - 99$

89. $(2x-5)(x+4)$
$= 2x(x+4) - 5(x+4)$

$= 2x^2 + 8x - 5x - 20$
$= 2x^2 + 3x - 20$

91. $(y - 13)(3y - 4)$
$= y(3y - 4) - 13(3y - 4)$
$= 3y^2 - 4y - 39y + 52$
$= 3y^2 - 43y + 52$

93. $(3y - 2)(5y - 4)$
$= 3y(5y - 4) - 2(5y - 4)$
$= 15y^2 - 12y - 10y + 8$
$= 15y^2 - 22y + 8$

95. $(2x + 3)(2x - 3)$
$= 2x(2x - 3) + 3(2x - 3)$
$= 4x^2 - 6x + 6x - 9$
$= 4x^2 - 9$

97. $(y + 1)(y^2 + 2y + 3)$
$= y(y^2 + 2y + 3) + 1(y^2 + 2y + 3)$
$= y^3 + 2y^2 + 3y + y^2 + 2y + 3$
$= y^3 + 3y^2 + 5y + 3$

99. $(y - 3)(y^2 - 3y + 4)$
$= y(y^2 - 3y + 4) - 3(y^2 - 3y + 4)$
$= y^3 - 3y^2 + 4y - 3y^2 + 9y - 12$
$= y^3 - 6y^2 + 13y - 12$

101. $(2a - 3)(a^2 - 3a + 5)$
$= 2a(a^2 - 3a + 5) - 3(a^2 - 3a + 5)$
$= 2a^3 - 6a^2 + 10a - 3a^2 + 9a - 15$
$= 2a^3 - 9a^2 + 19a - 15$

103. $(z - 4)(-2z^2 - 3z + 2)$
$= z(-2z^2 - 3z + 2) - 4(-2z^2 - 3z + 2)$
$= -2z^3 - 3z^2 + 2z + 8z^2 + 12z - 8$
$= -2z^3 + 5z^2 + 14z - 8$

105. $(2y - 5)(-2y^2 + 4y - 3)$
$= 2y(-2y^2 + 4y - 3) - 5(-2y^2 + 4y - 3)$
$= -4y^3 + 8y^2 - 6y + 10y^2 - 20y + 15$
$= -4y^3 + 18y^2 - 26y + 15$

107.

$$\begin{array}{r} x^2 - 5x + 3 \\ \underline{x + 8} \\ 8x^2 - 40x + 24 \\ \underline{x^3 - 5x^2 + 3x } \\ x^3 + 3x^2 - 37x + 24 \end{array}$$

109.

$$\begin{array}{r} x^2 - 3x + 9 \\ \underline{2x - 3} \\ -3x^2 + 9x - 27 \\ \underline{2x^3 - 6x^2 + 18x } \\ 2x^3 - 9x^2 + 27x - 27 \end{array}$$

111.

$$\begin{array}{r} 2x^3 + x^2 + 2x + 3 \\ \underline{x + 4} \\ 8x^3 + 4x^2 + 8x + 12 \\ \underline{2x^4 + x^3 + 2x^2 + 3x } \\ 2x^4 + 9x^3 + 6x^2 + 11x + 12 \end{array}$$

113.

$$\begin{array}{r} 4z^3 - 2z^2 + 5z - 4 \\ \underline{3z - 2} \\ -8z^3 + 4z^2 - 10z + 8 \\ \underline{12z^4 - 6z^3 + 15z^2 - 12z } \\ 12z^4 - 14z^3 + 19z^2 - 22z + 8 \end{array}$$

115.

$$\begin{array}{r} 7x^3 - 5x^2 + 6x \\ \underline{3x^2 - 4x} \\ -28x^4 + 20x^3 - 24x^2 \\ \underline{21x^5 - 15x^4 + 18x^3 } \\ 21x^5 - 43x^4 + 38x^3 - 24x^2 \end{array}$$

117.

$$\begin{array}{r} 2y^5 \qquad - 3y^3 + y^2 - 2y + 3 \\ \underline{2y - 1} \\ -2y^5 \qquad + 3y^3 - y^2 + 2y - 3 \\ \underline{4y^6 - \qquad 6y^4 + 2y^3 - 4y^2 + 6y } \\ 4y^6 - 2y^5 - 6y^4 + 5y^3 - 5y^2 + 8y - 3 \end{array}$$

119. $(x+5)(2x-3)$
$= x(2x-3)+5(2x-3)$
$= 2x^2-3x+10x-15$
$= 2x^2+7x-15$

121. Area $= (2x+1)(x+2) = 2x^2+5x+2$
Or: Area $= 2x(x)+x(1)+2x(2)+2(1)$
$= 2x^2+5x+2$

123. b is true;
$(5y^2)^3(2y-1)$
$= 125y^6(2y-1)$
$= 250y^7-125y^6$

125. $(x+1)(x-3)$
$= x(x-3)+1(x-3)$
$= x^2-3x+x-3$
$= x^2-2x-3$

127–129. Answers may vary.

131. $(3x+4)(2x-1)-(2x+1)(x-2)$
$= 3x(2x-1)+4(2x-1)$
$\quad -[2x(x-2)+1(x-2)]$
$= 6x^2-3x+8x-4-(2x^2-4x+x-2)$
$= 6x^2+5x-4-(2x^2-3x-2)$
$= 6x^2+5x-4-2x^2+3x+2$
$= 8x^2+8x-2$

133. Let x equal Annie's present age.
$5x$ equals Warbuck's present age.
$x+6$ equals Annie's age 6 years from now.
$5x+6$ equals Warbuck's age 6 years from now.
$(x+6)(5x+6)$
$= 5x^2+6x+30x+36$
$= 5x^2+36x+36$

135. a. $(x-1)(x+1)$
$= x^2+x-x-1$
$= x^2-1$

b. $(x-1)(x^2+x+1)$
$= x^3+x^2+x-x^2-x-1$
$= x^3-1$

c. $(x-1)(x^3+x^2+x+1)$
$= x^4+x^3+x^2+x-x^3-x^2-x-1$
$= x^4-1$

d. $(x-1)(x^4+x^3+x^2+x+1)$
$= x^5-1$

Review Problems

137. $3x+5y=9$
$4x+3y=1$

$$\begin{aligned} -9x-15y &= -27 \\ 20x+15y &= 5 \\ \hline 11x &= -22 \\ x &= -2 \\ -8+3y &= 1 \\ 3y &= 9 \\ y &= 3 \end{aligned}$$

$(-2, 3)$

138. $5x-4y \ge -20$

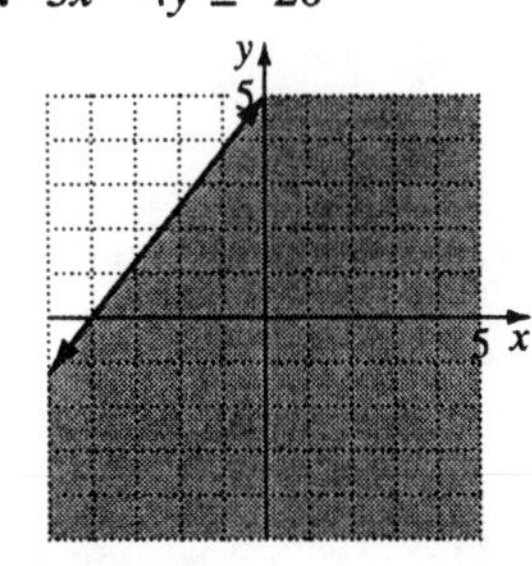

139. Let x equal the numerator.
$3x-2$ equals the denominator.
$x+(3x-2)=79$
$4x=81$
$x=\frac{81}{4}$

$3x-2=\frac{243-8}{4}=\frac{235}{4}$

$\frac{\frac{81}{4}}{\frac{235}{4}}=\frac{81}{4}\cdot\frac{4}{235}=\frac{81}{235}$

Problem Set 6.3

1. $(x+3)(x+5)$
$=x^2+5x+3x+15$
$=x^2+8x+15$

3. $(y-5)(y+3)$
$=y^2+3y-5y-15$
$=y^2-2y-15$

5. $(2b-1)(b+2)$
$=2b^2+4b-b-2$
$=2b^2+3b-2$

7. $(2x-3)(x+1)$
$=2x^2+2x-3x-3$
$=2x^2-x-3$

9. $(2y-3)(5y+3)$
$=10y^2+6y-15y-9$
$=10y^2-9y-9$

11. $(3y-7)(4y-5)$
$=12y^2-15y-28y+35$
$=12y^2-43y+35$

13. $(x^2-5)(x^2-3)$
$=x^4-3x^2-5x^2+15$
$=x^4-8x^2+15$

15. $(3y^3+2)(y^3+4)$
$=3y^6+12y^3+2y^3+8$
$=3y^6+14y^3+8$

17. $(3y^6-5)(2y^6-2)$
$=6y^{12}-6y^6-10y^6+10$
$=6y^{12}-16y^6+10$

19. $(x^2-3)(x+2)$
$=x^3+2x^2-3x-6$

21. $(4+5y)(5-4y)$
$=20-16y+25y-20y^2$
$=20+9y-20y^2$

23. $(-3+2y)(4+y)$
$=-12-3y+8y+2y^2$
$=-12+5y+2y^2$

25. $(-3+r)(-5-2r)$
$=15+6r-5r-2r^2$
$=15+r-2r^2$

27. $(6x^{10}-4)(3x^{10}+7)$
$=18x^{20}+42x^{10}-12x^{10}-28$
$=18x^{20}+30x^{10}-28$

29. $(x+5)(x^2-3)$
$=x^3-3x+5x^2-15$
$=x^3+5x^2-3x-15$

31. $(2x^2-3)(4x^3+1)$
$=8x^5+2x^2-12x^3-3$
$=8x^5-12x^3+2x^2-3$

33. $(x+3)(x-3)$
$=x^2-3^2=x^2-9$

35. $(3x+2)(3x-2)$
$=(3x)^2-2^2=9x^2-4$

37. $(3r-4)(3r+4)$
$=(3r)^2-4^2=9r^2-16$

39. $(3+r)(3-r)$
$=3^2-r^2=9-r^2$

41. $(5-7x)(5+7x)$
$=5^2-(7x)^2=25-49x^2$

43. $\left(2x+\frac{1}{2}\right)\left(2x-\frac{1}{2}\right)$
$=(2x)^2-\left(\frac{1}{2}\right)^2=4x^2-\frac{1}{4}$

45. $(y^2+1)(y^2-1)=y^4-1$

47. $(r^3+2)(r^3-2)=r^6-4$

49. $(1-y^4)(1+y^4)=1-y^8$

51. $(x+2)^2$
$=x^2+2(2x)+2^2=x^2+4x+4$

53. $(y-3)^2$
$=y^2-2(3y)+3^2=y^2-6y+9$

55. $(2x^2+3)^2=4x^4+12x^2+9$

57. $(4x^2-1)^2=16x^4-8x^2+1$

59. $\left(2x+\frac{1}{2}\right)^2$
$=4x^2+2(2x)\left(\frac{1}{2}\right)+\frac{1}{4}=4x^2+2x+\frac{1}{4}$

61. $\left(4y-\frac{1}{4}\right)^2$
$=16y^2-2(4y)\left(\frac{1}{4}\right)+\frac{1}{16}=16y^2-2y+\frac{1}{16}$

63. $(7-2x)^2=49-28x+4x^2$

65. $(7-12y^3)^2=49-168y^3+144y^6$

67. $(-3x-7)(x+5)$
$=-3x^2-15x-7x-35$
$=-3x^2-22x-35$

69. $(2x-5)^2$
$=4x^2-2(10x)+25$
$=4x^2-20x+25$

71. $(3x+11)(3x-11)=9x^2-121$

73. $(7m^4+m^2)(m^2+m)$
$=7m^6+7m^5+m^4+m^3$

75. $(y-5)(y^2+5y+25)$
$=y^3+5y^2+25y-5y^2-25y-125$
$=y^3-125$

77. $\left(\frac{4}{5}-2x^3\right)\left(\frac{4}{5}+2x^3\right)=\frac{16}{25}-4x^6$

79. $(4x^2-11)(2x^2+3)$
$=8x^4+12x^2-22x^2-33$
$=8x^4-10x^2-33$

81. $A=(x+1)^2=x^2+2x+1$

83. $A=(2x-3)(2x+3)=4x^2-9$

85. Let x equal an integer.
$x+1$ equals consecutive integer.
$x^2+(x+1)^2$
$=x^2+x^2+2x+1$
$=2x^2+2x+1$

87. $A=3x+3x+x^2+9$
$=x^2+6x+9=(x+3)^2$,
the square of a binomial.

89. a.

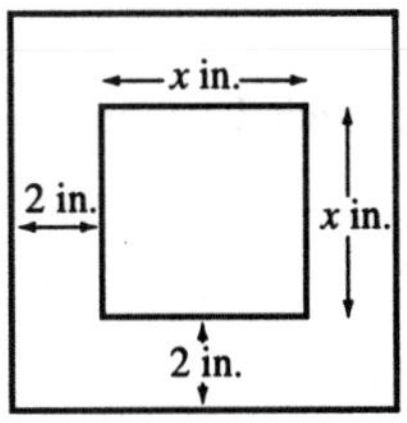

x in.
2 in.
x in.
2 in.

$(x+4)^2 - x^2$
$= x^2 + 8x + 16 - x^2$
$= 8x + 16$ square inches

b. $g(x) = 8x + 16$

c.

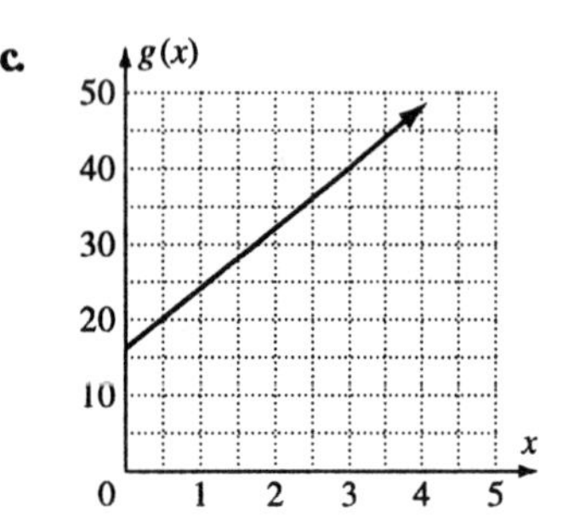

d. 1 inch

91. $(5d+3)(4d-2)$
$= 20d^2 - 10d + 12d - 6$
$= 20d^2 + 2d - 6$ desks
$f(d) = 20d^2 + 2d - 6$
$f(7) = 20(7)^2 + 2(7) - 6 = 988$
There are 988 desks in the room.

93. $(x+4)(x+3) - (x+1)(x+2)$
$= x^2 + 7x + 12 - (x^2 + 3x + 2)$
$= x^2 + 7x + 12 - x^2 - 3x - 2$
$= 4x + 10$
$f(x) = 4x + 10$
$f(3) = 4(3) + 10 = 22$
The area of the shaded region is 22 square units.

95. $f(x) = 3x$
$f(3) = 3(3) = 9$
The area of the shaded region is 9 square units.

97. a is true;
$(40+1)(40-1)$
$= (40)^2 - (1)^2 = 1600 - 1 = 1599$

99. a. $(x-1)^2 = x^2 - 2x + 1$

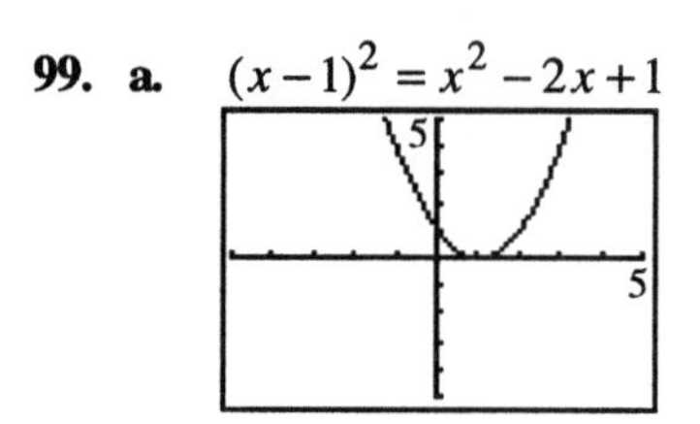

b. $(x+3)^2 = x^2 + 6x + 9$

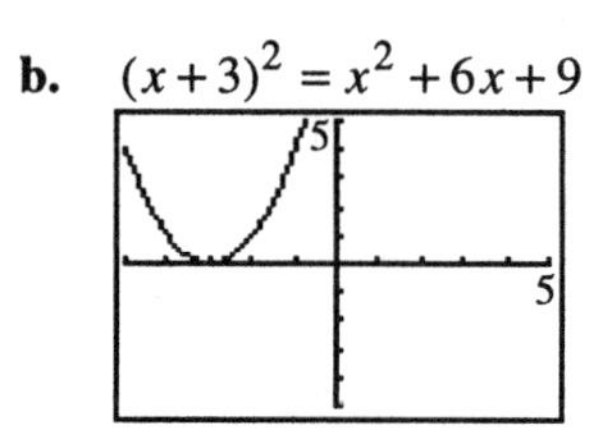

c. $(x+2)^2 = x^2 + 4x + 4$

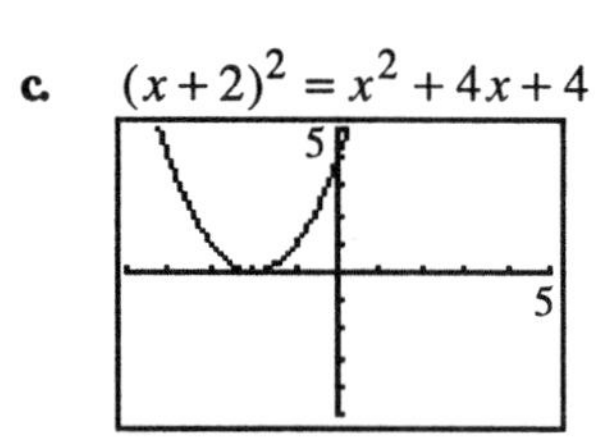

101–103. Answers may vary.

105. $(2x+3)^2 - (7-2x)^2$
$= 4x^2 + 12x + 9 - (49 - 28x + 4x^2)$
$= 4x^2 + 12x + 9 - 49 + 28x - 4x^2$
$= 40x - 40$

107. area of the region inside the right triangle and outside the square
= area of triangle – area of square
$= \frac{1}{2}(2x-3)(2x+3) - (x+1)^2$
$= \frac{1}{2}(4x^2 - 9) - (x^2 + 2x + 1)$
$= 2x^2 - \frac{9}{2} - x^2 - 2x - 1$
$= x^2 - 2x - \frac{11}{2}$

109. $(x+5)(x+3) = x^2 + 8x + 15;\ 8x$

111. $x^2 - 7x + 10 = (x-5)(x-2)$

113. Take a number: x
Add 2: $x+2$
Square the sum: $(x+2)^2$
Add 25 to the result: $(x+2)^2+25$
Subtract the product of the original number times four more than the original number:
$(x+2)^2+25-x(x+4)$
Add 6 to the difference:
$(x+2)^2+25-x(x+4)+6$
Using polynomial operations:
$(x+2)^2+25-x(x+4)+6=35$
$x^2+4x+4+25-x^2-4x+6=35$
$35=35$
The answer is 35, regardless of what number is originally chosen.

115. Group Activity

Review Problems

116. $y=-\frac{1}{2}x+3$

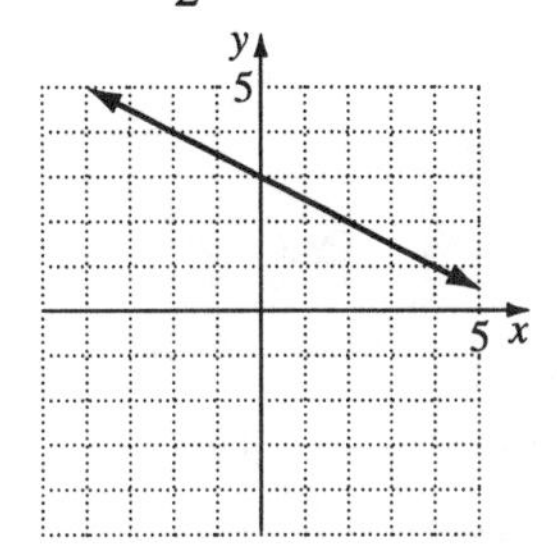

117. $|-5|=5$
$|-8|=8$
$|-5|<|-8|$

118. $7-2x+5x=-2(4-3x)$
$7+3x=-8x+6x$
$7=-2x-3x$
$7=-5x$
$-\frac{7}{5}=x$
The solution is $-\frac{7}{5}$.

Problem Set 6.4

1. $x^2+2xy-y^2$
$=(4)^2+2(4)(-3)-(-3)^2$
$=16-24-9=-17$

3. $4xy^3+x^2y^2-3y+6$
$=4(4)(-3)^3+(4)^2(-3)^2-3(-3)+6$
$=4(4)(-27)+16(9)-3(-3)+6$
$=-432+144+9+6=-273$

5. $yz-2xy+4xz$
$=(3)(-2)-2(-1)(3)+4(-1)(-2)$
$=-6+6+8=8$

7. $x^3y+4x^2yz-3xyz^2$
$=(-1)^3(3)+4(-1)^2(3)(-2)-3(-1)(3)(-2)^2$
$=(-1)(3)+4(1)(3)(-2)-3(-1)(3)(4)$
$=-3-24+36=9$

9.

Term	Coefficient	Degree
x^3y^2	1	$3+2=5$
$-5x^2y^7$	-5	$2+7=9$
$6y^2$	6	2
-3	-3	0

The degree of the polynomial is 9.

11.

Term	Coefficient	Degree
$4x^2yz$	4	$2+1+1=4$
$-5xyz$	-5	$1+1+1=3$
$12z^3$	12	3

The degree of the polynomial is 4.

13. $(5x^2y-3xy)+(2x^2y-xy)$
$=(5x^2y+2x^2y)+(-3xy-xy)$
$=7x^2y-4xy$

15. $(4y^2z+8yz+11)+(-2y^2z+5yz+2)$
$=(4y^2z-2y^2z)+(8yz+5yz)+(11+2)$
$=2y^2z+13yz+13$

17. $(x^3+7xy-5y^2)-(6x^3-xy+4y^2)$
$=(x^3+7xy-5y^2)+(-6x^3+xy-4y^2)$
$=(x^3-6x^3)+(7xy+xy)+(-5y^2-4y^2)$
$=-5x^3+8xy-9y^2$

19. $(3a^4b^2+5a^3b-3b)$
$-(2a^4b^2-3a^3b-4b+6a)$
$=(3a^4b^2+5a^3b-3b)$
$+(-2a^4b^2+3a^3b+4b-6a)$
$=(3a^4b^2-2a^4b^2)+(5a^3b+3a^3b)$
$+(-3b+4b)-6a$
$=a^4b^2+8a^3b+b-6a$

21. Add:

$$\begin{array}{r} 5x^2y^2-4xy^2+6y^2 \\ \underline{-8x^2y^2+5xy^2-\ \ y^2} \\ -3x^2y^2+\ \ xy^2+5y^2 \end{array}$$

23. Subtract:

$$\begin{array}{r} 3a^2b^4-5ab^2+7ab \\ \underline{-(-5a^2b^4-8ab^2-\ \ ab)} \end{array}$$

Add:

$$\begin{array}{r} 3a^2b^4-5ab^2+7ab \\ \underline{5a^2b^4+8ab^2+\ \ ab} \\ 8a^2b^4+3ab^2+8ab \end{array}$$

25. $(7a+13b)+(-26a+19b)-(11a-5b)$
$=(7a-26a)+(13b+19b)+(-11a+5b)$
$=(-19a+32b)+(-11a+5b)$
$=(-19a-11a)+(32b+5b)$
$=-30a+37b$

27. $(6x^2y)(3xy)=(6)(3)x^{2+1}y^{1+1}=18x^3y^2$

29. $(-7x^3y^4)(2x^2y^5)$
$=(-7)(2)x^{3+2}y^{4+5}$
$=-14x^5y^9$

31. $(-7a^{11}b^4c)(12a^3bc^5)$
$=(-7)(12)a^{11+3}b^{4+1}c^{1+5}$
$=-84a^{14}b^5c^6$

33. $5xy(2x+3y)$
$=5xy(2x)+5xy(3y)$
$=10x^2y+15xy^2$

35. $3ab^2(6a^2b^3+5ab)$
$=3ab^2(6a^2b^3)+3ab^2(5ab)$
$=18a^3b^5+15a^2b^3$

37. $-4y^2z(3y^3z^5-14y^2z+1)$
$=-4y^2z(3y^3z^5)-4y^2z(-14y^2z)-4y^2z(1)$
$=-12y^5z^6+56y^4z^2-4y^2z$

39. $(x+5y)(7x+3y)$
$=x(7x)+x(3y)+5y(7x)+5y(3y)$
$=7x^2+3xy+35xy+15y^2$
$=7x^2+38xy+15y^2$

41. $(a-3b)(2a+7b)$
$=a(2a)+a(7b)-3b(2a)-3b(7b)$
$=2a^2+7ab-6ab-21b^2$
$=2a^2+ab-21b^2$

43. $(xy+8)(xy-7)$
$=xy(xy)+xy(-7)+8(xy)+8(-7)$
$=x^2y^2-7xy+8xy-56$
$=x^2y^2+xy-56$

45. $(3ab-1)(5ab+2)$
$=3ab(5ab)+3ab(2)-1(5ab)-1(2)$
$=15a^2b^2+6ab-5ab-2$
$=15a^2b^2+ab-2$

47. $(7a+5b)^2$
$= (7a)^2 + 2(7a)(5b) + (5b)^2$
$= 49a^2 + 70ab + 25b^2$

49. $(x^2y^2 - 3)^2$
$= (x^2y^2)^2 - 2(x^2y^2)(3) + (-3)^2$
$= x^4y^4 - 6x^2y^2 + 9$

51. $(x^2 + y^2z^2)^2$
$= (x^2)^2 + 2(x^2)(y^2z^2) + (y^2z^2)^2$
$= x^4 + 2x^2y^2z^2 + y^4z^4$

53. $(x^2 + yz)(x^2 - yz)$
$= (x^2)^2 - (yz)^2$
$= x^4 - y^2z^2$

55. $(x - y)(x^2 + xy + y^2)$
$= x^3 + x^2y + xy^2 - x^2y - xy^2 - y^3$
$= x^3 - y^3$

57. $(a^2 - b^2)(a+b)$
$= a^2(a) + a^2(b) - b^2(a) - b^2(b)$
$= a^3 + a^2b - ab^2 - b^3$

59. $(m - n^3)(2m^3 + n)$
$= m(2m^3) + m(n) - n^3(2m^3) - n^3(n)$
$= 2m^4 + mn - 2m^3n^3 - n^4$

61. $(r^2 - s)(r^2 + s)$
$= (r^2)^2 - (s)^2$
$= r^4 - s^2$

63. $(xy + ab)(xy - ab)$
$= (xy)^2 - (ab)^2$
$= x^2y^2 - a^2b^2$

65. $(x^2 + 1)(x^4y + x^2 + 1)$
$= x^2(x^4y) + x^2(x^2) + x^2(1)$
$\quad +1(x^4y) + 1(x^2) + 1(1)$
$= x^6y + x^4 + x^2 + x^4y + x^2 + 1$
$= x^6y + x^4 + 2x^2 + x^4y + 1$

67. $N = \dfrac{x^2y - 8xy + 16y}{4}$
$N = \dfrac{(10)^2(16) - 8(10)(16) + 16(16)}{4}$
$N = \dfrac{1600 - 1280 + 256}{4}$
$N = \dfrac{576}{4}$
$N = 144$
144 board feet from 1 tree
$20(144) = 2880$
2880 board feet from 20 trees
Yes, it is enough for the job.

69. $2\pi rh + 2\pi r^2$
$= 2\pi(2)(4) + 2\pi(2)^2$
$= 16\pi + 8\pi$
$= 24\pi$
$\approx 24(3.14) = 75.36$ square inches

71. $(3x + 5y)(x + y)$
$= 3x(x) + 3x(y) + 5y(x) + 5y(y)$
$= 3x^2 + 3xy + 5xy + 5y^2$
$= 3x^2 + 8xy + 5y^2$

73. $(8a - 3b)(8a + 3b)$
$= (8a)^2 - (3b)^2$
$= 64a^2 - 9b^2$

75. $(a + b)(a + b)$
$= (a+b)^2$
$= a^2 + 2ab + b^2$

77. c is true
$(2x + 3 - 5y)(2x + 3 + 5y)$
$= 4x^2 + 6x + 10xy + 6x + 9$
$\quad +15y - 10xy - 15y - 25y^2$
$= 4x^2 + 12x + 9 - 25y^2$

For problems 79-81, a TI-92 or equivalent calculator will be necessary.

79. $z = x^2 + y^2 - 2x - 6y + 14$

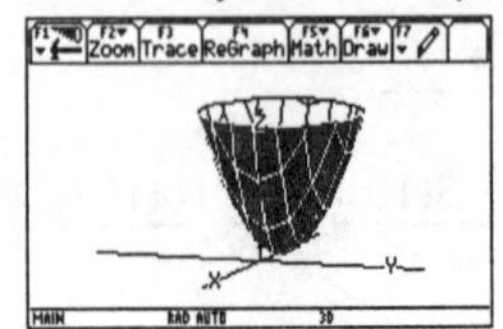

81. $z = \dfrac{x^3y - y^3x}{390}$

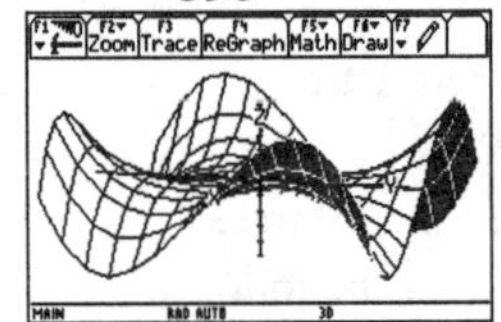

83. Answers may vary.

85. $2b(4b) - a(2a)$
$= 8b^2 - 2a^2$

87. $a^2 - (a - 2b)^2$
$= a^2 - (a^2 - 4ab + 4b^2)$
$= a^2 - a^2 + 4ab - 4b^2$
$= 4ab - 4b^2$

89. $a(2b) + \frac{1}{2}\pi b^2 = 2ab + \frac{1}{2}\pi b^2$

91. $\frac{1}{2}(2a)(7a + 4a) - \frac{1}{2}(3b)(2a)$
$= a(11a) - 3ab$
$= 11a^2 - 3ab$

Review Problems

93. $y = 5 - x$
$4x + 5y = 22$
$4x + 5(5 - x) = 22$
$4x + 25 - 5x = 22$
$25 - x = 22$
$25 - 22 = x$
$3 = x$
$y = 5 - 3 = 2$
$(3, 2)$

94. $A = \pi r^2 + 2\pi rh$
$A - \pi r^2 = 2\pi rh$
$\dfrac{A - \pi r^2}{2\pi r} = h$

95. $\frac{1}{3}x + \frac{2}{3} = \frac{1}{4}x - \frac{3}{4}$
$\frac{4}{4}\left(\frac{1}{3}x + \frac{2}{3}\right) = \frac{3}{3}\left(\frac{1}{4}x - \frac{3}{4}\right)$
$\frac{4}{12}x + \frac{8}{12} = \frac{3}{12}x - \frac{9}{12}$
$\frac{4}{12}x - \frac{3}{12}x = -\frac{9}{12} - \frac{8}{12}$
$\frac{1}{12}x = -\frac{17}{12}$
$x = -17$

Problem Set 6.5

1. $\dfrac{x^5}{x^2} = x^{5-2} = x^3$

3. $\dfrac{z^{13}}{z^5} = z^{13-5} = z^8$

5. $\dfrac{30y^{10}}{10y^5} = 3y^{10-5} = 3y^5$

7. $\dfrac{-8x^{22}}{4x^2} = -2x^{22-2} = -2x^{20}$

9. $\dfrac{-9a^8}{18a^5} = -\dfrac{1}{2}a^{8-5} = -\dfrac{a^3}{2}$

11. $\dfrac{7x^{17}}{5x^5} = \dfrac{7x^{17-5}}{5} = \dfrac{7}{5}x^{12}$

13. $7^0 = 1$

15. $-3^0 = -1(3^0) = -1(1) = -1$

17. $(-3)^0 = 1$

19. $4x^0 = 4(1) = 4$

21. $(4x)^0 = 1$

23. $-5^0 + (-5)^0 = -1 + 1 = 0$

25. $\left(\frac{x}{3}\right)^2 = \frac{x^2}{3^2} = \frac{x^2}{9}$

27. $\left(\frac{x^2}{4}\right)^3 = \frac{x^{2\cdot 3}}{4^3} = \frac{x^6}{64}$

29. $\left(\frac{2x^3}{5}\right)^2 = \frac{2^2 \cdot x^{3\cdot 2}}{5^2} = \frac{4x^6}{25}$

31. $\left(\frac{-3a^3}{4}\right)^3 = \frac{(-3)^3 \cdot a^{3\cdot 3}}{4^3} = -\frac{27a^9}{64}$

33. $\frac{6x^4 + 2x^3}{2} = \frac{6x^4}{2} + \frac{2x^3}{2} = 3x^4 + x^3$

35. $\frac{6x^4 - 2x^3}{2x} = \frac{6x^4}{2x} - \frac{2x^3}{2x} = 3x^3 - x^2$

37. $\frac{y^5 - 3y^2 + y}{y}$
$= \frac{y^5}{y} - \frac{3y^2}{y} + \frac{y}{y}$
$= y^4 - 3y + 1$

39. $\frac{15x^3 - 24x^2}{-3x}$
$= \frac{15x^3}{-3x} + \frac{-24x^2}{-3x}$
$= -5x^2 + 8x$

41. $\frac{18x^5 + 6x^4 + 9x^3}{3x^2}$
$= \frac{18x^5}{3x^2} + \frac{6x^4}{3x^2} + \frac{9x^3}{3x^2}$
$= 6x^3 + 2x^2 + 3x$

43. $\frac{12x^4 - 8x^3 + 40x^2}{4x}$
$= \frac{12x^4}{4x} - \frac{8x^3}{4x} + \frac{40x^2}{4x}$
$= 3x^3 - 2x^2 + 10x$

45. $(4x^2 - 6x) \div x = \frac{4x^2}{x} - \frac{6x}{x} = 4x - 6$

47. $\frac{30z^3 + 10z^2}{-5z} = \frac{30z^3}{-5z} + \frac{10z^2}{-5z} = -6z^2 - 2z$

49. $\frac{8x^3 + 3x^2 - 2x}{2x}$
$= \frac{8x^3}{2x} + \frac{3x^2}{2x} - \frac{2x}{2x} = 4x^2 + \frac{3}{2}x - 1$

51. $\frac{25x^7 - 15x^5 - 5x^4}{5x^3}$
$= \frac{25x^7}{5x^3} - \frac{15x^5}{5x^3} - \frac{5x^4}{5x^3}$
$= 5x^4 - 3x^2 - x$

53. $\frac{18x^7 - 9x^6 + 20x^5 - 10x^4}{-2x^4}$
$= \frac{18x^7}{-2x^4} - \frac{9x^6}{-2x^4} + \frac{20x^5}{-2x^4} - \frac{10x^4}{-2x^4}$
$= -9x^3 + \frac{9}{2}x^2 - 10x + 5$

55. $\frac{12x^2y^2 + 6x^2y - 15xy^2}{3xy}$
$= \frac{12x^2y^2}{3xy} + \frac{6x^2y}{3xy} - \frac{15xy^2}{3xy}$
$= 4xy + 2x - 5y$

57. $\frac{20x^7y^4 - 15x^3y^2 - 10x^2y}{-5x^2y}$
$= \frac{20x^7y^4}{-5x^2y} + \frac{-15x^3y^2}{-5x^2y} + \frac{-10x^2y}{-5x^2y}$
$= -4x^5y^3 + 3xy + 2$

59. d is true;
Example:
$$\frac{x^6+x^4}{x^2}=\frac{x^6}{x^2}+\frac{x^4}{x^2}=x^4-x^2$$

61. Graph $y=\frac{6x-7}{3}$ and $y=2x-\frac{7}{3}$
They are the same graph.

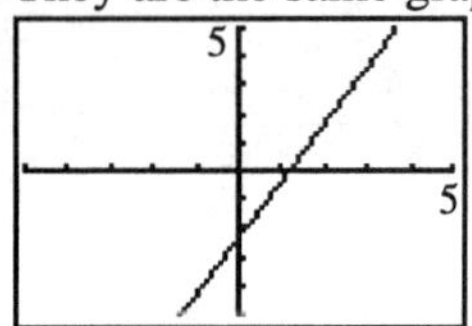

63. a. $\frac{x+2}{2}=\frac{x}{2}+\frac{2}{2}=\frac{x}{2}+1$

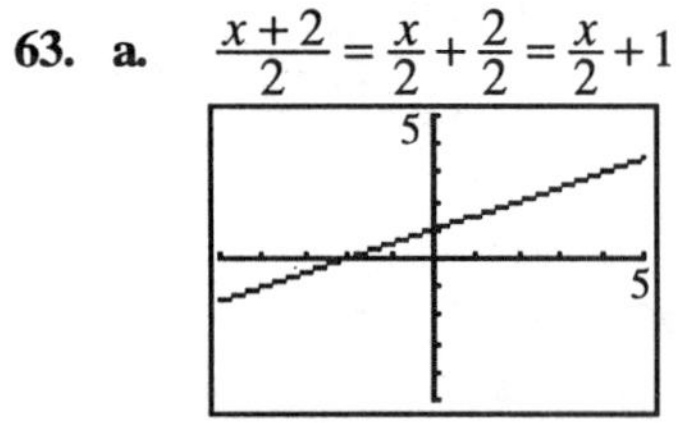

b. $\frac{x^2+2x}{x}=\frac{x^2}{x}+\frac{2x}{x}=x+2$

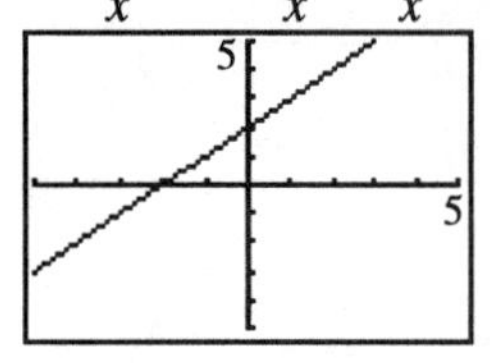

c. $\frac{x+2}{x}=\frac{x}{x}+\frac{2}{x}=1+\frac{2}{x}$

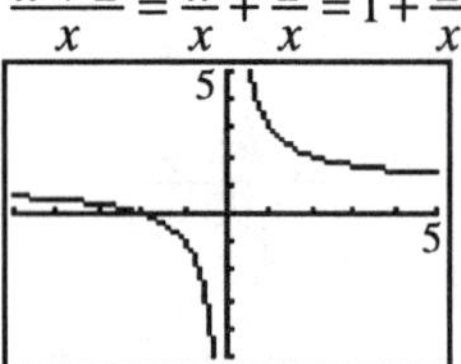

d. $\frac{x^6}{x^2}=x^{6-2}=x^4$

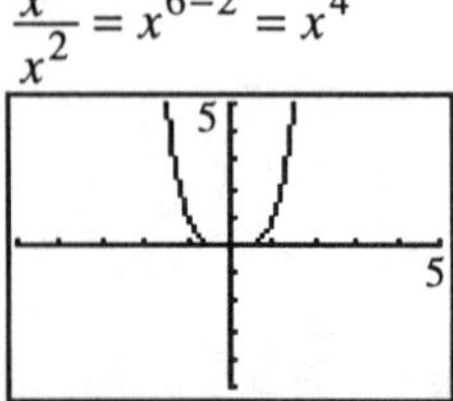

65. Answers may vary.

67. $\frac{6y^3(3y-1)+5y^2(6y-3)}{3y}$
$=\frac{18y^4-6y^3+30y^3-15y^2}{3y}$
$=\frac{18y^4+24y^3-15y^2}{3y}$
$=6y^3+8y^2-5y$

69. Let y equal the polynomial.
$\frac{y}{3x^2}=6x^6-9x^4+12x^2$
$y=(3x^2)(6x^6-9x^4+12x^2)$
$=18x^8-27x^6+36x^4$

71. $\frac{8x^4+4x^3+10x^2}{4x^2}$
$=\frac{8x^4}{4x^2}+\frac{4x^3}{4x^2}+\frac{10x^2}{4x^2}$
$=2x^2+x+\frac{5}{2}$
4, 2

73. $\frac{3x^{14}-6x^{12}-9x^7}{-3x^7}$
$=\frac{3x^{14}}{-3x^7}+\frac{-6x^{12}}{-3x^7}+\frac{-9x^7}{-3x^7}$
$=-x^7+2x^5+3$
9, –3, 7

Review Problems

74. $2x+y=11$
$x=18-3y$

$2(18-3y)+y=11$
$36-6y+y=11$
$36-5y=11$
$-5y=-25$
$y=5$
$x=18-3(5)=3$
(3, 5)

75. $2x - 3y > 6$

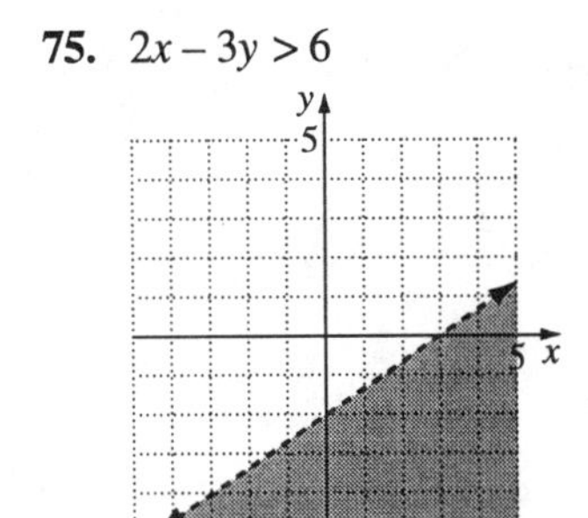

76. $R = \dfrac{L+3W}{2}$

$2R = L + 3W$

$2R - L = 3W$

$\dfrac{2R-L}{3} = W$

Problem Set 6.6

1. $\dfrac{x^2+6x+8}{x+2} = x+4$

$$\begin{array}{r}
x+4 \\
x+2\overline{\big)\,x^2+6x+8} \\
\underline{x^2+2x} \\
4x+8 \\
\underline{4x+8} \\
0
\end{array}$$

3. $\dfrac{2x^2+x-10}{x-2} = 2x+5$

$$\begin{array}{r}
2x+5 \\
x-2\overline{\big)\,2x^2+\ x-10} \\
\underline{2x^2-4x} \\
5x-10 \\
\underline{5x-10} \\
0
\end{array}$$

5. $\dfrac{x^2-5x+6}{x-3} = x-2$

$$\begin{array}{r}
x-2 \\
x-3\overline{\big)\,x^2-5x+6} \\
\underline{x^2-3x} \\
-2x+6 \\
\underline{-2x+6} \\
0
\end{array}$$

7. $\dfrac{2y^2+5y+2}{y+2} = 2y+1$

$$\begin{array}{r}
2y+1 \\
y+2\overline{\big)\,2y^2+5y+2} \\
\underline{2y^2+4y} \\
y+2 \\
\underline{y+2} \\
0
\end{array}$$

9. $\dfrac{x^2-5x+8}{x-3} = x-2+\dfrac{2}{x-3}$

$$\begin{array}{r}
x-2 \\
x-3\overline{\big)\,x^2-5x+8} \\
\underline{x^2-3x} \\
-2x+8 \\
\underline{-2x+6} \\
2
\end{array}$$

11. $\dfrac{5y+10+y^2}{y+2} = \dfrac{y^2+5y+10}{y+2} = y+3+\dfrac{4}{y+2}$

$$\begin{array}{r}
y+3 \\
y+2\overline{\big)\,y^2+5y+10} \\
\underline{y^2+2y} \\
3y+10 \\
\underline{3y+\ 6} \\
4
\end{array}$$

13. $\dfrac{x^3-6x^2+7x-2}{x-1} = x^2-5x+2$

$$\begin{array}{r}
x^2-5x+2 \\
x-1\overline{\big)\,x^3-6x^2+7x-2} \\
\underline{x^3-\ x^2} \\
-5x^2+7x \\
\underline{-5x^2+5x} \\
2x-2 \\
\underline{2x-2} \\
0
\end{array}$$

15. $\dfrac{12y^2-20y+3}{2y-3}=6y-1$

$$\begin{array}{r|l} & 6y-1 \\ \hline 2y-3 & 12y^2-20y+3 \\ & \underline{12y^2-18y} \\ & \quad -2y+3 \\ & \quad \underline{-2y+3} \\ & \qquad 0 \end{array}$$

17. $\dfrac{4a^2+4a-3}{2a-1}=2a+3$

$$\begin{array}{r|l} & 2a+3 \\ \hline 2a-1 & 4a^2+4a-3 \\ & \underline{4a^2-2a} \\ & \quad 6a-3 \\ & \quad \underline{6a-3} \\ & \qquad 0 \end{array}$$

19. $\dfrac{3y-y^2+2y^3+2}{2y+1}$

$=\dfrac{2y^3-y^2+3y+2}{2y+1}=y^2-y+2$

$$\begin{array}{r|l} & y^2-y+2 \\ \hline 2y+1 & 2y^3-y^2+3y+2 \\ & \underline{2y^3+y^2} \\ & \quad -2y^2+3y \\ & \quad \underline{-2y^2-\ y} \\ & \qquad 4y+2 \\ & \qquad \underline{4y+2} \\ & \qquad\quad 0 \end{array}$$

21. $\dfrac{2x^2-9x+8}{2x+3}=x-6+\dfrac{26}{2x+3}$

$$\begin{array}{r|l} & x-6 \\ \hline 2x+3 & 2x^2-9x+8 \\ & \underline{2x^2+3x} \\ & \quad -12x+\ 8 \\ & \quad \underline{-12x-18} \\ & \qquad 26 \end{array}$$

23. $\dfrac{x^3+4x-3}{x-2}=x^2+2x+8+\dfrac{13}{x-2}$

$$\begin{array}{r|l} & x^2+2x+8 \\ \hline x-2 & x^3+0x^2+4x-3 \\ & \underline{x^3-2x^2} \\ & \quad 2x^2+4x \\ & \quad \underline{2x^2-4x} \\ & \qquad 8x-\ 3 \\ & \qquad \underline{8x-16} \\ & \qquad\quad 13 \end{array}$$

25. $\dfrac{4y^3+8y^2+5y+9}{2y+3}=2y^2+y+1+\dfrac{6}{2y+3}$

$$\begin{array}{r|l} & 2y^2+y+1 \\ \hline 2y+3 & 4y^3+8y^2+5y+9 \\ & \underline{4y^3+6y^2} \\ & \quad 2y^2+5y \\ & \quad \underline{2y^2+3y} \\ & \qquad 2y+9 \\ & \qquad \underline{2y+3} \\ & \qquad\quad 6 \end{array}$$

27. $\dfrac{6y^3-5y^2+5}{3y+2}=2y^2-3y+2+\dfrac{1}{3y+2}$

$$\begin{array}{r|l} & 2y^2-3y+2 \\ \hline 3y+2 & 6y^3-5y^2+0y+5 \\ & \underline{6y^3+4y^2} \\ & \quad -9y^2+0y \\ & \quad \underline{-9y^2-6y} \\ & \qquad 6y+5 \\ & \qquad \underline{6y+4} \\ & \qquad\quad 1 \end{array}$$

29. $\dfrac{27x^3-1}{3x-1}=9x^2+3x+1$

$$\begin{array}{r|l} & 9x^2+3x+1 \\ \hline 3x-1 & 27x^3+0x^2+0x-1 \\ & \underline{27x^3-9x^2} \\ & \quad 9x^2+0x \\ & \quad \underline{9x^2-3x} \\ & \qquad 3x-1 \\ & \qquad \underline{3x-1} \\ & \qquad\quad 0 \end{array}$$

31. $\dfrac{81-12y^3+54y^2+y^4-108y}{y-3}$

$=\dfrac{y^4-12y^3+54y^2-108y+81}{y-3}$

$=y^3-9y^2+27y-27$

$$
\begin{array}{r}
y^3-9y^2+27y-27 \\
y-3\overline{)\,y^4-12y^3+54y^2-108y+81} \\
\underline{y^4-\ 3y^3} \qquad\qquad\qquad\qquad \\
-9y^3+54y^2 \qquad\qquad\quad \\
\underline{-9y^3+27y^2} \qquad\qquad\quad \\
27y^2-108y \qquad \\
\underline{27y^2-\ 81y} \qquad \\
-27y+81 \\
\underline{-27y+81} \\
0
\end{array}
$$

33. $\dfrac{4y^2+6y}{2y-1}=2y+4+\dfrac{4}{2y-1}$

$$
\begin{array}{r}
2y+4 \\
2y-1\overline{)\,4y^2+6y+0} \\
\underline{4y^2-2y} \qquad \\
8y+0 \\
\underline{8y-4} \\
4
\end{array}
$$

35. $\dfrac{y^4-2y^2+5}{y-1}=y^3+y^2-y-1+\dfrac{4}{y-1}$

$$
\begin{array}{r}
y^3+y^2-y-1 \\
y-1\overline{)\,y^4+0y^3-2y^2+0y+5} \\
\underline{y^4-\ y^3} \qquad\qquad\qquad\qquad \\
y^3-2y^2 \qquad\qquad\quad \\
\underline{y^3-\ y^2} \qquad\qquad\quad \\
-y^2+0y \qquad \\
\underline{-y^2+y} \qquad \\
-y+5 \\
\underline{-y+1} \\
4
\end{array}
$$

37. $\dfrac{y^4-4y^3+5y^2-3y+2}{y^2+3}$

$=y^2-4y+2+\dfrac{9y-4}{y^2+3}$

$$
\begin{array}{r}
y^2-4y+2 \\
y^2+3\overline{)\,y^4-4y^3+5y^2-3y+2} \\
\underline{y^4+0y^3+3y^2} \qquad\qquad \\
-4y^3+2y^2-3y \qquad \\
\underline{-4y^3+0y^2-12y} \qquad \\
2y^2+\ 9y+2 \\
\underline{2y^2+\ 0y+6} \\
9y-4
\end{array}
$$

39. $\dfrac{x^3+3x^2+5x+3}{x+1}=x^2+2x+3$ hours

$$
\begin{array}{r}
x^2+2x+3 \\
x+1\overline{)\,x^3+3x^2+5x+3} \\
\underline{x^3+\ x^2} \qquad\qquad\quad \\
2x^2+5x \qquad \\
\underline{2x^2+2x} \qquad \\
3x+3 \\
\underline{3x+3} \\
0
\end{array}
$$

41. b is true.

43. a. Yes;

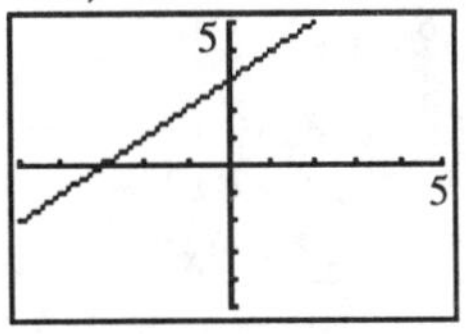

b. y is undefined for y_1, $y=2$ for y_2

c. $x=-1$

45. $\dfrac{2x^2+9x-35}{x+7}=2x-5$

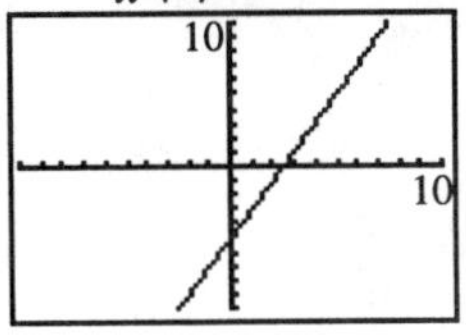

47. $\dfrac{6x^3+14x^2+10x+3}{3x+1}$

$= 2x^2+4x+2+\dfrac{1}{3x+1}$

49–51. Answers may vary.

53. Let y equal the polynomial.

$\dfrac{y}{2x+4} = x-3+\dfrac{17}{2x+4}$

$y = (2x+4)\left(x-3+\dfrac{17}{2x+4}\right)$

$y = (2x+4)(x-3)+(2x+4)\left(\dfrac{17}{2x+4}\right)$

$= 2x^2-2x-12+17$

$= 2x^2-2x+5$

55. Answers may vary.

$\dfrac{x^7-1}{x+1}$

$= x^6-x^5+x^4-x^3+x^2-x+1-\dfrac{2}{x+1}$

$$
\begin{array}{r}
x^6-x^5+x^4-x^3+x^2-x+1 \\
x+1\overline{\smash{)}x^7+0x^6+0x^5+0x^4+0x^3+0x^2+0x-1} \\
\underline{x^7+x^6} \qquad\qquad\qquad\qquad\qquad\qquad\qquad\quad \\
-x^6+0x^5 \qquad\qquad\qquad\qquad\qquad\qquad \\
\underline{-x^6-\ x^5} \qquad\qquad\qquad\qquad\qquad\qquad \\
x^5+0x^4 \qquad\qquad\qquad\qquad\qquad \\
\underline{x^5+\ x^4} \qquad\qquad\qquad\qquad\qquad \\
-x^4+0x^3 \qquad\qquad\qquad\qquad \\
\underline{-x^4-\ x^3} \qquad\qquad\qquad\qquad \\
x^3+0x^2 \qquad\qquad\qquad \\
\underline{x^3+\ x^2} \qquad\qquad\qquad \\
-x^2+0x \qquad\qquad \\
\underline{-x^2-\ x} \qquad\qquad \\
x-1 \\
\underline{x+1} \\
-2
\end{array}
$$

Review Problems

56. $x = 3-5y$

$x-2y = 10$

$$
\begin{array}{r}
x+5y=3 \\
\underline{-x+2y=-10} \\
7y=-7 \\
y=-1
\end{array}
$$

$x = 3-5y = 3+5 = 8$

$(8, -1)$

57. $52t+58t = 385$

$110t = 385$

$t = 3.5$

3.5 hours

58. $y \geq -2x+3$

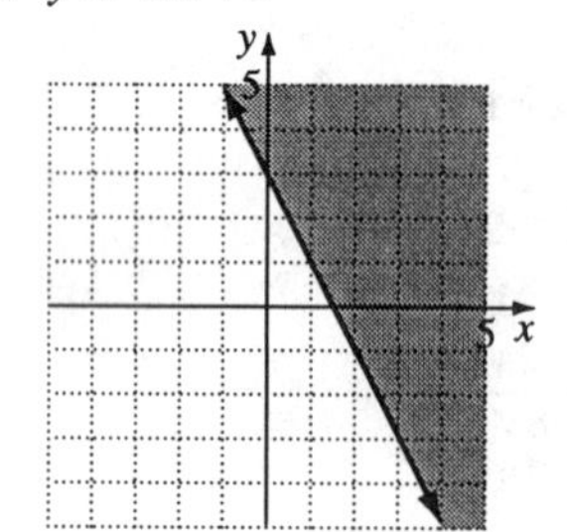

Problem Set 6.7

1. $5^{-2} = \dfrac{1}{5^2} = \dfrac{1}{25}$

3. $5^{-3} = \dfrac{1}{5^3} = \dfrac{1}{125}$

5. $\dfrac{1}{3^{-2}} = 3^2 = 9$

7. $2^{-1}+3^{-1} = \dfrac{1}{2}+\dfrac{1}{3} = \dfrac{5}{6}$

9. $\left(\dfrac{1}{4}\right)^{-2} = \dfrac{1}{\left(\frac{1}{4}\right)^2} = 4^2 = 16$

11. $-4^{-2} = -\dfrac{1}{4^2} = -\dfrac{1}{16}$

13. $(-4)^{-2} = \frac{1}{(-4)^2} = \frac{1}{16}$

15. $\frac{2^{-3}}{8^{-2}} = \frac{8^2}{2^3} = \frac{64}{8} = 8$

17. $\frac{3}{(-5)^{-3}} = 3(-5)^3 = 3(-125) = -375$

19. $\frac{x^3}{x^9} = \frac{1}{x^{9-3}} = \frac{1}{x^6}$

21. $\frac{z^5}{z^{13}} = \frac{1}{z^{13-5}} = \frac{1}{z^8}$

23. $\frac{30y^5}{10y^{10}} = \frac{3}{y^{10-5}} = \frac{3}{y^5}$

25. $\frac{-8x^3}{2x^7} = \frac{-4}{x^{7-4}} = \frac{-4}{x^3}$

27. $\frac{-9a^5}{27a^8} = -\frac{1}{3a^3}$

29. $\frac{7w^5}{5w^{13}} = \frac{7}{5w^8}$

31. $\frac{15a^5b^3}{5a^2b^7} = \frac{3a^3}{b^4}$

33. $\frac{-20x^4y^7}{4x^2y^{13}} = \frac{-5x^2}{y^6}$

35. $\frac{-20xy^3z^4}{60x^4yz^{11}} = \frac{-y^2}{3x^3z^7}$

37. $\frac{6x^4 - 8x^3 + 20x}{2x^2}$

$$= \frac{6x^4}{2x^2} - \frac{8x^3}{2x^2} + \frac{20x}{2x^2}$$

$$= 3x^2 - 4x + \frac{10}{x}$$

39. $\frac{8y^4 - 20y^3 - 10y^2 + 8y - 6}{2y}$

$$= \frac{8y^4}{2y} - \frac{20y^3}{2y} - \frac{10y^2}{2y} + \frac{8y}{2y} - \frac{6}{2y}$$

$$= 4y^3 - 10y^2 - 5y + 4 - \frac{3}{y}$$

41. $\frac{x^6 - x^4 + 2x^3 - 5x^2 + 9x}{x^3}$

$$= \frac{x^6}{x^3} - \frac{x^4}{x^3} + \frac{2x^3}{x^3} - \frac{5x^2}{x^3} + \frac{9x}{x^3}$$

$$= x^3 - x + 2 - \frac{5}{x} + \frac{9}{x^2}$$

43. $\frac{8x^8 - 12x^4 - 16x^3 + 20x}{4x^4}$

$$= \frac{8x^8}{4x^4} - \frac{12x^4}{4x^4} - \frac{16x^3}{4x^4} + \frac{20x}{4x^4}$$

$$= 2x^4 - 3 - \frac{4}{x} + \frac{5}{x^3}$$

45. $\frac{9x^2y^2 + 3x^2y - 6x^3y^2}{3x^2y^3}$

$$= \frac{9x^2y^2}{3x^2y^3} + \frac{3x^2y}{3x^2y^3} - \frac{6x^3y^2}{3x^2y^3}$$

$$= \frac{3}{y} + \frac{1}{y^2} - \frac{2x}{y}$$

47. $\frac{4a^4b - 12a^6b^2 + 8a^8b^6}{-4a^4b}$

$$= \frac{4a^4b}{-4a^4b} - \frac{12a^6b^2}{-4a^4b} + \frac{8a^8b^6}{-4a^4b}$$

$$= -1 + 3a^2b - 2a^4b^5$$

49. $x^{-8} \cdot x^3 = x^{-8+3} = x^{-5} = \frac{1}{x^5}$

51. $(4x^{-5})(2x^2) = 8x^{-5+2} = 8x^{-3} = \frac{8}{x^3}$

53. $\frac{z^3}{(z^4)^2} = \frac{z^3}{z^8} = z^{3-8} = z^{-5} = \frac{1}{z^5}$

55. $\frac{z^{-3}}{(z^4)^2} = \frac{z^{-3}}{z^8} = z^{-3-8} = z^{-11} = \frac{1}{z^{11}}$

57. $\frac{(4x^3)^2}{x^8} = \frac{16x^6}{x^8} = 16x^{6-8} = 16x^{-2} = \frac{16}{x^2}$

59. $\frac{(6a^4)^3}{a^{-5}} = \frac{6^3a^{12}}{a^{-5}} = 216a^{12+5} = 216a^{17}$

61. $\left(\frac{y^4}{y^2}\right)^{-3} = (y^{4-2})^{-3} = (y^2)^{-3} = y^{-6} = \frac{1}{y^6}$

63. $\left(\frac{4x^5}{2x^2}\right)^{-4}$
$= (2x^{5-2})^{-4}$
$= 2^{-4}(x^3)^{-4} = 2^{-4}x^{-12} = \frac{1}{16x^{12}}$

65. $(-2z^{-1})^{-2} = (-2)^{-2}z^{(-1)(-2)} = \frac{z^2}{4}$

67. $\frac{2x^5 \cdot 3x^7}{15x^6} = \frac{2x^{5+7-6}}{5} = \frac{2x^6}{5}$

69. $(x^3)^5x^{-7} = x^{15}x^{-7} = x^{15-7} = x^8$

71. $(2y^3)^4y^{-6} = 2^4y^{12}y^{-6} = 16y^{12-6} = 16y^6$

73. $\frac{(y^3)^4}{(y^2)^7} = \frac{y^{12}}{y^{14}} = y^{12-14} = y^{-2} = \frac{1}{y^2}$

75. $(a^4b^5)^{-3} = a^{-12}b^{-15} = \frac{1}{a^{12}b^{15}}$

77. $(a^{-2}b^{-6})^{-4} = a^8b^{24}$

79. $(a^3b^{-4}c^{-5})(a^{-2}b^{-4}c^9)$
$= a^{3-2}b^{-4-4}c^{-5+9}$
$= a^1b^{-8}c^4$
$= \frac{ac^4}{b^8}$

81. $\left(\frac{x^2}{y^3}\right)^{-2} = \frac{x^{-4}}{y^{-6}} = \frac{y^6}{x^4}$

83. $\left(\frac{2m^2}{3n^4}\right)^{-3} = \frac{2^{-3}m^{-6}}{3^{-3}n^{-12}} = \frac{3^3n^{12}}{2^3m^6} = \frac{27n^{12}}{8m^6}$

85. $2.7 \times 10^2 = 270$ (move decimal point 2 places right)

87. $9.12 \times 10^5 = 912,000$ (move right 5)

89. $3.4 \times 10^0 = 3.4$

91. $7.9 \times 10^{-1} = 0.79$ (move left 1)

93. $2.15 \times 10^{-2} = 0.0215$ (move left 2)

95. $7.86 \times 10^{-4} = 0.000786$ (move left 4)

97. $32,400 = 3.24 \times 10^4$

99. $220,000,000 = 2.2 \times 10^8$

101. $713 = 7.13 \times 10^2$

103. $6751 = 6.751 \times 10^3$

105. $0.0027 = 2.7 \times 10^{-3}$

107. $0.000\,020\,2 = 2.02 \times 10^{-5}$

109. $0.005 = 5 \times 10^{-3}$

111. $3.141\,59 = 3.14159 \times 10^0$

113. $(2 \times 10^3)(3 \times 10^2)$
$= 6 \times 10^{3+2} = 6 \times 10^5 = 600,000$

115. $(2 \times 10^5)(8 \times 10^3)$
$= 16 \times 10^{5+3} = 1.6 \times 10 \times 10^8$
$= 1.6 \times 10^9 = 1,600,000,000$

117. $\frac{12 \times 10^6}{4 \times 10^2} = 3 \times 10^{6-2} = 3 \times 10^4 = 30,000$

119. $\frac{15\times 10^4}{5\times 10^{-2}} = 3\times 10^{4+2} = 3\times 10^6 = 3{,}000{,}000$

121. $\frac{15\times 10^{-4}}{5\times 10^2} = 3\times 10^{-4-2} = 3\times 10^{-6}$
$= 0.000\ 003$

123. $\frac{180\times 10^6}{2\times 10^3} = 90\times 10^{6-3} = 90\times 10^3$
$= 90{,}000$

125. $\frac{3\times 10^4}{12\times 10^{-3}} = 0.25\times 10^{4+3} = 0.25\times 10^7$
$= 2{,}500{,}000$

127. $(5\times 10^2)^3 = 5^3\times 10^{2(3)} = 125\times 10^6$
$= 125{,}000{,}000$

129. $(3\times 10^{-2})^4 = 3^4\times 10^{-2(4)} = 81\times 10^{-8}$
$= 0.000\ 000\ 81$

131. $(4\times 10^6)^{-1} = 4^{-1}\times 10^{6(-1)} = 0.25\times 10^{-6}$
$= 0.000\ 000\ 25$

133. $9.29\times 10^7 = 92{,}900{,}000$

135. $4\times 10^{-5} = 0.00004$

137. $650{,}000 = 6.5\times 10^5$

139. $9230 = 9.230\times 10^3$

141. $0.000\ 000\ 000\ 000\ 000\ 7 = 7\times 10^{-16}$

143. $0.000\ 001\ 54 = 1.54\times 10^{-6}$

145. $\frac{3\times 10^{10}}{7.5\times 10^9} = 4$

147. a. $(2\times 10^4)(42) = 8.4\times 10^5$ km

b. $\frac{8.4\times 10^5}{4\times 10^4} = 21$ times

c. $4(2\times 10^4) = 8\times 10^4$ hours

d. $\frac{8\times 10^4}{365\times 24} \approx 9.1324$
9.1 years

e. 21, 9

149.

Year	Projected Spending ($)
1995	1.78×10^{11}
1996	1.99×10^{11}
1997	2.19×10^{11}
1998	2.40×10^{11}
1999	2.63×10^{11}
2000	2.88×10^{11}
2001	3.15×10^{11}
2002	3.45×10^{11}
2003	3.79×10^{11}
2004	4.16×10^{11}
2005	4.58×10^{11}

151. b is true
$5^{-2} > 2^{-5}$
$\frac{1}{5^2} > \frac{1}{2^5}$
$\frac{1}{25} > \frac{1}{32}$
$\frac{32}{(25)(32)} > \frac{25}{(25)(32)}$
True

153. a is true; $\frac{x^3}{x^7} = x^{3-7} = x^{-4} = \frac{1}{x^4}$.
True, for any nonzero real number x.

155. d is true;
$\frac{8\times 10^{-9}}{4\times 10^{-5}} = 2\times 10^{-9-(-5)} = 2\times 10^{-4}$.

157. Students should verify results.

159. Students should verify results.

161. Answers may vary.

163. $2^{-1}+2^{-2}=\frac{1}{2}+\frac{1}{4}=\frac{3}{4}$, or 75% already gathered. Percentage to be gathered $= 100-75 = 25\%$.

Review Problems

165. Let x = number of years.
$4000 + 200x = 9400$
$200x = 5400$
$x = 27$
27 years

166. $6(3-x) < 2x + 12$
$18 - 6x < 2x + 12$
$-8x < -6$
$x > \frac{3}{4}$

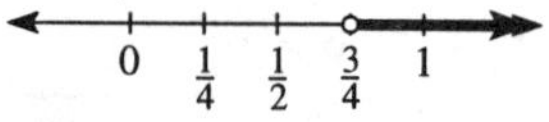

167. $(5x-2)(2x^2+3x-4)$
$= 5x(2x^2)+5x(3x)+5x(-4)-2(2x^2)-2(3x)-2(-4)$
$= 10x^3+15x^2-20x-4x^2-6x+8$
$= 10x^3+11x^2-26x+8$

Chapter 6 Review Problems

1. $7x^4+9x$ is a binomial of degree 4.

2. $3x+5x^2-2$ is a trinomial of degree 2.

3. $16x$ is a monomial of degree 1.

4. $f(x)=0.4x^2-40x+1039$
$f(20)=0.4(20)^2-40(20)+1039$
$f(20) = 399$
The average number of accidents per day in the U.S. involving drivers age 20 is 399.

5. a. $f(t)=-16t^2+16t+32$
$f(0)=-16(0)^2+16(0)+32=32$
$f(0.5)=-16(0.5)^2+16(0.5)+32=36$
$f(1)=-16(1)^2+16(1)+32=32$
$f(1.5)=-16(1.5)^2+16(1.5)+32=20$
$f(2)=-16(2)^2+16(2)+32=0$

b.

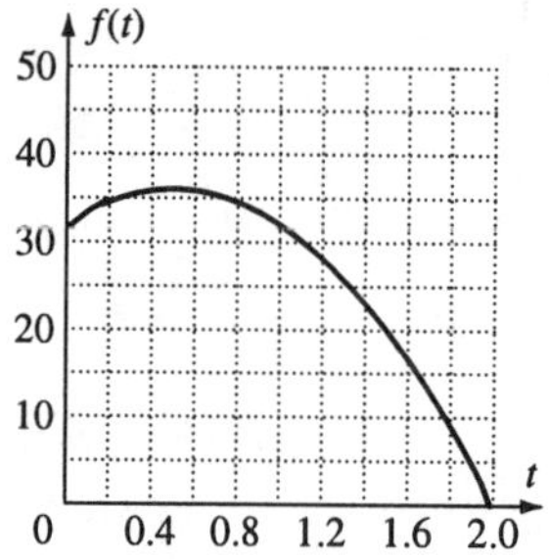

c. 0.5 seconds; 36 feet

d. 2 seconds; point on graph (2, 0)

e. Verify.

6. $(-6x^2+7x^2-9x+3)$
$+ (14x^3+3x^2-11x-7)$
$=(-6x^3+14x^3)+(7x^2+3x^2)$
$+(-9x-11x)+(3-7)$
$=8x^3+10x^2-20x-4$

7. $(-7a^2+4+9a^3)+(-13-8a^3+3a^2)$
$=(9a^3-8a^3)+(-7a^2+3a^2)+(4-13)$
$=a^3-4a^2-9$

8. $(5y^2-y-8)-(-6y^2+3y-4)$
$=5y^2-y-8+6y^2-3y+4$
$=(5y^2+6y^2)+(-y-3y)+(-8+4)$
$=11y^2-4y-4$

9. $(13x^4-8x^3+2x^2)-(5x^4-3x^3+2x^2-6)$
$=13x^4-8x^3+2x^2-5x^4+3x^3-2x^2+6$
$=(13x^4-5x^4)+(-8x^3+3x^3)$

$+(2x^2-2x^2)+6$
$=8x^4-5x^3+6$

10. $(-13x^4-6x^2+5x)-(x^4+7x^2-11x)$
$=-13x^4-6x^2+5x-x^4-7x^2+11x$
$=(-13x^4-x^4)+(-6x^2-7x^2)+(11x+5x)$
$=-14x^4-13x^2+16x$

11. Add:

$$\begin{array}{r} 7y^4-6y^3+4y^2-4y \\ \underline{y^3-\ y^2+3y-4} \\ 7y^4-5y^3+3y^2-\ y-4 \end{array}$$

12. Subtract:

$$\begin{array}{l} 7x^2-9x+2 \\ \underline{-(4x^2-2x-7)} \end{array}$$

Add:

$$\begin{array}{r} 7x^2-9x+2 \\ \underline{-4x^2+2x+7} \\ 3x^2-7x+9 \end{array}$$

13. Subtract:

$$\begin{array}{r} 5x^3-6x^2-\ 9x+14 \\ \underline{-(-5x^3+3x^2-11x+\ 3)} \end{array}$$

Add:

$$\begin{array}{r} 5x^3-6x^2-\ 9x+14 \\ \underline{5x^3-3x^2+11x-\ 3} \\ 10x^3-9x^2+2x+11 \end{array}$$

14. $7x(3x-9)=7x(3x)-(7x)(9)=21x^2-63x$

15. $-5x^3(4x^2-11x)=-20x^5+55x^4$

16. $3y^2(-7y^2+3y-6)$
$=3y^2(-7y^2)+3y^2(3y)+3y^2(-6)$
$=-21y^4+9y^3-18y^2$

17. $-2y^5(8y^3-4y^2-10y+6)$
$=-2y^5(8y^3)-2y^5(-4y^2)-2y^5(-10y)$
$-2y^5(6)$
$=-16y^8+8y^7+20y^6-12y^5$

18. $(x+3)(x^2-5x+2)$
$=x(x^2-5x+2)+3(x^2-5x+2)$
$=x(x^2)+x(-5x)+x(2)+3(x^2)+3(-5x)$
$+3(2)$
$=x^3-5x^2+2x+3x^2-15x+6$
$=x^3-5x^2+3x^2+2x-15x+6$
$=x^3-2x^2-13x+6$

19. $(3y-2)(4y^2+3y-5)$
$=3y(4y^2+3y-5)-2(4y^2+3y-5)$
$=3y(4y^2)+3y(3y)+3y(-5)-2(4y^2)$
$-2(3y)-2(-5)$
$=12y^3+9y^2-15y-8y^2-6y+10$
$=12y^3+9y^2-8y^2-15y-6y+10$
$=12y^3+y^2-21y+10$

20. $(x-6)(x+2)=x^2+2x-6x-12$
$=x^2-4x-12$

21. $(3y-5)(2y+1)=6y^2+3y-10y-5$
$=6y^2-7y-5$

22. $(4x^3-2x^2)(x^2-3)$
$=4x^3(x^2)+4x^3(-3)-2x^2(x^2)-2x^2(-3)$
$=4x^5-12x^3-2x^4+6x^2$
$=4x^5-2x^4-12x^3+6x^2$

23.

$$\begin{array}{r} y^2-4y+\ 7 \\ \underline{3y-\ 5} \\ -5y^2+20y-35 \\ \underline{3y^3-12y^2+21y} \\ 3y^3-17y^2+41y-35 \end{array}$$

24.

$$\begin{array}{r} 4x^3-\ 2x^2-6x-1 \\ \underline{2x+3} \\ 12x^3-\ 6x^2-18x-3 \\ \underline{8x^4-4x^3-12x^2-\ 2x} \\ 8x^4+8x^3-18x^2-20x-3 \end{array}$$

25. $(3x^3-2)(x^3+4)$
$=3x^6+12x^3-2x^3-8$
$=3x^6+10x^3-8$

26. $(x+3)^2=x^2+2x(3)+3^2=x^2+6x+9$

27. $(3y-4)^2$
$=(3y)^2-2(3y)(4)+4^2=9y^2-24y+16$

28. $(4x+5)(4x-5)=(4x)^2-5^2=16x^2-25$

29. $(2z+9)(2z-9)=(2z)^2-9^2=4z^2-81$

30. a. $(x+20)(x+30)$
$=x^2+30x+20x+600$
$=x^2+50x+600$ square yards

b. $f(x)=x^2+50x+600$

c. $f(5)=5^2+50(5)+600$
$= 25 + 250 + 600 = 875$
If the garage is increased by 5 yards in length and width, the new area is 875 square yards.

31. a. $2x(x+8)-x(2x-8)$
$=2x^2+16x-2x^2+8x$
$= 24x$ square inches

b. $g(x)=24x$

c.

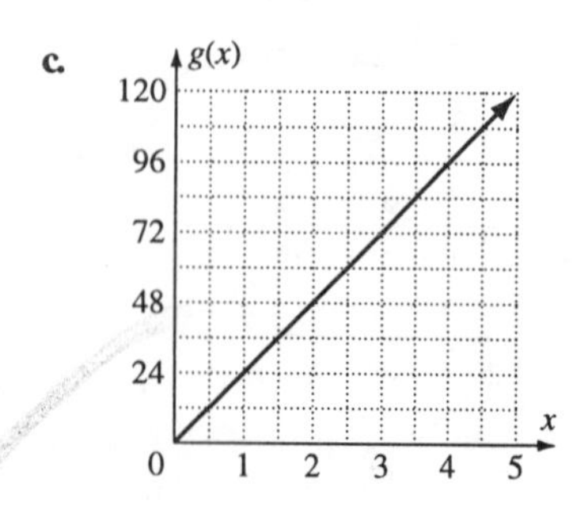

32. $(x+3)(x+4)$
$=x^2+7x+12$
$f(6)=6^2+7(6)+12=90$
If $x = 6$, the area of the shaded region is 90 sq. cm.

33. $(x+4)(x+5)-(4)(5)$
$=x^2+9x+20-20$
$=x^2+9x$
$f(6)=6^2+9(6)=90$
If $x = 6$, the area of the shaded region is 90 sq. cm.

34. $3-4xy+2y^2-5xy^3+x^8$
$=3-(4)(-1)(2)+2(2)^2-5(-1)(2)^3+(-1)^8$
$= 3 + 8 + 8 + 40 + 1 = 60$

35. $V=\pi r^2h+\frac{4}{3}\pi r^3$
$V=\pi(4)^2(21)+\frac{4}{3}\pi(4)^3$
$=336\pi+\frac{256\pi}{3}$
$=\frac{1264\pi}{3}$
$=\frac{1264(3.14)}{3}$
≈ 1323 cubic meters

36.

Term	Coefficient	Degree
$4x^2y$	4	$2 + 1 = 3$
$9x^3y^2$	9	$3 + 2 = 5$
$-17x^4$	−17	4
−12	−12	0

Degree of the polynomial = 5

37. $(7a^2-8ab+b^2)+(-8a^2-9ab-4b^2)$
$=(7a^2-8a^2)+(-8ab-9ab)+(b^2-4b^2)$
$=-a^2-17ab-3b^2$

38. $(13x^3y^2 - 5x^2y - 9x^2)$
$\quad - (-11x^3y^2 - 6x^2y + 3x^2 - 4)$
$= (13x^3y^2 - 5x^2y - 9x^2)$
$\quad + (11x^3y^2 + 6x^2y - 3x^2 + 4)$
$= (13x^3y^2 + 11x^3y^2) + (-5x^2y + 6x^2y)$
$\quad + (-9x^2 - 3x^2) + 4$
$= 24x^3y^2 + x^2y - 12x^2 + 4$

39. $(-7x^2y^3)(5x^4y^6)$
$= (-7)(5)x^{2+4}y^{3+6}$
$= -35x^6y^9$

40. $5ab^2(3a^2b^3 - 4ab)$
$= 5ab^2(3a^2b^3) + 5ab^2(-4ab)$
$= 15a^3b^5 - 20a^2b^3$

41. $(x + 7y)(3x - 5y)$
$= x(3x) + x(-5y) + 7y(3x) + 7y(-5y)$
$= 3x^2 - 5xy + 21xy - 35y^2$
$= 3x^2 + 16xy - 35y^2$

42. $(4xy - 3)(9xy - 1)$
$= 4xy(9xy) + 4xy(-1) - 3(9xy) - 3(-1)$
$= 36x^2y^2 - 4xy - 27xy + 3$
$= 36x^2y^2 - 31xy + 3$

43. $(3x - 5y)^2$
$= (3x)^2 - 2(3x)(5y) + (-5y)^2$
$= 9x^2 - 30xy + 25y^2$

44. $(3a^4 + 2b^3)^2$
$= (3a^4)^2 + 2(3a^4)(2b^3) + (2b^3)^2$
$= 9a^8 + 12a^4b^3 + 4b^6$

45. $(7x + 4y)(7x - 4y)$
$= (7x)^2 - (4y)^2$
$= 49x^2 - 16y^2$

46. $(a - b)(a^2 + ab + b^2)$
$= (a)^3 - (b)^3$
$= a^3 - b^3$

47. $\dfrac{-15y^8}{3y^2} = -5y^{8-2} = -5y^6$

48. $\dfrac{18y^4 - 12y^2 + 36y}{6y}$
$= \dfrac{18y^4}{6y} - \dfrac{12y^2}{6y} + \dfrac{36y}{6y}$
$= 3y^{4-1} - 2y^{2-1} + 6y^{1-1}$
$= 3y^3 - 2y + 6y^0$
$= 3y^3 - 2y + 6$

49. $(30x^8 - 25x^7 + 3x^6 - 40x^5) \div (-5x^5)$
$= \dfrac{30x^8}{-5x^5} - \dfrac{25x^7}{-5x^5} + \dfrac{3x^6}{-5x^5} - \dfrac{40x^5}{-5x^5}$
$= -6x^{8-5} + 5x^{7-5} - \dfrac{3}{5}x^{6-5} + 8x^{5-5}$
$= -6x^3 + 5x^2 - \dfrac{3}{5}x + 8$

50. $\dfrac{2z^3 - 6z^2 + 5z}{2z^2}$
$= \dfrac{2z^3}{2z^2} - \dfrac{6z^2}{2z^2} + \dfrac{5z}{2z^2}$
$= z^{3-2} - 3z^{2-2} + \dfrac{5}{2}z^{1-2}$
$= z - 3 + \dfrac{5}{2z}$

51. $\dfrac{20x^7 - 8x^6 - 16x^4 + 12x^2 - 2}{4x^5}$
$= \dfrac{20x^7}{4x^5} - \dfrac{8x^6}{4x^5} - \dfrac{16x^4}{4x^5} + \dfrac{12x^2}{4x^5} - \dfrac{2}{4x^5}$
$= 5x^{7-5} - 2x^{6-5} - 4x^{4-5} + 3x^{2-5} - \dfrac{1}{2x^5}$
$= 5x^2 - 2x - \dfrac{4}{x} + \dfrac{3}{x^3} - \dfrac{1}{2x^5}$

52. $\dfrac{27x^3y-9x^2y-18xy^2}{3xy}$
$=\dfrac{27x^3y}{3xy}-\dfrac{9x^2y}{3xy}-\dfrac{18xy^2}{3xy}$
$=9x^2-3x-6y$

53. $\dfrac{2x^2+3x-14}{x-2}=2x+7$

$$\begin{array}{r|l} & 2x+7 \\ x-2 & 2x^2+3x-14 \\ & \underline{2x^2-4x} \\ & \quad 7x-14 \\ & \quad \underline{7x-14} \\ & \qquad 0 \end{array}$$

54. $\dfrac{2y^3-5y^2+7y+5}{2y+1}=y^2-3y+5$

$$\begin{array}{r|l} & y^2-3y+5 \\ 2y+1 & 2y^3-5y^2+7y+5 \\ & \underline{2y^3+y^2} \\ & \quad -6y^2+7y \\ & \quad \underline{-6y^2-3y} \\ & \qquad 10y+5 \\ & \qquad \underline{10y+5} \\ & \qquad\quad 0 \end{array}$$

55. $\dfrac{z^3-2z^2-33z-7}{z-7}=z^2+5z+2+\dfrac{7}{z-7}$

$$\begin{array}{r|l} & z^2+5z+2 \\ z-7 & z^3-2z^2-33z-7 \\ & \underline{z^3-7z^2} \\ & \quad 5z^2-33z \\ & \quad \underline{5z^2-35z} \\ & \qquad 2z-7 \\ & \qquad \underline{2z-14} \\ & \qquad\quad 7 \end{array}$$

56. $(3y^6)(-2y^4)=-6y^{6+4}=-6y^{10}$

57. $(3x^3)^4=3^4x^{3(4)}=81x^{12}$

58. $4(2y^5)^3=4(2^3y^{5(3)})=4(8)y^{15}=32y^{15}$

59. $(2x)(4x)^2+15x^3=(2x)(16x^2)+15x^3$
$=32x^3+15x^3=47x^3$

60. $\dfrac{x^3}{x^9}=x^{3-9}=x^{-6}=\dfrac{1}{x^6}$

61. $\dfrac{30y^6}{5y^8}=6y^{6-8}=6y^{-2}=\dfrac{6}{y^2}$

62. $(5y^{-7})(6y^2)=30y^{-7+2}=30y^{-5}=\dfrac{30}{y^5}$

63. $\dfrac{x^4\cdot x^{-2}}{x^{-6}}=x^{4-2+6}=x^8$

64. $\dfrac{(3y^3)^4}{y^{10}}=\dfrac{3^4y^{3(4)}}{y^{10}}=81y^{12-10}=81y^2$

65. $\dfrac{y^{-7}}{(y^4)^3}=\dfrac{y^{-7}}{y^{12}}=y^{-7-12}=y^{-19}=\dfrac{1}{y^{19}}$

66. $\left(\dfrac{x^7}{x^4}\right)^{-4}=(x^{7-4})^{-4}=x^{3(-4)}=x^{-12}=\dfrac{1}{x^{12}}$

67. $\dfrac{(y^3)^4y^{-3}}{(y^{-2})^4}=\dfrac{y^{3(4)}y^{-3}}{y^{(-2)(4)}}=\dfrac{y^{12}y^{-3}}{y^{-8}}=y^{12-3+8}$
$=y^{17}$

68. $(2x^2y^{-3})^{-4}=2^{-4}x^{-8}y^{12}=\dfrac{y^{12}}{16x^8}$

69. $(4x^{-2}y^3)(-3x^4y^{-6})=(4)(-3)x^{-2+4}y^{3-6}$
$=-12x^2y^{-3}=\dfrac{-12x^2}{y^3}$

70. $\left(\dfrac{a^3}{b^2}\right)^{-4}=\dfrac{a^{-12}}{b^{-8}}=\dfrac{b^8}{a^{12}}$

71. $2.3\times10^4=23{,}000$
(move decimal point to right 4 places)

72. $1.76\times10^{-3}=0.00176$ (move left 3 places)

73. $9.84\times10^{-1}=0.984$ (move left 1 place)

74. $7^{-2}=\frac{1}{7^2}=\frac{1}{49}$

75. $2^{-1}+4^{-1}=\frac{1}{2}+\frac{1}{4}=\frac{3}{4}$

76. $(2^3)^{-2}=2^{3(-2)}=2^{-6}=\frac{1}{2^6}=\frac{1}{64}$

77. $\frac{5^{-5}}{5^{-3}}=5^{-5-(-3)}=5^{-2}=\frac{1}{5^2}=\frac{1}{25}$

78. $73,900,000=7.39\times10^7$

79. $0.000\ 089\ 4=8.94\times10^{-5}$

80. $0.000\ 972\ 5=9.725\times10^{-4}$

81. $0.38=3.8\times10^{-1}$

82. $8.639=8.639\times10^0$

83. $37,000=3.7\times10^4$

84. $(6\times10^{-3})(1.5\times10^6)=6(1.5)\times10^{-3+6}$
$=9\times10^3=9000$

85. $\frac{2\times10^2}{4\times10^{-3}}=\frac{10^{2+3}}{2}=0.5\times10^5$
$=5\times10^{-1}\times10^5=5.0\times10^4=50,000$

86. $(4\times10^{-2})^2=4^2\times10^{-2(2)}=16\times10^{-4}$
$=1.6\times10\times10^{-4}=1.6\times10^{1-4}=1.6\times10^{-3}$
$=0.0016$

87. $\frac{10^{-6}}{10^{-9}}=\frac{10^9}{10^6}=10^{9-6}=10^3=1000$

88. $\frac{3\times10^{10}}{6.5\times10^7}\approx4.615\times10^2$ or 461.5

89. $2(5.4\times10^9)=1.08\times10^{9+1}=1.08\times10^{10}$

90. $\frac{6\times10^{27}}{1.1\times10^{-6}}\approx5.4545\times10^{33}$ tons

Chapter 6 Test

1. Trinomial of degree two

2.
$$\begin{array}{r} 7x^3+3x^2-5x-11 \\ +6x^3-2x^2+4x-13 \\ \hline 13x^3+\ x^2-\ x-24 \end{array}$$

3.
$$\begin{array}{r} 9x^3-6x^2-11x-4 \\ +-4x^3+8x^2+13x-5 \\ \hline 5x^3+2x^2+\ 2x-9 \end{array}$$

4. $-6x^2(8x^2-7x-4)$
$=-48x^4+42x^3+24x^2$

5.
$$\begin{array}{r} 2x^2+\ 4x-\ 3 \\ \times\qquad\qquad 3x-\ 5 \\ \hline -10x^2-20x+15 \\ 6x^3+12x^2-\ 9x\qquad \\ \hline 6x^3+\ 2x^2-29x+15 \end{array}$$

6. $(3y+7)(2y-9)=6y^2+14y-27y-63$
$=6y^2-13y-63$

7. $(5x-3)(5x-3)=25x^2-2(15x)+9$
$=25x^2-30x+9$

8. $(4x^3-2)(5x^2-1)=20x^5-4x^3-10x^2+2$

9. $(3x+4y)(3x+4y)$
$=9x^2+2(12xy)+16y^2$
$=9x^2+24xy+16y^2$

10. $(7x+11)(7x-11)$
$=(7x)^2-(11)^2$
$=49x^2-121$

11. $\frac{12x^9}{-3x^5}=\frac{12}{-3}\cdot\frac{x^9}{x^5}=-4x^{9-5}=-4x^4$

12.
$$\begin{array}{r} 3x^3-2x^2+5x \\ 5x\overline{)15x^4-10x^3+25x^2} \\ \underline{15x^4\qquad\qquad\qquad} \\ -10x^3\qquad \\ \underline{-10x^3\qquad} \\ 25x^2 \\ \underline{25x^2} \\ 0 \end{array}$$

13. $\frac{20x^4-8x^3+12x^2-4}{4x^3}$
$=\frac{20x^4}{4x^3}-\frac{8x^3}{4x^3}+\frac{12x^2}{4x^3}-\frac{4}{4x^3}$
$=5x-2+\frac{3}{x}-\frac{1}{x^3}$

14.
$$\begin{array}{r} x^2-2x+3 \\ 2x+1\overline{)2x^3-3x^2+4x+4} \\ \underline{2x^3+\ x^2\qquad\qquad} \\ -4x^2+4x\qquad \\ \underline{-4x^2-2x\qquad} \\ 6x+4 \\ \underline{6x+3} \\ 1 \end{array}$$

$\frac{2x^3-3x^2+4x+4}{2x+1}=x^2-2x+3+\frac{1}{2x+1}$

15. $(-7x^3)(5x^8)=-35x^{11}$

16. $(-3x^2)^3=(-3)^3(x^2)^3=-27x^6$

17. $\frac{20x^3}{5x^8}=\frac{4}{x^5}$

18. $(-7x^{-8})(3x^2)=-21x^{-6}=-\frac{21}{x^6}$

19. $\frac{(2x^3)^4}{x^8}=\frac{2^4(x^3)^4}{x^8}=\frac{16x^{12}}{x^8}=16x^4$

20. $(3x^3)^2(-2x^3)^5=(9x^6)(-32x^{15})$
$=-288x^{21}$

21. $\left(\frac{x^{11}}{x^5}\right)^{-3}=(x^6)^{-3}=x^{-18}=\frac{1}{x^{18}}$

22. $4^{-3}=\frac{1}{4^3}=\frac{1}{64}$

23. $3.7\times10^{-4}=0.00037$

24. $(3^2)^{-2}=9^{-2}=\frac{1}{9^2}=\frac{1}{81}$

25. $\frac{2^{-5}}{2^{-3}}=\frac{1}{2^2}=\frac{1}{4}$

26. $7{,}600{,}000{,}000{,}000=7.6\times10^{12}$

27. $\frac{3.5\times10^4}{1.4\times10^{-13}}=\frac{3.5}{1.4}\times10^{4+13}=2.5\times10^{17}$

28. $(3.4\times10^6)(5\times10^{13})=3.4\times5\times10^6\times10^{13}$
$=17\times10^{19}=1.7\times10^{20}$

29. $(4x)(3x)-[x\cdot(x+2)]=12x^2-[x^2+2x]$
$=12x^2-x^2-2x=11x^2-2x$

30. $(x+8)(x+2)=x^2+8x+2x+16$
$=x^2+10x+16$

Cumulative Review Problems (Chapters 1–6)

1. $2(x+3)+2x=x+4$
$2x+6+2x=x+4$
$4x+6=x+4$
$3x=-2$
$x=-\frac{2}{3}$

2. $\frac{\$4.7\text{ billion}}{\$13\text{ billion}}\approx0.36=36\%$

3. $3-\frac{1}{4}x \le 2+\frac{3}{8}x$

$-\frac{1}{4}x-\frac{3}{8}x \le 2-3$

$-\frac{5}{8}x \le -1$

$x \ge -\frac{8}{5}(-1)$

$x \ge \frac{8}{5}$

$\left\{x \middle| x \ge \frac{8}{5}\right\}$

4. $5x-2y=-10$

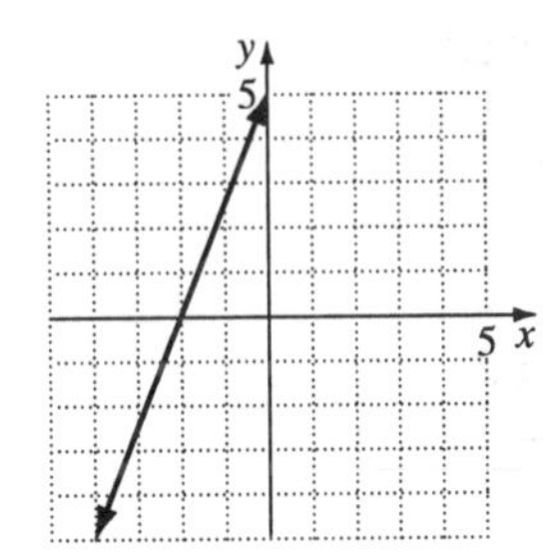

5. $y \ge -\frac{2}{5}x+2$

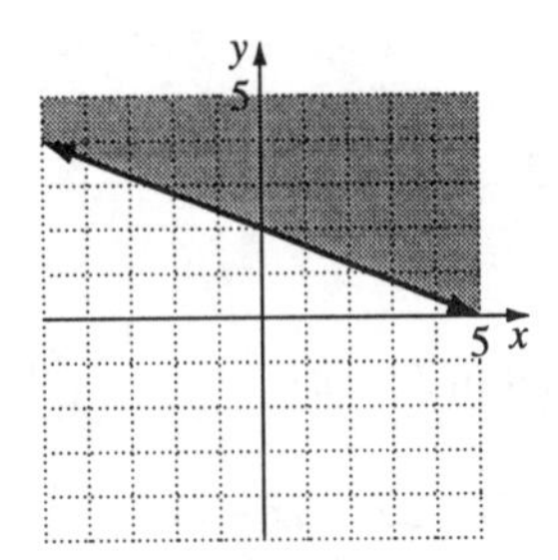

6. $3x-6y=1$

$x=2y+3$

$3(2y+3)-6y=1$

$6y+9-6y=1$

$9=1$ False

No solution; the system is inconsistent; $\varnothing$

7. $f(x)=0.00011x^4-0.013x^3+0.44x^2-3.6x+87$

$f(10)=0.00011(10)^4-0.013(10)^3+0.44(10)^2-3.6(10)+87$

$=1.1-13+44-36+87=83.1$

Ten years after 1930, or in 1940, the fertility rate in the U.S. was 83.1 live births per 1000 women. The bar for 1940 slightly lower, so the actual fertility rate in 1940 is less than that obtained from the model.

8. $x+y=-1$

$-2x+y=5$

Solution: $(-2, 1)$

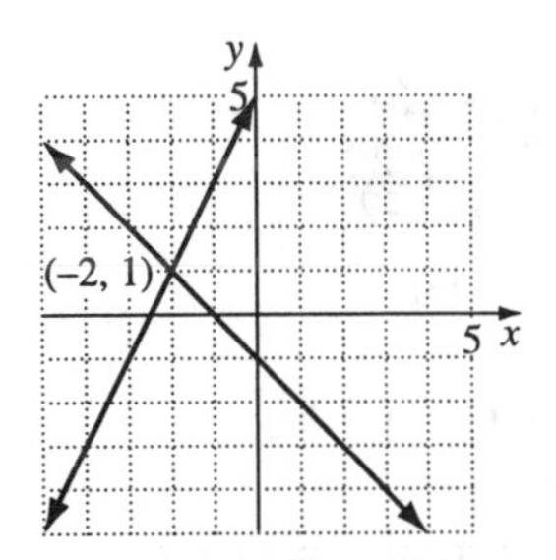

9. Let x = amount spent on education.

Then defense = $5x + 9000$

and welfare = $x - 6300$

$x+5x+9000+x-6300=103{,}500$

$7x+2700=103{,}500$

$7x=100{,}800$

$x=14{,}400$

$5x+9000=81{,}000$

$x-6300=8100$

Defense = \$81,000

Education = \$14,400

Welfare = \$8100

Social security ≈ \$70,000

Interest on the debt ≈ \$40,000

Medicare and Medicaid ≈ \$35,000

10. $f(x) = 0.25x + 12.75$
Slope = 0.25
y-intercept = 12.75
The number of serious crimes in 1987 ($x = 0$) was 12.75 million and this number increases by 0.25 million each year.

11. Let w = width in feet. Then $l = 2w - 7$.
$2w + 2(2w - 7) = 280$
$2w + 4w - 14 = 280$
$6w - 14 = 280$
$6w = 294$
$w = 49$
$2w - 7 = 91$
Width = 49 feet; length = 91 feet

12. Points (–1, 3), (–3, 5)
Slope $\frac{5-3}{-3-(-1)} = \frac{2}{-2} = -1$
$y - 3 = -1[x - (-1)]$
Point-slope form:
$y - 3 = -1(x + 1)$ or $y - 5 = -1(x + 3)$
$y - 3 = -x - 1$
Slope-intercept form: $y = -x + 2$

13. Points (1970, 2200) and (2000, 5400)
Slope $\frac{5400-2200}{2000-1970} = \frac{3200}{30} = 106\frac{2}{3}$
The rate of change per pupil is increasing at \$106.67 per year.

14. $\left(\frac{2}{3} + \frac{6}{11}\right) - \left(-\frac{1}{4} + \frac{5}{12}\right)$
$= \left(\frac{22}{33} + \frac{18}{33}\right) + \left(\frac{3}{12} - \frac{5}{12}\right)$
$= \frac{40}{33} - \frac{2}{12}$
$= \frac{160}{132} - \frac{22}{132}$
$= \frac{138}{132}$
$= \frac{23}{22}$ or $1\frac{1}{22}$

15. $\frac{x^3 + 3x^2 + 5x + 3}{x+1} = x^2 + 2x + 3$

$$\begin{array}{r} x^2 + 2x + 3 \\ x+1 \overline{)\, x^3 + 3x^2 + 5x + 3} \\ \underline{x^3 + x^2 \quad\quad\quad\;\;} \\ 2x^2 + 5x \quad\;\; \\ \underline{2x^2 + 2x \quad\;\;} \\ 3x + 3 \\ \underline{3x + 3} \\ 0 \end{array}$$

16. $0.3x - 4 = 0.1(x + 10)$
$0.3x - 4 = 0.1x + 1$
$0.3x - 0.1x = 4 + 1$
$0.2x = 5$
$x = 25$

17.
$$\begin{array}{r} 9x^5 - 3x^3 + 2x - 7 \\ \underline{-(9x^5 + 3x^3 - 7x - 9)} \end{array}$$

$$\begin{array}{r} 9x^5 - 3x^3 + 2x - 7 \\ \underline{-9x^5 - 3x^3 + 7x + 9} \\ -6x^3 + 9x + 2 \end{array}$$

18. Let t equal the time (in hours).
$13t + 11t = 72$
$24t = 72$
$t = 3$
They will meet in 3 hours.

19. $(2x)(x)(3x - 7) = 2x^2(3x - 7)$
$= 6x^3 - 14x^2$
$f(x) = 6x^3 - 14x^2$ cubic in.
$f(4) = 6(4)^3 - 14(4)^2 = 160$
If $x = 4$ in., the volume is 160 cubic in.

20. $(3x - 2)(4x^2 - 5x + 1)$
$= 3x(4x^2) + 3x(-5x) + 3x(1) - 2(4x^2) - 2(-5x) - 2(1)$
$= 12x^3 - 15x^2 + 3x - 8x^2 + 10x - 2$
$= 12x^3 - 23x^2 + 13x - 2$

21. $\frac{5000}{20} = 250; \frac{26,000}{250} = 104$ pounds

$\frac{104}{20} = 5.2$

5.2 or approximately 6 bags

22. Let x = length of middle-sized piece.
Then $2x$ = length of longest piece and
$x - 10$ = length of shortest piece.

$x + 2x + x - 10 = 70$
$4x - 10 = 70$
$4x = 80$
$x = 20$
$2x = 40$
$x - 10 = 10$

The lengths are 10 cm, 20 cm, and 40 cm.

23. $3x + 2y = 10 \quad (\times 3)$
$4x - 3y = -15 \quad (\times 2)$

$$\begin{aligned} 9x + 6y &= 30 \\ 8x - 6y &= -30 \\ \hline 17x &= 0 \\ x &= 0 \\ 3(0) + 2y &= 10 \\ 2y &= 10 \\ y &= 5 \end{aligned}$$

(0, 5)

24. $2x + 5y \le 10$
$x - y \ge 4$

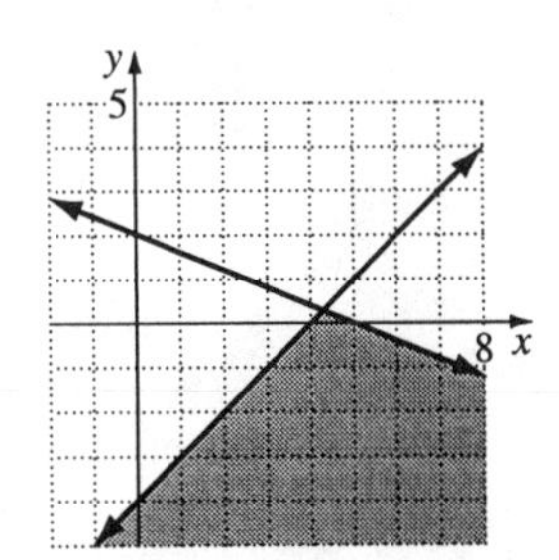

25. $f(x) = -0.000625x^2 + 0.025x + 0.501$
$f(10) = -0.000625(10)^2 + 0.025(10) + 0.501$
$f(10) = 0.6885$
In 1980 (10 years after 1970), there were 68.85% of women age 20–34 in the labor force. The point on the graph is (10, 68.85). The graph increases until about 1990, then starts to decrease.

26. Points (–1, 0) and (0, –3)

Slope $= \frac{-3-0}{0-(-1)} = \frac{-3}{1} = -3$

$y - 0 = -3[x - (-1)]$
$y = -3(x + 1)$
$y = -3x - 3$

27. $\frac{(8-10)^3 - (-4)^2}{2 + 8(2) \div 4} = \frac{(-2)^3 - 16}{2 + 16 \div 4}$

$= \frac{-8-16}{2+4}$

$= \frac{-24}{6} = -4$

28. $12 \div 2 - 3 = 3$
$x = 12, y = 2, z = 3$

29. Let x equal the speed of boat in still water.
y equals the speed of current.

	R	$\times$	T	$= D$
with the current	$x + y$		1 hour	16 miles
against the current	$x - y$		8 hours	16 miles

$1(x + y) = 16$
$8(x - y) = 16 \quad (\div 8)$

$$\begin{aligned} x + y &= 16 \\ x - y &= 2 \\ \hline 2x &= 18 \\ x &= 9 \\ 9 + y &= 16 \\ y &= 7 \end{aligned}$$

Speed of boat in still water, 9 mph;
speed of current, 7 mph

30.

	10	–5	1	Score
1	3	0	0	30
2	2	1	0	15
3	2	0	1	21
4	1	2	0	0
5	1	0	2	12
6	1	1	1	6
7	0	3	0	–15
8	0	2	1	–9
9	0	1	2	–3
10	0	0	3	3

10 possible different scores

Chapter 7

Problem Set 7.1

1. Possible answers:
$8x^3 = 2x \cdot 4x^2$
$8x^3 = x \cdot 8x^2$
$8x^3 = 2x \cdot 2x \cdot 2x$

3. Possible answers:
$-12x^5 = -3x^3 \cdot 4x^2$
$-12x^5 = -x \cdot 12x^4$
$-12x^5 = -2x^2 \cdot 6x^3$

5. Possible answers:
$36x^4 = x \cdot 36x^3$
$36x^4 = 2x^2 \cdot 18x^2$
$36x^4 = 6x^2 \cdot 6x^2$

7. $5x + 5 = 5(x) + 5(1) = 5(x + 1)$

9. $3z - 3 = 3(z) - 3(1) = 3(z - 1)$

11. $8x + 16 = 8(x) + 8(2) = 8(x + 2)$

13. $25x - 10 = 5(5x) - 5(2) = 5(5x - 2)$

15. $y^2 + y = y(y) + y(1) = y(y + 1)$

17. $18x^2 - 24 = 6(3x^2) - 6(4) = 6(3x^2 - 4)$

19. $25y^2 - 13y = y(25y) - y(13) = y(25y - 13)$

21. $36x^3 + 24x^2 = 12x^2(3x) + 12x^2(2)$
$= 12x^2(3x + 2)$

23. $27y^6 + 9y^4 = 9y^4(3y^2) + 9y^4(1)$
$9y^4(3y^2 + 1)$

25. $8x^2 - 4x^4 = 4x^2(2) - 4x^2(x^2)$
$= 4x^2(2 - x^2)$

27. $12x^2 - 13y^3 = (1)(12x^2) - (1)(13y^3)$
$= 1(12x^2 - 13y^3)$

29. $12y^2 + 16y - 8 = 4(3y^2) + 4(4y) + 4(-2)$
$= 4(3y^2 + 4y - 2)$

31. $100 + 75y - 50y^2$
$= 25(4) + 25(3y) + 25(-2y^2)$
$= 25(4 + 3y - 2y^2)$

33. $9y^4 + 18y^3 + 6y^2$
$= 3y^2(3y^2) + 3y^2(6y) + 3y^2(2)$
$= 3y^2(3y^2 + 6y + 2)$

35. $100y^5 - 50y^3 + 100y^2$
$= 50y^2(2y^3) + 50y^2(-y) + 50y^2(2)$
$= 50y^2(2y^3 - y + 2)$

37. $10x - 20x^2 + 5x^3$
$= 5x(2) + 5x(-4x) + 5x(x^2)$
$= 5x(2 - 4x + x^2)$

39. $-2y^2 - 3y^3 + 6y^5$
$= y^2(-2) + y^2(-3y) + y^2(6y^3)$
$= y^2(-2 - 3y + 6y^3)$

41. $6x^3y^2 + 9xy = 3xy(2xy + 3)$

43. $30x^3y^2 - 10x^3y + 20x^2y$
$= 10x^2y(3xy - x + 2)$

45. $16a^5b^3 - 48a^4b^4 + 8a^3b^2 - 56a^3b^3$
$= 8a^3b^2(2a^2b - 6ab^2 + 1 - 7b)$

47. $54a^2b^3c - 6a^2b^2c^2 + 12abc$
$= 6abc(9ab^2 - abc + 2)$

49. $-2x^2+8x-10=2(-x^2+4x-5)$
$=-2(x^2-4x+5)$

51. $3a-15=3(a-5)=-3(-a+15)$

53. $-4x+12x^2=4x(-1+3x)=-4x(1-3x)$

55. $-y^3+7y^2=y^2(-y+7)=-y^2(y-7)$

57. $y^2+y=y(y+1)=-y(-y-1)$

59. $3-x=1(3-x)=-1(-3+x)$

61. $x(x+5)+3(x+5)=(x+5)(x+3)$

63. $7x(x-3)-4(x-3)=(7x-4)(x-3)$

65. $3x(2x+5)+2x+5=3x(2x+5)+1(2x+5)$
$=(3x+1)(2x+5)$

67. $x^2(x+7)+2(x+7)=(x^2+2)(x+7)$

69. $4x^2(3x^3+2)-7(3x^3+2)$
$=(4x^2-7)(3x^3+2)$

71. $y^2(y+7)+y+7=y^2(y+7)+1(y+7)$
$=(y^2+1)(y+7)$

73. $x^3-3x^2+2x-6=x^2(x-3)+2(x-3)$
$=(x^2+2)(x-3)$

75. $3x^3+6x^2+2x+4=3x^2(x+2)+2(x+2)$
$=(3x^2+2)(x+2)$

77. $x^3+5x^2+x+5=x^2(x+5)+1(x+5)$
$=(x^2+1)(x+5)$

79. $10y^3-25y^2+4y-10$
$=5y^2(2y-5)+2(2y-5)$
$=(5y^2+2)(2y-5)$

81. $y^3+8y^2-3y-24=y^2(y+8)-3(y+8)$
$=(y^2-3)(y+8)$

83. $8y^5+12y^2-10y^3-15$
$=4y^2(2y^3+3)-5(2y^3+3)$
$=(4y^2-5)(2y^3+3)$

85. x = the width of the rectangular painting
$44x+x^2=x(44+x)$
length = $(44+x)$ in.

87. P = Principal amount; t = number of years;
r = simple interest rate
$P+Prt=P(1+rt)$

89. d is true.

91. $x^2-2x+5x-10\neq(x-2)(x-5)$
$x(x-2)+5(x-2)=(x+5)(x-2)$
Correct factorization: $(x+5)(x-2)$

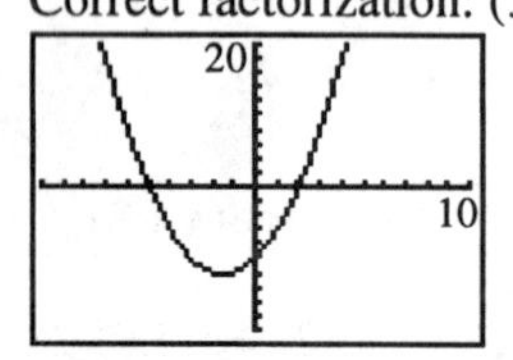

93. $-3x-6\neq-3(x-2)$

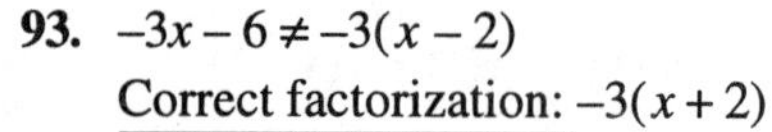
Correct factorization: $-3(x+2)$

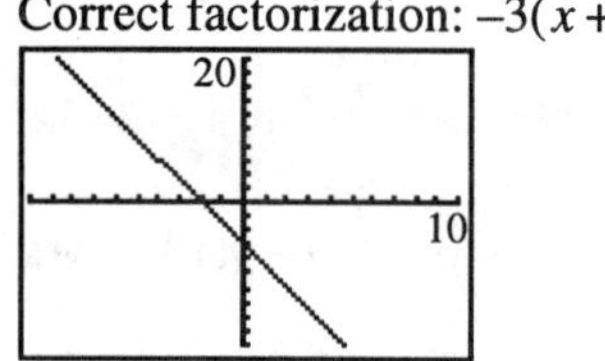

95. Answers may vary.

97. Answers may vary.

99. Answers may vary.

101. area of shaded region = area of large circle − area of small circle
$A=\pi(3x)^2-\pi x^2=\pi x^2(9-1)=8\pi x^2$

103. $A = (3x)(2x) + \frac{1}{2}\pi x^2 = \frac{1}{2}x^2(12 + \pi)$

Review Problems

105. Let x = amount invested at 6%.
$x - 350$ = amount invested at 8%.
$x(0.06) + (x - 350)(0.08) = 147$
$0.14x - 28 = 147$
$0.14x = 175$
$x = \$1250$
$x - 350 = 1250 - 350 = 900$
\$1250 at 6%; \$900 at 8%

106. $2x - y = -4$
$x - 3y = 3$
Lines intersect at $x = -3$, $y = -2$:
$(-3, -2)$

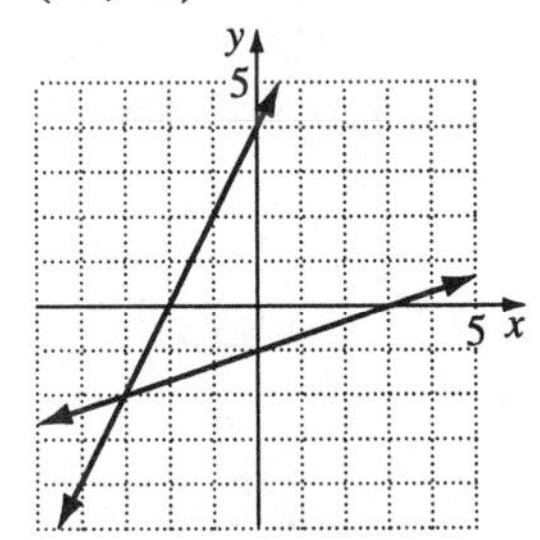

107. slope $= \frac{5-2}{-4-(-7)} = \frac{3}{3} = 1$
$y - 2 = 1[x - (-7)]$
$y - 2 = x + 7$ or $y - 5 = x + 4$
$y = x + 7 + 2$
$y = x + 9$

Problem Set 7.2

1. $x^2 + 3x + 2 = (x + 2)(\quad)$
Using FOIL, we must find a number y such that $2y = 2$ and $2 + y = 3$. The value of y is 1. Hence the missing factor is $x + 1$.
Check: $(x + 2)(x + 1) = x^2 + 3x + 2$

3. $y^2 + y - 6 = (y + 3)(\quad) = (y + 3)(y - 2)$
Pairs of integers whose products are −6 are:
(−6)(1)
(6)(−1)
(3)(−2) $\leftarrow 3 - 2 = 1$
(−3)(2)
Only 3 and −2 have a sum of 1.

5. $x^2 + x - 12 = (x - 3)(\quad) = (x - 3)(x + 4)$
Pairs of integers whose products are −12:
(1)(−12)
(−1)(12)
(2)(−6)
(−2)(6)
(3)(−4)
(−3)(4) ← only pair with a sum of 1.
$-3 + 4 = 1$

7. $y^2 - 5y + 4 = (y - 1)(\quad) = (y - 1)(y - 4)$
Pairs of integers whose products are 4:
(−1)(−4) ← only pair with a sum of −5.
(1)(4) $\quad -1 + (-4) = -5$
(−2)(−2)
(2)(2)

9. $y^2 - 2y - 3 = (y + 1)(\quad) = (y + 1)(y - 3)$
Pairs of integers whose products are −3:
(3)(−1)
(−3)(1) ← only pair with a sum of −2.
$(-3) + 1 = -2$

11. $r^2 - 6r + 8 = (r - 2)(\quad) = (r - 2)(r - 4)$
Pairs of integers whose products are 8:
(1)(8)
(−1)(−8)
(2)(4)
(−2)(−4) ← only pair with a sum of −6.
$-2 + (-4) = -6$

13. $x^2 + 5x + 6 = (x+2)(x+3)$
$(2)(3) = 6,\ 2 + 3 = 5$

15. $r^2 + 13r + 12 = (r+1)(r+12)$
$(1)(12) = 12,\ 1 + 12 = 13$

17. $x^2 + 9x + 8 = (x+1)(x+8)$
$(1)(8) = 8,\ 1 + 8 = 9$

19. $y^2 - 2y - 15 = (y+3)(y-5)$
$(3)(-5) = -15,\ 3 - 5 = -2$

21. $x^2 - 5x - 6 = (x-6)(x+1)$
$(-6)(1) = -6,\ -6 + 1 = -5$

23. $y^2 - 14y + 45 = (y-5)(y-9)$
$(-5)(-9) = 45,\ -5 + (-9) = -14$

25. $r^2 + 12r + 27 = (r+3)(r+9)$
$3(9) = 27,\ 3 + 9 = 12$

27. $n^2 - 11n - 42 = (n+3)(n-14)$
$3(-14) = -42,\ 3 - 14 = -11$

29. $y^2 - 9y - 36 = (y+3)(y-12)$
$3(-12) = -36,\ 3 - 12 = -9$

31. $x^2 + 10x - 75 = (x+15)(x-5)$
$15(-5) = -75,\ 15 - 5 = 10$

33. $x^2 - 8x + 32$ is prime.
cannot be factored

35. $y^2 + 30y + 200 = (y+10)(y+20)$
$(10)(20) = 200,\ 10 + 20 = 30$

37. $x^2 - 6x + 8 = (x-2)(x-4)$
$(-2)(-4) = 8,\ -2 + (-4) = -6$

39. $r^2 + 17r + 16 = (r+1)(r+16)$
$(16)(1) = 16,\ 1 + 16 = 17$

41. $m^2 - 15m + 36 = (m-3)(m-12)$
$(-3)(-12) = 36,\ -3 + (-12) = -15$

43. $y^2 + y - 56 = (y-7)(y+8)$
$(-7)(8) = -56,\ -7 + 8 = 1$

45. $r^2 + 4r + 12$ is prime.
cannot be factored

47. $y^2 - 4y - 21 = (y-7)(y+3)$
$(-7)(3) = -21,\ -7 + 3 = -4$

49. $x^2 + 8x - 105 = (x+15)(x-7)$
$(15)(-7) = -105,\ 15 - 7 = 8$

51. $r^2 + 27r + 72 = (r+3)(r+24)$
$(3)(24) = 72,\ 3 + 24 = 27$

53. $a^2 + 5ab + 6b^2 = (a+2b)(a+3b)$
$(2)(3) = 6,\ 2 + 3 = 5$

55. $x^2 + 5xy - 24y^2 = (x-3y)(x+8y)$
$(-3)(8) = -24,\ -3 + 8 = 5$

57. $3x^2 + 15x + 18 = 3(x^2 + 5x + 6)$
$= 3(x+2)(x+3)$
$(2)(3) = 6,\ 2 + 3 = 5$

59. $4y^2 - 4y - 8 = 4(y^2 - y - 2)$
$= 4(y+1)(y-2)$
$(1)(-2) = -2,\ 1 - 2 = -1$

61. $10x^2 - 40x - 600 = 10(x^2 - 4x - 60)$
$= 10(x+6)(x-10)$
$(6)(-10) = -60,\ 6 - 10 = -4$

63. $3x^2 - 33x + 54 = 3(x^2 - 11x + 18)$
$= 3(x-2)(x-9)$
$(-2)(-9) = 18,\ -2 - 9 = -11$

65. $2r^3 + 6r^2 + 4r = 2r(r^2 + 3r + 2)$
$= 2r(r+1)(r+2)$

67. $4x^3+12x^2-72x=4x(x^2+3x-18)$
$=4x(x+6)(x-3)$

69. $2r^3+8r^2-64r=2r(r^2+4r-32)$
$=2r(r+8)(r-4)$

71. $y^4+2y^3-80y^2=y^2(y^2+2y-80)$
$=y^2(y+10)(y-8)$

73. $x^4-3x^3-10x^2=x^2(x^2-3x-10)$
$=x^2(x-5)(x+2)$

75. $2w^4-26w^3-96w^2=2w^2(w^2-13w-48)$
$=2w^2(w-16)(w+3)$

77. $-2x^2+14x-24=-2(x^2-7x+12)$
$=-2(x-3)(x-4)$
$(-3)(-4)=12, -3-4=-7$

79. $-x^3-11x^2+42x=-x(x^2+11x-42)$
$=-x(x-3)(x+14)$
$(-3)(14)=-42, -3+14=11$

81. $2x^2-10xy-28y^2=2(x^2-5xy-14y^2)$
$2(x+2y)(x-7y)$
$(2)(-7)=-14, 2-7=-5$

83. $x^3+8x^2y+15xy^2=x(x^2+8xy+15y^2)$
$=x(x+3y)(x+5y)$
$(3)(5)=15, 3+5=8$

85. $x^2+3x-10=(x-2)(x+5)$
width $=x-2$

87. Statement b is true.
$x^2-10x+9=(x-1)(x-9)$
$(-1)(-9)=9, -1+(-9)=-10$

89. Statement d is true.
$(y-2)(y-1)=y^2-3y+2$, not
y^2-3y-2.
None is correct.

91. $2x^2+2x-12=2(x^2+x-6)$
$=2(x-2)(x+3)$
$(-2)(3)=-6, -2+3=1$

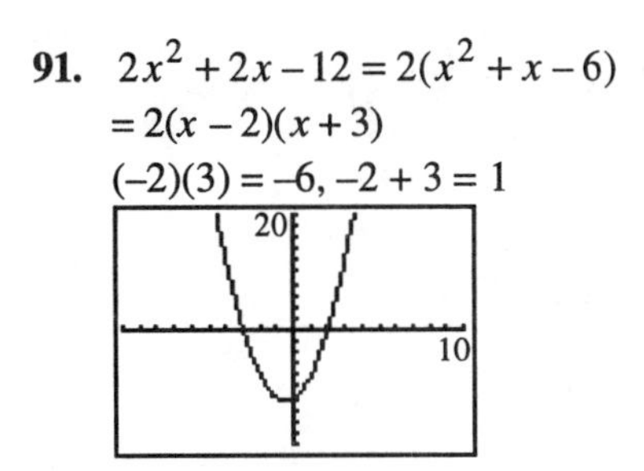

93. $x^3-x^2-2x=x(x^2-x-2)$
$=x(x+1)(x-2)$
$(1)(-2)=-2, 1-2=-1$

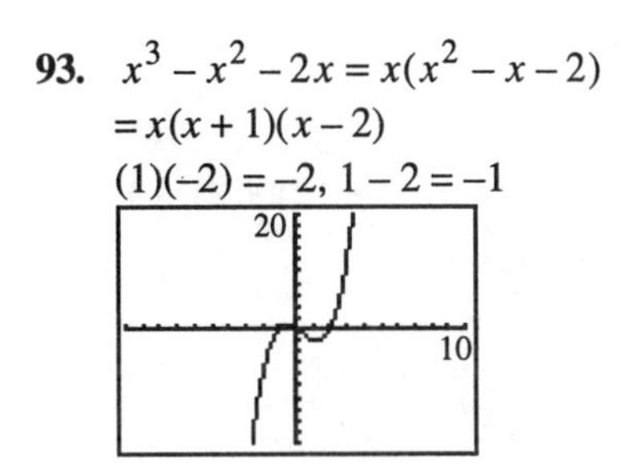

95. Answers may vary.

97. Answers may vary.

99. $A=x^2+7x+12=(x+3)(x+4)$
length of shaded area $=x+4-x=4$
width of shaded area $=x+3-x=3$
Area of shaded region
$=(x+3-x)(x+4-x)=3\cdot 4=12$

101. $x^2+bx+15$ can be factored if $b=8$ or $b=16$ or $b=-8$ or $b=-16$.
$x^2+8x+15=(x+5)(x+3)$
$x^2+16x+15=(x+1)(x+15)$
$x^2-8x+15=(x-5)(x-3)$
$x^2-16x+15=(x-1)(x-15)$

103. $x^2+4x+c=(x+2)(x+2)$, $(2)(2)=4$, $2+2=4$
$x^2+4x+c=(x+1)(x+3)$, $(1)(3)=3$, $1+3=4$
Possible values of c are 3 and 4.

105. $x^2 - \frac{10}{3}x - \frac{8}{3} = \left(x + \frac{2}{3}\right)(x - 4)$
$\left(\frac{2}{3}\right)(-4) = -\frac{8}{3},\ \frac{2}{3} - 4 = -\frac{10}{3}$

Review Problems

106. $179.5 + 2.35x = 282.9$
$2.35x = 103.4$
$x = 44$
$1960 + 44 = 2004$

107. $(4y+1)(2y-3) = 8y^2 - 12y + 2y - 3$
$= 8y^2 - 10y - 3$

108. $(3x+4)(3x+1) = 9x^2 + 3x + 12x + 4$
$= 9x^2 + 15x + 4$

Problem Set 7.3

1. $5x^2 + 6x + 1 = (5x+1)(x+1)$
The factors of 5 are 5 and 1 and of 1 are 1 and 1.

3. $5y^2 + 29y - 6 = (y+6)(5y-1)$
Possible factors:
$(5y + 6)(y - 1), -5y + 6y = xy$
$(5y - 6)(y + 1), 5y - 6y = -y$
$(y + 6)(5y - 1), -y + 30y = 29y$
This gives the only correct middle term.
$(y - 6)(5y + 1), y - 30y = -29y$
$(5y - 3)(y + 2), 10y - 3y = 7y$
$(5y + 3)(y - 2), -10y + 3y = -7y$
$(5y - 2)(y + 3), 15y - 2y = 13y$
$(5y + 2)(y - 3), -15y + 2y = -13y$

5. $24r^2 - 22r - 35 = (6r+5)(4r-7)$
Possible factors of $24r^2$:
$24r$ and r, $2r$ and $12r$, $3r$ and $8r$, $4r$ and $6r$
Possible factors of –35:
35 and –1, –1 and 35, 5 and –7, –5 and 7
$6r(-7) + 5(4r) = -42r + 20r = -22r$
This is the only correct possibility.

7. $6y^2 - 31y + 5 = (y-5)(6y-1)$
$-y - 30y = -31y$

9. $7y^2 - 40y - 63 = (y-7)(7y+9)$
$9y - 49y = -40y$

11. $15m^2 + 7m - 22 = (m-1)(15m+22)$
$22m - 15m = 7m$

13. $2x^2 + 7x + 3 = (2x+1)(x+3)$
$6x + x = 7x$

15. $2x^2 + 17x + 35 = (2x+7)(x+5)$
$10x + 7x = 17x$

17. $2y^2 - 17y + 30 = (2y-5)(y-6)$
$-12y - 5y = -17y$

19. $4x^2 - 11x + 7 = (x-1)(4x-7)$
$-7x - 4x = -11x$

21. $5y^2 - 12y + 6$ is prime

23. $3x^2 - x - 2 = (3x+2)(x-1)$
$-3x + 2x = -x$

25. $3y^2 + y - 10 = (3y-5)(y+2)$
$6y - 5y = y$

27. $3r^2 - 25r - 28 = (3r-28)(r+1)$
$3r - 28r = -25r$

29. $6y^2 - 11y + 4 = (2y-1)(3y-4)$
$-8y - 3y = -11y$

31. $8t^2 + 33t + 4 = (8t+1)(t+4)$
$32t + t = 33t$

33. $5x^2 + 33x - 14 = (5x-2)(x+7)$
$35x - 2x = 33x$

35. $14y^2 + 15y - 9 = (7y-3)(2y+3)$
$21y - 6y = 15y$

37. $25r^2 - 30r + 9 = (5r - 3)(5r - 3) = (5r - 3)^2$
$(-15r) + (-15r) = -30r$

39. $6x^2 - 7x + 3$ is prime

41. $10y^2 + 43y - 9 = (5y - 1)(2y + 9)$
$45y - 2y = 43y$

43. $8r^2 - 38r - 21 = (4r - 21)(2r + 1)$
$4r - 42r = -38r$

45. $15y^2 - y - 2 = (5y - 2)(3y + 1)$
$5y - 6y = -y$

47. $8m^2 - 2m - 1 = (4m + 1)(2m - 1)$
$-4m + 2m = -2m$

49. $35z^2 + 43z - 10 = (7z + 10)(5z - 1)$
$-7z + 50z = 43z$

51. $9y^2 - 9y + 2 = (3y - 2)(3y - 1)$
$-3y - 6y = -9y$

53. $20x^2 - 41x + 20 = (4x - 5)(5x - 4)$
$-16x - 25x = -41x$

55. $-4x^2 - x + 3 = (-4x + 3)(x + 1)$
$= -(4x - 3)(x + 1)$
$-4x + 3x = -x$

57. $-4y^2 + 5y + 6 = (-4y - 3)(y - 2)$
$= -(4y + 3)(y - 2)$
$8y - 3y = 5y$

59. $2 + 7y + 6y^2 = (2 + 3y)(1 + 2y)$
$4y + 3y = 7y$

61. $38 - 67x + 15x^2 = (2 - 3x)(19 - 5x)$
$-10x - 57x = -67x$

63. $2x^2 + 3xy + y^2 = (2x + y)(x + y)$
$2xy + xy = 3xy$

65. $15x^2 + 11xy - 14y^2 = (3x - 2y)(5x + 7y)$
$21xy - 10xy = 11xy$

67. $2x^2 - 9xy + 9y^2 = (x - 3y)(2x - 3y)$
$-3xy - 6xy = -9xy$

69. $2x^2 + 7xy + 5y^2 = (x + y)(2x + 5y)$
$5xy + 2xy = 7xy$

71. $6a^2 - 5ab - 6b^2 = (2a - 3b)(3a + 2b)$
$4ab - 9ab = -5ab$

73. $3a^2 - ab - 14b^2 = (a + 2b)(3a - 7b)$
$-7ab + 6ab = -ab$

75. $12r^2 - 25rs + 12s^2 = (3r - 4s)(4r - 3s)$
$-9rs - 16rs = -25rs$

77. $18x^2 + 48x + 32 = 2(9x^2 + 24x + 16)$
$= 2(3x + 4)(3x + 4) = 2(3x + 4)^2$

79. $4y^2 + 2y - 30 = 2(2y^2 + y - 15)$
$= 2(2y - 5)(y + 3)$
$6y - 5y = y$

81. $9r^2 + 33r - 60 = 3(3r^2 + 11r - 20)$
$= 3(3r - 4)(r + 5)$
$15r - 4r = 11r$

83. $2y^3 - 3y^2 - 5y = y(2y^2 - 3y - 5)$
$= y(2y - 5)(y + 1)$
$2y - 5y = -3y$

85. $9r^3 - 39r^2 + 12r = 3r(3r^2 - 13r + 4)$
$= 3r(r - 4)(3r - 1)$
$-r - 12r = -13r$

87. $14m^3 + 94m^2 - 28m$
$= 2m(7m^2 + 47m - 14)$
$= 2m(m + 7)(7m - 2)$
$-2m + 49m = 47m$

89. $15x^4 - 39x^3 + 18x^2 = 3x^2(5x^2 - 13x + 6)$
$= 3x^2(5x - 3)(x - 2)$
$-10x - 3x = -13x$

91. $10x^5 - 17x^4 + 3x^3 = x^3(10x^2 - 17x + 3)$
$= x^3(5x - 1)(2x - 3)$
$-15x - 2x = -17x$

93. $36x^2 + 54xy - 70y^2$
$= 2(18x^2 + 27xy - 35y^2)$
$= 2(3x + 7y)(6x - 5y)$
$-15xy + 42xy = 27xy$

95. $12a^2b - 34ab^2 + 14b^3$
$= 2b(6a^2 - 17ab + 7b^2)$
$= 2b(2a - b)(3a - 7b)$
$-14ab - 3ab = -17ab$

97. $-15a^2b^2 + 7ab^2 + 4b^2$
$= (-b^2)(15a^2 - 7a - 4)$
$= -b^2(3a + 1)(5a - 4)$

99. Statement a is true.
$18y^2 - 6y + 6$
$= 6(3y^2 - y + 1)$
$3y^2 - y + 1$ is prime.

101. $2x^2 + 5x + 3 = (2x + 3)(x + 1)$
correct factorization

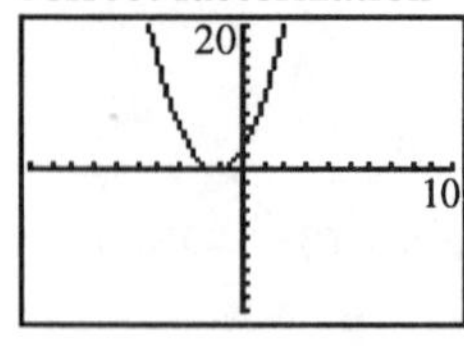

103. $18x^3 - 21x^2 - 9x = 3x(2x - 3)(3x + 1)$
correct factorization

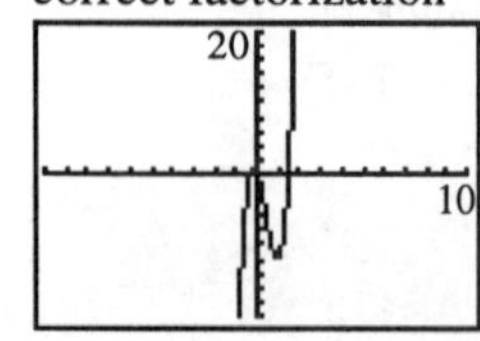

105. Answers may vary.

107. Answers may vary.

109. $2x^{2n} - 7x^n - 4 = (2x^n + 1)(x^n - 4)$
$-8x^n + x^n = -7x^n$

111. $3x^2 + bx + 2 = (3x + 2)(x + 1)$ if $b = 5$
or $(3x + 1)(x + 2)$ if $b = 7$
$= (3x - 2)(x - 1)$ if $b = -5$
or $(3x - 1)(x - 2)$ if $b = -7$.

113. $3(x + 2)^2 - (x + 2) - 4$
$= 3(x^2 + 4x + 4) - x - 6$
$= 3x^2 + 12x + 12 - x - 6$
$= 3x^2 + 11x + 6 = (3x + 2)(x + 3)$

115. Answers may vary.

Review Problems

116. $(9x + 7)(9x - 7) = 81x^2 + 63x - 63x - 49$
$= 81x^2 - 49$

117. $(5x - 6)^2 = (5x - 6)(5x - 6)$
$= 25x^2 - 60x + 36$

118. $(x + 2)(x^2 - 2x + 4)$
$= x^3 - 2x^2 + 4x + 2x^2 - 4x + 8$
$= x^3 + 8$

Problem Set 7.4

1. $x^2 - 25 = x^2 - 5^2 = (x + 5)(x - 5)$

3. $y^2 - 1 = y^2 - 1^2 = (y + 1)(y - 1)$

5. $4x^2 - 1 = (2x)^2 - 1^2 = (2x + 1)(2x - 1)$

7. $x^2 - 7$ is prime.

9. $9y^2 - 4 = (3y)^2 - 2^2 = (3y + 2)(3y - 2)$

11. $9x^2+4$ is prime.

13. $1-49x^2=1^2-(7x)^2=(1+7x)(1-7x)$

15. $25a^2-16b^2=(5a)^2-(4b)^2$
$=(5a-4b)(5a+4b)$

17. x^2+9 is prime.

19. $16z^2-y^2=(4z)^2-y^2=(4z+y)(4z-y)$

21. $9-121a^2=3^2-(11a)^2$
$=(3+11a)(3-11a)$

23. $(x+1)^2-16=(x+1)^2-4^2$
$=[(x+1)+4][(x+1)-4]=(x+5)(x-3)$

25. $(2x+3)^2-49=(2x+3)^2-7^2$
$=[(2x+3)+7][(2x+3)-7]$
$=(2x+10)(2x-4)=4(x+5)(x-2)$

27. $(3x-1)^2-64=(3x-1)^2-8^2$
$=[(3x-1)+8][(3x-1)-8]$
$=(3x+7)(3x-9)=3(3x+7)(x-3)$

29. $25-(x+3)^2=5^2-(x+3)^2$
$=[5-(x+3)][5+(x+3)]=(2-x)(8+x)$

31. $2y^2-18=2(y^2-9)=2(y^2-3^2)$
$=2(y+3)(y-3)$

33. $2x^3-72x=2x(x^2-36)=2x(x^2-6^2)$
$=2x(x-6)(x+6)$

35. $50-2y^2=2(25-y^2)=2(5^2-y^2)$
$=2(5-y)(5+y)$

37. $8y^3-2y=2y(4y^2-1)=2y[(2y)^2-1^2]$
$=2y(2y-1)(2y+1)$

39. $2x^3-2x=2x(x^2-1)=2x(x+1)(x-1)$

41. $x^4-16=(x^2+4)(x^2-4)$
$=(x^2+4)(x-2)(x+2)$

43. $16y^4-81=(4y^2-9)(4y^2+9)$
$=(4y^2+9)(2y-3)(2y+3)$

45. $1-y^4=(1+y^2)(1-y^2)$
$=(1+y^2)(1+y)(1-y)$

47. $x^8-1=(x^4+1)(x^4-1)$
$=(x^4+1)(x^2+1)(x^2-1)$
$=(x^4+1)(x^2+1)(x-1)(x+1)$

49. $16a^4-b^4=(4a^2+b^2)(4a^2-b^2)$
$=(4a^2+b^2)(2a-b)(2a+b)$

51. $x^2+2x+1=(x+1)(x+1)=(x+1)^2$

53. $x^2-14x+49=(x-7)(x-7)=(x-7)^2$

55. $x^2-2x+1=(x-1)(x-1)=(x-1)^2$

57. $x^2+24x+144=(x+12)(x+12)$
$=(x+12)^2$

59. $4y^2+4y+1=(2y)^2+2(2y)+1$
$=(2y+1)^2$

61. $9r^2-6r+1=(3r)^2-2(3r)+1=(3r-1)^2$

63. $16t^2+1+8t=16t^2+8t+1$
$=(4t)^2+2(4t)+1=(4t+1)^2$

65. $9b^2-42b+49=(3b)^2-2(3b)(7)+7^2$
$=(3b-7)^2$

67. $x^2-10x+100$ is prime.

69. $12k^2-12k+3=3(4k^2-4k+1)$
$=3[(2k)^2-2(2k)+1]=3(2k-1)^2$

71. $9x^3 + 6x^2 + x = x(9x^2 + 6x + 1)$
$= x(3x+1)(3x+1)$
$= x(3x+1)^2$

73. $2y^2 - 4y + 2 = 2(y^2 - 2y + 1)$
$= 2(y-1)(y-1)$
$= 2(y-1)^2$

75. $2y^3 + 28y^2 + 98y = 2y(y^2 + 14y + 49)$
$= 2(y+7)(y+7)$
$= 2(y+7)^2$

77. $25x^2 + 20xy + 4y^2 = (5x+2y)(5x+2y)$
$= (5x+2y)^2$

79. $a^2 - 6ab + 9b^2 = (a-3b)(a-3b)$
$= (a-3b)^2$

81. $4a^2 - 12ab + 9b^2 = (2a-3b)(2a-3b)$
$= (2a-3b)^2$

83. $32x^2 + 80xy + 50y^2$
$= 2(16x^2 + 40xy + 25y^2)$
$= 2(4x+5y)(4x+5y)$
$= 2(4x+5y)^2$

85. $x^3 + 27 = x^3 + 3^3 = (x+3)(x^2 - 3x + 9)$

87. $x^3 - 64 = x^3 - 4^3 = (x-4)(x^2 + 4x + 16)$

89. $8y^3 - 1 = (2y)^3 - 1$
$= (2y-1)[(2y)^2 + 2y + 1]$
$= (2y-1)(4y^2 + 2y + 1)$

91. $64x^3 + 125 = (4x)^3 + 5^3$
$= (4x+5)[(4x)^2 - (4x)(5) + 5^2]$
$= (4x+5)(16x^2 - 20x + 25)$

93. $2x^4 + 16x = 2x(x^3 + 8) = 2x(x^3 + 2^3)$
$= 2x(x+2)(x^2 - 2x + 4)$

95. $27y^4 - 8y = y(27y^3 - 8)$
$= y[(3y)^3 - (2)^3]$
$= y(3y-2)[(3y)^2 + 3y \cdot 2 + (2)^2]$
$= y(3y-2)(9y^2 + 6y + 4)$

97. $54 - 16y^3 = 2(27 - 8y^3)$
$= 2(3-2y)(9 + 6y + 4y^2)$

99. $64x^3 + 27y^3$
$= (4x+3y)(16x^2 - 12xy + 9y^2)$

101. $125x^3 - 64y^3$
$= (5x-4y)(25x^2 + 20xy + 16y^2)$

103. Statement d is true.
$2x^2 - 18 = 2(x^2 - 9) = 2(x-3)(x+3)$
Three distinct factors.

105. $4x^2 - 9 = (2x-3)(2x+3)$

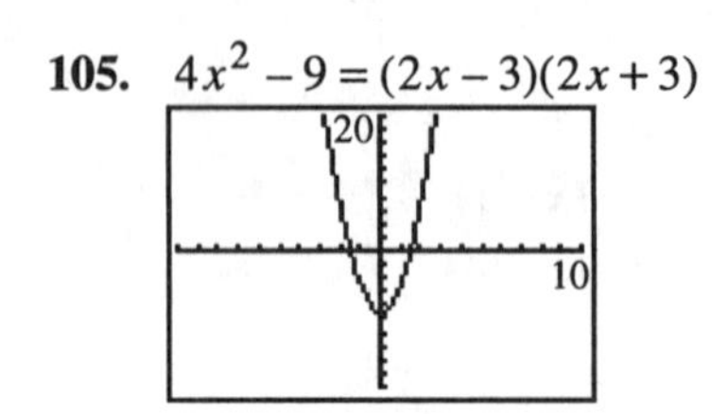

107. $4x^2 - 4x + 1 = (2x-1)(2x-1)$ or $(2x-1)^2$

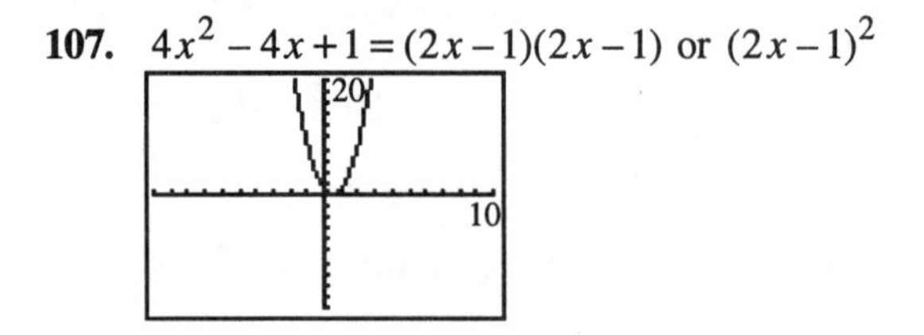

109. Answers may vary.

111. $A = x^2 - 5^2 = (x-5)(x+5)$

113. $A = x^2 - 4 \cdot 3^2 = x^2 - 6^2 = (x-6)(x+6)$

115. $1000^2 - 999^2 = (1000-999)(1000+999)$
$= (1)(1999) = 1999$

117. $1000^2 - 990^2 = (1000 - 990)(1000 + 990)$
$= 10(1990) = 19{,}900$

119. $9x^2 + kx + 1 = (3x + 1)(3x + 1)$
$3x + 3x = 6x,\ k = 6$

121. $x^5 - 1 = (x - 1)(x^4 + x^3 + x^2 + x + 1)$

123. Answers may vary.

Review Problems

125. $\left(\frac{3x^2}{2}\right)^4 = \left(\frac{3^4 x^{2\cdot 4}}{2^4}\right) = \frac{81x^8}{16}$

126. $6 - 2x > 4x - 12$
$6 + 12 > 2x + 4x$
$18 > 6x$
$3 > x$

1 2 3 4 5

127. $2x + 5 = 12 - 6x + 3(2x + 3)$
$2x + 5 = 12 - 6x + 6x + 9$
$2x + 5 = 21$
$2x = 16$
$x = 8$

Problem Set 7.5

1. $3x^3 - 3x = 3x(x^2 - 1) = 3x(x - 1)(x + 1)$

3. $3x^3 + 3x = 3x(x^2 + 1)$

5. $4x^2 - 4x - 24 = 4(x^2 - x - 6)$
$= 4(x - 3)(x + 2)$

7. $2x^4 - 162 = 2(x^4 - 81) = 2(x^2 + 9)(x^2 - 9)$
$= 2(x^2 + 9)(x - 3)(x + 3)$

9. $x^3 + 2x^2 - 9x - 18$
$= (x^3 + 2x^2) + (-9x - 18)$
$= x^2(x + 2) - 9(x + 2) = (x^2 - 9)(x + 2)$
$= (x - 3)(x + 3)(x + 2)$

11. $3x^3 - 30x^2 + 75x = 3x(x^2 - 10x + 25)$
$= 3x(x - 5)^2$

13. $2x^5 + 54x^2 = 2x^2(x^3 + 27)$
$= 2x^2(x + 3)(x^2 - 3x + 9)$

15. $6x^2 + 8x = 2x(3x + 4)$

17. $2y^2 - 2y - 112 = 2(y^2 - y - 56)$
$= 2(y + 7)(y - 8)$

19. $7y^4 + 14y^3 + 7y^2 = 7y^2(y^2 + 2y + 1)$
$= 7y^2(y + 1)^2$

21. $y^2 + 8y - 16$ is prime.

23. $16y^2 - 4y - 2 = 2(8y^2 - 2y - 1)$
$= 2(2y - 1)(4y + 1)$

25. $r^2 - 25r = r(r - 25)$

27. $4w^2 + 8w - 5 = (2w + 5)(2w - 1)$

29. $x^3 - 4x = x(x^2 - 4) = x(x + 2)(x - 2)$

31. $x^2 + 64$ is prime.

33. $9y^2 + 13y + 4 = (9y + 4)(y + 1)$

35. $y^3 + 2y^2 - 4y - 8 = y^2(y + 2) - 4(y + 2)$
$= (y + 2)(y^2 - 4) = (y + 2)(y - 2)(y + 2)$
$= (y - 2)(y + 2)^2$

37. $9y^2 + 24y + 16 = (3y + 4)^2$

39. $5y^3 - 45y^2 + 70y = 5y(y^2 - 9y + 14)$
$= 5y(y - 2)(y - 7)$

41. $y^5 - 81y = y(y^4 - 81) = y(y^2 - 9)(y^2 + 9)$
$= y(y^2 + 9)(y - 3)(y + 3)$

43. $20a^4 - 45a^2 = 5a^2(4a^2 - 9)$
$= 5a^2(2a-3)(2a+3)$

45. $12y^2 - 11y + 2 = (3y-2)(4y-1)$

47. $9y^2 - 64 = (3y-8)(3y+8)$

49. $9y^2 + 64$
The polynomial is prime.

51. $2y^3 + 3y^2 - 50y - 75$
$= (2y^3 + 3y^2) + (-50y - 75)$
$= y^2(2y+3) - 25(2y+3)$
$= (y^2 - 25)(2y+3)$
$= (y-5)(y+5)(2y+3)$

53. $-6x^2 - x + 1 = (-3x+1)(2x+1)$
$= -(3x-1)(2x+1)$

55. $2r^3 + 30r^2 - 68r = 2r(r^2 + 15r - 34)$
$= 2r(r+17)(r-2)$

57. $8x^5 - 2x^3 = 2x^3(4x^2 - 1)$
$= 2x^3(2x-1)(2x+1)$

59. $3x^2 + 243 = 3(x^2 + 81)$

61. $x^4 + 8x = x(x^3 + 8)$
$= x(x+2)(x^2 - 2x + 4)$

63. $2y^5 - 2y^2 = 2y^2(y^3 - 1)$
$= 2y^2(y-1)(y^2 + y + 1)$

65. $6x^2 + 8xy = 2x(3x + 4y)$

67. $xy - 7x + 3y - 21 = x(y-7) + 3(y-7)$
$= (x+3)(y-7)$

69. $x^2 - 3xy - 4y^2 = (x-4y)(x+y)$

71. $72a^3b^2 + 12a^2 - 24a^4b^2$
$= 12a^2(6ab^2 + 1 - 2a^2b^2)$

73. $3a^2 + 27ab + 54b^2 = (a+6b)(3a+9b)$

75. $48x^4y - 3x^2y = 3x^2y(16x^2 - 1)$
$= 3x^2y(4x+1)(4x-1)$

77. $6a^2b + ab - 2b = b(6a^2 + a - 2)$
$= b(3a+2)(2a-1)$

79. $7x^5y - 7xy^5 = 7xy(x^4 - y^4)$
$= 7xy(x^2 - y^2)(x^2 + y^2)$
$= 7xy(x^2 + y^2)(x-y)(x+y)$

81. $24a^2b + 6a^3b - 45a^4b$
$= 3a^2b(8 + 2a - 15a^2)$
$= 3a^2b(2+3a)(4-5a)$

83. $2bx^2 + 44bx + 242b = 2b(x^2 + 22x + 121)$
$= 2b(x+11)^2$

85. $15a^2 + 11ab - 14b^2 = (3a-2b)(5a+7b)$

87. $36x^3y - 62x^2y^2 + 12xy^3$
$= 2xy(18x^2 - 31xy + 6y^2)$
$= 2xy(2x-3y)(9x-2y)$

89. $a^2y - b^2y - a^2x + b^2x$
$= y(a^2 - b^2) - x(a^2 - b^2)$
$= (y-x)(a^2 - b^2) = (y-x)(a-b)(a+b)$

91. $9ax^3 + 15ax^2 - 14ax = ax(9x^2 + 15x - 14)$
$= ax(3x+7)(3x-2)$

93. $81x^4y - y^5 = y(81x^4 - y^4)$
$= y(9x^2 - y^2)(9x^2 + y^2)$
$= y(9x^2 + y^2)(3x+y)(3x-y)$

95. height $= x$
$x^3 - 60x^2 + 900x = x(x^2 - 60x + 900)$
$= x(x-30)^2$
x ft by $(x-30)$ ft by $(x-30)$ ft

97. t = time in seconds
$-16t^2 + 256 = -16(t^2 - 16)$
$= -16(t+4)(t-4)$
$t - 4 = 0$ when $t = 4$
4 seconds

99. b is true.

101. $3x^3 - 12x^2 - 15x = 3x(x^2 - 4x - 5)$
$= 3x(x-5)(x+1)$

103. $x^4 - 16 = (x^2 + 4)(x-2)(x+2)$

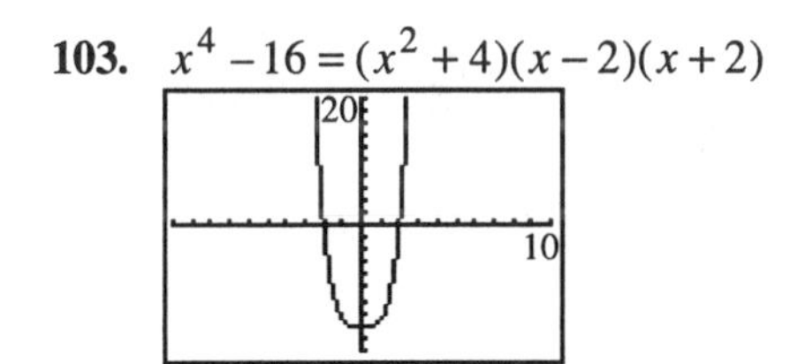

105. Answers may vary.

107. Answers may vary.

109. $5y^5 - 5y^4 - 20y^3 + 20y^2$
$= 5y^2(y^3 - y^2 - 4y + 4)$
$= 5y^2[y^2(y-1) - 4(y-1)]$
$= 5y^2(y^2 - 4)(y-1)$
$= 5y^2(y-1)(y+2)(y-2)$

111. $(x+5)^2 - 20(x+5) + 100 = [(x+5) - 10]^2$
$= (x-5)^2$

113. Answers may vary.

Review Problems

114. $5x - 2y > 10$
$-2y > 10 - 5x$
$y < \frac{5}{2}x - 5$

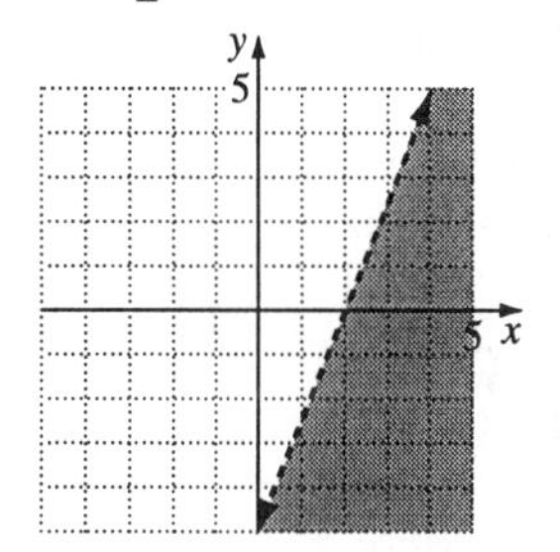

115. $x = 13 - 3y; x + y = 5$
$(13 - 3y) + y = 5$
$13 - 2y = 5$
$-2y = -8$
$y = 4$
$x = 13 - 3(4) = 1$
$(1, 4)$

116. Let x = the measure of 1st angle. Then
$3x$ = the measure of the 2nd angle and
$x + 30$ = the measure of the 3rd angle.
$x + 3x + (x + 30) = 180$
$5x + 30 = 180$
$5x = 150$
$x = 30$
$3x = 90$
$x + 30 = 60$
The measures of the angles are 30°, 60°, and 90°.

Problem Set 7.6

1. $(x-2)(x+3) = 0$
$x - 2 = 0$ or $x + 3 = 0$
(Apply zero-product principle)
$x = 2$ or $x = -3$
Check: $x = 2$:
$(2-2)(2+3) = 0$
$0(5) = 0$

$0 = 0$ True
$x = -3$:
$(-3 - 2)(-3 + 3) = 0$
$(-5)(0) = 0$
$0 = 0$ True
The solutions are $-3, 2$.

3. $(3x + 4)(2x - 1) = 0$
$3x + 4 = 0$ or $2x - 1 = 0$
$x = -\frac{4}{3}$ or $x = \frac{1}{2}$
Solution: $-\frac{4}{3}, \frac{1}{2}$

5. $(2y - 1)(4y + 1) = 0$
$2y - 1 = 0$ or $4y + 1 = 0$
$y = \frac{1}{2}$ or $y = -\frac{1}{4}$
$-\frac{1}{4}, \frac{1}{2}$

7. $(2y + 7)(3y - 1) = 0$
$2y + 7 = 0$ or $3y - 1 = 0$
$y = -\frac{7}{2}$ or $y = \frac{1}{3}$
$-\frac{7}{2}, \frac{1}{3}$

9. $(z - 2)(3z + 7) = 0$
$z - 2 = 0$ or $3z + 7 = 0$
$z = 2$ or $z = -\frac{7}{3}$
$-\frac{7}{3}, 2$

11. $(4w - 9)(2w + 5) = 0$
$4w - 9 = 0$ or $2w + 5 = 0$
$w = \frac{9}{4}$ or $w = -\frac{5}{2}$
$-\frac{5}{2}, \frac{9}{4}$

13. $x^2 + 8x + 15 = 0$
Factor the left-hand side and apply the zero-product principle.
$(x + 3)(x + 5) = 0$
Apply the zero-product principle
$x + 3 = 0$ or $x + 5 = 0$
$x = -3$ or $x = -5$
Check:
$x = -3$:
$(-3)^2 + 8(-3) + 15 = 9 - 24 + 15 = 0$ True
$x = -5$:
$(-5)^2 + 8(-5) + 15 = 25 - 40 + 15 = 0$ True
Solution: $-5, -3$

15. $y^2 - 2y - 15 = 0$
$(y - 5)(y + 3) = 0$
$y - 5 = 0$ or $y + 3 = 0$
$y = 5$ or $y = -3$
$-3, 5$

17. $m^2 - 4m = 21$
$m^2 - 4m - 21 = 0$
$(m - 7)(m + 3) = 0$
$m - 7 = 0$ or $m + 3 = 0$
$m = 7$ or $m = -3$
$-3, 7$

19. $z^2 + 9z = -8$
$z^2 + 9z + 8 = 0$
$(z + 8)(z + 1) = 0$
$z + 8 = 0$ or $z + 1 = 0$
$z = -8$ or $z = -1$
$-8, -1$

21. $y^2 + 4y = 0$
$y(y + 4) = 0$
$y = 0$ or $y + 4 = 0$
$y = -4$
$-4, 0$

23. $x^2 - 5x = 0$
$x(x - 5) = 0$
$x = 0$ or $x - 5 = 0$
$x = 5$
$0, 5$

25. $x^2 = 4x$
$x^2 - 4x = 0$
$x(x-4) = 0$
$x = 0$ or $x - 4 = 0$
$x = 4$
0, 4

27. $2x^2 = 5x$
$2x^2 - 5x = 0$
$x(2x-5) = 0$
$2x - 5 = 0$ or $x = 0$
$2x = 5$
$x = \frac{5}{2}$
$0, \frac{5}{2}$

29. $3x^2 = -5x$
$3x^2 + 5x = 0$
$x(3x+5) = 0$
$3x + 5 = 0$ or $x = 0$
$3x = -5$
$x = -\frac{5}{3}$
$0, -\frac{5}{3}$

31. $x^2 + 4x + 4 = 0$
$(x+2)^2 = 0$
$x + 2 = 0$
$x = -2$
-2

33. $x^2 - 12x = 36$
$x^2 - 12x - 36$ is prime.

35. $4x^2 - 12x = 9$
$4x^2 - 12x - 9$ is prime.

37. $2x^2 = 7x + 4$
$2x^2 - 7x - 4 = 0$
$(2x+1)(x-4) = 0$
$2x + 1 = 0$ or $x - 4 = 0$
$x = -\frac{1}{2}$ or $x = 4$
$-\frac{1}{2}, 4$

39. $5x^2 + x = 18$
$5x^2 + x - 18 = 0$
$(5x-9)(x+2) = 0$
$5x - 9 = 0$ or $x + 2 = 0$
$x = \frac{9}{5}$ or $x = -2$
$-2, \frac{9}{5}$

41. $x(6x+32) + 7 = 0$
$6x^2 + 23x + 7 = 0$
$(2x+7)(3x+1) = 0$
$2x + 7 = 0$ or $3x + 1 = 0$
$x = -\frac{7}{2}$ or $x = -\frac{1}{3}$
$-\frac{7}{2}, -\frac{1}{3}$

43. $3s^2 + 4s = -1$
$3s^2 + 4s + 1 = 0$
$(3s+1)(s+1) = 0$
$3s + 1 = 0$ or $s + 1 = 0$
$s = -\frac{1}{3}$ or $s = -1$
$-1, -\frac{1}{3}$

45. $4x(x+1) = 15$
$4x^2 + 4x - 15 = 0$
$(2x+5)(2x-3) = 0$
$2x + 5 = 0$ or $2x - 3 = 0$
$x = -\frac{5}{2}$ or $x = \frac{3}{2}$
$-\frac{5}{2}, \frac{3}{2}$

47. $12r^2 + 31r + 20 = 0$
$(3x + 4)(4x + 5) = 0$
$3x + 4 = 0$ or $4x + 5 = 0$
$r = -\frac{4}{3}$ or $r = -\frac{5}{4}$
$-\frac{4}{3}, -\frac{5}{4}$

49. $12s^2 + 28s - 24 = 0$
$4(3s^2 + 7s - 6) = 0$
$4(3s - 2)(s + 3) = 0$
$3s - 2 = 0$ or $s + 3 = 0$
$s = \frac{2}{3}$ or $s = -3$
$-3, \frac{2}{3}$

51. $w^2 - 5w = 18 + 2w$
$w^2 - 7w - 18 = 0$
$(w - 9)(w + 2) = 0$
$w - 9 = 0$ or $w + 2 = 0$
$w = 9$ or $w = -2$
$-2, 9$

53. $z(z + 8) = 16(z - 1)$
$z^2 + 8z - 16z + 16 = 0$
$z^2 - 8z + 16 = 0$
$(z - 4)^2 = 0$
$z - 4 = 0$
$z = 4$
4

55. $16x^2 - 49 = 0$
$(4x - 7)(4x + 7) = 0$
$4x - 7 = 0$ or $4x + 7 = 0$
$4x = 7$ or $4x = -7$
$x = \frac{7}{4}$ or $x = -\frac{7}{4}$
$-\frac{7}{4}, \frac{7}{4}$

57. $(y - 3)(y + 8) = -30$
$y^2 + 5y - 24 = -30$
$y^2 + 5y + 6 = 0$
$(y + 2)(y + 3) = 0$
$y + 2 = 0$ or $y + 3 = 0$
$y = -2$ or $y = -3$
$-3, -2$

59. $(z + 1)(2z + 5) = -1$
$2z^2 + 7z + 5 = -1$
$2z^2 + 7z + 6 = 0$
$(2z + 3)(z + 2) = 0$
$2z + 3 = 0$ or $z + 2 = 0$
$z = -\frac{3}{2}$ or $z = -2$
$-2, -\frac{3}{2}$

61. $4y^2 + 20y + 25 = 0$
$(2y + 5)^2 = 0$
$2y + 5 = 0$
$y = -\frac{5}{2}$
$-\frac{5}{2}$

63. $64w^2 - 48w + 9 = 0$
$(8w - 3)^2 = 0$
$8w - 3 = 0$
$w = \frac{3}{8}$
$\frac{3}{8}$

65. $f(t) = -16t^2 + 128t$

a. $-16t^2 + 128t = 0$
$-16(t^2 - 8t) = 0$
$-16t(t - 8) = 0$
$t = 0$ or $t - 8 = 0$
$t = 8$ seconds (reject $t = 0$)

b.

Time t	Height $f(t) = -16t^2 + 128t$
0	0
1	112
2	192
3	240
4	256
5	240
6	192
7	112
8	0

c.

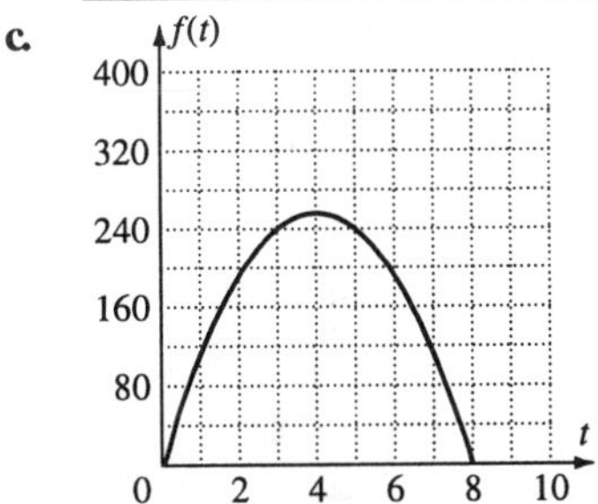

d. The maximum height is reached at $t = 4$ seconds, and the height is 256 ft.

e. The height is 0 at $t = 8$ seconds when the projectile hits the ground. (8, 0) is a point on the graph.

67. a. $3500 + 475t - 10t^2 = 7250$
$-3750 + 475t - 10t^2 = 0$
$-5(750 - 95t + 2t^2) = 0$
$-5(10 - t)(75 - 2t) = 0$
$10 - t = 0$ or $75 - 2t = 0$
$-t = -10$ or $-2t = -75$
$t = 10$ or $t = 37.5$
(not valid since $0 < t \leq 20$)
10 years

b. (10, 7250)

69. $S = \frac{n^2 + n}{2} = 91$
$n^2 + n - 182 = 0$
$(n + 14)(n - 13) = 0$
$n = 13$ numbers (reject $n = -14$)

71. x = number of VCR's
$13 + x^2 = 14x$
$x^2 - 14x + 13 = 0$
$(x - 13)(x - 1) = 0$
$x = 13$ or $x = 1$
13 years

73. $xy = 110$
$y = x + 1$
$x(x + 1) = 110$
$x^2 + x - 110 = 0$
$(x + 11)(x - 10) = 0$
$x + 11 = 0$ or $x - 10 = 0$
$x = -11$ or $x = 10$
Pages 10 and 11

75. w = width of lot
length = $w + 3$
$(w + 3)w = 180$
$w^2 + 3w - 180 = 0$
$(w + 15)(w - 12) = 0$
$w + 15 = 0$ or $w - 12 = 0$
$w = -15$ or $w = 12$
$w + 3 = 15$
width = 12 yds, length = 15 yds

77. w = width of rug
$w + 5$ = length or rug
$w(w + 5) = 2w + 2(w + 5) + 10$
$w^2 + 5w = 2w + 2w + 10 + 10$
$w^2 + 5w = 4w + 20$
$w^2 + w - 20 = 0$

$(w + 5)(w - 4) = 0$
$w + 5 = 0$ or $w - 4 = 0$
$w = -5$ or $w = 4$
$4 + 5 = 9$
The rug is 4 ft by 9 ft.

79. $x =$ base
$2x - 1 =$ height
$\frac{1}{2}x(2x-1) = 60$
$x(2x-1) = 120$
$2x^2 - x - 120 = 0$
$(2x + 15)(x - 8) = 0$
$x = -\frac{15}{2}$ or $x = 8$
$2(8) - 1 = 15$
base = 8 in., height = 15 in.

81. **a.** $(2x + 12)(2x + 15)$
$= 4x^2 + 54x + 180$

b. $f(x) = 4x^2 + 54x + 180$

c. $4x^2 + 54x + 180 = 378$
$4x^2 + 54x - 198 = 0$
$2(2x^2 + 27x - 99) = 0$
$2(2x + 33)(x - 3) = 0$
$2x + 33 = 0$ or $x - 3 = 0$
$2x = -33$ or $x = 3$
$x = -\frac{33}{2}$
3 m

d.

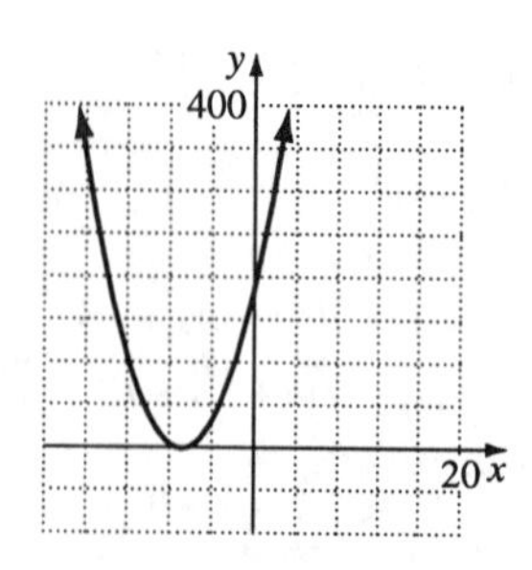

83. $x^2 + (x-7)^2 = (x+1)^2$
$x^2 + x^2 - 14x + 49 = x^2 + 2x + 1$
$2x^2 - 14x + 49 - x^2 - 2x - 1 = 0$
$x^2 - 16x + 48 = 0$
$(x - 4)(x - 12) = 0$
$x - 4 = 0$ or $x - 12 = 0$
$x = 4$ or $x = 12$
12 m and 5 m

85. $2W + 2L = 34$
$W + L = 17$
$W = 17 - L$
$W^2 + L^2 = 13^2$
$(17 - L)^2 + L^2 = 169$
$289 - 34L + L^2 + L^2 = 169$
$289 - 34L + 2L^2 = 169$
$2L^2 - 34L + 120 = 0$
$2(L^2 - 17L + 60) = 0$
$2(L - 12)(L - 5) = 0$
$L - 12 = 0$ or $L - 5 = 0$
$L = 12$ or $L = 5$
The lake is 12 ft × 5 ft.

87. a is true

89.

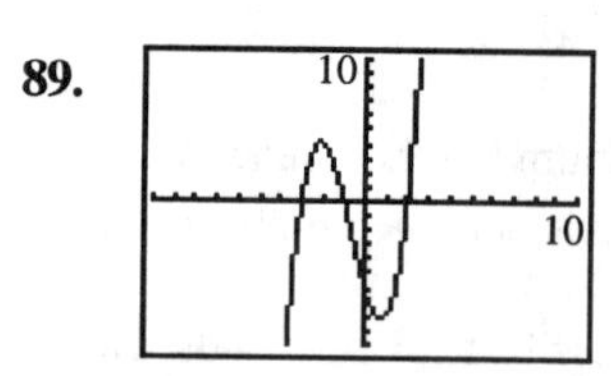

$y = x^3 + 2x^2 - 5x - 6$
$y = (x + 3)(x + 1)(x - 2)$
$x = -3$ or $x = -1$ or $x = 2$

91. Students should verify.

93. Students should verify.

95. Answers may vary.

97. $x^3+3x^2-10x=0$
$x(x+5)(x-2)=0$
$x=0$ or $x+5=0$ or $x-2=0$
$x=0$ or $x=-5$ or $x=2$
$-5, 0, 2$

99. $(x^2-5x+5)^3=1$
$x^2-5x+5=1$
since $1^3=1$
$x^2-5x+4=0$
$(x-1)(x-4)=0$
$x-4=0$ or $x-4=0$
$x=1$ or $x=4$
$1, 4$

101. Let x = the width of the rectangle.
$x+3$ = length of the rectangle
$x+2$ = width of new rectangle
$x+3+2=x+5$ = length of new rectangle
$(x+2)(x+5)=54$
$x^2+7x+10=54$
$x^2+7x-44=0$
$(x+11)(x-4)=0$
$x=4$ (reject $x=-11$)
$x+3=4+3=7$
dimensions of original rectangle:
width, 4ft; length, 7 ft

103. Area of trapezoid $=\frac{1}{2}h(b_1+b_2)$
$30=\frac{x}{2}[(3+2x-2)+2x-2]$
$60=x(4x-1)$
$60=4x^2-x$
$4x^2-x-60=0$
$(4x+15)(x-4)=0$
$x=4$ $\left(\text{reject } x=-\frac{15}{4}\right)$
Area of triangle $=\frac{3x}{2}=\frac{3(4)}{2}$
= 6 square inches

105. x = length of garden
$x(x-2)=63$
$x^2-2x-63=0$
$(x+7)(x-9)=0$
$x+7=0$ or $x-9=0$
$x=-7$ or $x=9$
The garden is 7 yds by 9 yds.

Review Problems

107. $y>-\frac{2}{3}x+1$

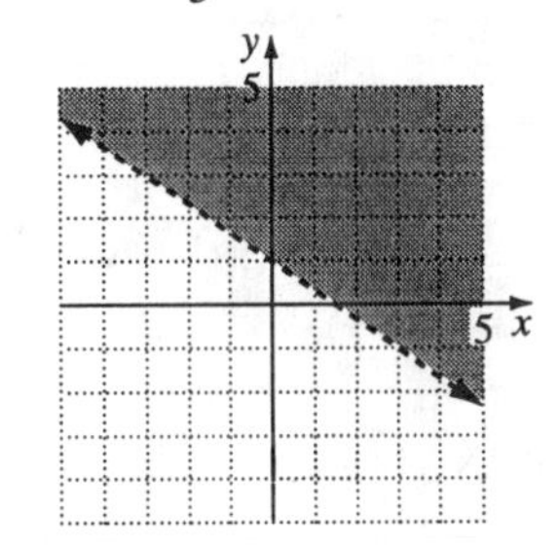

108. $\left(\frac{8x^4}{4x^7}\right)^{-2}=\left(\frac{2}{x^3}\right)^{-2}=\left(\frac{x^3}{2}\right)^2=\frac{x^6}{4}$

109.
$$\begin{array}{r} y^2-2y+5 \\ 3y-5\overline{)3y^3-11y^2+25y-25} \\ \underline{3y^3-5y^2} \qquad\qquad \\ -6y^2+25y \qquad \\ \underline{-6y^2+10y} \qquad \\ 15y-25 \\ \underline{15y-25} \\ 0 \end{array}$$

$\frac{3y^3-11y^2+25y-25}{3y-5}=y^2-2y+5$

Chapter 7 Review Problems

1. $9y^2-18y=9y(y-2)$

2. $x^2-11x+28=(x-4)(x-7)$

3. $y^3-8y^2+7y=y(y-7)(y-1)$

4. $10r^2+9r+2=(5r+2)(2r+1)$

5. $15z^2 - z - 2 = (5z - 2)(3z + 1)$

6. $x^2 - 144 = (x + 12)(x - 12)$

7. $64 - y^2 = (8 - y)(8 + y)$

8. $9r^2 + 6r + 1 = (3r + 1)^2$

9. $20a^7 - 36a^3 = 4a^3(5a^4 - 9)$

10. $8x^5 + 6x^2 - 20x^3 - 15$
$= 2x^2(4x^3 + 3) - 5(4x^3 + 3)$
$= (2x^2 - 5)(4x^3 + 3)$

11. $x^3 - 3x^2 - 9x + 27 = x^2(x - 3) - 9(x - 3)$
$= (x^2 - 9)(x - 3) = (x + 3)(x - 3)^2$

12. $12y^2 + 11y - 5 = (4y + 5)(3y - 1)$

13. $16x^2 - 40x + 25 = (4x - 5)^2$

14. $r^2 + 16$ is prime.

15. $2x^3 + 19x^2 + 35x = x(2x^2 + 19x + 35)$
$= x(2x + 5)(x + 7)$

16. $3x^3 - 30x^2 + 75x = 3x(x^2 - 10x + 25)$
$= 3x(x - 5)^2$

17. $10z^2 + 37z + 7 = (5z + 1)(2z + 7)$

18. $3x^5 - 24x^2 = 3x^2(x^3 - 8)$
$= 3x^2(x - 2)(x^2 + 2x + 4)$

19. $4y^4 - 36y^2 = 4y^2(y^2 - 9)$
$= 4y^2(y - 3)(y + 3)$

20. $36y^2 - 59y - 7 = (9y + 1)(4y - 7)$

21. $5x^2 + 20x - 105 = 5(x^2 + 4x - 21)$
$= 5(x + 7)(x - 3)$

22. $9r^2 + 8r - 3$ is prime.

23. $10x^5 - 44x^4 + 16x^3$
$= 2x^3(5x^2 - 22x + 8) = 2x^3(5x - 2)(x - 4)$

24. $40x^2 + 17x - 12 = (5x + 4)(8x - 3)$

25. $486z^2 - 24 = 6(81z^2 - 4)$
$= 6(9z - 2)(9z + 2)$

26. $48r^2 - 120r + 75 = 3(16r^2 - 40r + 25)$
$= 3(4r - 5)^2$

27. $3y^4 - 9y^3 - 30y^2 = 3y^2(y^2 - 3y - 10)$
$= 3y^2(y - 5)(y + 2)$

28. $100y^2 - 49 = (10y - 7)(10y + 7)$

29. $256x^4 - 1 = (16x^2 - 1)(16x^2 + 1)$
$= (4x - 1)(4x + 1)(16x^2 + 1)$

30. $9x^5 - 18x^4 = 9x^4(x - 2)$

31. $3w^2 + w - 5$ is prime.

32. $64y^2 - 144y + 81 = (8y - 9)^2$

33. $x^2 + x + 1$ is prime.

34. $x^4 - 16 = (x^2 - 4)(x^2 + 4)$
$= (x - 2)(x + 2)(x^2 + 4)$

35. $y^3 - 8 = (y - 2)(y^2 + 2y + 4)$

36. $x^3 + 64 = (x + 4)(x^2 - 4x + 16)$

37. $-10y^2 + 31y - 15 = -(5y - 3)(2y - 5)$

38. $6x^2 + 11x - 10 = (2x + 5)(3x - 2)$

39. $3x^4 - 12x^2 = 3x^2(x^2 - 4)$
$= 3x^2(x - 2)(x + 2)$

40. $3r^4 + 12r^2 = 3r^2(r^2 + 4)$

41. $56y^3 - 70y^2 + 21y = 7y(8y^2 - 10y + 3)$
$= 7y(2y - 1)(4y - 3)$

42. $a^2 + 4a + 16$ is prime.

43. $s^2 - s - 90 = (s - 10)(s + 9)$

44. $x^2 - 6x - 27 = (x - 9)(x + 3)$

45. $8y^2 - 14y - 5$ is prime.

46. $25x^2 + 25x + 6 = (5x + 2)(5x + 3)$

47. $p^4 + 125p = p(p^3 + 125)$
$= p(p + 5)(p^2 - 5p + 25)$

48. $32y^3 + 32y^2 + 6y = 2y(16y^2 + 16y + 3)$
$= 2y(4y + 1)(4y + 3)$

49. $16x^5 - 25x^7 = x^5(16 - 25x^2)$
$= x^5(4 + 5x)(4 - 5x)$

50. $2y^2 - 16y + 32 = 2(y^2 - 8y + 16)$
$= 2(y - 4)^2$

51. $12x^4y^3 - 9x^3y^2 + 15x^2y$
$= 3x^2y(4x^2y^2 - 3xy + 5)$

52. $x^2 - 2xy - 35y^2 = (x + 5y)(x - 7y)$

53. $a^2b^2 + ab - 12 = (ab + 4)(ab - 3)$

54. $15x^2 - 11xy + 2y^2 = (3x - y)(5x - 2y)$

55. $x^2 + 7x + xy + 7y = x(x + 7) + y(x + 7)$
$= (x + y)(x + 7)$

56. $9a^2 + 24ab + 16b^2 = (3a + 4b)^2$

57. $4x^2 - 20xy + 25y^2 = (2x - 5y)^2$

58. $20a^7b^2 - 36a^3b^4 = 4a^3b^2(5a^4 - 9b^2)$

59. $4x^2 - 20x + 2xy - 10y$
$= 4x(x - 5) + 2y(x - 5)$
$= (4x + 2y)(x - 5)$
$= 2(2x + y)(x - 5)$

60. $2x^4y - 2x^2y = 2x^2y(x^2 - 1)$
$= 2x^2y(x + 1)(x - 1)$

61. $39a^2b - 52a + 13ab^4 = 13a(3ab - 4 + b^4)$

62. $100y^2 - 49z^2 = (10y + 7z)(10y - 7z)$

63. $9x^5y^2 - 18x^4y^5 = 9x^4y^2(x - 2y^3)$

64. $x^2 + xy + y^2$ is prime.

65. $a^2q + a^2z - p^2q - p^2z$
$= a^2(q + z) - p^2(q + z)$
$= (a^2 - p^2)(q + z) = (a - p)(a + p)(q + z)$

66. $x^2y^2 - 16x^2 - 4y^2 + 64$
$= x^2(y^2 - 16) - 4(y^2 - 16)$
$= (x^2 - 4)(y^2 - 16)$
$= (x - 2)(x + 2)(y - 4)(y + 4)$

67. $3x^4y^2 - 12x^2y^4 = 3x^2y^2(x^2 - 4y^2)$
$= 3x^2y^2(x - 2y)(x + 2y)$

68. $125x^3 - 8y^3$
$= (5x - 2y)(25x^2 + 10xy + 4y^2)$

69. $y^2 + 5y = 14$
$y^2 + 5y - 14 = 0$
$(y + 7)(y - 2) = 0$
$y = -7$ or $y = 2$
$-7, 2$

70. $x(x-4)=32$
$x^2-4x-32=0$
$(x-8)(x+4)=0$
$x=8$ or $x=-4$
$-4, 8$

71. $8w^2-37w+20=0$
$(8w-5)(w-4)=0$
$w=\frac{5}{8}$ or $w=4$
$\frac{5}{8}, 4$

72. $2x^2+15x=8$
$2x^2+15x-8=0$
$(2x-1)(x+8)=0$
$2x-1=0$ or $x+8=0$
$x=\frac{1}{2}$ or $x=-8$
$\frac{1}{2}, -8$

73. $5x^2+20x=0$
$5x(x+4)=0$
$5x=0$ or $x+4=0$
$x=0$ or $x=-4$
$0, -4$

74. $3x^2=-21x-30$
$3x^2+21x+30=0$
$3(x^2+7x+10)=0$
$3(x+5)(x+2)=0$
$x+5=0$ or $x+2=0$
$x=-5$ or $x=-2$
$-5, -2$

75. $a^2-3^2=(a+3)(a-3)$

76. $a^2-4b^2=(a+2b)(a-2b)$

77. $\pi a^2-\pi(a-b)^2=\pi[a^2-(a-b)^2]$
$=\pi[a+(a-b)][a-(a-b)]$
$=\pi(2a-b)(b)$
$=\pi b(2a-b)$

78. $\frac{4}{3}\pi a^3-\frac{4}{3}\pi b^3=\frac{4}{3}\pi(a^3-b^3)$
$=\frac{4}{3}\pi(a-b)(a^2+2ab+b^2)$

79. $f(t)=-16t^2+16t+32$

a. $f(t)=0$:
$-16(t^2-t-2)=0$
$-16(t-2)(t+1)=0$
2 seconds (reject $t=-1$)

b.

Time t	Height $f(t)=-16t^2+16t+32$
0	32
$\frac{1}{2}$	36
1	32
$1\frac{1}{2}$	20
2	0

c.

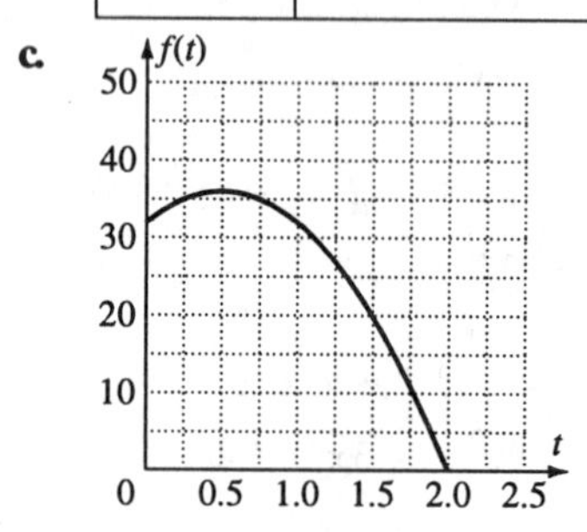

d. Maximum height of 36 ft is reached at $\frac{1}{2}$ second.

e. (2, 0) is a point on the graph.

80. t = number of truck routes;
n = number of cities
$21=\frac{n^2-n}{2}$
$42=n^2-n$
$n^2-n-42=0$
$(n-7)(n+6)=0$

$n - 7 = 0$ or $n + 6 = 0$
$n = 7$ or $n = -6$
7 cities

81. x = robbery victimization rate for African-Americans
$(x + 2)(x - 6) = 180$
$x^2 - 4x - 12 = 180$
$x^2 - 4x - 192 = 0$
$(x + 12)(x - 16) = 0$
$x + 12 = 0$ or $x - 16 = 0$
$x = -12$ or $x = 16$
There are 16 African American victims per 1000 whereas there are 11 Hispanic victims and 5 white victims per 1000.

82. x = width of calculator
$x(x + 5) = 84$
$x^2 + 5x - 84 = 0$
$(x + 12)(x - 7) = 0$
$x + 12 = 0$ or $x - 7 = 0$
$x = -12$ or $x = 7$
width = 7 cm, length = 12 cm

83. x = length of a garden side
$x(x - 3) = 88$
$x^2 - 3x - 88 = 0$
$(x + 8)(x - 11) = 0$
$x + 8 = 0$ or $x - 11 = 0$
$x = -8$ or $x = 11$
11 m by 11 m

84. b = base of sail
$\frac{1}{2}b(b + 4) = 30$
$b^2 + 4b = 60$
$b^2 + 4b - 60 = 0$
$(b + 10)(b - 6) = 0$
$b + 10 = 0$ or $b - 6 = 0$
$b = -10$ or $b = 6$
base = 6 m, height = 10 m

85. $x^2 + (x + 2)^2 = 10^2$
$x^2 + x^2 + 4x + 4 = 100$
$2x^2 + 4x + 4 - 100 = 0$
$2x^2 + 4x - 96 = 0$
$2(x^2 + 2x - 48) = 0$
$2(x + 8)(x - 6) = 0$
$x + 8 = 0$ or $x - 6 = 0$
$x = -8$ or $x = 6$
$x + 2 = 8$
6 ft and 8 ft

Chapter 7 Test

1. $x^2 - 9x + 18 = (x - 3)(x - 6)$

2. $x^2 - 14x + 49 = (x - 7)(x - 7) = (x - 7)^2$

3. $15y^4 - 35y^3 + 10y^2 = 5y^2(3y^2 - 7y + 2)$
$= 5y^2(3y - 1)(y - 2)$

4. $x^3 + 2x^2 + 3x + 6 = x^2(x + 2) + 3(x + 2)$
$= (x^2 + 3)(x + 2)$

5. $x^2 - 9x = x(x - 9)$

6. $x^3 + 6x^2 - 7x = x(x^2 + 6x - 7)$
$= x(x + 7)(x - 1)$

7. $14x^2 + 64x - 30 = 2(7x^2 + 32x - 15)$
$= 2(x + 5)(7x - 3)$

8. $25x^2 - 9 = (5x - 3)(5x + 3)$

9. $x^3 + 8 = x^3 + 2^3 = (x + 2)(x^2 - 2x + 2^2)$
$= (x + 2)(x^2 - 2x + 4)$

10. $x^2 - 4x - 21 = (x + 3)(x - 7)$

11. $x^2 + 4$ is prime.

12. $6y^3 + 9y^2 + 3y = 3y(2y^2 + 3y + 1)$
$= 3y(y + 1)(2y + 1)$

13. $4y^2 - 36 = 4(y^2 - 9)$
$= 4(y^2 - 3^2) = 4(y+3)(y-3)$

14. $16x^2 + 48x + 36 = 4(4x^2 + 12x + 9)$
$= 4(2x+3)(2x+3) = 4(2x+3)^2$

15. $2x^4 - 32 = 2(x^4 - 16) = 2[(x^2)^2 - (2^2)^2]$
$= 2(x^2 + 2^2)(x^2 - 2^2)$
$= 2(x^2 + 4)(x+2)(x-2)$

16. $36x^2 - 84x + 49 = (6x-7)(6x-7)$
$= (6x-7)^2$

17. $7x^2 - 50x + 7 = (7x-1)(x-7)$

18. $x^4 + 2x^3 - 5x - 10 = (x^3 - 5)(x+2)$

19. $12y^3 - 12y^2 - 45y = 3y(4y^2 - 4y - 15)$
$= 3y(2y+3)(2y-5)$

20. $y^3 - 125 = y^3 - 5^3 = (y-5)(y^2 + 5y + 5^2)$
$= (y-5)(y^2 + 5y + 25)$

21. $5x^2 - 5xy - 30y^2 = 5(x^2 - xy - 6y^2)$
$= 5(x+2y)(x-3y)$

22. $x^2 + 2x - 24 = 0$
$(x+6)(x-4) = 0$
$x + 6 = 0$ or $x - 4 = 0$
$x = -6$ or $x = 4$
$-6, 4$

23. $3x^2 - 5x = 2$
$3x^2 - 5x - 2 = 0$
$(3x+1)(x-2) = 0$
$3x + 1 = 0$ or $x - 2 = 0$
$3x = -1$
$x = -\frac{1}{3}$ or $x = 2$
$-\frac{1}{3}, 2$

24. $x(x-6) = 16$
$x^2 - 6x - 16 = 0$
$(x+2)(x-8) = 0$
$x + 2 = 0$ or $x - 8 = 0$
$x = -2$ or $x = 8$
$-2, 8$

25. $x^2 - 4(2^2) = x^2 - 16$
$= x^2 - 4^2$
$= (x-4)(x+4)$

26. Solve $f(t) = 0$ for t.
$-5t^2 + 29t + 6 = 0$
$(5t+1)(-t+6) = 0$
$5t + 1 = 0$ or $-t + 6 = 0$
$t = \frac{-1}{5}$ or $t = 6$
Since t cannot be negative, $t = 6$.
The ball will hit the ground in 6 seconds.

27. Let w be the width.
Then length is $2w + 3$.
$w(2w+3) = 90$
$2w^2 + 3w - 90 = 0$
$(2w+15)(w-6) = 0$
$2w + 15 = 0$ or $w - 6 = 0$
$w = -\frac{15}{2}$ or $w = 6$
$2(6) + 3 = 15$
width = 6 yd
length = 15 yd

Cumulative Review Problems (Chapters 1–7)

1. a. Natural numbers: $\{1, 9\}$

b. Whole numbers: $\{0, 1, 9\}$

c. Integers: $\{-3, -2, 0, 1, 9\}$

d. Rational numbers:
$\left\{-3, -2, \frac{1}{7}, 0, 1, 9, 11.3\right\}$

e. Irrational numbers: $\left\{\sqrt{7}, 8\pi\right\}$

f. Real numbers:
$\left\{-3, -2, \frac{1}{7}, 0, 1, 9, 11.3, \sqrt{7}, 8\pi\right\}$

2. $6[5 + 2(3 - 8) - 3] = 6[5 + 2(-5) - 3]$
$= 6[5 - 10 - 3] = 6(-8) = -48$

3. $4(x - 2) = 2(x - 4) + 3x$
$4x - 8 = 2x - 8 + 3x$
$4x - 8 = 5x - 8$
$4x - 5x = -8 + 8$
$-x = 0$
$x = 0$

4. $\frac{x}{2} - 1 = \frac{x}{3} + 1$
$6\left(\frac{x}{2} - 1\right) = 6\left(\frac{x}{3} + 1\right)$
$3x - 6 = 2x + 6$
$3x - 2x = 6 + 6$
$x = 12$

5. Let x = the measure of the base angles.
$3x - 10$ = the measure of third angle
$2x + 3x - 10 = 180$
$5x = 190$
$x = 38$
$3x - 10 = 3(38) - 10 = 104$
The measures of the angles are 38°, 38° and 104°.

6. $5x + 6y > -30$
$6y > -5x - 30$
$y > -\frac{5}{6}x - 5$

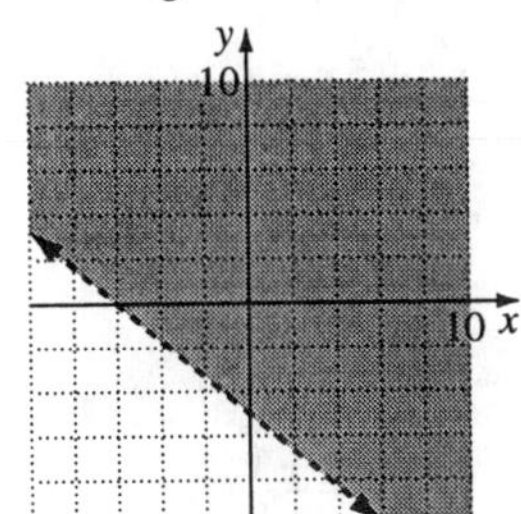

7. $5x + 2y = 14$
$y = 2x - 11$
Substitute for y.
$5x + 2(2x - 11) = 14$
$5x + 4x - 22 = 14$
$9x = 36$
$x = 4$
$y = 2x - 11 = 2(4) - 11 = -3$
$(4, -3)$

8. Let x = cost of pen.
y = cost of pad
$4x + 7y = 6.40$
$19x + 2y = 5.40$
Multiply equation 1 by –2.
Multiply equation 2 by 7.
$$\begin{aligned} -8x - 14y &= -12.80 \\ 133x + 14y &= 37.80 \\ \hline 125x &= 25.00 \\ x &= 0.20 \end{aligned}$$
$y = \frac{6.40 - 4x}{7} = \frac{6.40 - 0.80}{7} = 0.80$
cost of pen, \$0.20; cost of pad, \$0.80

9. $\frac{6x^5 - 3x^4 + 9x^2 + 27}{-3x}$
$= -2x^4 + x^3 - 3x - \frac{9}{x}$ $(x \neq 0)$

10. $\left(\frac{4y^{-1}}{2y^{-3}}\right)^3 = 8y^{-3-(-9)} = 8y^6$

11. x = total number of nuns
$\frac{15,040}{x} = \frac{16}{100}$
$16x = 1,504,000$
$x = \frac{1,504,000}{16} = 94,000$
$94,000 \cdot 40\% = 37,600$
$94,000 \cdot 44\% = 41,360$
94,000 means total
Over 70 years = 37,600 nuns
51 to 70 years = 41,360 nuns

12. $y(5y+17)=12$
$5y^2+17y-12=0$
$(5y-3)(y+4)=0$
$y=\frac{3}{5}$ or $y=-4$
$-4, \frac{3}{5}$

13. $5-5x>2(5-x)+1$
$5-5x>10-2x+1$
$-5x+2x>-5+11$
$-3x>6$
$x<-2$
$\{x|x<-2\}$

14. $f(x)=0.025x^3-0.7x^2+4.43x+16.77$
$f(10)=0.025(10)^3-0.7(10)^2+4.43(10)+16.77$
$f(10)=16.07$
Approximately 16.07% of 18–25 yr olds used hallucinogens in 1984.

15. Teacher salary = t
Surgeon salary = $6t + 18{,}600$
$t+(6t+18{,}600)=258{,}000$
$7t+18{,}600=258{,}000$
$7t=239{,}400$
$t=34{,}200$
Teacher = \$34,200
Surgeon = \$223,800
Ped = \$115,000
Att. = \$62,000

16. (2, –4), (3, 1)
slope = $\frac{1+4}{3-2}=\frac{5}{1}=5$
point-slope: $y-1=5(x-3)$ or
$y+4=5(x-2)$
$y-1=5x-15$
slope-intercept: $y=5x-14$

17. a. $(x+1)(x+2)=x^2+3x+2$

b. $f(x)=x^2+3x+2$

c. $f(8)=8^2+3(8)+2=90$
When the width of the original garden is 8 m, the area of the new garden is 90 m^2.

18. $y<-\frac{2}{5}x+2$

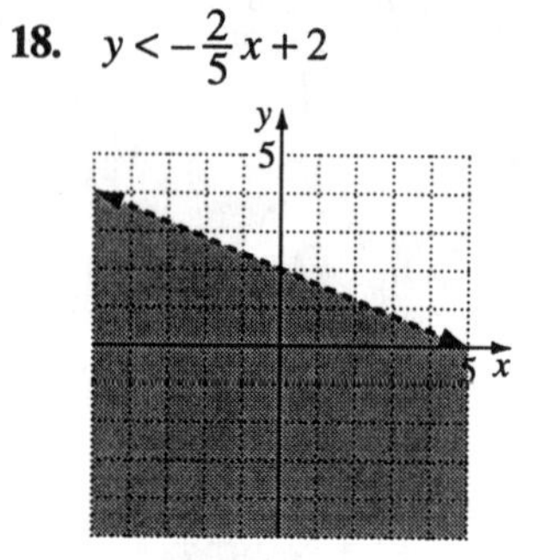

19. $2x+3y=5$
$3x-2y=-4$
Multiply equation 1 by 3.
Multiply equation 2 by –2.
$6x+9y=15$
$-6x+4y=8$
$13y=23$
$y=\frac{23}{13}$
$x=\frac{-4+2y}{3}=\frac{-4+\frac{46}{13}}{3}=\frac{-52+46}{39}=-\frac{6}{39}$
$=-\frac{2}{13}$
$\left(-\frac{2}{13}, \frac{23}{13}\right)$

20. $\frac{6x^3+5x^2-34x+13}{3x-5}$
$=2x^2+5x-3-\frac{2}{3x-5}$

$$\begin{array}{r} 2x^2+5x-3 \\ 3x-5\overline{)6x^3+5x^2-34x+13} \\ \underline{6x^3-10x^2} \\ 15x^2-34x \\ \underline{15x^2-25x} \\ -9x+13 \\ \underline{-9x+15} \\ -2 \end{array}$$

quotient $= 2x^2 + 5x - 3$

remainder $= -\dfrac{2}{3x-5}$

21. Let x = the first integer.
Then $x + 1$ and $x + 2$ are the next two integers.
$x + x + 1 + x + 2 = 48$
$3x + 3 = 48$
$3x = 45$
$x = 15$
$x = 15,\ x + 1 = 16,\ x + 2 = 17$
The integers are 15, 16, and 17.

22. $x\overline{)133}$ with quotient $1y$
Since $7(19) = 133$, $x = 7$ and $y = 9$.

23. $3x^2 + 11x + 6 = (3x + 2)(x + 3)$

24. $f(t) = -16t^2 + 64t$

a. $f(t) = 0 = -16t^2 + 64t$
$-16(t^2 - 4t) = 0$
$-16t(t - 4) = 0$
$t = 0$ or $t - 4 = 0$
$t = 4$ seconds (reject $t = 0$)

b.

Time t	Height $f(t) = -16t^2 + 64t$
0	0
1	48
2	64
3	48
4	0

c.

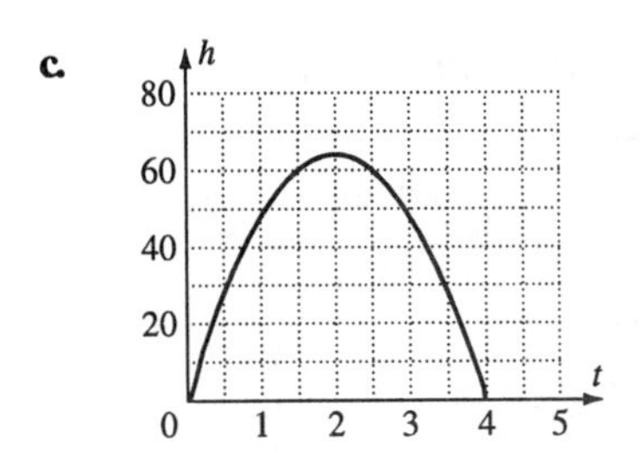

d. The maximum height is reached at 2 seconds, and the height is 64 ft.

e. The height is 0 at 4 seconds when the projectile is dropping. (4, 0) is a point on the graph.

25. $y^5 - 16y = y(y^4 - 16) = y(y^2 + 4)(y^2 - 4)$
$= y(y^2 + 4)(y - 2)(y + 2)$

26. $x(x + 2) = 24$
$x^2 + 2x - 24 = 0$
$(x + 6)(x - 4) = 0$
$x + 6 = 0$ or $x - 4 = 0$
$x = -6$ or $x = 4$
Width = 4 ft, length = 6 ft

27. Let x = cost of dinner before tax
$x + 0.07x = 160$
$1.07x = 160$
$x \approx 149.53$
$149.53

28. $\frac{4}{5} - \frac{9}{8} = \left(\frac{4}{5}\cdot\frac{8}{8}\right) - \left(\frac{9}{8}\cdot\frac{5}{5}\right) = \frac{32}{40} - \frac{45}{40} = -\frac{13}{40}$

29. $A = \dfrac{B + C}{2}$
$2A = B + C$
$B = 2A - C$

30. 5 nickels, 3 dimes, 2 quarters

Nickels	Dimes	Quarters
	2	1
2	1	1
4		1
3	3	
5	2	

5 ways

Chapter 8

Problem Set 8.1

1. $\frac{7}{2x}$
   $2x = 0$
   $x = 0$
   Since 0 will make the denominator zero, the rational expression is undefined for $x = 0$.

3. $\frac{x}{x-7}$
   $x - 7 = 0$
   $x = 7$

5. $\frac{5y^2}{5y-15} = \frac{5y^2}{5(y-3)}$
   $5(y - 3) = 0$
   $y = 3$

7. $\frac{x+4}{(x+7)(x-3)}$
   $(x + 7)(x - 3) = 0$
   $x = -7$ or $x = 3$

9. $\frac{13z}{(3z-15)(z+2)} = \frac{13z}{3(z-5)(z+2)}$
   $3(z - 5)(z + 2) = 0$
   $z = 5$ or $z = -2$

11. $\frac{x+5}{x^2+x-12} = \frac{x+5}{(x+4)(x-3)}$
    $(x + 4)(x - 3) = 0$
    $x = -4$ or $x = 3$

13. $\frac{y+3}{4y^2+y-3} = \frac{y+3}{(4y-3)(y+1)}$
    $(4y - 3)(y + 1) = 0$
    $y = \frac{3}{4}$ or $y = -1$

15. $\frac{7x}{x^2+4}$
    $x^2 + 4 = 0$
    No values of x for which the expression is undefined.

17. $\frac{y^2-16}{8}$
    $8 \neq 0$
    No values of y for which the expression is undefined.

19. $\frac{14x^2}{7x} = \frac{2 \cdot 7 \cdot x \cdot x}{7 \cdot x} = 2x$

21. $\frac{60x^4}{10x^6} = \frac{2 \cdot 2 \cdot 3 \cdot 5 \cdot x \cdot x \cdot x \cdot x}{2 \cdot 5 \cdot x \cdot x \cdot x \cdot x \cdot x \cdot x} = \frac{6}{x^2}$

23. $\frac{5x-15}{25} = \frac{5(x-3)}{5 \cdot 5} = \frac{x-3}{5}$

25. $\frac{-2x+8}{-4x} = \frac{-2(x-4)}{-2(2x)} = \frac{x-4}{2x}$

27. $\frac{3}{3x-9} = \frac{3}{3(x-3)} = \frac{1}{x-3}$

29. $\frac{-15}{3x-5}$ is not reducible.

31. $\frac{3y+9}{y+3} = \frac{3(y+3)}{y+3} = 3$

33. $\frac{x+5}{x^2-25} = \frac{x+5}{(x+5)(x-5)} = \frac{1}{x-5}$

35. $\frac{2y-10}{3y-6} = \frac{2(y-5)}{3(y-2)}$

37. $\frac{s+1}{s^2-2s-3} = \frac{s+1}{(s-3)(s+1)} = \frac{1}{s-3}$

39. $\frac{4b-8}{b^2-4b+4} = \frac{4(b-2)}{(b-2)^2} = \frac{4}{b-2}$

41. $\dfrac{y^2-3y+2}{y^2+7y-18}=\dfrac{(y-1)(y-2)}{(y+9)(y-2)}=\dfrac{y-1}{y+9}$

43. $\dfrac{2y^2-7y+3}{2y^2-5y+2}=\dfrac{(2y-1)(y-3)}{(2y-1)(y-2)}=\dfrac{y-3}{y-2}$

45. $\dfrac{2x+3}{2x-5}$ is irreducible.

47. $\dfrac{x^2+5x+2x+10}{x^2-25}$
$=\dfrac{x(x+5)+2(x+5)}{x^2-25}$
$=\dfrac{(x+5)(x+2)}{(x+5)(x-5)}$
$=\dfrac{x+2}{x-5}$

49. $\dfrac{x^3+5x^2-6x}{x^3-x}$
$=\dfrac{x(x^2+5x-6)}{x(x^2-1)}$
$=\dfrac{(x+6)(x-1)}{(x+1)(x-1)}$
$=\dfrac{x+6}{x+1}$

51. $\dfrac{2y^8+y^7}{2y^6+y^5}=\dfrac{y^7(2y+1)}{y^5(2y+1)}=y^2$

53. $\dfrac{x-5}{5-x}=\dfrac{-(5-x)}{5-x}=-1$

55. $\dfrac{2x-2}{1-x}=\dfrac{2(x-1)}{1-x}=-2$

57. $\dfrac{-2x-8}{x^2-16}=\dfrac{-2(x+4)}{(x+4)(x-4)}=-\dfrac{2}{x-4}=\dfrac{2}{4-x}$

59. $\dfrac{4-6y}{3y^2-2y}=\dfrac{2(2-3y)}{y(3y-2)}=\dfrac{-2(3y-2)}{y(3y-2)}=-\dfrac{2}{y}$

61. $\dfrac{9-x^2}{x^2-x-6}$
$=\dfrac{(3-x)(3+x)}{(x-3)(x+2)}$
$=\dfrac{-(x-3)(3+x)}{(x-3)(x+2)}$
$=-\dfrac{3+x}{x+2}$

63. $\dfrac{y^2-9y+18}{y^3-27}$
$=\dfrac{(y-3)(y-6)}{(y-3)(y^2+3y+9)}$
$=\dfrac{y-6}{y^2+3y+9}$

65. $\dfrac{b^2-b-12}{4-b}$
$=\dfrac{(b-4)(b+3)}{4-b}$
$=\dfrac{(b-4)(b+3)}{-(b-4)}$
$=-(b+3)$ or $-b-3$

67. $y=\dfrac{x^2-25}{x-5}=\dfrac{(x+5)(x-5)}{x-5}=x+5\ (x\neq 5)$

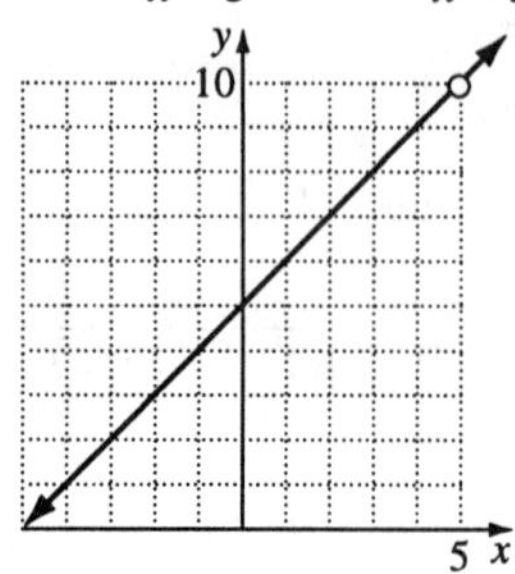

69. $f(x)=\dfrac{9x-18}{3x-6}=\dfrac{9(x-2)}{3(x-2)}=3\ (x\neq 2)$

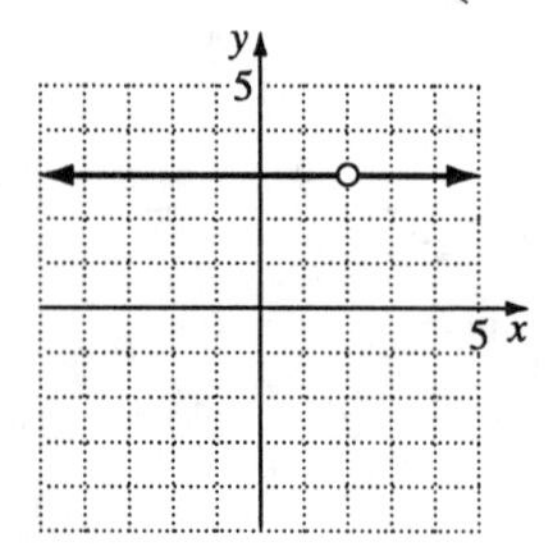

71. $\dfrac{10x^3y}{5xy^2}=\dfrac{2x^2}{y}$

73. $\frac{7x+2y}{14x+4y}=\frac{7x+2y}{2(7x+2y)}=\frac{1}{2}$

75. $\frac{x^2-4y^2}{x+2y}=\frac{(x-2y)(x+2y)}{x+2y}=x-2y$

77. $\frac{6a^2}{2a(a-3b)}=\frac{2a(3a)}{2a(a-3b)}=\frac{3a}{a-3b}$

79. $\frac{x^2+3xy-10y^2}{3x^2-7xy+2y^2}$
$=\frac{(x+5y)(x-2y)}{(3x-y)(x-2y)}$
$=\frac{x+5y}{3x-y}$

81. $\frac{6a^2-11ab+4b^2}{9a^2-16b^2}$
$=\frac{(3a-4b)(2a-b)}{(3a-4b)(3a+4b)}$
$=\frac{2a-b}{3a+4b}$

83. $f(x)=\frac{130x}{100-x}$

a. $f(40)=\frac{130(40)}{100-40}\approx 86.67$ is the cost in millions of dollars to inoculate 40% of the population
$f(80)=\frac{130(80)}{100-80}=520$ is the cost to inoculate 80% of the population
$f(90)=\frac{130(90)}{100-90}=1170$ is the cost to inoculate 90% of the population

b. $x=100$

c. The cost becomes infinitely large as x approaches 100%.

d. $y=\frac{130x}{100-x}$

x	0	10	20	40	50	60	70	80	90	95	99	100
$f(x)$	0	14	33	87	130	195	303	520	1170	2470	12870	undefined

e.

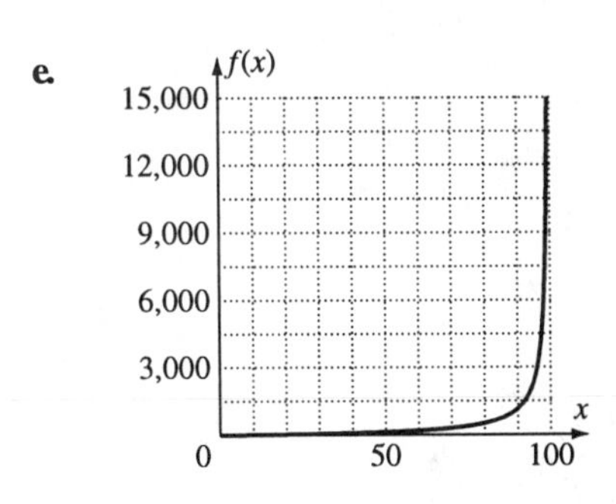

f. The values of the cost increase rapidly as the values of x (% inoculated) approach 100.

85. $f(x)=\frac{5000}{x}+2x$
minimum = (50, 200)
minimum perimeter is 200 square feet
50 by 50
square

87. Statement c is true.

89. $\frac{2x^2-x-1}{x-1}=2x+1$

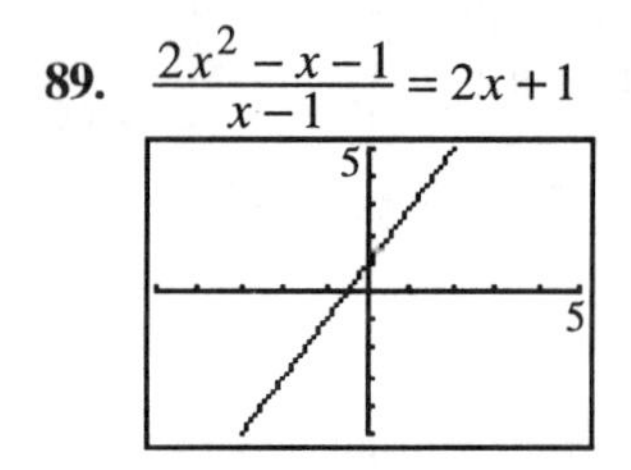

91. $f(x)$ goes to infinity.

93. Answers may vary.

95. $\frac{1}{x+4}$ approaches 0 as x becomes large.

x	0	10	100
$\frac{1}{x+4}$	0.25	0.07	0.01

97. $\frac{2x}{x-1}$ approaches 2 as x becomes large.

x	0	10	100	1000
$\frac{2x}{x-1}$	0	2.22	2.02	2.00

99. $\frac{x+2}{x^2}$ approaches 0 as x becomes large.

x	0	10	100
$\frac{x+2}{x^2}$	undefined	0.12	0.01

101. $\frac{x}{x^2-1}$ approaches 0 as x becomes large.

x	0	10	100	1000
$\frac{x}{x^2-1}$	0	0.10	0.01	0.001

103. Possible answer:
$\frac{x^2+7x+12}{x+3}=\frac{(x+3)(x+4)}{x+3}=x+4$

105. Possible answer:
$\frac{x^2+7x+12}{x+4}$ is undefined when $x=-4$

107. Others are $\frac{26}{65}=\frac{2}{5}$ and $\frac{49}{98}=\frac{4}{8}$.

Review Problems

109. $\frac{17}{68}=0.25$; 25%

110. $2x-5y=-2$
$3x+4y=20$

Multiply equation 1 by 3.
Multiply equation 2 by –2.

$$\begin{aligned} 6x-15y &= -6 \\ -6x-8y &= -40 \\ \hline -23y &= -46 \\ y &= 2 \end{aligned}$$

$x=\frac{5y-2}{2}=\frac{10-2}{2}=4$
$y=2$
$\{(4, 2)\}$

111. $\frac{8.5\times10^{-3}}{1.7\times10^{-7}}$
$=\frac{8.5}{1.7}\times10^{-3-(-7)}$
$=\frac{8.5}{1.7}\times10^{4}=5\times10^{4}=50,000$

Problem Set 8.2

1. $\frac{5}{x+2} \cdot \frac{x-3}{7} = \frac{5(x-3)}{7(x+2)} \ (x \neq -2)$

3. $\frac{3x}{7} \cdot \frac{x+1}{x-2}$
$= \frac{3x(x+1)}{7(x-2)}$
$= \frac{3x^2+3x}{7x-14} \ (x \neq 2)$

5. $\frac{x}{2} \cdot \frac{4}{x+1} = \frac{4x}{2(x+1)} = \frac{2x}{x+1} \ (x \neq -1)$

7. $\frac{3}{x} \cdot \frac{2x}{9} = \frac{3 \cdot 2x}{9x} = \frac{2}{3} \ (x \neq 0)$

9. $\frac{x-2}{3x+9} \cdot \frac{2x+6}{2x-4} = \frac{x-2}{3(x+3)} \cdot \frac{2(x+3)}{2(x-2)} = \frac{1}{3}$
$(x+3)(x-2) \neq 0 \ (x \neq -3, 2)$

11. $\frac{y-5}{y+2} \cdot \frac{6y-8}{2y+4}$
$= \frac{y-5}{y+2} \cdot \frac{2(3y-4)}{2(y+2)}$
$= \frac{3y^2-19y+20}{y^2+4y+4}$
$(y+2)(y+2) \neq 0 \ (y \neq -2)$

13. $\frac{4y+30}{y^2-3y} \cdot \frac{y-3}{2y+15}$
$= \frac{2(2y+15)}{y(y-3)} \cdot \frac{y-3}{2y+15} = \frac{2}{y}$
$y(y-3)(2y+15) \neq 0 \ \left(y \neq 0,\ 3,\ -\frac{15}{2}\right)$

15. $\frac{r^2-9}{r^2} \cdot \frac{r^2-3r}{r^2+r-12}$
$= \frac{(r-3)(r+3)}{r^2} \cdot \frac{r(r-3)}{(r+4)(r-3)}$
$= \frac{(r-3)(r+3)}{r(r+4)}$
$r^2(r+4)(r-3) \neq 0$
$(r \neq 0, 3, -4)$

17. $\frac{y^2-7y-30}{y^2-6y-40} \cdot \frac{2y^2+5y+2}{2y^2+7y+3}$
$= \frac{(y-10)(y+3)}{(y-10)(y+4)} \cdot \frac{(2y+1)(y+2)}{(2y+1)(y+3)}$
$= \frac{y+2}{y+4}$
$(y-10)(y+4)(2y+1)(y+3) \neq 0$
$\left(y \neq -4,\ 10,\ -\frac{1}{2},\ -3\right)$

19. $(y^2-9)\left(\frac{4}{y-3}\right)$
$= \frac{(y-3)(y+3)(4)}{y-3} = 4(y+3) \ (y \neq 3)$

21. $\frac{x^2-2x+4}{x^2-4} \cdot \frac{(x+2)^3}{2x+4}$
$= \frac{x^2-2x+4}{(x-2)(x+2)} \cdot \frac{(x+2)^3}{2(x+2)}$
$= \frac{(x^2-2x+4)(x+2)}{2(x-2)}$
$(x-2)(x+2)2(x+2) \neq 0$
$(x \neq -2, 2)$

23. $\frac{x^2-x-6}{3x-9} \cdot \frac{x^2-9}{x^2+6x+9}$
$= \frac{(x-3)(x+2)}{3(x-3)} \cdot \frac{(x-3)(x+3)}{(x+3)^2}$
$= \frac{(x+2)(x-3)}{3(x+3)}$
$3(x-3)(x+3)^2 \neq 0$
$(x \neq -3, 3)$

25. $\frac{y^2+10y+25}{y-4} \cdot \frac{y^2-y-12}{y+5} \cdot \frac{1}{y+3}$
$= \frac{(y+5)^2}{y-4} \cdot \frac{(y-4)(y+3)}{y+5} \cdot \frac{1}{y+3} = y+5$
$(y-4)(y+5)(y+3) \neq 0$
$(y \neq -3, 4, -5)$

27. $\frac{(x-2)^3}{(x-1)^3}\cdot\frac{x^2-2x+1}{x^2-4x+4}$

$=\frac{(x-2)^3}{(x-1)^3}\cdot\frac{(x-1)^2}{(x-2)^2}=\frac{x-2}{x-1}$

$(x-1)^3(x-2)^2\neq 0$

$(x\neq 1, 2)$

29. $\frac{25-y^2}{y^2-2y-35}\cdot\frac{y^2-8y-20}{y^2-3y-10}$

$=\frac{(5-y)(5+y)}{(y-7)(y+5)}\cdot\frac{(y-10)(y+2)}{(y-5)(y+2)}$

$=-\frac{y-10}{y-7}$

$(y-7)(y+5)(y-5)(y+2)\neq 0$

$(y\neq -2, -5, 5, 7)$

31. $\frac{x}{7}\div\frac{5}{3}=\frac{x}{7}\cdot\frac{3}{5}=\frac{3x}{35}$

33. $\frac{3}{x}\div\frac{12}{x}=\frac{3}{x}\cdot\frac{x}{12}=\frac{1}{4}\ (x\neq 0)$

35. $\frac{15}{x}\div\frac{3}{2x}=\frac{15}{x}\cdot\frac{2x}{3}=10\ (x\neq 0)$

37. $\frac{2}{x+1}\div\frac{3}{x-1}$

$=\frac{2}{x+1}\cdot\frac{x-1}{3}$

$=\frac{2x-2}{3x+3}$

$(x+1)\neq 0$

$(x\neq -1)$

39. $\frac{x}{y^2}\div\frac{x^4}{y^3}=\frac{x}{y^2}\cdot\frac{y^3}{x^4}=\frac{y}{x^3}$

$(y^2)(x^4)\neq 0$

$(x\neq 0, y\neq 0)$

41. $\frac{x+3}{x-4}\div\frac{x-3}{x+4}$

$=\frac{x+3}{x-4}\cdot\frac{x+4}{x-3}$

$=\frac{x^2+7x+12}{x^2-7x+12}$

$(x-4)(x-3)\neq 0$

$(x\neq 4, 3)$

43. $\frac{x+1}{3}\div\frac{3x+3}{7}=\frac{x+1}{3}\cdot\frac{7}{3(x+1)}=\frac{7}{9}\ (x\neq -1)$

45. $\frac{7}{x-5}\div\frac{28}{3x-15}$

$=\frac{7}{x-5}\cdot\frac{3(x-5)}{28}=\frac{3}{4}\ (x\neq 5)$

47. $\frac{x^2-4}{x}\div\frac{x+2}{x-2}$

$=\frac{(x-2)(x+2)}{x}\cdot\frac{x-2}{x+2}$

$=\frac{(x-2)^2}{x}=\frac{x^2-4x+4}{x}$

$(x)(x+2)=0$

$(x\neq 0, -2)$

49. $(y^2-16)\div\frac{y^2+3y-4}{y^2+4}$

$=(y-4)(y+4)\cdot\frac{y^2+4}{(y+4)(y-1)}$

$=\frac{(y-4)(y^2+4)}{y-1}$

$(y+4)(y-1)\neq 0$

$(y\neq -4, 1)$

51. $\frac{y^2-y}{15}\div\frac{y-1}{5}=\frac{y(y-1)}{15}\cdot\frac{5}{y-1}=\frac{y}{3}$

$(y-1)\neq 0$

$(y\neq 1)$

53. $\frac{x^2+2x+1}{6x^2}\div\frac{x+1}{12x^3}$

$=\frac{(x+1)^2}{6x^2}\cdot\frac{12x^3}{x+1}=2x(x+1)$

$6x^2(x+1)\neq 0$

$(x\neq 0, -1)$

55. $\frac{y^3+y}{y^2-y} \div \frac{y^3-y^2}{y^2-2y+1}$

$= \frac{y(y^2+1)}{y(y-1)} \cdot \frac{(y-1)^2}{y^2(y-1)}$

$= \frac{y^2+1}{y^2}$

$y(y-1)y^2(y-1) \neq 0$

$(y \neq 0, 1)$

57. $\frac{m^2+5m+4}{m^2+12m+32} \div \frac{m^2-12m+35}{m^2+3m-40}$

$= \frac{(m+4)(m+1)}{(m+4)(m+8)} \cdot \frac{(m+8)(m-5)}{(m-7)(m-5)}$

$= \frac{m+1}{m-7}$

$(m+4)(m+8)(m-7)(m-5) \neq 0$

$(m \neq -4, -8, 7, 5)$

59. $\frac{2y^2-128}{y^2+16y+64} \div \frac{y^2-6y-16}{3y^2+30y+48}$

$= \frac{2(y-8)(y+8)}{(y+8)^2} \cdot \frac{3(y+8)(y+2)}{(y-8)(y+2)} = 6$

$(y+8)^2(y-8)(y+2) \neq 0$

$(y \neq -2, -8, 8)$

61. $\frac{\frac{x^2+5}{3}}{7} = \frac{x^2+5}{3} \div 7 = \frac{x^2+5}{3} \cdot \frac{1}{7} = \frac{x^2+5}{21}$

63. $\frac{\frac{7x}{9}}{x} = \frac{7x}{9} \div x = \frac{7x}{9} \cdot \frac{1}{x} = \frac{7}{9}\ (x \neq 0)$

65. $\frac{\frac{x^3}{6}}{\frac{x}{3}} = \frac{x^3}{6} \div \frac{x}{3} = \frac{x^3}{6} \cdot \frac{3}{x} = \frac{x^2}{2}\ (x \neq 0)$

67. $\frac{\frac{3x+12}{4}}{\frac{x+4}{2}}$

$= \frac{3x+12}{4} \div \frac{x+4}{2}$

$= \frac{3(x+4)}{4} \cdot \frac{2}{x+4} = \frac{3}{2}\ (x \neq -4)$

69. $\frac{\frac{x}{y-7}}{\frac{4}{7-y}}$

$= \frac{x}{y-7} \div \frac{4}{7-y}$

$= \frac{x}{y-7} \cdot \frac{7-y}{4} = -\frac{x}{4}\ (y \neq 7)$

71. $\frac{\frac{x^2-9x+18}{x^2-9}}{\frac{2x^3-11x^2-6x}{2x^2+x}}$

$= \frac{x^2-9x+18}{x^2-9} \div \frac{2x^3-11x^2-6x}{2x^2+x}$

$= \frac{(x-6)(x-3)}{(x+3)(x-3)} \cdot \frac{x(2x+1)}{x(2x+1)(x-6)} = \frac{1}{x+3}$

$\left(x \neq 3, -3, 0, -\frac{1}{2}, 6\right)$

73. $\frac{\left(\frac{7x}{3}\right)^2}{\left(\frac{7x}{2}\right)^3}$

$= \left(\frac{7x}{3}\right)^2 \div \left(\frac{7x}{2}\right)^3$

$= \frac{(7x)^2}{9} \cdot \frac{8}{(7x)^3} = \frac{8}{63x}\ (x \neq 0)$

75. $\frac{\frac{4}{x^2-3x-28}}{\frac{2}{x-7}}$

$= \frac{4}{x^2-3x-28} \div \frac{2}{x-7}$

$= \frac{4}{(x-7)(x+4)} \cdot \frac{x-7}{2}$

$= \frac{2}{x+4}$

$(x \neq -4, 7)$

77. $\frac{3x}{y^2} \cdot \frac{y}{12x^3} = \frac{1}{4x^2y}$

$(y^2)(12x^3) \neq 0$

$(x \neq 0, y \neq 0)$

79. $\dfrac{x^2-y^2}{x}\cdot\dfrac{x^2+xy}{x+y}$

$=\dfrac{(x+y)(x-y)(x)(x+y)}{x(x+y)}$

$=(x+y)(x-y)$

$(x)(x+y)\neq 0$

$(x\neq 0, x\neq -y)$

81. $\dfrac{a^2-b^2}{a+b}\cdot\dfrac{a+2b}{2a^2-ab-b^2}$

$=\dfrac{(a+b)(a-b)(a+2b)}{(a+b)(2a+b)(a-b)}$

$=\dfrac{a+2b}{2a+b}$

$(a+b)(2a+b)(a-b)\neq 0$

$\left(a\neq -b,\ a\neq -\frac{b}{2},\ a\neq b\right)$

83. $\dfrac{12x^2}{4yz}\div\dfrac{3x^2}{yz}=\dfrac{12x^2}{4yz}\cdot\dfrac{yz}{3x^2}=1$

$(x\neq 0, y\neq 0, z\neq 0)$

85. $\dfrac{4a+4b}{ab^2}\div\dfrac{3a+3b}{a^2b}$

$=\dfrac{4(a+b)}{ab^2}\cdot\dfrac{a^2b}{3(a+b)}$

$=\dfrac{4a}{3b}$

$(ab^2)(3)(a+b)\neq 0$

$(a\neq 0,\ b\neq 0,\ a\neq -b)$

87. $\dfrac{4x^2-y^2}{x^2+4xy+4y^2}\div\dfrac{4x-2y}{3x+6y}$

$=\dfrac{(2x-y)(2x+y)(3)(x+2y)}{(x+2y)(x+2y)(2)(2x-y)}$

$=\dfrac{3(2x+y)}{2(x+2y)}$

$(x+2y)(x+2y)(2)(2x-y)\neq 0$

$\left(x\neq -2y,\ x\neq \frac{y}{2}\right)$

89. $\dfrac{a^2-4b^2}{a^2+3ab+2b^2}\div\dfrac{a^2-4ab+4b^2}{a+b}$

$=\dfrac{(a-2b)(a+2b)(a+b)}{(a+2b)(a+b)(a-2b)(a-2b)}$

$=\dfrac{1}{a-2b}$

$(a+2b)(a+b)(a-2b)^2\neq 0$

$(a\neq -2b,\ a\neq -b,\ a\neq 2b)$

91. Statement c is true.

93. Correct

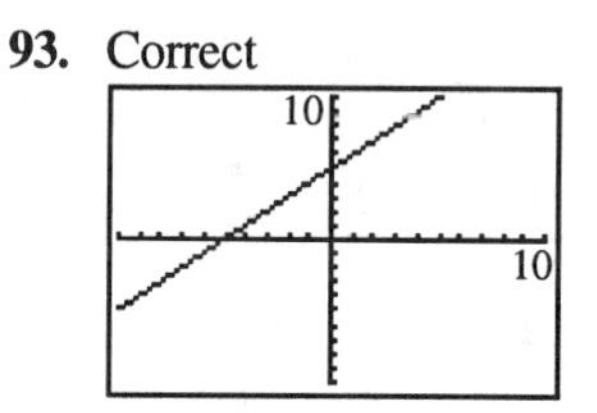

95. Answers may vary.

97. $\dfrac{x}{x-4}\cdot\dfrac{3(x-4)}{2x}=\dfrac{3}{2}$

99. $\dfrac{\text{area of trapezoid}}{\text{area of rectangle}}$

$=\dfrac{\frac{1}{2}(3x+x)\left(\frac{1}{8x+4}\right)}{\left(\frac{1}{14x+7}\right)(3)}$

$=\dfrac{2x}{4(2x+1)}\cdot\dfrac{7(2x+1)}{3}$

$=\dfrac{14x}{12}=\dfrac{7x}{6}$

Review Problems

101. $\dfrac{(3x^{-3})^4}{9x^5}=\dfrac{3^4x^{-12}}{9x^5}=\dfrac{9\cdot 9}{9x^{5+12}}=\dfrac{9}{x^{17}}$

102. $2x+3<3(x-5)$

$2x+3<3x-15$

$2x-3x<-15-3$

$-x<-18$

$x>18$

103. $y(2y+9)=5$
$2y^2+9y-5=0$
$(2y-1)(y+5)=0$
$2y-1=0$ or $y+5=0$
$y=\frac{1}{2}$ or $y=-5$
$\frac{1}{2}, -5$

Problem Set 8.3

1. $\frac{4x}{9}+\frac{2x}{9}=\frac{6x}{9}=\frac{2x}{3}$

3. $\frac{5}{x}+\frac{3}{x}=\frac{8}{x}\ (x\neq 0)$

5. $\frac{7}{9x}+\frac{5}{9x}=\frac{12}{9x}=\frac{4}{3x}\ (x\neq 0)$

7. $\frac{m}{5}+\frac{2m}{5}=\frac{3m}{5}$

9. $\frac{7}{4-y}+\frac{3}{4-y}=\frac{10}{4-y}\quad (y\neq 4)$

11. $\frac{3x+2}{3x+4}+\frac{3x+6}{3x+4}$
$=\frac{6x+8}{3x+4}=\frac{2(3x+4)}{3x+4}=2\quad \left(x\neq -\frac{4}{3}\right)$

13. $\frac{y^2-2y}{y^2+3y}+\frac{y^2+y}{y^2+3y}$
$=\frac{2y^2-y}{y^2+3y}=\frac{y(2y-1)}{y(y+3)}=\frac{2y-1}{y+3}\ (y\neq 0, -3)$

15. $\frac{y+2}{6y^3}+\frac{3y-2}{6y^3}=\frac{4y}{6y^3}=\frac{2}{3y^2}\quad (y\neq 0)$

17. $\frac{y^2+9y}{4y^2-11y-3}+\frac{3y-5y^2}{4y^2-11y-3}$
$=\frac{-4y^2+12y}{4y^2-11y-3}$
$=\frac{-4y(y-3)}{(4y+1)(y-3)}$
$=-\frac{4y}{4y+1}\quad \left(y\neq 3, -\frac{1}{4}\right)$

19. $\frac{y}{2y+7}-\frac{2}{2y+7}=\frac{y-2}{2y+7}\quad \left(y\neq -\frac{7}{2}\right)$

21. $\frac{x}{x-1}-\frac{1}{x-1}=\frac{x-1}{x-1}=1\quad (x\neq 1)$

23. $\frac{2y+1}{3y-7}-\frac{y+8}{3y-7}$
$=\frac{2y+1-y-8}{3y-7}=\frac{y-7}{3y-7}\quad \left(y\neq \frac{7}{3}\right)$

25. $\frac{2y+3}{3y-6}-\frac{3-y}{3y-6}$
$=\frac{2y+3-3+y}{3y-6}$
$=\frac{3y}{3y-6}$
$=\frac{3y}{3(y-2)}=\frac{y}{y-2}\quad (y\neq 2)$

27. $\frac{y^3-3}{2y^4}-\frac{7y^3-3}{2y^4}$
$=\frac{y^3-3-7y^3+3}{2y^4}$
$=\frac{-6y^3}{2y^4}=-\frac{3}{y}\quad (y\neq 0)$

29. $\frac{y^2+3y}{y^2+y-12}-\frac{y^2-12}{y^2+y-12}$
$=\frac{y^2+3y-y^2+12}{y^2+y-12}$
$=\frac{3y+12}{y^2+y-12}$
$=\frac{3(y+4)}{(y+4)(y-3)}=\frac{3}{y-3}\quad (y\neq -4, 3)$

31. $\frac{16r^2+3}{16r^2+16r+3}-\frac{3-4r}{16r^2+16r+3}$
$=\frac{16r^2+3-3+4r}{16r^2+16r+3}$
$=\frac{16r^2+4r}{16r^2+16r+3}$
$=\frac{4r(4r+1)}{(4r+1)(4r+3)}=\frac{4r}{4r+3}\quad \left(r\neq -\frac{1}{4}, -\frac{3}{4}\right)$

33. $\frac{9x}{10}-\frac{7x}{10}+\frac{3x}{10}=\frac{5x}{10}=\frac{x}{2}$

35. $\frac{6y^2+y}{2y^2-9y+9}-\frac{2y+9}{2y^2-9y+9}-\frac{4y-3}{2y^2-9y+9}$
$=\frac{6y^2+y-2y-9-4y+3}{2y^2-9y+9}$
$=\frac{6y^2-5y-6}{2y^2-9y+9}$
$=\frac{(2y-3)(3y+2)}{(2y-3)(y-3)}$
$=\frac{3y+2}{y-3}\ \left(y\neq\frac{3}{2},\ 3\right)$

37. $\frac{2y+7}{y-6}+\frac{3y}{6-y}$
$=\frac{2y+7}{y-6}-\frac{3y}{y-6}=\frac{7-y}{y-6}\ (y\neq 6)$

39. $\frac{5x-2}{3x-4}+\frac{2x-3}{4-3x}$
$=\frac{5x-2}{3x-4}-\frac{2x-3}{3x-4}$
$=\frac{5x-2-2x+3}{3x-4}$
$=\frac{3x+1}{3x-4}\ \left(x\neq\frac{4}{3}\right)$

41. $\frac{y^2}{y-2}+\frac{4}{2-y}$
$=\frac{y^2}{y-2}-\frac{4}{y-2}$
$=\frac{y^2-4}{y-2}$
$=\frac{(y-2)(y+2)}{y-2}=y+2\ (y\neq 2)$

43. $\frac{b-3}{b^2-25}+\frac{b-3}{25-b^2}$
$=\frac{b-3}{b^2-25}-\frac{b-3}{b^2-25}=0\ (b\neq -5,5)$

45. $\frac{y}{y-1}-\frac{1}{1-y}=\frac{y}{y-1}+\frac{1}{y-1}=\frac{y+1}{y-1}\ (y\neq 1)$

47. $\frac{3-a}{a-7}-\frac{2a-5}{7-a}$
$=\frac{3-a}{a-7}+\frac{2a-5}{a-7}$
$=\frac{a-2}{a-7}$
$(a\neq 7)$

49. $\frac{z-2}{z^2-25}-\frac{z-2}{25-z^2}$
$=\frac{z-2}{z^2-25}+\frac{z-2}{z^2-25}$
$=\frac{2(z-2)}{z^2-25}$
$(z\neq -5,5)$

51. $\frac{3(m-2)}{2m-3}+\frac{3(m-1)}{3-2m}+\frac{5(2m+1)}{2m-3}$
$=\frac{3(m-2)}{2m-3}-\frac{3(m-1)}{2m-3}+\frac{5(2m+1)}{2m-3}$
$=\frac{3m-6-3m+3+10m+5}{2m-3}$
$=\frac{10m+2}{2m-3}$
$=\frac{2(5m+1)}{2m-3}\ \left(m\neq\frac{3}{2}\right)$

53. $\frac{2x-y}{3}+\frac{x+4y}{3}$
$=\frac{2x-y+x+4y}{3}$
$=\frac{3x+3y}{3}$
$=\frac{3(x+y)}{3}=x+y$

55. $\frac{27x+18y}{(3x-2y)(3x+4y)(3x+2y)}$
$-\frac{18x+24y}{(3x-2y)(3x+4y)(3x+2y)}$
$=\frac{27x+18y-18x-24y}{(3x-2y)(3x+4y)(3x+2y)}$
$=\frac{9x-6y}{(3x-2y)(3x+4y)(3x+2y)}$
$=\frac{3(3x-2y)}{(3x-2y)(3x+4y)(3x+2y)}$
$=\frac{3}{(3x+4y)(3x+2y)}$
$\left(x\neq\frac{2}{3}y,\ x\neq -\frac{4}{3}y,\ x\neq -\frac{2}{3}y\right)$

57. $\dfrac{2(x-2y)}{(x+2y)(x+y)(x-2y)}+\dfrac{4(x+2y)}{(x+2y)(x+y)(x-2y)}-\dfrac{3(x+y)}{(x+2y)(x+y)(x-2y)}$

$=\dfrac{2(x-2y)+4(x+2y)-3(x+y)}{(x+2y)(x+y)(x-2y)}$

$=\dfrac{2x-4y+4x+8y-3x-3y}{(x+2y)(x+y)(x-2y)}$

$=\dfrac{3x+y}{(x+2y)(x+y)(x-2y)}$ $(x\neq -2y, x\neq -y, x=2y)$

59. $\dfrac{2a-b}{a-b}+\dfrac{a-2b}{b-a}=\dfrac{2a-b-(a-2b)}{a-b}=\dfrac{2a-b-a+2b}{a-b}=\dfrac{a+b}{a-b}$ $(a\neq b)$

61. $\dfrac{a+b}{a^2-b^2}+\dfrac{a-b}{a^2-b^2}-\dfrac{2a}{a^2-b^2}=\dfrac{a+b+a-b-2a}{(a+b)(a-b)}=\dfrac{0}{(a+b)(a-b)}=0$ $(a\neq b, -b)$

63. Statement c is true.

$\dfrac{2y+1}{y-7}+\dfrac{3y+1}{y-7}-\dfrac{5y+2}{y-7}=\dfrac{2y+1+3y+1-5y-2}{y-7}=\dfrac{0}{y-7}=0,$ $(y\neq 7)$

65. Correct

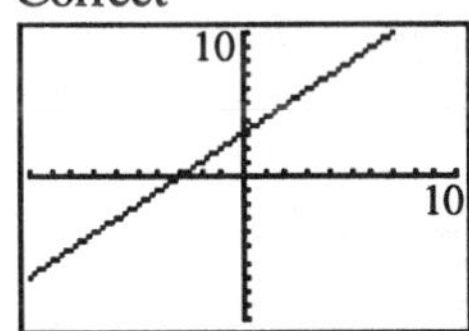

67. $\dfrac{x^2-13}{x+4}-\dfrac{3}{x+4}=\dfrac{x^2-13-3}{x+4}=\dfrac{x^2-16}{x+4}=\dfrac{(x+4)(x-4)}{x+4}=x-4$ $(x\neq -4)$

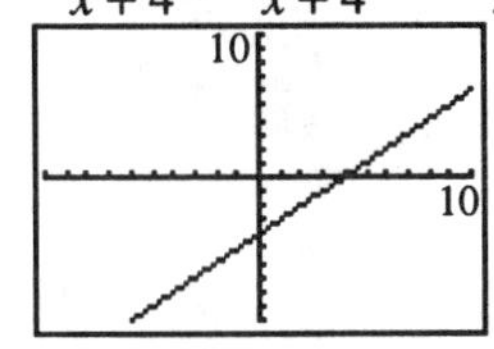

69. Answers may vary.

71. $2\left(\dfrac{2x}{x+1}+\dfrac{3}{x+1}\right)-2\left(\dfrac{x}{x+1}+\dfrac{2}{x+1}\right)=\dfrac{2(2x+3)}{x+1}-\dfrac{2(x+2)}{x+1}=\dfrac{4x+6-2x-4}{x+1}=\dfrac{2x+2}{x+1}=\dfrac{2(x+1)}{x+1}=2$

$(x\neq -1)$

73. $\dfrac{3x}{x+2}-\dfrac{20x-6}{x+2}=\dfrac{3x-20x+6}{x+2}=\dfrac{6-17x}{x+2}$

75. $\dfrac{a^2}{a-4}-\dfrac{a+12}{a-4}=\dfrac{a^2-a-12}{a-4}=\dfrac{(a-4)(a+3)}{a-4}=a+3$

Review Problems

77. $81y^4 - 1 = (9y^2 - 1)(9y^2 + 1)$
$= (3y - 1)(3y + 1)(9y^2 + 1)$

78. Let t = the time (in hours)
$550t + 475t = 2050$
$1025t = 2050$
$t = 2$ hours
time = noon + 2 hours = 2:00 P.M.

79. $\frac{3x^3 + 2x^2 - 26x - 15}{x + 3} = 3x^2 - 7x - 5$
$(x \neq -3)$

$$\begin{array}{r} 3x^2 - 7x - 5 \\ x + 3 \overline{\big) 3x^3 + 2x^2 - 26x - 15} \\ \underline{3x^3 + 9x^2 } \\ -7x^2 - 26x \\ \underline{-7x^2 - 21x } \\ -5x - 15 \\ \underline{-5x - 15} \\ 0 \end{array}$$

Problem Set 8.4

1. LCM (12, 10):
$12 = 2^2 \cdot 3$
$10 = 2 \cdot 5$
LCM $= 2^2 \cdot 3 \cdot 5 = 60$

3. LCM($3x$, x^3):
$3x = 3 \cdot x$
$x^3 = x^3$
LCM $= 3x^3$

5. LCM($15x^2$, $24x^5$):
$15x^2 = 3 \cdot 5 \cdot x^2$
$24x^5 = 2^3 \cdot 3 \cdot x^5$
LCM $= 2^3 \cdot 3 \cdot 5 \cdot x^5 = 120x^5$

7. LCM($100y$, $120y$)
$100y = 2^2 \cdot 5^2 \cdot y$
$120y = 2^3 \cdot 3 \cdot 5 \cdot y$
LCM $= 2^3 \cdot 3 \cdot 5^2 \cdot y = 600y$

9. LCM ($15x^2$, $6x^5$):
$15x^2 = 3 \cdot 5 \cdot x^2$
$6x^5 = 2 \cdot 3 \cdot x^5$
LCM $= 2 \cdot 3 \cdot 5 \cdot x^5 = 30x^5$

11. LCM($y - 3$, $y + 1$):
$y - 3 = y - 3$
$y + 1 = y + 1$
LCM $= (y - 3)(y + 1) = y^2 - 2y - 3$

13. LCM(x, $7(x + 2)$):
$x = x$
$7(x + 2) = 7 \cdot (x + 2)$
LCM $= 7x(x + 2)$

15. LCM ($18x^2$, $27x(x - 5)$):
$18x^2 = 2 \cdot 3^2 \cdot x^2$
$27x(x - 5) = 3^3 \cdot x \cdot (x - 5)$
LCM $= 2 \cdot 3^3 \cdot x^2 \cdot (x - 5) = 54x^2(x - 5)$

17. LCM($x + 3$, $x^2 - 9$):
$x + 3 = x + 3$
$x^2 - 9 = (x + 3) \cdot (x - 3)$
LCM $= (x + 3)(x - 3) = x^2 - 9$

19. LCM ($y^2 - 4$, $y(y + 2)$):
$y^2 - 4 = (y + 2) \cdot (y - 2)$
$y(y + 2) = y \cdot (y + 2)$
LCM $= y(y - 2)(y + 2) = y(y^2 - 4)$

21. LCM ($y^2 - 25$, $y^2 - 10y + 25$):
$y^2 - 25 = (y - 5) \cdot (y + 5)$
$y^2 - 10y + 25 = (y - 5)^2$
LCM $= (y + 5)(y - 5)^2$

23. LCM $(2x^2+7x-4,\ x^2+2x-8)$:
$2x^2+7x-4=(2x-1)\cdot(x+4)$
$x^2+2x-8=(x+4)\cdot(x-2)$
LCM $=(2x-1)(x-2)(x+4)$

25. $\frac{3}{x}+\frac{5}{x^2}=\frac{3x}{x^2}+\frac{5}{x^2}=\frac{3x+5}{x^2}\quad (x\neq 0)$

27. $\frac{2}{9w}-\frac{11}{6w}=\frac{4}{18w}-\frac{33}{18w}=-\frac{29}{18w}\quad (w\neq 0)$

29. $\frac{x-1}{6}-\frac{x+2}{3}$
$=\frac{x-1}{6}-\frac{2x+4}{6}$
$=\frac{-x-5}{6}=-\frac{x+5}{6}$

31. $\frac{2}{x-1}+\frac{3}{x+2}$
$=\frac{2(x+2)+3(x-1)}{(x-1)(x+2)}$
$=\frac{5x+1}{(x-1)(x+2)}\quad (x\neq -2, 1)$

33. $\frac{2}{r+5}+\frac{3}{4r}$
$=\frac{2(4r)+3(r+5)}{4r(r+5)}=\frac{11r+15}{4r(r+5)}\quad (r\neq 0, -5)$

35. $\frac{4y-9}{3y}-\frac{3y-8}{4y}$
$=\frac{4(4y-9)-3(3y-8)}{12y}$
$=\frac{7y-12}{12y}\quad (y\neq 0)$

37. $\frac{5a+3}{2a^2}-\frac{3a-4}{a}$
$=\frac{5a+3}{2a^2}-\frac{2a(3a-4)}{2a(a)}$
$=\frac{5a+3-6a^2+8a}{2a^2}$
$=\frac{-6a^2+13a+3}{2a^2}\quad (a\neq 0)$

39. $\frac{7}{x+5}-\frac{4}{x-5}$
$=\frac{7(x-5)-4(x+5)}{(x+5)(x-5)}$
$=\frac{3x-55}{(x+5)(x-5)}\quad (x\neq -5, 5)$

41. $\frac{2z}{z^2-16}+\frac{z}{z-4}$
$=\frac{2z}{(z+4)(z-4)}+\frac{z}{z-4}$
$=\frac{2z+z(z+4)}{(z+4)(z-4)}$
$=\frac{z^2+6z}{(z+4)(z-4)}$
$=\frac{z(z+6)}{(z+4)(z-4)}\quad (z\neq -4, 4)$

43. $\frac{5y}{y^2-9}-\frac{4}{y+3}$
$=\frac{5y}{(y-3)(y+3)}-\frac{4}{y+3}$
$=\frac{5y-4(y-3)}{(y-3)(y+3)}$
$=\frac{y+12}{(y-3)(y+3)}\quad (y\neq -3, 3)$

45. $\frac{7}{y-1}-\frac{3}{(y-1)^2}$
$=\frac{7(y-1)-3}{(y-1)^2}=\frac{7y-10}{(y-1)^2}\quad (y\neq 1)$

47. $\frac{3r}{4r-20}+\frac{9r}{6r-30}$
$=\frac{3r}{4(r-5)}+\frac{9r}{6(r-5)}$
$=\frac{3r}{4(r-5)}+\frac{3r}{2(r-5)}$
$=\frac{3r+6r}{4(r-5)}$
$=\frac{9r}{4(r-5)}\quad (r\neq 5)$

49. $\frac{y+4}{y}-\frac{y}{y+4}$
$=\frac{(y+4)(y+4)-y(y)}{y(y+4)}$
$=\frac{y^2+8y+16-y^2}{y(y+4)}$
$=\frac{8(y+2)}{y(y+4)}\ (y\neq 0,-4)$

51. $\frac{z}{z^2+2z+1}+\frac{4}{z^2+5z+4}$
$=\frac{z}{(z+1)^2}+\frac{4}{(z+1)(z+4)}$
$=\frac{z(z+4)+4(z+1)}{(z+1)^2(z+4)}$
$=\frac{z^2+8z+4}{(z+1)^2(z+4)}\ (z\neq -1,-4)$

53. $\frac{y-5}{y+3}+\frac{y+3}{y-5}$
$=\frac{(y-5)^2+(y+3)^2}{(y+3)(y-5)}$
$=\frac{2y^2-4y+34}{(y+3)(y-5)}$
$=\frac{2(y^2-2y+17)}{(y+3)(y-5)}\ (y\neq -3, 5)$

55. $\frac{5}{2y^2-2y}-\frac{3}{2y-2}$
$=\frac{5}{2y(y-1)}-\frac{3}{2(y-1)}$
$=\frac{5-3y}{2y(y-1)}\ (y\neq 0, 1)$

57. $\frac{4r+3}{r^2-9}-\frac{r+1}{r-3}$
$=\frac{4r+3}{(r+3)(r-3)}-\frac{r+1}{r-3}$
$=\frac{4r+3-(r+1)(r+3)}{(r+3)(r-3)}$
$=\frac{4r+3-r^2-4r-3}{(r+3)(r-3)}$
$=\frac{-r^2}{(r+3)(r-3)}\ (r\neq -3, 3)$

59. $\frac{y^2-39}{y^2+3y-10}-\frac{y-7}{y-2}$
$=\frac{y^2-39}{(y-2)(y+5)}-\frac{y-7}{y-2}$
$=\frac{y^2-39-(y+5)(y-7)}{(y-2)(y+5)}$
$=\frac{y^2-39-y^2+2y+35}{(y-2)(y+5)}$
$=\frac{2(y-2)}{(y-2)(y+5)}$
$=\frac{2}{y+5}\ (y\neq -5, 2)$

61. $\frac{w^2-11}{3w^2+5w-2}-\frac{w-5}{3w-1}$
$=\frac{w^2-11}{(3w-1)(w+2)}-\frac{w-5}{3w-1}$
$=\frac{w^2-11-(w+2)(w-5)}{(3w-1)(w+2)}$
$=\frac{w^2-11-w^2+3w+10}{(3w-1)(w+2)}$
$=\frac{3w-1}{(3w-1)(w+2)}$
$=\frac{1}{w+2}\ \left(w\neq -2, \frac{1}{3}\right)$

63. $4+\frac{1}{x-3}=\frac{4(x-3)+1}{x-3}=\frac{4x-11}{x-3}\ (x\neq 3)$

65. $3-\frac{3y}{y+1}=\frac{3(y+1)-3y}{y+1}=\frac{3}{y+1}\ (y\neq -1)$

67. $\frac{9x+3}{x^2-x-6}+\frac{x}{3-x}$
$=\frac{3(3x+1)}{(x-3)(x+2)}+\frac{x}{3-x}$
$=\frac{3(3x+1)}{(x-3)(x+2)}-\frac{x}{x-3}$
$=\frac{3(3x+1)-x(x+2)}{(x-3)(x+2)}$
$=\frac{-x^2+7x+3}{(x-3)(x+2)}$
$(x\neq -2, 3)$

69. $\frac{y+3}{5y^2}-\frac{y-5}{15y}$
$=\frac{3(y+3)}{15y^2}-\frac{y(y-5)}{15y^2}$
$=\frac{3y+9-y^2+5y}{15y^2}$
$=\frac{-y^2+8y+9}{15y^2}$
$=\frac{(-y+9)(y+1)}{15y^2}\ (y\neq 0)$

71. $\frac{x+1}{4x^2+4x-15}-\frac{4x+5}{8x^2-10x-3}$
$=\frac{x+1}{(2x+5)(2x-3)}-\frac{4x+5}{(4x+1)(2x-3)}$
$=\frac{(x+1)(4x+1)-(4x+5)(2x+5)}{(2x+5)(2x-3)(4x+1)}$
$=\frac{4x^2+5x+1-8x^2-30x-25}{(2x+5)(2x-3)(4x+1)}$
$=\frac{-4x^2-25x-24}{(2x+5)(2x-3)(4x+1)}$
$\left(x\neq -\frac{5}{2},\ \frac{3}{2},\ -\frac{1}{4}\right)$

73. $\frac{4x}{x^2-1}-\frac{2}{x}-\frac{2}{x+1}$
$=\frac{4x}{(x-1)(x+1)}-\frac{2}{x}-\frac{2}{x+1}$
$=\frac{4x(x)-2(x-1)(x+1)-2x(x-1)}{x(x-1)(x+1)}$
$=\frac{4x^2-2x^2+2-2x^2+2x}{x(x-1)(x+1)}$
$=\frac{2(x+1)}{x(x-1)(x+1)}$
$=\frac{2}{x(x-1)}$
$(x\neq 0, -1, 1)$

75. $\frac{7}{3x^2}+\frac{4}{x^2}-\frac{10}{7x}$
$=\frac{7\cdot 7+4\cdot 21-10\cdot 3\cdot x}{21x^2}$
$=\frac{-30x+133}{21x^2}\ (x\neq 0)$

77. $\frac{2}{x}+\frac{3}{y}=\frac{2}{x}\cdot\frac{y}{y}+\frac{3}{y}\cdot\frac{x}{x}=\frac{2y+3x}{xy}$
$(x\neq 0, y\neq 0)$

79. $\frac{5}{4x^2y}-\frac{2}{5xy^2}$
$=\frac{5}{4x^2y}\cdot\frac{5y}{5y}-\frac{2}{5xy^2}\cdot\frac{4x}{4x}$
$=\frac{25y-8x}{20x^2y^2}$
$(x\neq 0, y\neq 0)$

81. $\frac{x+2}{y}+\frac{y-2}{x}$
$=\frac{x+2}{y}\cdot\frac{x}{x}+\frac{y-2}{x}\cdot\frac{y}{y}$
$=\frac{x(x+2)+y(y-2)}{xy}$
$=\frac{x^2+2x+y^2-2y}{xy}\ (x\neq 0, y\neq 0)$

83. $\frac{y}{xy-x^2}-\frac{x}{y^2-xy}$
$=\frac{y}{x(y-x)}-\frac{x}{y(y-x)}$
$=\frac{y}{x(y-x)}\cdot\frac{y}{y}-\frac{x}{y(y-x)}\cdot\frac{x}{x}$
$=\frac{y^2-x^2}{xy(y-x)}$
$=\frac{(y-x)(y+x)}{xy(y-x)}$
$=\frac{y+x}{xy}\ (x\neq 0, y\neq 0, x\neq y)$

85. $\frac{1}{a-b}-\frac{a}{a^2-ab}+\frac{a^2}{a^3-a^2b}$
$=\frac{1}{a-b}-\frac{a}{a(a-b)}+\frac{a^2}{a^2(a-b)}$
$=\frac{1}{a-b}-\frac{1}{a-b}+\frac{1}{a-b}$
$=\frac{1}{a-b}\ (a\neq b, a\neq 0)$

87. $\frac{a}{b}-\frac{c}{d}=\frac{a}{b}\cdot\frac{d}{d}-\frac{c}{d}\cdot\frac{b}{b}$
$=\frac{ad-cb}{bd}\quad (b\neq 0,\ d\neq 0)$

89. $2\left(\frac{2x+1}{2x}\right)+2\left(\frac{x+5}{3x^2}\right)$
$=\frac{2x+1}{x}+\frac{2x+10}{3x^2}$
$=\frac{2x+1}{x}\cdot\frac{3x}{3x}+\frac{2x+10}{3x^2}$
$=\frac{3x(2x+1)+2x+10}{3x^2}$
$=\frac{6x^2+3x+2x+10}{3x^2}$
$=\frac{6x^2+5x+10}{3x^2}\quad (x\neq 0)$

91. $C=\frac{DA}{A+12}$
$\frac{7D}{19}-\frac{3D}{15}$
$=\frac{7D}{19}-\frac{D}{5}$
$=\frac{7D}{19}\cdot\frac{5}{5}-\frac{D}{5}\cdot\frac{19}{19}$
$=\frac{35D-19D}{95}=\frac{16D}{95}$
The difference between a 7-year-old's dose and a 3-year-old's dose is $\frac{16}{95}$ times the adult dose.

93. Statement d is true.
d. $y-3,\ y+3$
LCD = $(y-3)(y+3)=y^2-9$ True

95. $\frac{x}{5}+\frac{1}{3}=\frac{x}{5}\cdot\frac{3}{3}+\frac{1}{3}\cdot\frac{5}{5}=\frac{3x+5}{15}$

97–99. Answers may vary.

101. $\frac{2}{x-1}+\frac{3}{x^2}$
$=\frac{2}{x-1}\cdot\frac{x^2}{x^2}+\frac{3}{x^2}\cdot\frac{x-1}{x-1}$
$=\frac{2x^2+3(x-1)}{x^2(x-1)}=\frac{2x^2+3x-3}{x^2(x-1)}$

103. Fraction left to paint is
$1-\frac{1}{x}-\frac{1}{x+3}$
$=\frac{x(x+3)-(x+3)-x}{x(x+3)}$
$=\frac{x^2+3x-x-3-x}{x(x+3)}$
$=\frac{x^2+x-3}{x(x+3)}$

Review Problems

104. $(3y+5)(2y-7)$
$=6y^2-21y+10y-35$
$=6y^2-11y-35$

105. $3x-y<3$
$y>3x-3$

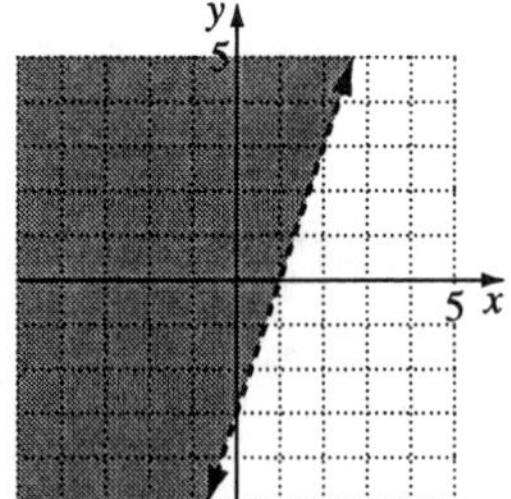

106. $(-3,-4),\ (1,0)$: $m=\frac{-4-0}{-3-1}=1$
using (1, 0):
$y-0=1(x-1)$
$y=x-1$

Problem Set 8.5

1. $\frac{\frac{1}{2}+\frac{1}{4}}{\frac{1}{2}+\frac{1}{3}}=\frac{\frac{4+2}{8}}{\frac{3+2}{6}}=\frac{3}{4}\cdot\frac{6}{5}=\frac{18}{20}=\frac{9}{10}$

3. $\dfrac{3+\frac{1}{2}}{4-\frac{1}{4}}=\dfrac{\frac{6+1}{2}}{\frac{16-1}{4}}=\dfrac{7}{2}\cdot\dfrac{4}{15}=\dfrac{28}{30}=\dfrac{14}{15}$

5. $\dfrac{\frac{2}{5}-\frac{1}{3}}{\frac{2}{3}-\frac{3}{4}}=\dfrac{\left(\frac{2}{5}-\frac{1}{3}\right)}{\left(\frac{2}{3}-\frac{3}{4}\right)}\cdot\dfrac{60}{60}=\dfrac{4(6-5)}{5(8-9)}=\dfrac{4}{-5}=-\dfrac{4}{5}$
(Using LCM of denominators)

7. $\dfrac{\frac{3}{4}-x}{\frac{3}{4}+x}=\dfrac{\frac{3-4x}{4}}{\frac{3+4x}{4}}=\dfrac{3-4x}{3+4x}$

9. $\dfrac{5-\frac{2}{x}}{3+\frac{1}{x}}=\dfrac{\frac{5x-2}{x}}{\frac{3x+1}{x}}=\dfrac{5x-2}{3x+1},\ \left(x\neq 0,\ -\frac{1}{3}\right)$

11. $\dfrac{2+\frac{3}{y}}{1-\frac{7}{y}}=\dfrac{\frac{2y+3}{y}}{\frac{y-7}{y}}$
$=\dfrac{2y+3}{y}\cdot\dfrac{y}{y-7}=\dfrac{2y+3}{y-7}\ \ (y\neq 0,7)$

13. $\dfrac{\frac{1}{y}-\frac{3}{2}}{\frac{1}{y}+\frac{3}{4}}=\dfrac{\frac{2-3y}{2y}}{\frac{4+3y}{4y}}$
$=\dfrac{2-3y}{2y}\cdot\dfrac{4y}{4+3y}=\dfrac{2(2-3y)}{4+3y}\ \ \left(y\neq 0,\ -\frac{4}{3}\right)$

15. $\dfrac{\frac{x}{5}-\frac{5}{x}}{\frac{1}{5}+\frac{1}{x}}=\dfrac{\frac{x^2-25}{5x}}{\frac{x+5}{5x}}$
$=\dfrac{(x+5)(x-5)}{5x}\cdot\dfrac{5x}{x+5}=x-5\ \ (x\neq 0,-5)$

17. $\dfrac{1+\frac{1}{x}}{1-\frac{1}{x^2}}=\dfrac{\frac{x+1}{x}}{\frac{x^2-1}{x^2}}$
$=\dfrac{x+1}{x}\cdot\dfrac{x^2}{(x+1)(x-1)}=\dfrac{x}{x-1}\ \ (x\neq 0,1,-1)$

19. $\dfrac{\frac{1}{7}-\frac{1}{y}}{\frac{7-y}{7}}=\dfrac{y-7}{7y}\cdot\dfrac{7}{7-y}=-\dfrac{1}{y}\ \ (y\neq 0,7)$

21. $\dfrac{\frac{12}{y^2}-\frac{3}{y}}{\frac{15}{y}-\frac{9}{y^2}}=\dfrac{\frac{12-3y}{y^2}}{\frac{15y-9}{y^2}}\cdot\dfrac{y^2}{y^2}$
$=\dfrac{12-3y}{15y-9}=\dfrac{3(4-y)}{3(5y-3)}=\dfrac{4-y}{5y-3}\ \ \left(y\neq 0,\ \frac{3}{5}\right)$

23. $\dfrac{\frac{1}{w}+\frac{2}{w^2}}{\frac{2}{w}+1}=\dfrac{\frac{w+2}{w^2}}{\frac{w(2+w)}{w^2}}\cdot\dfrac{w^2}{w^2}$
$=\dfrac{w+2}{w(2+w)}=\dfrac{1}{w}\ \ (w\neq 0,-2)$

25. $\dfrac{\frac{9}{5}+\frac{4}{5s}}{\frac{4}{s^2}+\frac{9}{5}}=\dfrac{9s+4}{5s}\cdot\dfrac{5s^2}{20+9s^2}$
$=\dfrac{s(9s+4)}{9s^2+20}\ \ (s\neq 0)$

27. $\dfrac{\frac{7}{x^3}+\frac{11}{x^2}}{\frac{7}{x^4}+\frac{11}{x^3}}=\dfrac{7+11x}{x^3}\cdot\dfrac{x^4}{7+11x}=x$
$\left(x\neq 0,\ -\frac{7}{11}\right)$

29. $\dfrac{\frac{7}{6x^3}-\frac{5}{12x}}{\frac{7}{2x}+\frac{3}{2x^3}}$
$=\dfrac{\frac{14-5x^2}{12x^3}}{\frac{7x^2+3}{2x^3}}$
$=\dfrac{14-5x^2}{12x^3}\cdot\dfrac{2x^3}{7x^2+3}=\dfrac{14-5x^2}{6(7x^2+3)}\ \ (x\neq 0)$

31. $\dfrac{x-5+\frac{3}{x}}{x-7+\frac{2}{x}}=\dfrac{\frac{x^2-5x+3}{x}}{\frac{x^2-7x+2}{x}}$
$=\dfrac{x^2-5x+3}{x}\cdot\dfrac{x}{x^2-7x+2}$
$=\dfrac{x^2-5x+3}{x^2-7x+2}\ \ \left(x\neq 0,\ \dfrac{7\pm\sqrt{41}}{2}\right)$

33. $\dfrac{\frac{1}{y+2}}{1+\frac{1}{y+2}}=\dfrac{\frac{1}{y+2}}{\frac{y+3}{y+2}}=\dfrac{1}{y+3}\ \ (y\neq -2,-3)$

35. $\dfrac{\frac{1}{x}+\frac{1}{x^2}+\frac{1}{x^3}}{\frac{1}{x^4}+\frac{1}{x^5}+\frac{1}{x^6}}=\dfrac{\frac{1}{x^3}(x^2+x+1)}{\frac{1}{x^6}(x^2+x+1)}=\dfrac{x^6}{x^3}=x^3$

$x=1$: $\dfrac{\frac{1}{x}+\frac{1}{x^2}+\frac{1}{x^3}}{\frac{1}{x^4}+\frac{1}{x^5}+\frac{1}{x^6}}=x^3=1$

$x=2$: $\dfrac{\frac{1}{x}+\frac{1}{x^2}+\frac{1}{x^3}}{\frac{1}{x^4}+\frac{1}{x^5}+\frac{1}{x^6}}=x^3=8$

$x=3$: $\dfrac{\frac{1}{x}+\frac{1}{x^2}+\frac{1}{x^3}}{\frac{1}{x^4}+\frac{1}{x^5}+\frac{1}{x^6}}=x^3=27$

$x=4$: $\dfrac{\frac{1}{x}+\frac{1}{x^2}+\frac{1}{x^3}}{\frac{1}{x^4}+\frac{1}{x^5}+\frac{1}{x^6}}=x^3=64$

$x=5$: $\dfrac{\frac{1}{x}+\frac{1}{x^2}+\frac{1}{x^3}}{\frac{1}{x^4}+\frac{1}{x^5}+\frac{1}{x^6}}=x^3=125$

37. $\dfrac{\frac{4}{a}-\frac{8}{b}}{2}=\dfrac{\frac{4b-8a}{ab}}{2}=\dfrac{4(b-2a)}{ab}\cdot\dfrac{1}{2}=\dfrac{2(b-2a)}{ab}$

$(a\neq 0), b\neq 0)$

39. $\dfrac{a-\frac{a}{b}}{\frac{1+a}{b}}=\dfrac{ab-a}{b}\cdot\dfrac{b}{1+a}=\dfrac{a(b-1)}{1+a}$

$(a\neq -1,\ b\neq 0)$

41. $\dfrac{1}{\frac{1}{a}+b}=\dfrac{1}{\frac{1+ab}{a}}=1\cdot\dfrac{a}{1+ab}=\dfrac{a}{1+ab}$

$(a\neq 0,\ ab\neq -1)$

43. $\dfrac{\frac{1}{x}+\frac{1}{y}}{\frac{1}{xy}}=\dfrac{y+x}{xy}\cdot\dfrac{xy}{1}=y+x\quad(x\neq 0, y\neq 0)$

45. $\dfrac{\frac{x}{y}+\frac{1}{x}}{\frac{y}{x}+\frac{1}{x}}=\dfrac{\frac{x^2+y}{xy}}{\frac{1+y}{x}}=\dfrac{x^2+y}{xy}\cdot\dfrac{x}{1+y}=\dfrac{x^2+y}{y(1+y)}$

$(x\neq 0, y\neq 0, -1)$

47. $\dfrac{\frac{a}{10b^3}-\frac{3}{5b}}{\frac{a}{10b}+\frac{3}{b^4}}=\dfrac{\frac{a-6b^2}{10b^3}}{\frac{ab^3+30}{10b^4}}=\dfrac{a-6b^2}{10b^3}\cdot\dfrac{10b^4}{ab^3+30}$

$=\dfrac{b(a-6b^2)}{ab^3+30}\quad(ab^3\neq -30,\ b\neq 0)$

49. $\dfrac{\frac{2}{x^3y}+\frac{5}{xy^4}}{\frac{5}{x^3y}-\frac{3}{xy}}=\dfrac{\frac{2y^3+5x^2}{x^3y^4}}{\frac{5-3x^2}{x^3y}}=\dfrac{2y^3+5x^2}{x^3y^4}\cdot\dfrac{x^3y}{5-3x^2}$

$=\dfrac{2y^3+5x^2}{y^3(5-3x^2)}\quad\left(x\neq 0,\ y\neq 0,\ x\neq\pm\sqrt{\frac{5}{3}}\right)$

51. $\dfrac{7-\frac{2}{xy^3}}{\frac{x+2}{x^2y}}=\dfrac{7xy^3-2}{xy^3}\cdot\dfrac{x^2y}{x+2}=\dfrac{x(7xy^3-2)}{y^2(x+2)}$

$(x\neq 0, y\neq 0, x\neq -2)$

53. $\dfrac{\frac{3}{x+y}-\frac{3}{x-y}}{\frac{5}{x^2-y^2}}$

$=\dfrac{3(x-y)-3(x+y)}{(x+y)(x-y)}\cdot\dfrac{(x+y)(x-y)}{5}$

$=\dfrac{3x-3y-3x-3y}{5}=-\dfrac{6y}{5}\quad(x\neq y, x\neq -y)$

55. $r_{\text{average}}=\dfrac{2d}{\frac{d}{r_1}+\frac{d}{r_2}}=\dfrac{2d}{\frac{dr_2+dr_1}{r_1r_2}}=\dfrac{2dr_1r_2}{d(r_1+r_2)}$

$=\dfrac{2r_1r_2}{r_1+r_2}$

$r_1=30,\ r_2=20$

$r_{\text{average}}=\dfrac{2(30)(20)}{30+20}=\dfrac{1200}{50}$

$=24$ miles per hour

The answer is not 25 miles per hour, which would be $\dfrac{r_1+r_2}{2}$, but is the total distance divided by the total time.

57. Statement b is true.

b. $\dfrac{y-\frac{1}{2}}{y+\frac{3}{4}}=\dfrac{\frac{2y-1}{2}}{\frac{4y+3}{4}}=\dfrac{2y-1}{2}\cdot\dfrac{4}{4y+3}$

$=\dfrac{2(2y-1)}{4y+3}=\dfrac{4y-2}{4y+3}$

True for all values exccept $y=-\frac{3}{4}$.

59. Correct

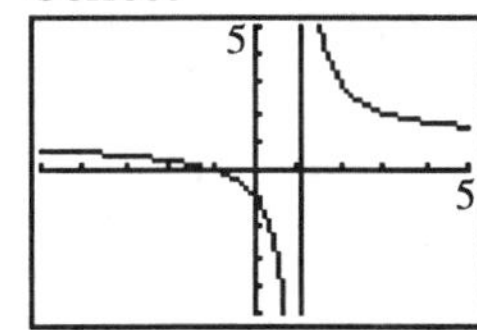

61. $\dfrac{1+\frac{1}{y}-\frac{6}{y^2}}{1-\frac{5}{y}+\frac{6}{y^2}}-\dfrac{1-\frac{1}{y}}{1-\frac{2}{y}-\frac{3}{y^2}}$

$=\dfrac{\frac{y^2+y-6}{y^2}}{\frac{y^2-5y+6}{y^2}}-\dfrac{\frac{y-1}{y}}{\frac{y^2-2y-3}{y^2}}$

$=\dfrac{y^2+y-6}{y^2-5y+6}-\dfrac{y(y-1)}{y^2-2y-3}$

$=\dfrac{(y+3)(y-2)}{(y-2)(y-3)}-\dfrac{y(y-1)}{(y-3)(y+1)}$

$=\dfrac{y+3}{y-3}-\dfrac{y(y-1)}{(y-3)(y+1)}$

$=\dfrac{(y+3)(y+1)-y(y-1)}{(y-3)(y+1)}$

$=\dfrac{y^2+4y+3-y^2+y}{(y-3)(y+1)}$

$=\dfrac{5y+3}{(y-3)(y+1)}$

$(y \neq 0, -1, 3, 2)$

Review Problems

63. $x^2-4x+9-(3x^2-7x-5)$

$=x^2-3x^2-4x+7x+9+5$

$=-2x^2+3x+14$

64. $h=-16t^2+32t \quad h=16$ feet

$16=-16t^2+32t$

$1=-t^2+2t$

$t^2-2t+1=0$

$(t-1)^2=0$

$t=1$ second

65. $f(x)=4x-3$

$f(-2)=4(-2)-3=-8-3=-11$

$3f(4)=3[4(4)-3]=3(16-3)=39$

$f(-2)+3f(4)=-11+39=28$

Problem Set 8.6

1. $\frac{x}{3}=\frac{x}{2}-2$

$6\left(\frac{x}{3}\right)=6\left(\frac{x}{2}-2\right)$

$2x=3x-12$

$12=3x-2x$

$12=x$

12

3. $\frac{x}{9}-\frac{3}{5}=\frac{2}{3}$

$45\left(\frac{x}{9}-\frac{3}{5}\right)=45\left(\frac{2}{3}\right)$

$5x-27=30$

$5x=57$

$x=\frac{57}{5}$

$\frac{57}{5}$

5. $2-\frac{8}{x}=6$

$x\left(2-\frac{8}{x}\right)=6x$

$2x-8=6x$

$-8=4x$

$x=-2$

-2

7. $\frac{2}{3}-\frac{5}{6}=\frac{1}{y}$

$6y\left(\frac{2}{3}-\frac{5}{6}\right)=6y\left(\frac{1}{y}\right) \ (y \neq 0)$

$6y\left(\frac{-1}{6}\right)=6$

$-y=6$

$y=-6$

-6

9. $\frac{4}{y}+\frac{1}{2}=\frac{5}{y}$

$\frac{8+y}{2y}=\frac{5}{y}$

$2y\left(\frac{8+y}{2y}\right)=2y\left(\frac{5}{y}\right)$

$8+y=10$

$y=10-8=2$

2

11. $\frac{2}{y}+3=\frac{5}{2y}+\frac{13}{4}$

$4y\left(\frac{2+3y}{y}\right)=4y\left(\frac{10+13y}{4y}\right)$

$4(2+3y)=10+13y$

$8+12y=10+13y$

$y=-2$

–2

13. $\frac{1}{z-1}+5=\frac{11}{z-1}$

$5=\frac{11-1}{z-1}=\frac{10}{z-1}$

$5z-5=10$

$z=3$

3

15. $\frac{8y}{y+1}=4-\frac{8}{y+1}$

$y+1\left(\frac{8y}{y+1}\right)=y+1\left(4-\frac{8}{y+1}\right)$

$8y=4y+4-8$

$4y=-4$

$y=-1$

Since y is not permitted to be –1, there is no solution.

$\varnothing$

17. $\frac{4}{r^2-4}+\frac{2}{r-2}=\frac{1}{r+2}$

$\text{LCM}=r^2-4$

$(r^2-4)\left[\frac{4}{r^2-4}+\frac{2}{r-2}\right]=(r^2-4)\left(\frac{1}{r+2}\right)$

$4+2(r+2)=r-2$

$4+2r+4=r-2$

$r=-8-2=-10$

–10

19. $\frac{2}{y+1}-\frac{1}{y-1}=\frac{2y}{y^2-1}$

$(y^2-1)\left(\frac{2}{y+1}-\frac{1}{y-1}\right)=\left(\frac{2y}{y^2-1}\right)(y^2-1)$

$2(y-1)-1(y+1)=2y$

$2y-2-y-1=2y$

$y-3=2y$

$-3=y$

–3

21. $\frac{4}{y-3}-\frac{2}{y-2}=\frac{7-y}{y^2-5y+6}$

$\text{LCM}=y^2-5y+6=(y-2)(y-3)$

$(y-3)(y-2)\left[\frac{4}{y-3}-\frac{2}{y-2}\right]$

$=(y^2-5y+6)\left[\frac{7-y}{y^2-5y+6}\right]$

$(y\neq 2, 3)$

$4(y-2)-2(y-3)=7-y$

$4y-8-2y+6=7-y$

$3y=9$

$y=3$

but $y\neq 3$

Since y is not permitted to be 3, there is no solution.

$\varnothing$

23. $\frac{5}{2x+6}-\frac{1}{x+3}=\frac{1}{x+1}$

$\text{LCM} = 2(x+3)(x+1)$

$2(x+3)(x+1)\left(\frac{5}{2x+6}-\frac{1}{x+3}\right)$

$=2(x+3)(x+1)\left(\frac{1}{x+1}\right)$

$5(x+1)-2(x+1)=2(x+3)$

$3(x+1)=2(x+3)$

$3x+3=2x+6$

$x=3$

3

25. $\frac{3y}{y-4}-5=\frac{12}{y-4}$

$(y-4)\left(\frac{3y}{y-4}-5\right)=(y-4)\left(\frac{12}{y-4}\right)$

$3y-5(y-4)=12$

$-2y+20=12$

$-2y=-8$

$y=4$

but $y \neq 4$

Since $y=4$ is not permitted, there is no solution.

$\varnothing$

27. $\frac{4}{w}-\frac{w}{2}=\frac{7}{2}$

$(2w)\left(\frac{4}{w}-\frac{w}{2}\right)=2w\left(\frac{7}{2}\right)$

$8-w^2=7w$

$w^2+7w-8=0$

$(w+8)(w-1)=0$

$w=1$ or $w=-8$

Check:

$w=1$: $\frac{4}{1}-\frac{1}{2}=\frac{7}{2}$ true

$w=-8$: $\frac{4}{-8}-\frac{-8}{2}=-\frac{1}{2}+\frac{8}{2}=\frac{7}{2}$ true

Solution is –8, 1

29. $\frac{5}{3y-8}=\frac{y}{y+2}$

$5(y+2)=3y^2-8y$

$\left(y \neq -2, \frac{8}{3}\right)$

$5y+10=3y^2-8y$

$3y^2-13y-10=0$

$(3y+2)(y-5)=0$

$y=-\frac{2}{3}$ or $y=5$

Check:

$y=-\frac{2}{3}$: $\frac{5}{-2-8}=\frac{-\frac{2}{3}}{\frac{4}{3}}$

$-\frac{1}{2}=-\frac{1}{2}$ true

$y=5$: $\frac{5}{15-8}=\frac{5}{5+2}$

$\frac{5}{7}=\frac{5}{7}$ true

Solution is $-\frac{2}{3}$, 5

31. $\frac{3}{z-1}+\frac{8}{z}=3$

$z(z-1)\left(\frac{3}{z-1}+\frac{8}{z}\right)=z(z-1)(3)$

$3z+8(z-1)=3z(z-1)$

$3z+8z-8=3z^2-3z$

$3z^2-14z+8=0$

$(3z-2)(z-4)=0$

$z=\frac{2}{3}$ or $z=4$

Check:

$z=\frac{2}{3}$: $\frac{3}{-\frac{1}{3}}+\frac{8}{\frac{2}{3}}=-9+12=3$

$3=3$ true

$z=4$: $\frac{3}{3}+\frac{8}{4}=1+2=3$

$3=3$ true

Solution is $\frac{2}{3}$, 4.

33. $\frac{2}{y-2}+\frac{y}{y+2}=\frac{y+6}{y^2-4}$

LCM = y^2-4

$(y^2-4)\left(\frac{2}{y-2}+\frac{y}{y+2}\right)=(y^2-4)\left(\frac{y+6}{y^2-4}\right)$

$(y \neq -2, 2)$

$2(y+2)+y(y-2)=y+6$

$2y+4+y^2-2y=y+6$

$y^2-y-2=0$

$(y-2)(y+1)=0$

$y=-1$ or $y=2$ (reject $y=2$ since $y \neq 2$)

Solution is –1.

35. $x+\frac{6}{x}=-5$

$x\left(x+\frac{6}{x}\right)=-5x$

$x^2+6=-5x$

$x^2+5x+6=0$

$(x+3)(x+2)=0$

$x=-3$ or $x=-2$

Solution is –3, –2.

37. $\frac{1}{x}+\frac{1}{x-3}=\frac{x-2}{x-3}$

$x(x-3)\left(\frac{1}{x}+\frac{1}{x-3}\right)=x(x-3)\left(\frac{x-2}{x-3}\right)$

$x-3+x=x^2-2x$

$x^2-4x+3=0$

$(x-3)(x-1)=0$

$x=3$ or $x=1$

$x \neq 3$ so the solution is 1.

39. $P=\frac{500(1+3t)}{5+t}$

$P=1000$

$1000=\frac{500(1+3t)}{5+t}$

$2(5+t)=1+3t$

$(t \neq -5)$

$10+2t=1+3t$

$t=9$ hours

It will take 9 hours for the population to increase to 1000 insects.

41. $y=\frac{0.8x}{1+0.03x}$

$20=\frac{0.8x}{1+0.03x}$

$20+0.6x=0.8x$

$20=0.2x$

$x=100$ prey per unit area

43. $Q=\frac{PV}{P-V}$

$85=\frac{80,000V}{80,000-V}$

$85(80,000)-85V=80,000V$

$V=\frac{85(80,000)}{80,000+85}\approx 84.91$

Last year's volume, approximately 84.91 thousands = 84,910

45. Let

x = number of marijuana/hashish incidents.

Then

$x+94,151$ = number of cocaine incidents.

$\frac{x+94,151}{x}=4+\frac{6653}{x}$

$x\left(\frac{x+94,151}{x}\right)=x\left(4+\frac{6653}{x}\right)$

$x+94,151=4x+6653$

$-3x=-87,498$

$x=29,166$

$x+94,151=123,317$

Marijuana/hashish = 29,166

Cocaine = 123,317

47. Let x = the numerator of the fraction.

$x+5$ = the denominator of the fraction.

$\frac{x+1}{x+5+2}=\frac{1}{3}$

$3x+3=x+7$

$2x=4$

$x=2$

$x+5=7$

fraction = $\frac{2}{7}$

Check: $\frac{2+1}{7+2}=\frac{3}{9}=\frac{1}{3}$

49. Let x = the number.

$\frac{1}{x}$ = the reciprocal of the number.

$x+\frac{1}{x}=\frac{25}{12}$

$\frac{x^2+1}{x}=\frac{25}{12}$

$12x^2+12=25x$

$12x^2-25x+12=0$

$(4x-3)(3x-4)=0$

$x=\frac{3}{4}$ or $x=\frac{4}{3}$

The number is $\frac{3}{4}$ or $\frac{4}{3}$.

51. Let x = the speed of current.

	D	R	$T=\frac{D}{R}$
with the current	33 miles	$18+x$	$\frac{33}{18+x}$
against the current	21 miles	$18-x$	$\frac{21}{18-x}$

$\text{time}=\frac{\text{distance}}{\text{speed}}=\frac{33}{18+x}=\frac{21}{18-x}$

$33(18-x)=21(18+x)$

$18(33-21)=x(21+33)$

$x=\frac{216}{54}=4$

The speed of the current is 4 miles per hour.

53. Let x = rate of walking.

$9x$ = rate of driving

$\frac{90}{9x}+\frac{5}{x}=3$

$90+45=27x$

$x=\frac{135}{27}=5$ miles per hour = walking rate

55. Let t = number of hours to complete job when working together.

$\frac{t}{55}+\frac{t}{66}=1$

$\frac{6t+5t}{330}=1$

$11t=330$

$t=30$

30 hours to complete job.

57. Let t = time to fill tub with both pipes.

$\frac{t}{15}+\frac{t}{10}=1$

$\frac{2t+3t}{30}=1$

$5t=30$

$t=6$

6 minutes to fill tub.

59. Let x = additional consecutive hits.

$\text{batter average}=\frac{\text{hits}}{\text{number of times at bat}}$

$\frac{12+x}{40+x}=0.440$

$12+x=17.6+0.440x$

$0.560x=5.6$

$x=10$

The player must hit the ball 10 additional consecutive times.

61. Area of first rectangle
 – area of second rectangle
= area of third rectangle

$9\left(\frac{1}{x+5}\right)-\frac{1}{x-5}(1)=3x\left(\frac{1}{x^2-25}\right)$

$9(x-5)-(x+5)=3x$

$8x-50=3x$

$5x=50$

$x=10$

First rectangle: 9 by $\frac{1}{15}$

Second rectangle: 1 by $\frac{1}{5}$

Third rectangle: 30 by 75

63. Statement b is true.

b. $\frac{y+7}{2y-6}=\frac{5}{y-3}-1$

$\frac{y+7}{2(y-3)}=\frac{5-y+3}{y-3}$

$\frac{y+7}{2(y-3)}=\frac{8-y}{(y-3)}$

$y+7=16-2y$

$3y=9$

$y=3$ but $y \neq 3$

No solution; true

65. Statement d is true. $\frac{1}{x}$ is the reciprocal of x.

$\frac{1}{3}\left(\frac{1}{x}\right)=\frac{1}{3x}$

67. $\frac{50}{x}=2x, \quad x=\pm 5$

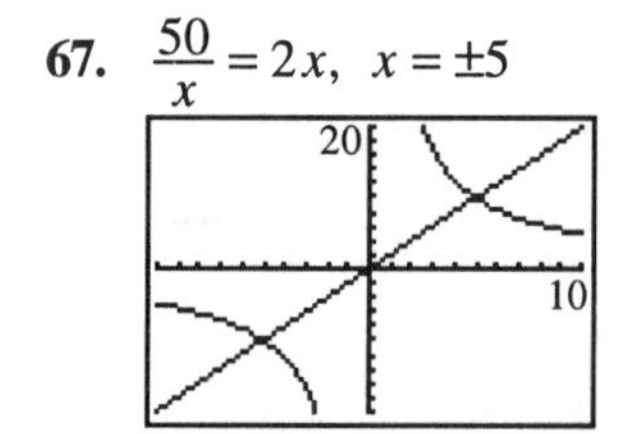

69. $\frac{2}{x+2}=\frac{1}{x^2-4}+1, \quad x=1$

71. $y=\frac{40}{x}+\frac{40}{x+30}$

73. $x=30$

30; 2

75. $\frac{1}{y^2+3y+2}+\frac{1}{y-1}=\frac{2}{y^2-1}$

$(y+2)(y+1)(y-1)\left[\frac{1}{(y+2)(y+1)}+\frac{1}{y-1}\right]$

$=\left[\frac{2}{(y+1)(y-1)}\right](y+2)(y+1)(y-1)$

$y-1+(y+2)(y+1)=2(y+2)$

$y-1+y^2+3y+2=2y+4$

$y^2+4y+1=2y+4$

$y^2+2y-3=0$

$(y+3)(y-1)=0$

$y=-3, y=1$

Since y cannot equal 1, the solution is -3.

77. Let x = speed of car.

$\frac{100}{x}-1=\frac{100}{x+5}$

$100(x+5)-x(x+5)=100x$

$100x+500-x^2-5x=100x$

$x^2+5x-500=0$

$x=20$ (reject $x=-25$)

Car's speed is 20 miles per hour

79. Let x = number of questions on test after first 30 questions.

$\frac{20+x}{30+x}=0.75$

$20+x=22.5+0.75x$

$0.25x=2.5$

$x=10$

Total number of questions = 30 + 10 = 40

Review Problems

81. Let x = the unknown calories.

$\frac{28.4}{110}=\frac{42.6}{x}$

$x=\frac{110(42.6)}{28.4}=165$ calories

82. $(5, -2), (3, 8)$: $m=\frac{8-(-2)}{3-5}=\frac{10}{-2}=-5$

83. $2x^3+8x^2-42x=2x(x^2+4x-21)$

$=2x(x+7)(x-3)$

Problem Set 8.7

1. $C=\frac{4x}{100-x}$

$100C-Cx=4x$

$(x \neq 100)$

$(4+C)x=100C$

$x = \frac{100C}{4+C}$

If $C = 16$, $x = \frac{1600}{4+16} = \frac{1600}{20} = 80\%$.

3. $P = 30 - \frac{9}{t+1} = \frac{30t+30-9}{t+1} = \frac{30t+21}{t+1}$

$Pt + P = 30t + 21$

$(P-30)t = 21 - P$

$t = \frac{21-P}{P-30}$

$t = \frac{P-21}{30-P}$

If $P = 29$, $t = \frac{29-21}{30-29} = 8$ years from 1990, or the year 1998.

5. $B = \frac{F}{S-V}$

a. $BS - BV = F$

$S = \frac{BV+F}{B}$

$S = V + \frac{F}{B}$

b. If $F = 20{,}000$, $V = 60$, $B = 100$, then

$S = 60 + \frac{20{,}000}{100} = 60 + 200 = \$260$.

7. $\frac{1}{R} = \frac{1}{R_1} + \frac{1}{R_2}$

a. If $R = 4$ ohms, $R_1 = 12$ ohms, then,

$\frac{1}{4} = \frac{1}{12} + \frac{1}{R_2}$

$\frac{1}{R_2} = \frac{1}{4} - \frac{1}{12}$

$\frac{1}{R_2} = \frac{1}{6}$

$R_2 = 6$ ohms

b. $\frac{1}{R} = \frac{R_2 + R_1}{R_1 R_2}$

$R = \frac{R_1 R_2}{R_1 + R_2}$

9. $P = \frac{DN}{N+2}$

$PN + 2P = DN$

$N(P-D) = -2P$

$N = \frac{2P}{D-P}$

11. $A = P + Prt = P(1 + rt)$

$P = \frac{A}{1+rt}$

13. $A = \frac{1}{2}bh$ for h

$2A = bh$

$h = \frac{2A}{b}$

Area of a triangle (A), base (b), height (h)

15. $s = \frac{1}{2}at^2$ for a

$2s = at^2$

$a = \frac{2s}{t^2}$

17. $F = \frac{mv^2}{r}$ for r

$rF = mv^2$

$r = \frac{mv^2}{F}$

19. $\frac{P_1V_1}{T_1} = \frac{P_2V_2}{T_2}$ for T_2

$T_2P_1V_1 = T_1P_2V_2$

$T_2 = \frac{T_1P_2V_2}{P_1V_1}$

21. $S = \frac{a}{1-r}$ for r

$1 - r = \frac{a}{S}$

$r = 1 - \frac{a}{S}$

23. $f = \frac{f_1f_2}{f_1+f_2}$ for f_2

$f(f_1 + f_2) = f_1f_2$

$ff_1 + ff_2 = f_1f_2$

$ff_2 - f_1f_2 = -ff_1$

$f_2(f - f_1) = -ff_1$

$f_2 = -\frac{ff_1}{f - f_1}$

25. $V = \frac{4}{3}\pi r^3$ for r^3

$\frac{3}{4}V = \pi r^3$

$r^3 = \frac{3V}{4\pi}$

Volume of a sphere (V), radius (r)

27. $A = \frac{rs}{r+s}$ for s

$Ar + As = rs$

$s(A - r) = -Ar$

$s = -\frac{Ar}{A-r} = \frac{Ar}{r-A}$

29. $\frac{b}{y} = 1 + c$ for y

$b = y(1 + c)$

$y = \frac{b}{1+c}$

31. a. $P = kH$

$425 = k(25)$

$k = \frac{425}{25}$

$k = 17$

$P = 17H$

b. $P = 17(40)$

$P = \$680$

c.

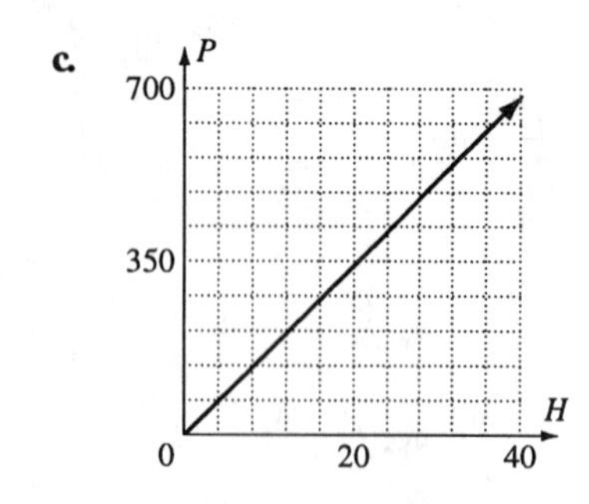

33. $A = kE$

$126 = k(1800)$

$k = 0.07$

$A = 0.07E$

$A = 0.07(2600)$

$A = \$182$

35. a.

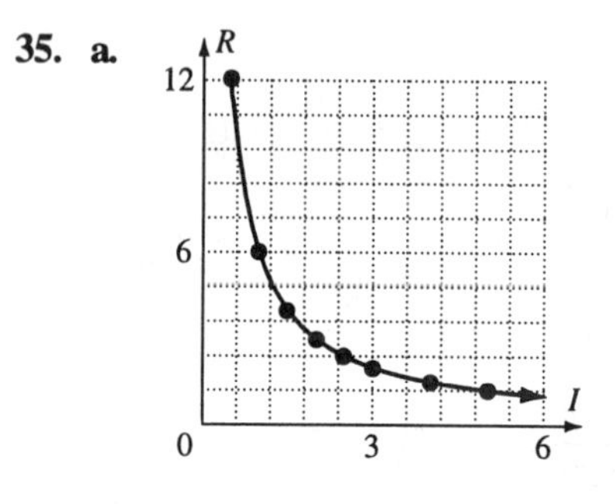

b. Inversely; the graph is not linear.

c. $R = \frac{k}{I}$

$k = RI$

$k = 0.5(12) = 6$

$k = 6$

$R = \frac{6}{I}$

37. $L = \frac{k}{W}$

$27 = \frac{k}{8}$

$k = 216$

$L = \frac{216}{W}$

$L = \frac{216}{12}$

$L = 18$ yards.

39. b is true.

$nt - 6n = 32 - 2t$

$n(t - 6) = 32 - 2t$

$n = \frac{32 - 2t}{t - 6}$

41. Students should verify solution.

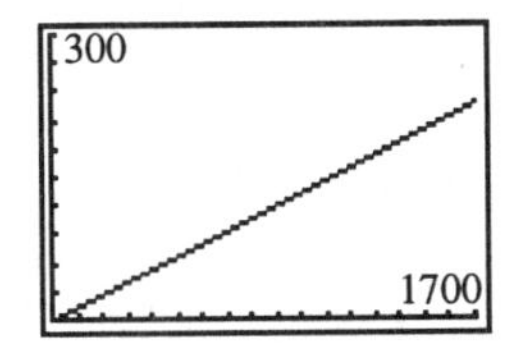

43. Students should verify solution.

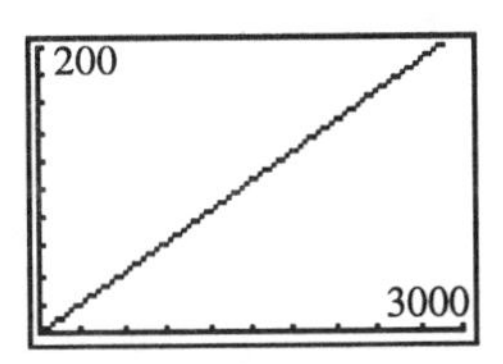

45. Answers may vary.

47. 17; $17 more for each additional hour.

49. $I = \frac{kW}{L}$
$75 = \frac{k(6)}{8}$
$k = \frac{8(75)}{6} = 100$
$I = \frac{100W}{L}$
$I = \frac{100(7)}{10}$
$I = 70$

Review Problems

50. $25x^2 - 81 = (5x+9)(5x-9)$

51. $x^2 - 12x + 36 = 0$
$(x-6)(x-6) = 0$
$x - 6 = 0$
$x = 6$

52. $y = -\frac{2}{3}x + 4$

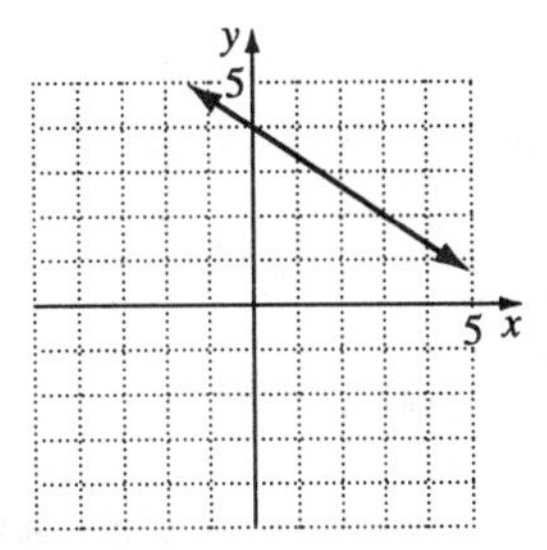

Chapter 8 Review Problems

1. $\frac{5x}{6x-24}$
$6x - 24 \neq 0$
$6x \neq 24$
$x \neq 4$
4

2. $\frac{x+3}{(x-2)(x+5)}$
$(x-2)(x+5) \neq 0$
$x - 2 \neq 0,\ x + 5 \neq 0$
$x \neq 2,\ x \neq -5$
2, –5

3. $\frac{x^2+3}{x^2-3x+2}$
$x^2 - 3x + 2 \neq 0$
$(x-2)(x-1) \neq 0$
$x - 2 \neq 0, \quad x - 1 \neq 0$
$x \neq 2 \qquad x \neq 1$
2, 1

4. $\frac{5}{x^2+1}$
$x^2 + 1$ is always positive, therefore no value of x is excluded.
$\varnothing$

5. $f(x)=\dfrac{80,000x}{100-x}$

a. $f(20)=\dfrac{80,000(20)}{100-20}=\$20,000$
$=$ cost to remove 20% of pollutants
$f(50)=\dfrac{80,000(50)}{100-50}=\$80,000$
$=$ cost to remove 50% of pollutants
$f(90)=\dfrac{80,000(90)}{100-90}=\$720,000$
$=$ cost to remove 90% of pollutants
$f(98)=\dfrac{80,000(98)}{100-98}=\$3,920,000$
$=$ cost to remove 98% of pollutants

b. $x=100$

c. The cost increases rapidly as x approaches 100%. The difficulty of removing the pollutants increases greatly as higher levels of purity are demanded.

6. $f(x)=\dfrac{80,000x}{100-x}$

x	0	10	20	50	60	80	90	98	99	100
$f(x)$	0	8889	20,000	80,000	120,000	320,000	720,000	3,920,000	7,920,000	undefined

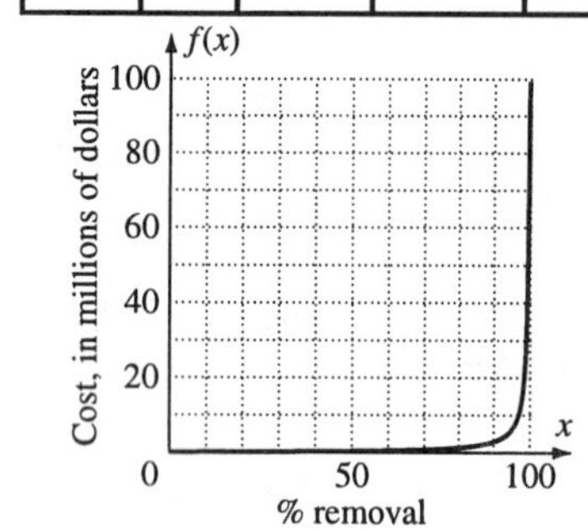

7. $\dfrac{16x^2}{12x}=\dfrac{4x^2(4)}{4x(3)}=\dfrac{4x}{3}\ (x\neq 0)$

8. $\dfrac{x^3+2x^2}{x+2}=\dfrac{x^2(x+2)}{x+2}=x^2\quad (x\neq -2)$

9. $\dfrac{x^2+3x-18}{x^2-36}$
$=\dfrac{(x+6)(x-3)}{(x+6)(x-6)}$
$=\dfrac{x-3}{x-6}\ (x\neq \pm 6)$

10. $\frac{x^2-4x-5}{x^2+8x+7}$
$=\frac{(x-5)(x+1)}{(x+7)(x+1)}$
$=\frac{x-5}{x+7}\ (x\neq -1, -7)$

11. $\frac{y^2+2y}{y^2+4y+4}=\frac{y(y+2)}{(y+2)^2}=\frac{y}{y+2}\ (y\neq -2)$

12. $\frac{3a^2-5a-2}{4-a^2}$
$=\frac{(3a+1)(a-2)}{(2-a)(2+a)}$
$=-\frac{3a+1}{a+2}\ (a\neq -2, 2)$

13. $y=\frac{x^2-4}{x-2}=\frac{(x-2)(x+2)}{x-2}=x+2$
$y=x+2\ (x\neq 2)$

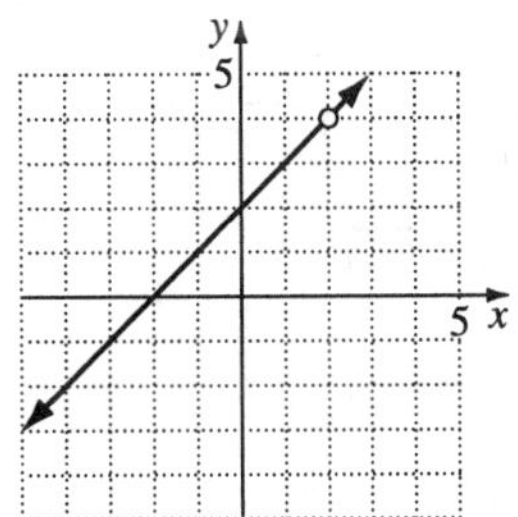

14. $\frac{5y+5}{6}\cdot\frac{3y}{y^2+y}$
$=\frac{5(y+1)}{6}\cdot\frac{3y}{y(y+1)}=\frac{15}{6}=\frac{5}{2}\ (y\neq 0, -1)$

15. $\frac{x^2+6x+9}{x^2-4}\cdot\frac{x+3}{x-2}$
$=\frac{(x+3)^2}{(x-2)(x+2)}\cdot\frac{x+3}{x-2}$
$=\frac{(x+3)^3}{(x+2)(x-2)^2}\ (x\neq -2, 2)$

16. $\frac{2y^2+y-3}{4y^2-9}\cdot\frac{3y+3}{5y-5y^2}$
$=\frac{(2y+3)(y-1)}{(2y+3)(2y-3)}\cdot\frac{3(y+1)}{5y(1-y)}$
$=-\frac{3(y+1)}{5y(2y-3)}\ \left(y\neq -\frac{3}{2}, \frac{3}{2}, 0, 1\right)$

17. $\frac{x^2+x-6}{x^2+6x+9}\cdot\frac{x+2}{x-3}\cdot\frac{x^2-7x+12}{x^2-x-2}$
$=\frac{(x+3)(x-2)}{(x+3)^2}\cdot\frac{x+2}{x-3}\cdot\frac{(x-3)(x-4)}{(x-2)(x+1)}$
$=\frac{(x+2)(x-4)}{(x+3)(x+1)}\ (x\neq -3, 3, 2, -1)$

18. $\frac{y^2+y-2}{10}\div\frac{2y+4}{5}$
$=\frac{(y+2)(y-1)}{10}\cdot\frac{5}{2(y+2)}$
$=\frac{y-1}{4}\ (y\neq -2)$

19. $\frac{6y+2}{y^2-1}\div\frac{3y^2+y}{y-1}$
$=\frac{2(3y+1)}{(y-1)(y+1)}\cdot\frac{y-1}{y(3y+1)}$
$=\frac{2}{y(y+1)}\ \left(y\neq 0, -\frac{1}{3}, -1, 1\right)$

20. $\frac{y^2-5y-24}{2y^2-2y-24}\div\frac{y^2-10y+16}{4y^2+4y-24}$
$=\frac{(y-8)(y+3)}{2(y-4)(y+3)}\cdot\frac{4(y+3)(y-2)}{(y-2)(y-8)}$
$=\frac{2(y+3)}{y-4}\ (y\neq 4, -3, 2, 8)$

21. $\frac{z^2-10z+21}{7-z}\div(z+3)$
$=\frac{(z-3)(z-7)}{7-z}\cdot\frac{1}{z+3}$
$=\frac{3-z}{3+z}\ (z\neq 7, -3)$

22. $\frac{12x-5}{3x-1}+\frac{1}{3x-1}$
$\frac{12x-5+1}{3x-1}=\frac{12x-4}{3x-1}$
$=\frac{4(3x-1)}{3x-1}=4 \quad \left(x\neq\frac{1}{3}\right)$

23. $\frac{3y^2+2y}{y-1}-\frac{10y-5}{y-1}$
$=\frac{3y^2-8y+5}{y-1}$
$=\frac{(3y-5)(y-1)}{y-1}=3y-5 \quad y\neq 1$

24. $\frac{2y-1}{y^2+5y-6}-\frac{2y-7}{y^2+5y-6}$
$=\frac{6}{(y+6)(y-1)} \quad (y\neq 1, -6)$

25. $\frac{2x+7}{x^2-9}-\frac{x-4}{x^2-9}=\frac{x+11}{(x-3)(x+3)}$
$(x\neq 3, -3)$

26. $9x^3=3^2x^3$
$12x=2^2\cdot 3x$
$\text{LCM}=2^2\cdot 3^2x^3 \text{ or } 36x^3$

27. $8y^2(y-1)^2=2^3y^2(y-1)^2$
$10y^3(y-1)=2\cdot 5y^3(y-1)$
$\text{LCM}=2^3\cdot 5y^3(y-1)^2 \text{ or } 40y^3(y-1)^2$

28. $x^2+4x+3=(x+3)(x+1)$
$x^2+10x+21=(x+3)(x+7)$
$\text{LCM}=(x+1)(x+3)(x+7)$

29. $\frac{3}{10y^2}+\frac{7}{25y}=\frac{15}{50y^2}+\frac{14y}{50y^2}=\frac{15+14y}{50y^2}$

30. $\frac{6y}{y^2-4}-\frac{3}{y+2}$
$=\frac{6y}{y^2-4}-\frac{3(y-2)}{y^2-4}$
$=\frac{3y+6}{y^2-4}=\frac{3(y+2)}{y^2-4}=\frac{3}{y-2} \quad (y\neq -2, 2)$

31. $\frac{2}{3x}+\frac{5}{x+1}=\frac{2x+2+15x}{3x(x+1)}=\frac{17x+2}{3x(x+1)}$

32. $\frac{2y}{y^2+2y+1}+\frac{y}{y^2-1}$
$=\frac{2y}{(y+1)^2}+\frac{y}{(y+1)(y-1)}$
$=\frac{2y(y-1)}{(y+1)^2(y-1)}+\frac{y(y+1)}{(y+1)^2(y-1)}$
$=\frac{2y^2-2y+y^2+y}{(y+1)^2(y-1)}$
$=\frac{3y^2-y}{(y+1)^2(y-1)}$
$=\frac{y(3y-1)}{(y+1)^2(y-1)}$
$(y\neq -1, 1)$

33. $\frac{4z}{z^2+6z+5}-\frac{3}{z^2+5z+4}$
$=\frac{4z}{(z+1)(z+5)}-\frac{3}{(z+1)(z+4)}$
$=\frac{4z(z+4)-3(z+5)}{(z+1)(z+4)(z+5)}$
$=\frac{4z^2+13z-15}{(z+1)(z+4)(z+5)}$

34. $\frac{y}{y-2}-\frac{y-4}{2-y}$
$=\frac{y}{y-2}+\frac{y-4}{y-2}$
$=\frac{2y-4}{y-2}$
$=\frac{2(y-2)}{y-2}=2 \quad (y\neq 2)$

35. $\frac{4y-1}{2y^2+5y-3}-\frac{y+3}{6y^2+y-2}$
$=\frac{4y-1}{(2y-1)(y+3)}-\frac{y+3}{(3y+2)(2y-1)}$
$=\frac{(4y-1)(3y+2)-(y+3)(y+3)}{(2y-1)(y+3)(3y+2)}$

$$= \frac{12y^2 + 5y - 2 - y^2 - 6y - 9}{(2y-1)(y+3)(3y+2)}$$
$$= \frac{11y^2 - y - 11}{(2y-1)(y+3)(3y+2)}$$

36. $\frac{x+1}{5x} + 2$

$\frac{x+1}{5x} + 2\left(\frac{5x}{5x}\right)$

$= \frac{x+1+10x}{5x}$

$= \frac{11x+1}{5x} \quad (x \neq 0)$

37. $\frac{\frac{1}{x}}{1-\frac{1}{x}}$

$= \frac{\frac{1}{x}}{\frac{x}{x}-\frac{1}{x}}$

$= \frac{\frac{1}{x}}{\frac{x-1}{x}}$

$= \frac{1}{x} \cdot \frac{x}{x-1}$

$= \frac{1}{x-1} \quad (x \neq 0, 1)$

38. $\frac{\frac{1}{x}-\frac{1}{2}}{\frac{1}{3}-\frac{x}{6}} = \frac{2-x}{2x} \cdot \frac{6}{2-x} = \frac{3}{x} \quad (x \neq 0, 2)$

39. $\frac{3+\frac{12}{y}}{1-\frac{16}{y^2}}$

$= \frac{3y+12}{y} \cdot \frac{y^2}{y^2-16}$

$= \frac{3(y+4)y^2}{y(y+4)(y-4)}$

$= \frac{3y}{y-4} \quad (y \neq 0, -4, 4)$

40. $\frac{\frac{3}{5x^3}-\frac{1}{10x}}{\frac{3}{10x}+\frac{1}{x^2}}$

$= \frac{\frac{3}{5x^3}\cdot\frac{2}{2}-\frac{1}{10x}\cdot\frac{x^2}{x^2}}{\frac{3}{10x}\cdot\frac{x}{x}+\frac{1}{x^2}\cdot\frac{10}{10}}$

$= \frac{\frac{6}{10x^3}-\frac{x^2}{10x^3}}{\frac{3x}{10x^2}+\frac{10}{10x^2}}$

$= \frac{\frac{6-x^2}{10x^3}}{\frac{3x+10}{10x^2}}$

$= \frac{6-x^2}{10x^3} \cdot \frac{10x^2}{3x+10}$

$= \frac{6-x^2}{x(3x+10)} \quad \left(x \neq 0, -\frac{10}{3}\right)$

41. $\frac{2xy+2xz}{3x^2y+3x^2z} = \frac{2x(y+z)}{3x^2(y+z)} = \frac{2}{3x}$

$(x \neq 0, y \neq -z)$

42. $\frac{8a^2+2a^2b^2}{b^2+4b+4} = \frac{2a^2(4+b^2)}{(b+2)(b+2)} \quad (b \neq -2)$

43. $\frac{a^2-2ab+b^2}{a^2-b^2} \cdot \frac{a^2+ab}{3a^2b^2-3ab^3}$

$= \frac{(a-b)(a-b)}{(a-b)(a+b)} \cdot \frac{a(a+b)}{3ab^2(a-b)}$

$= \frac{1}{3b^2} \quad (a \neq b, a \neq -b, a \neq 0, b \neq 0)$

44. $\frac{x^2-y^2}{x-y} \div \frac{xy+x^2}{x+y}$

$= \frac{(x+y)(x-y)}{x-y} \cdot \frac{x+y}{x(y+x)}$

$= \frac{x+y}{x} \quad (x \neq y, x \neq -y, x \neq 0)$

45. $\frac{4a^2-16b^2}{9} \div \frac{(a+2b)^2}{12}$

$= \frac{4(a-2b)(a+2b)}{9} \cdot \frac{12}{(a+2b)(a+2b)}$

$= \frac{16(a-2b)}{3(a+2b)} \quad (a \neq -2b)$

46. $\frac{1}{4a} + \frac{6}{ab}$

$= \frac{1}{4a} \cdot \frac{b}{b} + \frac{6}{ab} \cdot \frac{4}{4}$

$= \frac{b}{4ab} + \frac{24}{4ab}$

$= \frac{b+24}{4ab} \quad (a, b \neq 0)$

47. $\frac{5}{3ab}-\frac{4}{a^2}$

$=\frac{5}{3ab}\cdot\frac{a}{a}-\frac{4}{a^2}\cdot\frac{3b}{3b}$

$=\frac{5a}{3a^2b}-\frac{12b}{3a^2b}$

$=\frac{5a-12b}{3a^2b}\ (a, b\neq 0)$

48. $\frac{x+y}{y}-\frac{x-y}{x}$

$=\frac{x+y}{y}\cdot\frac{x}{x}-\frac{x-y}{x}\cdot\frac{y}{y}$

$=\frac{x(x+y)-y(x-y)}{xy}$

$=\frac{x^2+xy-xy+y^2}{xy}$

$=\frac{x^2+y^2}{xy}\ (x, y\neq 0)$

49. $\frac{a-b}{ab}-\frac{c-b}{bc}$

$=\frac{a-b}{ab}\cdot\frac{c}{c}-\frac{c-b}{bc}\cdot\frac{a}{a}$

$=\frac{c(a-b)-a(c-b)}{abc}$

$=\frac{ac-bc-ac+ab}{abc}$

$=\frac{-bc+ab}{abc}$

$(a, b, c\neq 0)$

50. $\frac{a+\frac{1}{b}}{b^2}$

$=\frac{\frac{ab}{b}+\frac{1}{b}}{b^2}$

$=\frac{ab+1}{b}\cdot\frac{1}{b^2}$

$=\frac{ab+1}{b^3}\ (b\neq 0)$

51. $\frac{\frac{a}{3b}-\frac{1}{2}}{\frac{4}{3b}-\frac{2}{a}}$

$=\frac{\frac{a}{3b}\cdot\frac{2}{2}-\frac{1}{2}\cdot\frac{3b}{3b}}{\frac{4}{3b}\cdot\frac{a}{a}-\frac{2}{a}\cdot\frac{3b}{3b}}$

$=\frac{\frac{2a-3b}{6b}}{\frac{4a-6b}{3ab}}$

$=\frac{2a-3b}{6b}\cdot\frac{3ab}{4a-6b}$

$=\frac{2a-3b}{6b}\cdot\frac{3ab}{2(2a-3b)}$

$=\frac{a}{4}\ (a, b\neq 0, 2a\neq 3b)$

52. $\frac{2}{x}=\frac{2}{3}+\frac{x}{6}$

$6x\left(\frac{2}{x}\right)=6x\left(\frac{2}{3}+\frac{x}{6}\right)$

$12=4x+x^2$

$x^2+4x-12=0$

$(x+6)(x-2)=0$

$x=-6,\ x=2$

$-6, 2$

53. $\frac{13}{y-1}-3=\frac{1}{y-1}$

$\frac{13-1}{y-1}=3$

$12=3y-3$

$3y=15$

$y=5$

5

54. $\frac{3}{4x}-\frac{1}{x}=\frac{1}{4}$

$\frac{3-4}{4x}=\frac{1}{4}$

$-\frac{1}{4x}=\frac{1}{4}$

$-4=4x$

$x=-1$

-1

55. $\frac{5}{y+2}+\frac{y}{y+6}$

$=\frac{24}{y^2+8y+12}=\frac{24}{(y+6)(y+2)}$

$5(y+6)+y(y+2)=24$

$5y+30+y^2+2y=24$

$y^2+7y+6=0$

$(y+6)(y+1)=0$

$y=-1$

$(y\neq -2, -6)$

-1

56. $3-\frac{6}{y}=y+8$

$3y-6=y^2+8y$

$y^2+5y+6=0$

$(y+2)(y+3)=0$

$y=-2$ or $y=-3$ (Both answers check.)

$-2, -3$

57. $4-\frac{y}{y+5}=\frac{5}{y+5}$

$(y+5)\left(4-\frac{y}{y+5}\right)=(y+5)\left(\frac{5}{y+5}\right)$

$4(y+5)-y=5$

$4y+20-y=5$

$3y=-15$

$y=-5$

Equation undefined for $y=-5$.

No solution $\varnothing$

58. $P=30-\frac{9}{t+1}$

$27=30-\frac{9}{t+1}$

$-3=-\frac{9}{t+1}$

$3=\frac{9}{t+1}$

$3t+3=9$

$3t=6$

$t=2$

$1990+2=1992$

corresponding to 1992

59. $f(x)=\frac{20x+20,000}{x}$

a. $f(100)=\frac{20(100)+20,000}{100}=\$220$

$f(1000)=\frac{20(1000)+20,000}{1000}=\$40$

$f(10,000)=\frac{20(10,000)+20,000}{10,000}$

$=\$22$

b. The cost decreases, approaching \$20.

c. $20.20=\frac{20x+20,000}{x}$

$20.20x=20x+20,000$

$20.20x-20x=20,000$

$x(20.20-20)=20,000$

$x=\frac{20,000}{20.20-20}$

$x=100,000$

60. $C=\frac{200x}{100-x}$

$C(100-x)=200x$

$100C-Cx=200x$

$100C=200x+Cx$

$100C=x(200+C)$

$x=\frac{100C}{200+C}$

$x=\frac{100(300)}{200+300}=60\%$

61. $C=\frac{DA}{A+12}$

$80=\frac{200A}{A+12}$

$80(A+12)=200A$

$80A+960=200A$

$80A-200A=-960$

$A(80-200)=-960$

$A=\frac{-960}{80-200}$

$A=8$ yrs old

62. $C = \frac{DA}{A+12}$

$C(A+12) + DA$

$CA + 12C = DA$

$CA - DA = -12C$

$A(C-D) = -12C$

$A = \frac{-12C}{C-D}$

63. $\frac{1}{a} + \frac{1}{b} = \frac{1}{c}$ for a

$\frac{b+a}{ab} = \frac{1}{c}$

$bc + ac = ab$

$a(c-b) = -bc$

$a = \frac{bc}{b-c}$

64. $T = \frac{A-P}{Pr}$ for P

$PrT = A - P$

$P(rT+1) = A$

$P = \frac{A}{1+rT}$

65. Let x = numerator of fraction.

$x+6$ = denominator of fraction.

$\frac{x+3}{x+6+3} = \frac{2}{5}$

$\frac{x+3}{x+9} = \frac{2}{5}$

$5x + 15 = 2x + 18$

$3x = 3$

$x = 1$

$x + 6 = 7$

fraction = $\frac{1}{7}$

66. Let x = number in 1970

Then $x + 6515$ = number in 1993

$\frac{x+6515}{x} = 5 + \frac{639}{x}$

$x\left(\frac{x+6515}{x}\right) = x\left(5 + \frac{639}{x}\right)$

$x + 6515 = 5x + 639$

$6515 - 639 = 5x - x$

$5876 = 4x$

$x = 1469$

$x + 6515 = 7984$

Number of African-American officials elected in 1970 was 1469 and in 1993 was 7984.

67. Let x = speed of the boat in still water.

$\text{time} = \frac{\text{distance}}{\text{speed}}$

$\frac{11}{x+3} = \frac{9}{x-3}$

$11x - 33 = 9x + 27$

$2x = 60$

$x = 30$

The speed of the boat in still water is 30 miles/hour.

68. Let t = the hours to complete the job working together

$\frac{t}{6} + \frac{t}{12} = 1$

$12\left(\frac{t}{6} + \frac{t}{12}\right) = 1(12)$

$2t + t = 12$

$3t = 12$

$t = 4$ hours

69. Let t = minutes to fill tub with all three pipes

$\frac{t}{8} + \frac{t}{12} + \frac{t}{24} = 1$

$24\left(\frac{t}{8} + \frac{t}{12} + \frac{t}{24}\right) = 24(1)$

$3t + 2t + t = 24$

$6t = 24$

$t = 4$ minutes

70. a. $P = kH$

$210 = k(15)$

$k = 14$

$P = 14H$

b. $P = 14H$

$P = 14(40)$

$P = \$560$

c. Slope = 14
$14 more for each additional hour of work.

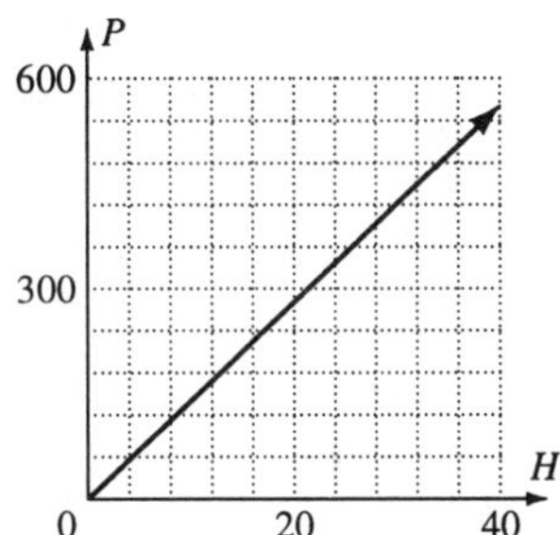

71. $B = kE$
$98 = k(1400)$
$k = 0.07$
$B = 0.07E$
$B = 0.07(2200)$
$B = \$154$

72. $I = \frac{k}{R}$
$24 = \frac{k}{4}$
$k = 96$
$I = \frac{96}{R}$
$I = \frac{96}{6}$
$I = 16$ amperes

Chapter 8 Test

1. $\frac{x+7}{x^2+5x-36}$ is undefined when
$x^2+5x-36=0.$
$x^2+5x-36=0$
$(x+9)(x-4)=0$
$x+9=0$ or $x-4=0$
$x=-9$ or $x=4$
$-9, 4$

2. $\frac{x^2+2x-3}{x^2-3x+2} = \frac{(x+3)(x-1)}{(x-2)(x-1)} = \frac{x+3}{x-2}$

3. $\frac{4y^2-20y}{y^2-4y-5} = \frac{4y(y-5)}{(y-5)(y+1)} = \frac{4y}{y+1}$

4. $\frac{x^2-16}{10} \cdot \frac{5}{x+4} = \frac{(x-4)(x+4)}{2} \cdot \frac{1}{(x+4)}$
$= \frac{x-4}{2}$

5. $\frac{y^2-7y+12}{y^2-4y} \cdot \frac{y^2}{y^2-9}$
$= \frac{(y-4)(y-3)}{y(y-4)} \cdot \frac{y^2}{(y-3)(y+3)} = \frac{y}{y+3}$

6. $\frac{2x+8}{x-3} \div \frac{x^2+5x+4}{x^2-9}$
$= \frac{2(x+4)}{x-3} \cdot \frac{(x-3)(x+3)}{(x+4)(x+1)} = \frac{2(x+3)}{x+1}$

7. $\frac{5y+5}{(y-3)^2} \div \frac{y^2-1}{y-3}$
$= \frac{5(y+1)}{(y-3)(y-3)} \cdot \frac{y-3}{(y-1)(y+1)}$
$= \frac{5}{(y-3)(y-1)}$

8. $\frac{2y^2+5}{y+3} + \frac{6y-5}{y+3} = \frac{2y^2+5+6y-5}{y+3}$
$= \frac{2y^2+6y}{y+3} = \frac{2y(y+3)}{y+3} = 2y$

9. $\frac{y^2-2y+3}{y^2+7y+12} - \frac{y^2-4y-5}{y^2+7y+12}$
$= \frac{y^2-2y+3-y^2+4y+5}{y^2+7y+12}$
$= \frac{2y+8}{y^2+7y+12} = \frac{2(y+4)}{(y+4)(y+3)} = \frac{2}{y+3}$

10. $\frac{x}{x+3} + \frac{5}{x-3}$
$= \frac{x(x-3)}{(x+3)(x-3)} + \frac{5(x+3)}{(x+3)(x-3)}$
$= \frac{x^2-3x+5x+15}{(x+3)(x-3)} = \frac{x^2+2x+15}{(x+3)(x-3)}$

11. $\frac{2}{y^2-4y+3}+\frac{6}{y^2+y-2}$
$=\frac{2}{(y-3)(y-1)}+\frac{6}{(y+2)(y-1)}$
$=\frac{2(y+2)}{(y-3)(y-1)(y+2)}$
$+\frac{6(y-3)}{(y-3)(y-1)(y+2)}$
$=\frac{2y+4+6y-18}{(y-3)(y-1)(y+2)}$
$=\frac{8y-14}{(y-3)(y-1)(y+2)}$
$=\frac{2(4y-7)}{(y-3)(y-1)(y+2)}$

12. $\frac{4}{y-3}+\frac{y+5}{3-y}=\frac{4-y-5}{y-3}=-\frac{y+1}{y-3}$

13. $6-\frac{3}{x-3}=\frac{6(x-3)-3}{x-3}$
$=\frac{6x-21}{x-3}=\frac{3(2x-7)}{x-3}$

14. $\frac{2y+3}{y^2-7y+12}-\frac{2}{y-3}$
$=\frac{2y+3}{(y-3)(y-4)}-\frac{2}{y-3}$
$=\frac{2y+3-2(y-4)}{(y-3)(y-4)}=\frac{11}{(y-3)(y-4)}$

15. $\frac{8y}{y^2-16}-\frac{4}{y-4}$
$=\frac{8y}{(y-4)(y+4)}-\frac{4(y+4)}{(y-4)(y+4)}$
$=\frac{8y-4y-16}{(y-4)(y+4)}=\frac{4(y-4)}{(y-4)(y+4)}=\frac{4}{y+4}$

16. $\frac{(x-y)^2}{x+y}\div\frac{x^2-xy}{3x+3y}$
$=\frac{(x+y)(x-y)}{x+y}\cdot\frac{3(x+y)}{x(x-y)}=\frac{3(x+y)}{x}$

17. $\frac{a+4b}{4b}-\frac{a+2b}{2a}=\frac{a(a+4b)}{4ab}-\frac{2b(a+2b)}{4ab}$
$=\frac{a^2+4ab-2ab-4b^2}{4ab}=\frac{a^2+2ab-4b^2}{4ab}$

18. $\frac{5+\frac{5}{x}}{2+\frac{1}{x}}=\frac{\frac{5x+5}{x}}{\frac{2x+1}{x}}=\frac{5x+5}{x}\cdot\frac{x}{2x+1}=\frac{5(x+1)}{2x+1}$

19. $\frac{\frac{1}{x}-\frac{1}{y}}{\frac{1}{x}}=\frac{\frac{y-x}{xy}}{\frac{1}{x}}=\frac{y-x}{xy}\cdot\frac{x}{1}=\frac{y-x}{y}$

20. $\frac{5}{y}+\frac{2}{3}=2-\frac{2}{y}-\frac{1}{6}$
Multiply each term by $6y$.
$30+4y=12y-12-y$
$-7y=-42$
$y=6$

21. $\frac{3}{y+5}-1=\frac{4-y}{2y+10}$
Multiply each term by $2(y+5)$.
$6-2(y+5)=4-y$
$6-2y-10=4-y$
$-y=8$
$y=-8$

22. $\frac{2}{x-1}=\frac{3}{x^2-1}+1$
Multiply each term by $(x-1)(x+1)$.
$2(x+1)=3+(x-1)(x+1)$
$2x+2=3+x^2-1$
$x^2-2x=0$
$x(x-2)=0$
$x=0$ or $x=2$

23. $\frac{1}{t}=\frac{1}{a}+\frac{1}{b}$ Multiply each term by tab.
$ab=tb+ta$
$ab=t(b+a)$
$\frac{ab}{b+a}=t$
$t=\frac{ab}{a+b}$

24. $12{,}000=\frac{12{,}840}{1+r}$
$1+r=\frac{12{,}840}{12{,}000}$
$r=1.07-1=0.07=7\%$

25. Let m be the number of endangered species of mammals and b be the number of endangered species of birds.
$b = 2m - 15$
$\frac{b}{m} = \frac{2m-15}{m} = 1 + \frac{21}{m}$
Multiply each term by m.
$2m - 15 = m + 21$
$m = 36$
$b = 2(36) - 15 = 57$
Mammals: 36, birds: 57

26. Let t be the time for both pipes to fill the tub.
$\frac{t}{20} + \frac{t}{30} = 1$
$3t + 2t = 60$
$5t = 60$
$t = 12$
It takes 12 minutes.

27. Let P = pressure in pounds per square inch (psi)
k = constant
d = distance under the surface
Then $P = kd$.
If $25 = k(60)$, then $k = \frac{25}{60} = \frac{5}{12}$
and $P = \frac{5}{12}(330) = 137.5$.
The submarine will experience 137.5 psi when 330 feet below the surface.

28. Let C = current in amperes
k = constant
r = resistance in ohms
Then $C = \frac{k}{r}$.
If $42 = \frac{k}{5}$, then $k = 42 \cdot 5 = 210$ and
$C = \frac{210}{4} = 52.5$
The current is 52.5 amperes when the resistance is 4 ohms.

Cumulative Review Problems (Chapters 1–8)

1. a. $8 + (7 + 3) = (8 + 7) + 3$
Associative property of addition

b. $-3(5 + 9) = (-3) \cdot 5 + (-3) \cdot 9$
Distributive property

c. $5(3 + 4) = 5(4 + 3)$
Commutative property of addition

2. $2x + 3 < 3(x - 5)$
$2x + 3 < 3x - 15$
$2x - 3x < -15 - 3$
$-x < -18$
$x > 18$
16 17 18 19 20

3. Let x = the number
$6x - 7 = 175$
$6x = 175 + 7$
$6x = 182$
$x = 30\frac{1}{3}$

4. $3x - 4y < 12$

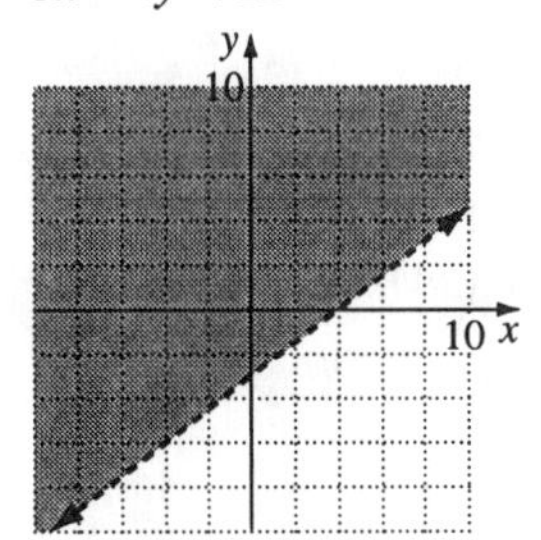

5. Let x = temperature in MN
Then $2x + 6$ = temperature in FL
$\frac{x + 2x + 6}{2} = 57$
$\frac{3x + 6}{2} = 57$
$3x + 6 = 2(57)$
$3x = 114 - 6$
$x = \frac{108}{3}$
$x = 36$

$2x + 6 = 78$
The average yearly temperature for America's coldest city is 36°F and warmest city is 78°F.

6. $3x^2 - 15x - 42$
$= 3(x^2 - 5x - 14)$
$= 3(x - 7)(x + 2)$

7. Area of circle $= \pi(\text{radius})^2$
$A = \pi r^2$
If the radius is doubled, $2r$, the area becomes:
$A = \pi(2r)^2$
$A = \pi 4r^2$
$A = 4\pi r^2$
The area is multiplied by 4.

8. Let w = the width
Then $3w + 1$ = the length
$w(3w + 1) = 14$
$3w^2 + w = 14$
$3w^2 + w - 14 = 0$
$(3w + 7)(w - 2) = 0$
$3w + 7 = 0$ or $w - 2 = 0$
$w = -\frac{7}{3}$ or $w = 2$
The width cannot be negative so width = 2 yards and length = 7 yards.

9. $2x^3 - 20x^2 + 50x$
$= 2x(x^2 - 10x + 25)$
$= 2x(x - 5)(x - 5)$
$= 2x(x - 5)^2$

10. $\frac{x^2 + 2x - 12}{x - 3} = x + 5 + \frac{3}{x - 3}$

$$\begin{array}{r} x+5 \\ x-3\overline{)x^2+2x-12} \\ \underline{x^2-3x} \\ 5x-12 \\ \underline{5x-15} \\ 3 \end{array}$$

11. $\left(\frac{4x^5}{2x^2}\right)^3 = \frac{4^3 x^{5\cdot 3}}{2^3 x^{2\cdot 3}} = \frac{64x^{15}}{8x^6} = 8x^9$

12. $5x + 2y = -1$
$2x - 5y = 1$
Multiply equation 1 by 5.
Multiply equation 2 by 2.

$$\begin{array}{r} 25x + 10y = -5 \\ \underline{4x - 10y = 2} \\ 29x = -3 \\ x = -\frac{3}{29} \end{array}$$

$5\left(-\frac{3}{29}\right) + 2y = -1$
$2y = -1 + \frac{15}{29}$
$2y = -\frac{14}{29}$
$y = -\frac{7}{29}$
$\left(-\frac{3}{29}, -\frac{7}{29}\right)$

13. $f(x) = -0.17x^3 + 6.8x^2 + 536x$
$f(10) = -0.17(10)^3 + 6.8(10)^2 + 536(10)$
$f(10) = -170 + 680 + 5360$
$f(10) = 5870$
An acre of 10 trees produced 5870 limes.

14. slope $= \frac{10 - (-5)}{-2 - 1} = -\frac{15}{3} = -5$
$y - (-5) = -5(x - 1)$
$y + 5 = -5(x - 1)$ or $y - 10 = -5(x + 2)$
$y = -5x + 5 - 5$
$y = -5x$

15. $f(t) = -16t^2 + 80t$
$f(0) = -16(0)^2 + 80(0) = 0$
$f(1) = -16(1)^2 + 80(1) = 64$
$f(2) = -16(2)^2 + 80(2) = 96$
$f(3) = -16(3)^2 + 80(3) = 96$
$f(4) = -16(4)^2 + 80(4) = 64$
$f(5) = -16(5)^2 + 80(5) = 0$
The height of the rocket above the ground after 0 sec, 1 sec, 2 sec, 3 sec, 4 sec, 5 sec is

0 ft, 64 ft, 96 ft, 96 ft, 64 ft, 0 ft, respectively.

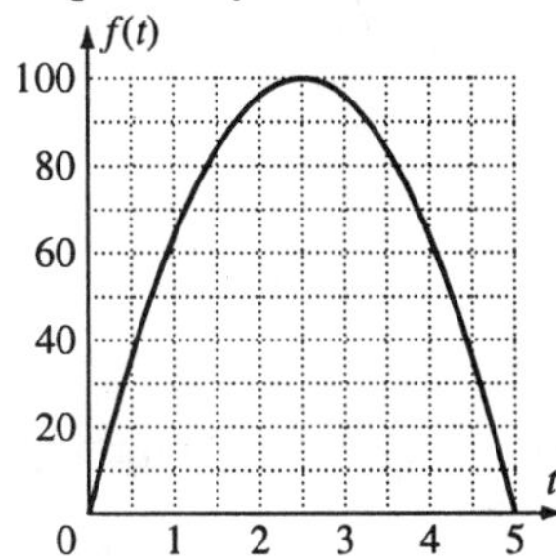

16. Let x = percent in Sweden
Then $2x + 14$ = percent in America
$x + 2x + 14 = 54.5$
$3x + 14 = 54.5$
$3x = 40.5$
$x = 13.5$
$2x + 14 = 41$
Sweden = 13.5%
America = 41%
New Zealand = 38.2%
Italy = 35.1%
England = 33%
Australia = 31.1%
Netherlands = 21.1%
Canada = 20%
Belgium = 19.6%
Finland = 17.9%

17. Let x = South Africa's population
$0.14x = 5.32$ million
$x = 38$ million
Black = 0.75(38 million) = 28.5 million
Indian = 0.02(38 million) = 0.76 million
Mixed = 0.09(38 million) = 3.42 million

18. Let h = height
Area = $\frac{1}{2}$(base)(height)
$33 = \frac{1}{2}(11)h$
$\frac{2(33)}{11} = h$
$h = 6$ feet

19. $2 - 3(x - 2) = 5(x + 5) - 1$
$2 - 3x + 6 = 5x + 25 - 1$
$-3x + 8 = 5x + 24$
$-3x - 5x = 24 - 8$
$-8x = 16$
$x = -2$

20. $2x + y = 4$
$x + y = 2$
Solution: (2, 0)

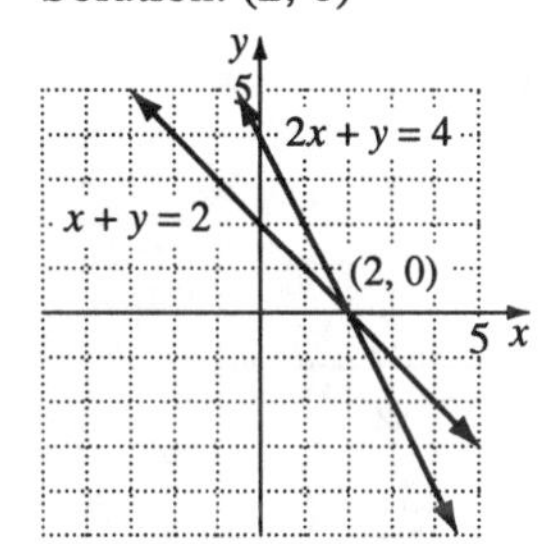

21. Let x = the number of cassettes.
y = the number of compact discs.
$5x + 2y = 65$
$3x + 4y = 81$
Multiply equation 1 by –2.
No change to equation 2.

$$\begin{array}{r} -10x - 4y = -130 \\ 3x + 4y = 81 \\ \hline -7x = -49 \end{array}$$

$x = 7$
$5(7) + 2y = 65$
$2y = 30$
$y = 15$
cassette, \$7 each; compact disc, \$15 each

22. Let x = the length of the shorter leg.
$x + 7$ = the length of the longer leg.
$2x + 1$ = the hypotenuse.
$x^2 + (x + 7)^2 = (2x + 1)^2$
$x^2 + x^2 + 14x + 49 = 4x^2 + 4x + 1$
$0 = 2x^2 - 10x - 48$
$2(x^2 - 5x - 24) = 0$
$2(x - 8)(x + 3) = 0$
$x = 8$ (reject $x = -3$)
$x + 7 = 15$

$2x + 1 = 17$
The lengths of the sides of the triangle are 8 meters, 15 meters, and 17 meters.

23. $x + \frac{12}{x} = -7$

$x\left(x + \frac{12}{x}\right) = -7x$

$x^2 + 12 = -7x$

$x^2 + 7x + 12 = 0$

$(x + 3)(x + 4) = 0$

$x + 3 = 0$ or $x + 4 = 0$

$x = -3$ or $x = -4$

$-3, -4$

24. $\frac{y}{y^2 + 5y + 6} - \frac{2}{y^2 + 3y + 2}$

$= \frac{y}{(y+2)(y+3)} - \frac{2}{(y+2)(y+1)}$

$= \frac{y(y+1) - 2(y+3)}{(y+1)(y+2)(y+3)}$

$= \frac{y^2 + y - 2y - 6}{(y+1)(y+2)(y+3)}$

$= \frac{y^2 - y - 6}{(y+1)(y+2)(y+3)}$

$= \frac{(y-3)(y+2)}{(y+1)(y+2)(y+3)}$

$= \frac{y-3}{(y+1)(y+3)}$

25. a. $(x + 100)(x + 100)$
$= x^2 + 200x + 10{,}000$

b. $f(x) = x^2 + 200x + 10{,}000$

c. $f(5) = 5^2 + 200(5) + 10{,}000$
$f(5) = 11{,}025$
If the lawn is increased by 5 feet on each side, the area of the expanded lawn is 11,025 square feet.

26. Let x = the speed of the boat in still water.
$x + 5$ = the speed of the boat with current.
$x - 5$ = the speed of the boat against current.

$\frac{240}{x+5} = \frac{160}{x-5}$

$\frac{3}{x+5} = \frac{2}{x-5}$

$3x - 15 = 2x + 10$

$x = 25$

Speed of boat in still water, 25 miles per hour.

27. Let y = yearly income and x = years of education

$y = 1600x + 6300$

$25{,}500 = 1600x + 6300$

$19200 = 1600x$

$x = 12$

12 years of education are needed to earn $25,500 per year.

28. $\frac{9 \times 10^{-4}}{3 \times 10^{-6}} = \frac{9}{3} \times 10^{-4+6} = 3 \times 10^2 = 300$

29. $\frac{80 + 96 + 88 + 92 + x}{5} \geq 90$

$356 + x \geq 450$

$x \geq 94$

94% or better

30. $\left(\frac{\$26{,}000}{82.4}\right)(148.2) = \$46{,}762$

Chapter 9

Section 9.1

1. The square roots of 36 are 6 and –6, since $6^2 = 36$ and $(-6)^2 = 36$.

3. Square roots of 144 are 12 and –12, since $12^2 = 144$ and $(-12)^2 = 144$.

5. Square roots of $\frac{9}{16}$ are $\frac{3}{4}$ and $-\frac{3}{4}$, since $\left(\frac{3}{4}\right)^2 = \frac{9}{16}$ and $\left(-\frac{3}{4}\right)^2 = \frac{9}{16}$.

7. Square roots of $\frac{49}{100}$ are $\frac{7}{10}$ and $-\frac{7}{10}$, since $\left(\frac{7}{10}\right)^2 = \frac{49}{100}$ and $\left(-\frac{7}{10}\right)^2 = \frac{49}{100}$.

9. $\sqrt{36} = 6$

11. $-\sqrt{36} = -6$

13. $\sqrt{-36}$ does not exist, not a real number

15. $\sqrt{\frac{1}{25}} = \frac{1}{5}$

17. $\sqrt{\frac{49}{25}} = \frac{7}{5}$

19. $-\sqrt{\frac{1}{9}} = -\frac{1}{3}$

21. $-\sqrt{\frac{49}{100}} = -\frac{7}{10}$

23. $\sqrt{0.04} = 0.2$

25. $\sqrt{33-8} = \sqrt{25} = 5$

27. $\sqrt{2 \cdot 32} = \sqrt{64} = 8$

29. $\sqrt{144+25} = \sqrt{169} = 13$

31. $\sqrt{144} + \sqrt{25} = 12 + 5 = 17$

33. $\sqrt{\frac{1}{225}} = \frac{1}{15}$, rational

35. $\sqrt{15} \approx 3.873$, irrational

37. $\sqrt{400} = 20$, rational

39. $-\sqrt{225} = -15$, rational

41. $\sqrt{-1}$ is not a real number

43. $-\sqrt{83} \approx -9.110$, irrational

45. $\sqrt{573} \approx 23.937$, irrational

47. $-\sqrt{1369} = -37$, rational

49. $\frac{9+\sqrt{144}}{3} = \frac{9+12}{3} = \frac{21}{3} = 7$, rational

51. $\frac{12+\sqrt{45}}{2} \approx 9.354$, irrational

53. $\frac{12+\sqrt{-45}}{2}$ is not a real number

55. $\sqrt[4]{1} = 1$

57. $\sqrt[3]{64} = 4$

59. $\sqrt[3]{-27} = -3$

61. $\sqrt[3]{125} = 5$

63. $\sqrt[4]{16} = 2$

65. $-\sqrt[4]{81} = -3$

67. $\sqrt[4]{-81}$ is not a real number.

69. $\sqrt[4]{256} = 4$

71. $\sqrt[5]{-32} = -2$

73. $\sqrt{\sqrt[3]{64}} = \sqrt{4} = 2$

75. $\sqrt[3]{\frac{8}{27}} = \frac{2}{3}$

77. $\sqrt[3]{-\frac{1}{64}} = -\frac{1}{4}$

79.

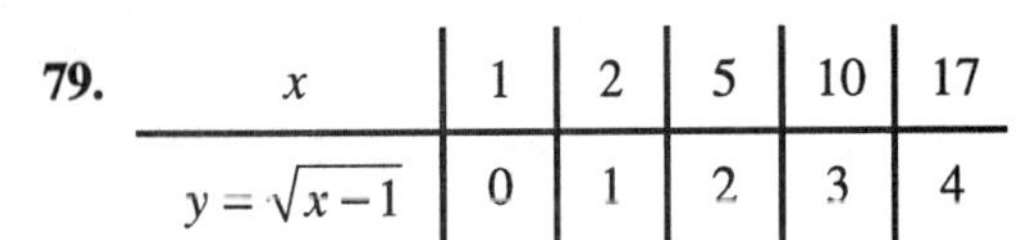

x	1	2	5	10	17
$y = \sqrt{x-1}$	0	1	2	3	4

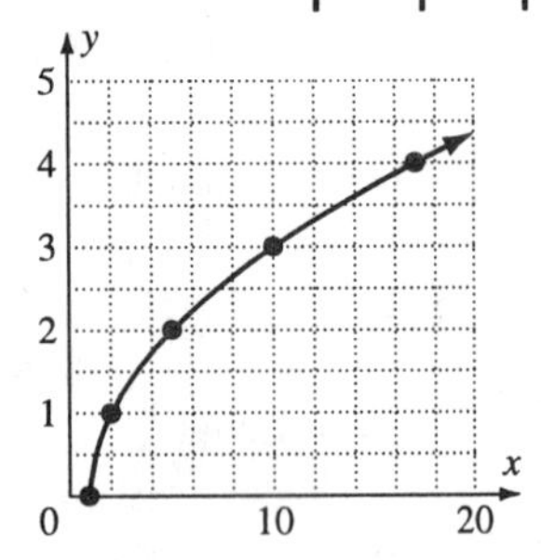

a. $x \geq 1$; $y = 0$ when $x = 1$, then the graph increases.

b. Verify graph.

81. $v = 4\sqrt{r}$
$v = 4\sqrt{9}$
$v = 4(3)$
$v = 12$
12 miles per hour

83. $t = \sqrt{\frac{d}{16}}$, $d = 144$ feet
$t = \sqrt{\frac{144}{16}} = \sqrt{9} = 3$ seconds

85. $H = (10.45 + \sqrt{100W} - W)(33 - t)$,
$W = 4$ meters per second, $t = 0$
$H = (10.45 + \sqrt{100 \cdot 4} - 4)(33 - 0)$
$= (10.45 + 20 - 4)(33) = (26.45)(33)$
$= 872.85$

Since $H > 200$, the exposed flesh will freeze in 1 minute or less, yes.

87. d is true.
$\sqrt{\frac{1}{4}} + \sqrt{\frac{1}{9}} = \sqrt{\frac{25}{36}}$
$\frac{1}{2} + \frac{1}{3} = \frac{5}{6}$
$\frac{3}{6} + \frac{2}{6} = \frac{5}{6}$
$\frac{5}{6} = \frac{5}{6}$

89. $d = 1.22\sqrt{x} = 1.22\sqrt{25000} \approx 192.90$ miles

91. $r = \frac{\sqrt{A} - \sqrt{P}}{\sqrt{P}} = \frac{\sqrt{900} - \sqrt{800}}{\sqrt{800}}$
≈ 0.061

93. $y_1 = \sqrt{x+4}$
$y_2 = \sqrt{x}$
$y_3 = \sqrt{x-3}$

Possible answer: They have the same shape, but different x- and y-intercepts.

95. $v = 4\sqrt{r}$

$v \approx 28.2843$ miles per hour

97. Yes

99. Yes

101-103. Answers may vary.

105. $-\sqrt{47} \approx -6.86$
-7 and -6

Review Problems

107. $4x - 5y = 20$

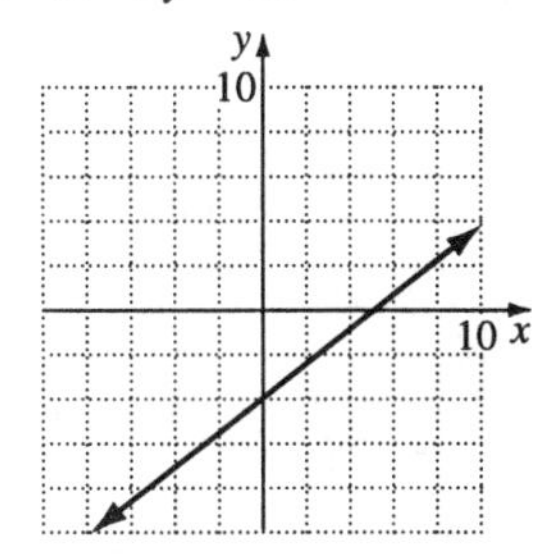

108. $\dfrac{1}{x^2 - 17x + 30} \div \dfrac{1}{x^2 + 7x - 18}$
$= \dfrac{1}{(x-2)(x-15)} \cdot \dfrac{(x+9)(x-2)}{1}$
$= \dfrac{x+9}{x-15}, (x \neq 2, 15)$

109. $2(x - 3) > 4x + 10$
$2x - 6 > 4x + 10$
$2x - 4x > 6 + 10$
$-2x > 16$
$x < -8$

−10 −9 −8 −7 −6

Problem Set 9.2

1. $\sqrt{7} \cdot \sqrt{6} = \sqrt{42}$

3. $\sqrt{6} \cdot \sqrt{6} = \sqrt{36} = 6$

5. $\sqrt{3} \cdot \sqrt{5y} = \sqrt{15y}$

7. $\sqrt{3x} \cdot \sqrt{6y} = \sqrt{18xy} = \sqrt{9 \cdot 2xy} = 3\sqrt{2xy}$

9. $\sqrt{\frac{1}{2}} \cdot \sqrt{\frac{5}{7}} = \sqrt{\frac{5}{14}}$

11. $\sqrt{50} = \sqrt{25 \cdot 2} = 5\sqrt{2}$

13. $\sqrt{45} = \sqrt{9 \cdot 5} = 3\sqrt{5}$

15. $\sqrt{80x} = \sqrt{16 \cdot 5x} = 4\sqrt{5x}$

17. $\sqrt{600xy} = \sqrt{100 \cdot 6xy} = 10\sqrt{6xy}$

19. $2\sqrt{27} = 2\sqrt{9 \cdot 3} = 6\sqrt{3}$

21. $7\sqrt{8a} = 7\sqrt{4 \cdot 2a} = 14\sqrt{2a}$

23. $\sqrt{27} \cdot \sqrt{18} = \sqrt{9 \cdot 3} \cdot \sqrt{9 \cdot 2} = 9\sqrt{6}$

25. $\sqrt{15} \cdot \sqrt{21x} = \sqrt{15 \cdot 21x} = \sqrt{9 \cdot 35x} = 3\sqrt{35x}$

27. $\sqrt{72a} \cdot \sqrt{50b} = (6\sqrt{2a})(5\sqrt{2b})$
$= 30 \cdot 2\sqrt{ab} = 60\sqrt{ab}$

29. $\sqrt{3} \cdot \sqrt{6} \cdot \sqrt{18} = \sqrt{18} \cdot \sqrt{18} = 18$

31. $\sqrt{y^3} = \sqrt{y^2 \cdot y} = y\sqrt{y}$

33. $\sqrt{50x^2} = \sqrt{50} \cdot \sqrt{x^2} = 5x\sqrt{2}$

35. $\sqrt{80x^4} = \sqrt{80} \cdot \sqrt{x^4} = 4x^2\sqrt{5}$

37. $\sqrt{72x^5} = \sqrt{36x^4 \cdot 2x} = \sqrt{36x^4} \cdot \sqrt{2x}$
$= 6x^2\sqrt{2x}$

39. $\sqrt{12x^{11}} = \sqrt{4x^{10}} \cdot \sqrt{3x} = 2x^5\sqrt{3x}$

41. $\sqrt{90p^{23}} = \sqrt{9p^{22}} \cdot \sqrt{10p} = 3p^{11}\sqrt{10p}$

43. $\sqrt{2x^2} \cdot \sqrt{6x} = \sqrt{12x^2}\sqrt{x} = 2x\sqrt{3} \cdot \sqrt{x}$
$= 2x\sqrt{3x}$

45. $\sqrt{2y^3} \cdot \sqrt{10y} = \sqrt{20y^4} = 2y^2\sqrt{5}$

47. $\sqrt{15r^2} \cdot \sqrt{5r^6} = \sqrt{75r^8} = 5r^4\sqrt{3}$

49. $\sqrt{x^2y} \cdot \sqrt{xy^5} = \sqrt{x^3y^6} = xy^3\sqrt{x}$

51. $\sqrt{50xy} \cdot \sqrt{4x^2y^4} = \sqrt{200x^3y^5}$
$= 10xy^2\sqrt{2xy}$

53. $\sqrt{\frac{49}{16}} = \frac{7}{4}$

55. $\sqrt{\frac{35}{4}} = \frac{\sqrt{35}}{2}$

57. $\sqrt{\frac{7}{x^4}} = \frac{\sqrt{7}}{x^2}$

59. $\sqrt{\frac{72}{x^6}} = \frac{6\sqrt{2}}{x^3}$

61. $\frac{\sqrt{54}}{\sqrt{6}} = \sqrt{\frac{54}{6}} = \sqrt{9} = 3$

63. $\frac{\sqrt{72}}{\sqrt{8}} = \sqrt{9} = 3$

65. $\frac{15\sqrt{10}}{3\sqrt{2}} = 5\sqrt{5}$

67. $\frac{30\sqrt{50}}{10\sqrt{5}} = 3\sqrt{10}$

69. $\sqrt{\frac{28y}{81}} = \frac{2\sqrt{7y}}{9}$

71. $\frac{\sqrt{96y^5}}{\sqrt{8y}} = \sqrt{12y^4} = 2y^2\sqrt{3}$

73. $\frac{\sqrt{8x^7}}{\sqrt{2x}} = \sqrt{4x^6} = 2x^3$

75. $\frac{\sqrt{24y^7}}{\sqrt{6}} = \sqrt{4y^7} = 2y^3\sqrt{y}$

77. $\sqrt[3]{32} = \sqrt[3]{8 \cdot 4} = 2\sqrt[3]{4}$

79. $\sqrt[3]{128} = \sqrt[3]{64 \cdot 2} = 4\sqrt[3]{2}$

81. $\sqrt[4]{80} = \sqrt[4]{16 \cdot 5} = 2\sqrt[4]{5}$

83. $\sqrt[3]{4} \cdot \sqrt[3]{2} = \sqrt[3]{8} = 2$

85. $\sqrt[3]{9} \cdot \sqrt[3]{6} = \sqrt[3]{54} = \sqrt[3]{27 \cdot 2} = 3\sqrt[3]{2}$

87. $\sqrt[5]{16} \cdot \sqrt[5]{4} = \sqrt[5]{64} = \sqrt[5]{32 \cdot 2} = 2\sqrt[5]{2}$

89. $\sqrt[3]{\frac{27}{8}} = \frac{3}{2}$

91. $\sqrt[4]{\frac{225}{81}} = \frac{\sqrt[4]{225}}{3} = \frac{\sqrt{15}}{3}$

(Note that $\left(\sqrt{15}\right)^4 = 225$.)

93. $\sqrt[3]{\frac{3}{8}} = \frac{\sqrt[3]{3}}{2}$

95. $A = (13\sqrt{2})(5\sqrt{6}) = 65\sqrt{12}$
$= 130\sqrt{3}$ square feet

97. $h = \frac{s}{2}\sqrt{3}$
$h = \frac{\sqrt{18}}{2}(\sqrt{3}) = \frac{(3\sqrt{2})\sqrt{3}}{2} = \frac{3}{2}\sqrt{6}$ feet

99. Statement c is true, since
$\sqrt{2x}\sqrt{6y} = \sqrt{12xy} = 2\sqrt{3xy}$.

101. Students should verify solution.

103. $\sqrt{x^4} = x^2$

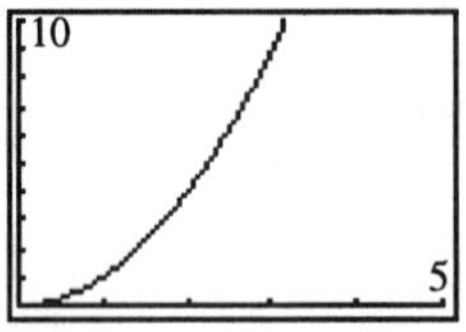

105. $\sqrt{18x^2} = 3x\sqrt{2}$

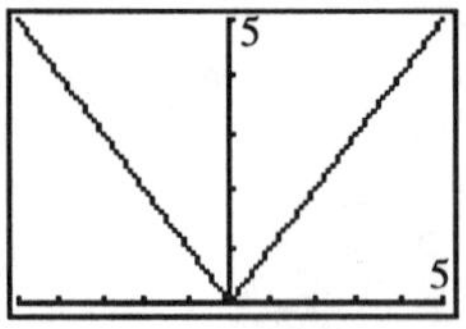

107. Answers may vary.

109. $\sqrt{x^{12n}} = x^{6n}$

111. $\sqrt{3a^3bc^6} \cdot \sqrt{6a^4b^5c^6} = \sqrt{18a^7b^6c^{12}}$
$= 3a^3b^3c^6\sqrt{2a}$

113. $\sqrt{2a^{12}b^5} \cdot \sqrt{8a^3b^7} = \sqrt{16a^{15}b^{12}}$
$= 4a^7b^6\sqrt{a}$; 12, 8, 7

Review Problems

114. $4x + 3y = 18$ $(\times 5)$
$5x - 9y = 48$ $(\times(-4))$

$$\begin{aligned} 20x + 15y &= 90 \\ -20x + 36y &= -192 \\ \hline 51y &= -102 \\ y &= -2 \\ 4x - 3(2) &= 18 \\ 4x &= 24 \\ x &= 6 \end{aligned}$$

(6, −2)

115. $\frac{2x+1}{6x+12} + \frac{x+1}{x^2+2x} - \frac{1}{6}$
$= \frac{2x+1}{6(x+2)} \cdot \frac{x}{x} + \frac{x+1}{x(x+2)} \cdot \frac{6}{6} - \frac{1}{6} \cdot \frac{x(x+2)}{x(x+2)}$
$= \frac{2x^2 + x + 6x + 6 - x^2 - 2x}{6x(x+2)}$
$= \frac{x^2 + 5x + 6}{6x(x+2)}, (x \neq -2, 0)$

116. Let b = the base of the triangle.
$b - 3$ = height of triangle
$A = \frac{1}{2}bh$
$35 = \frac{1}{2}b(b-3) = \frac{1}{2}(b^2 - 3b)$
$b^2 - 3b = 70$
$b^2 - 3b - 70 = 0$
$(b - 10)(b + 7) = 0$
$b = 10$ (reject $b = -7$)
$b - 3 = 7$
base, 10 cm; height, 7 cm

Problem Set 9.3

1. $7\sqrt{3} + 6\sqrt{3} = (7+6)\sqrt{3} = 13\sqrt{3}$

3. $4\sqrt{13} - 6\sqrt{13} = -2\sqrt{13}$

5. $\sqrt{5} + \sqrt{5} = 2\sqrt{5}$

7. $\sqrt{13x} + 2\sqrt{13x} = 3\sqrt{13x}$

9. $-4\sqrt{11y} - 8\sqrt{11y} = -12\sqrt{11y}$

11. $5\sqrt{6p} - \sqrt{6p} = 4\sqrt{6p}$

13. $4\sqrt{2} - 5\sqrt{2} + 8\sqrt{2} = 7\sqrt{2}$

15. $\sqrt{3} - 6\sqrt{7} - 12\sqrt{3} = -6\sqrt{7} - 11\sqrt{3}$

17. $6\sqrt[3]{4} - 5\sqrt[3]{4} = \sqrt[3]{4}$

19. $\sqrt{2} + \sqrt[3]{2}$ cannot be simplified

21. $\sqrt[4]{5} + \sqrt[3]{6} + 8\sqrt[4]{5} - 2\sqrt[3]{6} = 9\sqrt[4]{5} - \sqrt[3]{6}$

23. $\sqrt{8} + 3\sqrt{2} = 2\sqrt{2} + 3\sqrt{2} = 5\sqrt{2}$

25. $6\sqrt{3} - \sqrt{27} = 6\sqrt{3} - 3\sqrt{3} = 3\sqrt{3}$

27. $\sqrt{50a} + \sqrt{18a} = 5\sqrt{2a} + 3\sqrt{2a} = 8\sqrt{2a}$

29. $3\sqrt{18b} - 5\sqrt{50b} = 9\sqrt{2b} - 25\sqrt{2b}$
$= -16\sqrt{2b}$

31. $\frac{1}{4}\sqrt{12} - \frac{1}{2}\sqrt{48} = \frac{1}{4}\sqrt{12} - \frac{2}{2}\sqrt{12}$
$= -\frac{3}{4}\sqrt{12} = -\frac{3}{2}\sqrt{3}$

33. $3\sqrt{75} + 2\sqrt{12} - 2\sqrt{48}$
$= 15\sqrt{3} + 4\sqrt{3} - 8\sqrt{3}$
$= 11\sqrt{3}$

35. $6\sqrt{7} + 2\sqrt{28} - 3\sqrt{63}$
$= 6\sqrt{7} + 4\sqrt{7} - 9\sqrt{7} = \sqrt{7}$

37. $\frac{1}{6}\sqrt{72}-\frac{3}{8}\sqrt{8}+\frac{1}{5}\sqrt{50}$
$=\sqrt{2}-\frac{3}{4}\sqrt{2}+\sqrt{2}$
$=\frac{5}{4}\sqrt{2}$

39. $3\sqrt{54}-2\sqrt{20}+4\sqrt{45}-\sqrt{24}$
$=9\sqrt{6}-4\sqrt{5}+12\sqrt{5}-2\sqrt{6}$
$=7\sqrt{6}+8\sqrt{5}$

41. $\frac{1}{4}\sqrt{2x}+\frac{2}{3}\sqrt{8x}=\frac{1}{4}\sqrt{2x}+\frac{4}{3}\sqrt{2x}$
$=\frac{19}{12}\sqrt{2x}$

43. $\frac{\sqrt{45}}{4}-\sqrt{80}+\frac{\sqrt{20}}{3}=\frac{3}{4}\sqrt{5}-4\sqrt{5}+\frac{2}{3}\sqrt{5}$
$=\frac{9-48+8}{12}\sqrt{5}=-\frac{31}{12}\sqrt{5}$

45. $\sqrt[3]{81}+\sqrt[3]{24}=3\sqrt[3]{3}+2\sqrt[3]{3}=5\sqrt[3]{3}$

47. $5\sqrt[3]{54b}+2\sqrt[3]{16b}=15\sqrt[3]{2b}+4\sqrt[3]{2b}$
$=19\sqrt[3]{2b}$

49. $5\sqrt[3]{16}-2\sqrt[3]{54}=10\sqrt[3]{2}-6\sqrt[3]{2}=4\sqrt[3]{2}$

51. $\sqrt{2}(\sqrt{3}+4)=\sqrt{6}+4\sqrt{2}$
Distributive property

53. $\sqrt{7}(\sqrt{6}-5)=\sqrt{42}-5\sqrt{7}$

55. $\sqrt{5x}(\sqrt{3}+\sqrt{7})=\sqrt{15x}+\sqrt{35x}$

57. $\sqrt{2}(\sqrt{5}-\sqrt{2})=\sqrt{10}-\sqrt{4}=\sqrt{10}-2$

59. $\sqrt{3}(5\sqrt{2}+\sqrt{3})=5\sqrt{6}+3$

61. $\sqrt{3}(4\sqrt{3}+\sqrt{5})=4\cdot 3+\sqrt{15}=12+\sqrt{15}$

63. $5\sqrt{3}(4\sqrt{2}+6\sqrt{5})=20\sqrt{6}+30\sqrt{15}$

65. $3\sqrt{10a}(6\sqrt{2a}-4\sqrt{5b})$
$=18\sqrt{20a^2}-12\sqrt{50b}$
$=36a\sqrt{5}-60\sqrt{2b}$

67. $(\sqrt{5}+2)(\sqrt{5}+3)=5+5\sqrt{5}+6$
$=11+5\sqrt{5}$

69. $(\sqrt{2x}+6)(\sqrt{2x}-5)=2x+\sqrt{2x}-30$

71. $(\sqrt{2}+1)(\sqrt{3}-6)=\sqrt{6}-6\sqrt{2}+\sqrt{3}-6$

73. $(\sqrt{3}+\sqrt{a})(\sqrt{3}+2\sqrt{a})$
$=3+2\sqrt{3a}+\sqrt{3a}+2a$
$=3+3\sqrt{3a}+2a$

75. $(2\sqrt{7}+3)(4\sqrt{7}-5)$
$=56-10\sqrt{7}+12\sqrt{7}-15$
$=41+2\sqrt{7}$

77. $(\sqrt{5}+\sqrt{2})(\sqrt{5}+3\sqrt{2})$
$=5+3\sqrt{10}+\sqrt{10}+6$
$=11+4\sqrt{10}$

79. $(\sqrt{a}+\sqrt{b})(\sqrt{a}+3\sqrt{b})$
$=a+3\sqrt{ab}+\sqrt{ab}+3b$
$=a+4\sqrt{ab}+3b$

81. $(4\sqrt{3}+7\sqrt{2})(5\sqrt{3}-6\sqrt{2})$
$=60-24\sqrt{6}+35\sqrt{6}-84=-24+11\sqrt{6}$

83. $(\sqrt{5}+\sqrt{3})^2=5+2\sqrt{15}+3=8+2\sqrt{15}$

85. $(\sqrt{3}-1)^2=3-2\sqrt{3}+1=4-2\sqrt{3}$

87. $(2\sqrt{3}-4\sqrt{7})^2=12-16\sqrt{21}+112$
$=124-16\sqrt{21}$

89. $(\sqrt{a}+\sqrt{3})^2=a+2\sqrt{3a}+3$

91. $(\sqrt{y}-\sqrt{10})^2=y-2\sqrt{10y}+10$

93. $(4+\sqrt{7})(4-\sqrt{7})=16-7=9$

95. $(\sqrt{13}+\sqrt{5})(\sqrt{13}-\sqrt{5})=13-5=8$

97. $(2\sqrt{3}-7)(2\sqrt{3}+7)=12-49=-37$

99. $(2\sqrt{3}+\sqrt{5})(2\sqrt{3}-\sqrt{5})=12-5=7$

101. $(3\sqrt{7}-2\sqrt{3})(3\sqrt{7}+2\sqrt{3})=63-12=51$

103. perimeter $=2l+2w=\ 2(3\sqrt{75})+2(4\sqrt{18})$
$=6\sqrt{75}+8\sqrt{18}=30\sqrt{3}+24\sqrt{2}$ meters

105. $\sqrt{2}+\sqrt{8}=\sqrt{(2+8)+2\sqrt{2\cdot 8}}$
$=\sqrt{10+2\sqrt{16}}=\sqrt{10+8}=\sqrt{9\cdot 2}=3\sqrt{2}$
$\sqrt{2}+\sqrt{8}=\sqrt{2}+2\sqrt{2}=3\sqrt{2}$
Answers may vary.

107. Perimeter: $2(5\sqrt{2}+3+2\sqrt{3}-2)$
$=2(5\sqrt{2}+2\sqrt{3}+1)=10\sqrt{2}+4\sqrt{3}+2$ cm
Area: $(5\sqrt{2}+3)(2\sqrt{3}-2)$
$=10\sqrt{6}-10\sqrt{2}+6\sqrt{3}-6\text{ cm}^2$

109. Statement d is true.
a. $\sqrt{16}+\sqrt{9}=4+3=7$ *not* 5
b. $7\sqrt[3]{3}-4\sqrt{3};\ \sqrt[3]{3}\neq\sqrt{3}$
c. $\sqrt{5}+6\sqrt{5}=7\sqrt{5}$ *not* $7\sqrt{10}$
d. None of the above is true.

111. Statement c is correct.
$(\sqrt{5}-\sqrt{3})^2=(\sqrt{5})^2-2\sqrt{5}\sqrt{3}+(\sqrt{3})^2$
$=5-2\sqrt{15}+3=8-2\sqrt{15}$ True

113. $\sqrt{4x}+\sqrt{9x}=5\sqrt{x}$

115. $5\sqrt{x-2}-6\sqrt{x-2}=-\sqrt{x-2}$

117. $(\sqrt{x}-1)(\sqrt{x}-1)=x-2\sqrt{x}+1$

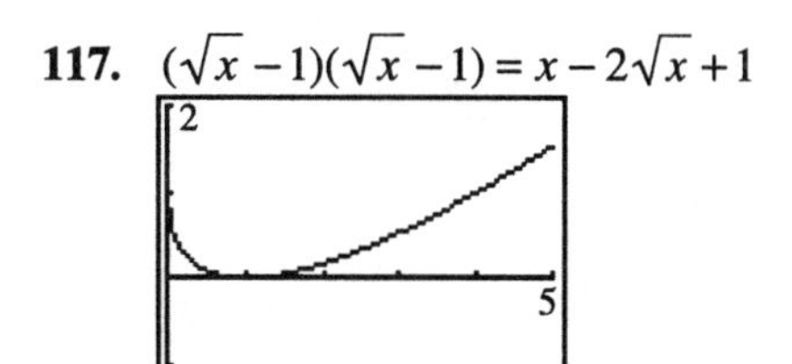

119. $\sqrt{x}(2\sqrt{x}+1)=2x+\sqrt{x}$

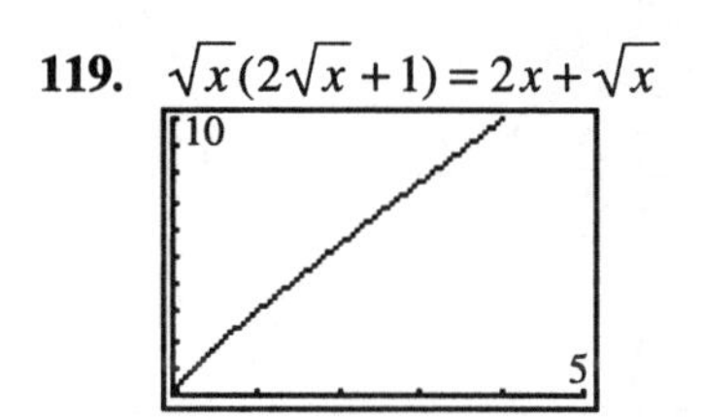

121-123. Answers may vary.

125. $\sqrt{5}\cdot\sqrt{15}+6\sqrt{3}=\sqrt{75}+6\sqrt{3}$
$=5\sqrt{3}+6\sqrt{3}=11\sqrt{3}$

127. $(\sqrt[3]{4}+1)(\sqrt[3]{2}-3)=\sqrt[3]{8}-3\sqrt[3]{4}+\sqrt[3]{2}-3$
$=2-3\sqrt[3]{4}+\sqrt[3]{2}-3=-1-3\sqrt[3]{4}+\sqrt[3]{2}$

129. $(4\sqrt{3x}+\sqrt{2y})(4\sqrt{3x}-\sqrt{2y})$
$=16(3x)-2y=48x-2y$

Review Problems

130. $64y^3-y=y(64y^2-1)=y(8y+1)(8y-1)$

131. $(3y-2)(4y-3)-(2y-5)^2$
$=12y^2-17y+6-4y^2+20y-25$
$=8y^2+3y-19$

132. $y=-\frac{1}{4}x+3$

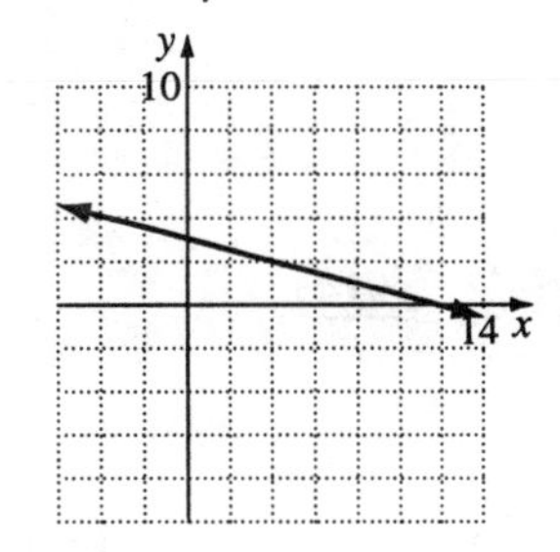

Problem Set 9.4

1. $\frac{2}{\sqrt{3}}=\frac{2}{\sqrt{3}}\cdot\frac{\sqrt{3}}{\sqrt{3}}=\frac{2\sqrt{3}}{3}$

3. $\frac{21}{\sqrt{7}}=\frac{21}{\sqrt{7}}\cdot\frac{\sqrt{7}}{\sqrt{7}}=\frac{21\sqrt{7}}{7}=3\sqrt{7}$

5. $\sqrt{\frac{2}{5}}=\frac{\sqrt{2}}{\sqrt{5}}\cdot\frac{\sqrt{5}}{\sqrt{5}}=\frac{\sqrt{2}\cdot\sqrt{5}}{5}=\frac{\sqrt{10}}{5}$

7. $\sqrt{\frac{7}{3}}=\frac{\sqrt{7}}{\sqrt{3}}\cdot\frac{\sqrt{3}}{\sqrt{3}}=\frac{\sqrt{21}}{3}$

9. $\sqrt{\frac{11}{x}}=\frac{\sqrt{11}}{\sqrt{x}}\cdot\frac{\sqrt{x}}{\sqrt{x}}=\frac{\sqrt{11x}}{x}$

11. $\sqrt{\frac{x}{y}}=\frac{\sqrt{x}}{\sqrt{y}}\cdot\frac{\sqrt{y}}{\sqrt{y}}=\frac{\sqrt{xy}}{y}$

13. $\frac{12}{\sqrt{32}}=\frac{12}{4\sqrt{2}}\cdot\frac{\sqrt{2}}{\sqrt{2}}=\frac{3\sqrt{2}}{2}$

15. $\frac{15}{\sqrt{12}}=\frac{15}{2\sqrt{3}}\cdot\frac{\sqrt{3}}{\sqrt{3}}=\frac{15\sqrt{3}}{2(3)}=\frac{5\sqrt{3}}{2}$

17. $\sqrt{\frac{5}{18}}=\frac{\sqrt{5}}{\sqrt{18}}\cdot\frac{\sqrt{18}}{\sqrt{18}}=\frac{\sqrt{90}}{18}=\frac{3\sqrt{10}}{18}=\frac{\sqrt{10}}{6}$

19. $\sqrt{\frac{20}{3}}=\frac{\sqrt{20}}{\sqrt{3}}\cdot\frac{\sqrt{3}}{\sqrt{3}}=\frac{\sqrt{60}}{3}=\frac{2\sqrt{15}}{3}$

21. $\sqrt{\frac{a}{32}}=\frac{\sqrt{a}}{4\sqrt{2}}\cdot\frac{\sqrt{2}}{\sqrt{2}}=\frac{\sqrt{2a}}{8}$

23. $\sqrt{\frac{x^2}{11}}=\frac{\sqrt{x^2}}{\sqrt{11}}\cdot\frac{\sqrt{11}}{\sqrt{11}}=\frac{x\sqrt{11}}{11}$

25. $\frac{\sqrt{7x}}{\sqrt{8}}=\frac{\sqrt{7x}}{2\sqrt{2}}\cdot\frac{\sqrt{2}}{\sqrt{2}}=\frac{\sqrt{14x}}{4}$

27. $\sqrt{\frac{7a}{12}}=\frac{\sqrt{7a}}{2\sqrt{3}}\cdot\frac{\sqrt{3}}{\sqrt{3}}=\frac{\sqrt{21a}}{6}$

29. $\sqrt{\frac{45}{x}}=\frac{3\sqrt{5}}{\sqrt{x}}\cdot\frac{\sqrt{x}}{\sqrt{x}}=\frac{3\sqrt{5x}}{x}$

31. $\sqrt{\frac{27}{a^3}}=\frac{3\sqrt{3}}{a\sqrt{a}}\cdot\frac{\sqrt{a}}{\sqrt{a}}=\frac{3\sqrt{3a}}{a^2}$

33. $\frac{\sqrt{50a^2}}{\sqrt{12a^3}}=\frac{5a\sqrt{2}}{2a\sqrt{3a}}\cdot\frac{\sqrt{3a}}{\sqrt{3a}}=\frac{5\sqrt{6a}}{6a}$

35. $\frac{5}{\sqrt{3}-1}=\frac{5}{\sqrt{3}-1}\cdot\frac{\sqrt{3}+1}{\sqrt{3}+1}$ use conjugate

$=\frac{5(\sqrt{3}+1)}{3-1}$

$=\frac{5(\sqrt{3}+1)}{2}$

37. $\frac{15}{\sqrt{7}+2}=\frac{15}{\sqrt{7}+2}\cdot\frac{\sqrt{7}-2}{\sqrt{7}-2}=\frac{15(\sqrt{7}-2)}{7-4}$

$=5(\sqrt{7}-2)$

39. $\frac{18}{3-\sqrt{3}}=\frac{18}{3-\sqrt{3}}\cdot\frac{3+\sqrt{3}}{3+\sqrt{3}}=\frac{18(3+\sqrt{3})}{9-3}$

$=3(3+\sqrt{3})$

41. $\frac{\sqrt{2}}{\sqrt{2}+1}=\frac{\sqrt{2}}{\sqrt{2}+1}\cdot\frac{\sqrt{2}-1}{\sqrt{2}-1}=\frac{\sqrt{2}(\sqrt{2}-1)}{2-1}$

$=2-\sqrt{2}$

43. $\frac{\sqrt{12}}{\sqrt{3}-1}=\frac{\sqrt{12}}{\sqrt{3}-1}\cdot\frac{\sqrt{3}+1}{\sqrt{3}+1}=\frac{2\sqrt{3}(\sqrt{3}+1)}{3-1}$

$=\sqrt{3}(\sqrt{3}+1)=3+\sqrt{3}$

45. $\frac{3\sqrt{2}}{\sqrt{10}+2}=\frac{3\sqrt{2}}{\sqrt{10}+2}\cdot\frac{\sqrt{10}-2}{\sqrt{10}-2}$

$=\frac{3\sqrt{2}(\sqrt{10}-2)}{10-4}=\frac{\sqrt{2}(\sqrt{10}-2)}{2}$

$=\frac{\sqrt{20}-2\sqrt{2}}{2}=\sqrt{5}-\sqrt{2}$

47. $\frac{\sqrt{3}+1}{\sqrt{2}-1}=\frac{\sqrt{3}+1}{\sqrt{2}-1}\cdot\frac{\sqrt{2}+1}{\sqrt{2}+1}$

$=\frac{\sqrt{6}+\sqrt{3}+\sqrt{2}+1}{2-1}=\sqrt{6}+\sqrt{3}+\sqrt{2}+1$

49. $\frac{\sqrt{2}-2}{2-\sqrt{3}} = \frac{\sqrt{2}-2}{2-\sqrt{3}} \cdot \frac{2+\sqrt{3}}{2+\sqrt{3}}$
$\frac{2\sqrt{2}+\sqrt{6}-4-2\sqrt{3}}{4-3}$
$= -4+2\sqrt{2}-2\sqrt{3}+\sqrt{6}$

51. $\frac{2\sqrt{3}+1}{\sqrt{6}-\sqrt{3}} = \frac{2\sqrt{3}+1}{\sqrt{6}-\sqrt{3}} \cdot \frac{\sqrt{6}+\sqrt{3}}{\sqrt{6}+\sqrt{3}}$
$= \frac{2\sqrt{18}+6+\sqrt{6}+\sqrt{3}}{6-3}$
$= \frac{6\sqrt{2}+6+\sqrt{6}+\sqrt{3}}{3}$
$= \frac{6+6\sqrt{2}+\sqrt{3}+\sqrt{6}}{3}$

53. $\frac{\sqrt{5}+\sqrt{6}}{\sqrt{5}+\sqrt{3}} = \frac{\sqrt{5}+\sqrt{6}}{\sqrt{5}+\sqrt{3}} \cdot \frac{\sqrt{5}-\sqrt{3}}{\sqrt{5}-\sqrt{3}}$
$= \frac{5-\sqrt{15}+\sqrt{30}-\sqrt{18}}{5-3}$
$= \frac{5-\sqrt{15}+\sqrt{30}-3\sqrt{2}}{2}$

55. $\frac{\sqrt{5}+\sqrt{2}}{\sqrt{5}-\sqrt{2}} = \frac{\sqrt{5}+\sqrt{2}}{\sqrt{5}-\sqrt{2}} \cdot \frac{\sqrt{5}+\sqrt{2}}{\sqrt{5}+\sqrt{2}}$
$= \frac{5+2\sqrt{10}+2}{5-2}$
$= \frac{7+2\sqrt{10}}{3}$

57. $\sqrt{56} = \sqrt{4}\sqrt{14} = 2\sqrt{14}$

59. $\sqrt[4]{32} = \sqrt[4]{16}\sqrt[4]{2} = 2\sqrt[4]{2}$

61. $8\sqrt{27}-3\sqrt{12} = 8\cdot 3\sqrt{3}-3\cdot 2\sqrt{3} = 18\sqrt{3}$

63. $7\sqrt{15}-2\sqrt{5}\cdot\sqrt{3} = 7\sqrt{15}-2\sqrt{15} = 5\sqrt{15}$

65. $\frac{9}{\sqrt{18}} = \frac{9}{3\sqrt{2}} = \frac{3}{\sqrt{2}} \cdot \frac{\sqrt{2}}{\sqrt{2}} = \frac{3\sqrt{2}}{2}$

67. $(2\sqrt{5}+\sqrt{3})(\sqrt{2}+\sqrt{7})$
$= 2\sqrt{10}+2\sqrt{35}+\sqrt{6}+\sqrt{21}$

69. $\frac{\sqrt{6}+1}{\sqrt{2}-4} = \frac{\sqrt{6}+1}{\sqrt{2}-4} \cdot \frac{\sqrt{2}+4}{\sqrt{2}+4}$
$= \frac{\sqrt{12}+4\sqrt{6}+\sqrt{2}+4}{2-16}$
$= -\frac{4+\sqrt{2}+2\sqrt{3}+4\sqrt{6}}{14}$

71. $\frac{w}{h} = \frac{2}{\sqrt{5}-1}$
$\frac{2}{\sqrt{5}-1} \cdot \frac{\sqrt{5}+1}{\sqrt{5}+1} = \frac{2(\sqrt{5}+1)}{5-1} = \frac{\sqrt{5}+1}{2}$
$\frac{w}{h} = \frac{\sqrt{5}+1}{2} \approx 1.62$

73. b is true; $\frac{3\sqrt{x}}{x\sqrt{6}} = \frac{3\sqrt{x}}{x\sqrt{6}}\frac{\sqrt{6}}{\sqrt{6}} = \frac{3\sqrt{6x}}{6x} = \frac{\sqrt{6x}}{2x}$
for $x > 0$, True.

75. $\sqrt{2}\cdot\sqrt{2x} = 2\sqrt{x}$

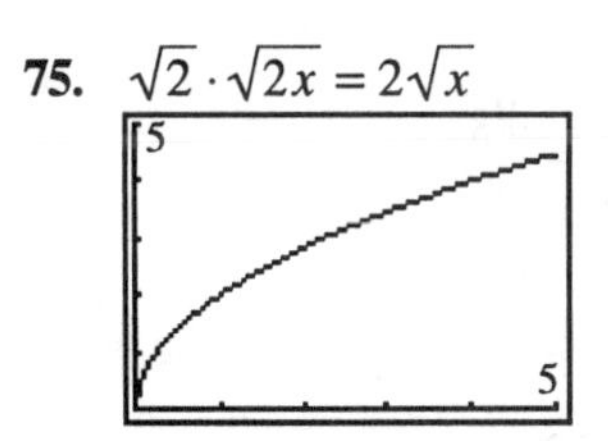

77. $\frac{x}{\sqrt{2}-1} = (\sqrt{2}+1)x$

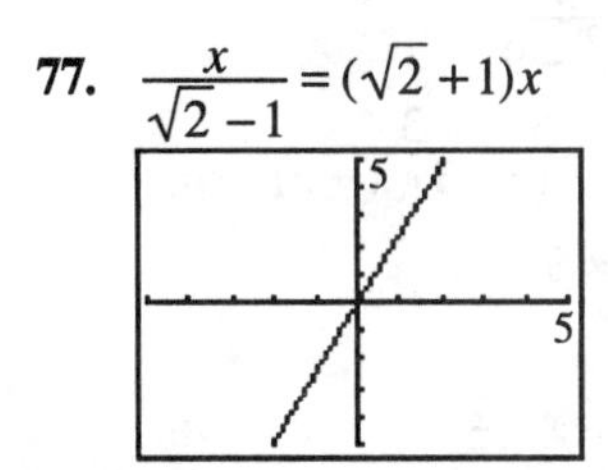

79-81. Answers may vary.

83. $\sqrt{13+\sqrt{2}+\frac{7}{3+\sqrt{2}}}$

$=\sqrt{13+\sqrt{2}+\frac{7(3-\sqrt{2})}{(3+\sqrt{2})(3-\sqrt{2})}}$

$=\sqrt{13+\sqrt{2}+\frac{21-7\sqrt{2}}{9-2}}$

$=\sqrt{13+\sqrt{2}+\frac{21-7\sqrt{2}}{7}}$

$=\sqrt{13+\sqrt{2}+3-\sqrt{2}}$

$=\sqrt{16}=4$

85. $\frac{2}{\sqrt[3]{4}}=\frac{2}{\sqrt[3]{4}}\cdot\frac{\sqrt[3]{2}}{\sqrt[3]{2}}=\frac{2\sqrt[3]{2}}{2}=\sqrt[3]{2}$

87. $\frac{1}{\sqrt[3]{3}}=\frac{1}{\sqrt[3]{3}}\cdot\frac{\sqrt[3]{9}}{\sqrt[3]{9}}=\frac{\sqrt[3]{9}}{3}$

89. $\frac{1}{\sqrt[4]{2}}=\frac{1}{\sqrt[4]{2}}\cdot\frac{\sqrt[4]{8}}{\sqrt[4]{8}}=\frac{\sqrt[4]{8}}{2}$

Review Problems

91. $\frac{x^2-6x+9}{12}\cdot\frac{3}{x^2-9}$

$=\frac{(x-3)(x-3)}{3\cdot 4}\cdot\frac{3}{(x-3)(x+3)}$

$=\frac{x-3}{4(x+3)}, (x\neq 3, -3)$

92. $\frac{1}{y-1}+\frac{1}{y+1}=\frac{3y-2}{y^2-1}$

$\frac{y+1+y-1}{y^2-1}=\frac{3y-2}{y^2-1}$

$2y=3y-2\quad (y\neq\pm 1)$

$y=2$

2

93. $(2x^2)^{-3}=2^{-3}x^{-6}=\frac{1}{2^3x^6}=\frac{1}{8x^6}, x\neq 0$

Problem Set 9.5

1. $\sqrt{x}=4$

$x=4^2=16$

Squaring Property of Equality

16

3. $\sqrt{x}=5$

$x=5^2=25$

25

5. $\sqrt{x+4}=2$

$x+4=4$

$x=0$

0

7. $\sqrt{x-4}=11$

$x-4=121$

$x=125$

125

9. $\sqrt{3y-2}=4$

$3y-2=16$

$3y=18$

$y=6$

6

11. $\sqrt{3x+5}=2$

$3x+5=4$

$3x=-1$

$x=-\frac{1}{3}$

$-\frac{1}{3}$

13. $3\sqrt{z}=\sqrt{8z+16}$

$9z=8z+16$

$z=16$

Check: $3\sqrt{16}=3\cdot 4=12$

$\sqrt{8\cdot 16+16}=\sqrt{144}=12$ √

$12=12$ True

16

15. $\sqrt{2y-3} = 2\sqrt{3y-2}$
$2y - 3 = 4(3y - 2)$
$2y - 3 = 12y - 8$
$-10y = -5$
$y = \frac{1}{2}$
$\frac{1}{2}$

17. $\sqrt{2y-3} = -5$
Since $\sqrt{2y-3}$ represents a nonnegative root, there is no solution of the equation.
solution: $\varnothing$

19. $\sqrt{3y+4} - 2 = 3$
$\sqrt{3y+4} = 5$
$3y + 4 = 25$
$3y = 21$
$y = 7$
7

21. $\sqrt{6x-8} - 3 = 1$
$\sqrt{6x-8} = 4$
$6x - 8 = 16$
$6x = 24$
$x = 4$
4

23. $3\sqrt{y-1} = \sqrt{3y+3}$
$9(y - 1) = 3y + 3$
$9y - 3y = 3 + 9$
$6y = 12$
$y = 2$
2

25. $\sqrt{y+3} = y - 3$
$y + 3 = y^2 - 6y + 9$
$y^2 - 7y + 6 = 0$
$(y - 6)(y - 1) = 0$
$y = 6$ or $y = 1$
Check: $y = 6$: $\sqrt{6+3} = \sqrt{9} = 3$; $6 - 3 = 3$
$y = 1$: $\sqrt{1+3} = \sqrt{4} = 2$; $1 - 3 = -2 \neq 2$
The solution is 6.

27. $\sqrt{2x+13} = x + 7$
$2x + 13 = x^2 + 14x + 49$
$x^2 + 12x + 36 = 0$
$(x + 6)^2 = 0$
$x = -6$
-6

29. $\sqrt{y^2+5} = y + 1$
$y^2 + 5 = y^2 + 2y + 1$
$2y = 4$
$y = 2$
2

31. $\sqrt{3y+3} + 5 = y$
$\sqrt{3y+3} = y - 5$
$3y + 3 = y^2 - 10y + 25$
$y^2 - 13y + 22 = 0$
$(y - 2)(y - 11) = 0$
$y = 2$ or $y = 11$
Check: $y = 2$: $\sqrt{6+3} + 5 = 3 + 5 = 8 \neq 2$
$y = 11$: $\sqrt{33+3} + 5 = 6 + 5 = 11$
The solution is 11.

33. $\sqrt{3z+7} - z = 3$
$\sqrt{3z+7} = z + 3$
$3z + 7 = z^2 + 6z + 9$
$z^2 + 3z + 2 = 0$
$(z + 1)(z + 2) = 0$
$z = -1$ or $z = -2$
Check: $z = -1$: $\sqrt{-3+7} - (-1) = 3$
$z = -2$: $\sqrt{-6+7} - (-2) = 3$
The solutions are -1, -2.

35. $\sqrt{3y} + 10 = y + 4$
$\sqrt{3y} = y - 6$
$3y = y^2 - 12y + 36$
$y^2 - 15y + 36 = 0$
$(y - 3)(y - 12) = 0$
$y = 3$ or $y = 12$

Check: $y = 3$: $\sqrt{9} + 10 = 13 \neq 3 + 4$

$y = 12$: $\sqrt{36} + 10 = 16 = 12 + 4$

The solution is 12.

37. $\sqrt{4z^2 + 3z - 2} - 2z = 0$

$\sqrt{4z^2 + 3z - 2} = 2z$

$4z^2 + 3z - 2 = 4z^2$

$3z = 2$

$z = \frac{2}{3}$

$\frac{2}{3}$

39. $\sqrt{3y^2 + 6y + 4} - 2 = 0$

$\sqrt{3y^2 + 6y + 4} = 2$

$3y^2 + 6y + 4 = 4$

$3y(y + 2) = 0$

$y = 0$ or $y = -2$

Check: $y = 0$: $\sqrt{4} = 2$; $2 - 2 = 0$

$y = -2$: $\sqrt{12 - 12 + 4} = \sqrt{4} = 2$; $2 - 2 = 0$

The solutions are -2, 0.

41. $3\sqrt{y} + 5 = 2$

$3\sqrt{y} = -3$

$\sqrt{y} = -1$

Since $\sqrt{y}$ represents a nonnegative value, there is no solution of the equation.

Solution: $\varnothing$

43. $N = 5000\sqrt{100 - x}$

$40000 = 5000\sqrt{100 - x}$

$8 = \sqrt{100 - x}$

$64 = 100 - x$

$x = 100 - 64 = 36$ years

45. $s = 30\sqrt{\frac{a}{p}}$

$90 = 30\sqrt{\frac{a}{100}}$

$3 = \sqrt{\frac{a}{100}}$

$9 = \frac{a}{100}$

$a = 900$ feet

47. $T = \frac{11}{7}\sqrt{\frac{L}{2}}$

$2 = \frac{11}{7}\sqrt{\frac{L}{2}}$

$\frac{14}{11} = \sqrt{\frac{L}{2}}$

$\frac{196}{121} = \frac{L}{2}$

$L = \frac{392}{121}$ feet ≈ 3.24 feet

49. $R = \sqrt{A^2 + B^2}$

$500 = \sqrt{(300)^2 + B^2}$

$250,000 = 90,000 + B^2$

$160,000 = B^2$

$400 = B$

400 pounds

51. Statement c is true. $\sqrt{x}$ is a non-negative number for $x \geq 0$. If a is negative, the solution is $\varnothing$.

53. $\sqrt{2x + 2} = \sqrt{3x - 5}$

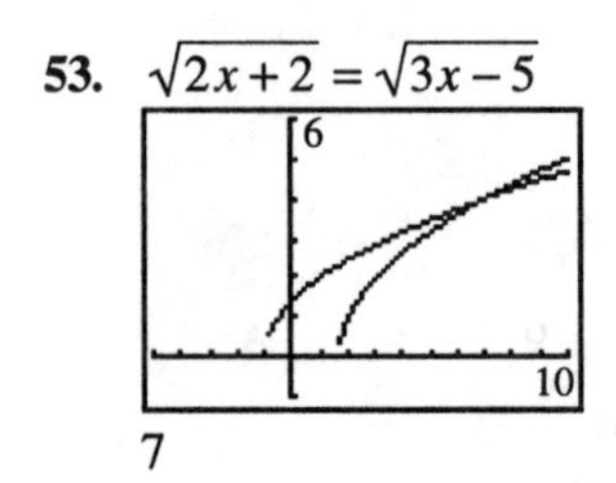

7

55. $\sqrt{x^2+3}=x+1$

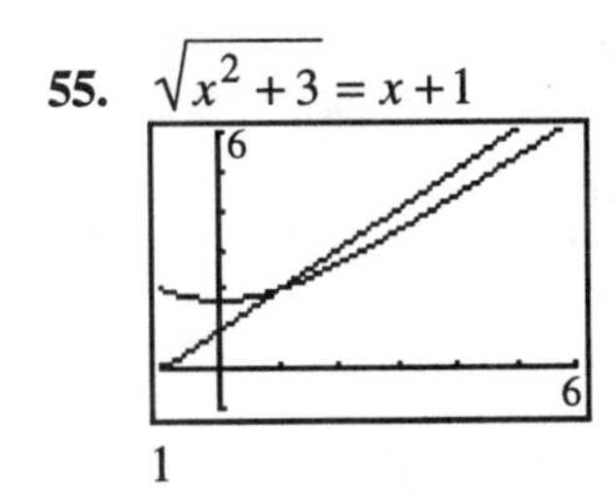

1

57. $\sqrt{x}+4=2$

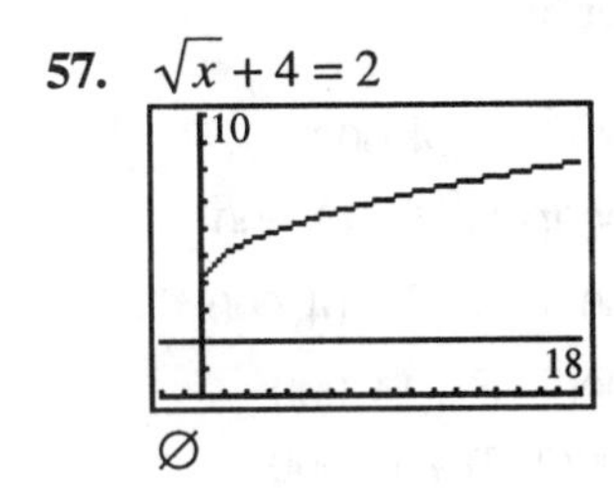

$\varnothing$

59. Answers may vary.

61. $\sqrt{x-8}=5-\sqrt{x+7}$

$x-8=25-10\sqrt{x+7}+x+7$

$10\sqrt{x+7}=40$

$\sqrt{x+7}=4$

$x+7=16$

$x=9$

9

63. $y=\sqrt{x-2}+2$

$z=\sqrt{y-2}+2$

$w=\sqrt{z-2}+2$

Since $w=2,\ 2=\sqrt{z-2}+2;\ z=2$

$2=\sqrt{y-2}+2;\ y=2$

$2=\sqrt{x-2}+2;\ x=2$

Review Problems

64. Let x be the part invested at 6%.
Then $9000-x$ = part invested at 4%.

$x(0.06)+(9{,}000-x)(0.04)=500$

$0.06x+360-0.04x=500$

$0.02x=140$

$x=7000$

$9000-x=2000$

\$7000 at 6%, \$2000 at 4%

65. Let x = orchestra seat price and y = mezzanine seat price.

$4x+2y=22$

$2x+3y=16$

$$\begin{array}{rl} 12x+6y &= 66 \\ -4x-6y &= -32 \\ \hline 8x \quad &= 34 \\ x \quad &= 4.25 \end{array}$$

Orchestra seats sell for \$4.25.

66. $2x+y=-4$

$x+y=-3$

Solution: $(-1,-2)$

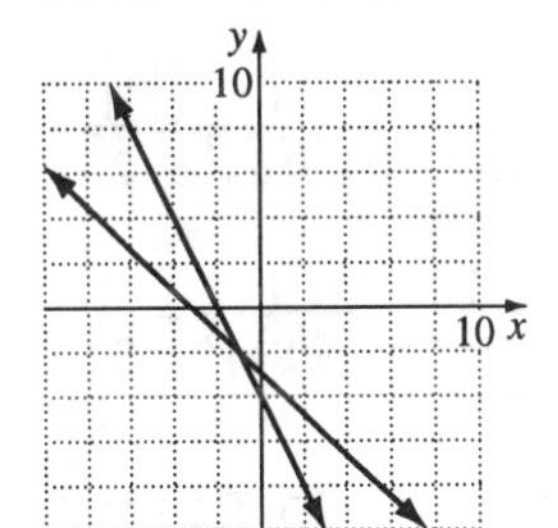

Problem Set 9.6

1. $49^{1/2}=\sqrt{49}=7$

3. $121^{1/2}=\sqrt{121}=11$

5. $100^{-1/2}=\frac{1}{100^{1/2}}=\frac{1}{\sqrt{100}}=\frac{1}{10}$

7. $16^{-1/2}=\frac{1}{16^{1/2}}=\frac{1}{\sqrt{16}}=\frac{1}{4}$

9. $27^{1/3}=\sqrt[3]{27}=3$

11. $125^{-1/3}=\frac{1}{125^{1/3}}=\frac{1}{\sqrt[3]{125}}=\frac{1}{5}$

13. $-125^{1/3}=-\sqrt[3]{125}=-5$

15. $16^{1/4}=\sqrt[4]{16}=2$

17. $\left(\frac{27}{64}\right)^{1/3}=\sqrt[3]{\frac{27}{64}}=\frac{3}{4}$

19. $32^{-1/5} = \frac{1}{32^{1/5}} = \frac{1}{\sqrt[5]{32}} = \frac{1}{2}$

21. $-32^{1/5} = -\sqrt[5]{32} = -2$

23. $81^{3/2} = \left(\sqrt{81}\right)^3 = 9^3 = 729$

25. $125^{2/3} = \left(\sqrt[3]{125}\right)^2 = 5^2 = 25$

27. $9^{3/2} = \sqrt{9^3} = \sqrt{(3^2)^3} = \sqrt{3^6} = 3^3 = 27$

29. $(-32)^{3/5} = \left(\sqrt[5]{-32}\right)^3 = (-2)^3 = -8$

31. $16^{-3/4} = \frac{1}{16^{3/4}} = \frac{1}{(\sqrt[4]{16})^3} = \frac{1}{2^3} = \frac{1}{8}$

33. $81^{-5/4} = \frac{1}{81^{5/4}} = \frac{1}{(\sqrt[4]{81})^5} = \frac{1}{3^5} = \frac{1}{243}$

35. $8^{-2/3} = \frac{1}{8^{2/3}} = \frac{1}{(\sqrt[3]{8})^2} = \frac{1}{2^2} = \frac{1}{4}$

37. $\left(\frac{4}{25}\right)^{-1/2} = \left(\frac{25}{4}\right)^{1/2} = \sqrt{\frac{25}{4}} = \frac{5}{2}$

39. $\left(\frac{8}{125}\right)^{-1/3} = \left(\frac{125}{8}\right)^{1/3} = \sqrt[3]{\frac{125}{8}} = \frac{5}{2}$

41. $(-8)^{-2/3} = \frac{1}{(-8)^{2/3}} = \frac{1}{\sqrt[3]{(-8)^2}} = \frac{1}{\sqrt[3]{64}} = \frac{1}{4}$

43. $27^{2/3} + 16^{3/4} = \left(\sqrt[3]{27}\right)^2 + \left(\sqrt[4]{16}\right)^3 = 3^2 + 2^3$
$= 9 + 8 = 17$

45. $25^{3/2} \cdot 81^{1/4} = \sqrt{25^3} \cdot \sqrt[4]{81} = 125 \cdot 3 = 375$

47. $v = \left(\frac{5r}{2}\right)^{1/2}$
$v = \left(\frac{5 \cdot 250}{2}\right)^{1/2}$
$v = (625)^{1/2}$
$v = \sqrt{625}$
$v = 25$
25 miles per hour

49. $f(t) = 1000t^{5/4} + 14,000$
$f(81) = 1000(81)^{5/4} + 14,000$
$f(81) = 1000(\sqrt[4]{81})^5 + 14,000$
$f(81) = 1000(3)^5 + 14,000$
$f(81) = 1000(243) + 14,000$
$f(81) = 243,000 + 14,000$
$f(81) = 257,000$
In 2051, 81 years after 1970, the average pollution will be 257,000 particles per cubic cm.

51. Statement a is true.
$2^{1/2} \cdot 2^{1/2} = 2^{1/2+1/2} = 2^1 = 2$
$4^{1/2} = 2$ True

53. a.

W	0	25	50	150	200	250	300
$T = W^{1.41}$	0	94	249	1170	1756	2405	3110

b.

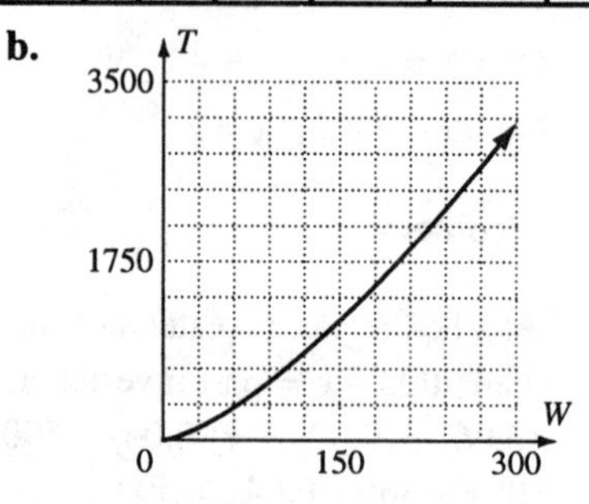

As weight increases, area increases.

c. Verify graph.

55. Answers may vary.

57. $25^{1/4} \cdot 25^{-3/4} = 25^{1/4-3/4} = 25^{-2/4}$
$= \frac{1}{25^{1/2}} = \frac{1}{5}$

Review Problems

59. $N = kt$
$6 = k(12)$
$k = \frac{1}{2}$
$N = \frac{1}{2}t$
$N = \frac{1}{2}(20)$
$N = 10$ inches

60. slope $= \frac{11-8}{7-6} = 3$
$y-8 = 3(x-6)$ or $y-11 = 3(x-7)$
$y-8 = 3x-18$
$y = 3x-10$

61. Let w = width and l = length
$l = 2w+3$
$wl = w(2w+3) = 44$
$2w^2 + 3w - 44 = 0$
$(2w+11)(w-4) = 0$
$w = 4 \left(\text{reject} = -\frac{11}{2}\right)$
$2w+3 = 11$
width, 4 meters; length, 11 meters

Chapter 9 Review

1. $\sqrt{64} = 8$
$-\sqrt{64} = -8$

2. $\sqrt{\frac{9}{25}} = \frac{3}{5}$
$-\sqrt{\frac{9}{25}} = -\frac{3}{5}$

3. $\sqrt{121} = 11$

4. $-\sqrt{121} = -11$

5. $\sqrt{-121}$ is not a real number

6. $\sqrt[3]{\frac{8}{125}} = \frac{2}{5}$

7. $\sqrt[5]{-32} = -2$

8. $-\sqrt[4]{81} = -3$

9. $\sqrt{\frac{8}{50}} = \frac{2\sqrt{2}}{5\sqrt{2}} = \frac{2}{5}$, rational

10. $\sqrt{1.21} = 1.1$, rational

11. $\sqrt{75} = 5\sqrt{3} \approx 8.660$, irrational

12. $\sqrt{-4}$ is not a real number

13. $\sqrt{300} = \sqrt{3} \cdot \sqrt{100} = 10\sqrt{3}$

14. $6\sqrt{20} = 6\sqrt{4} \cdot \sqrt{5} = 12\sqrt{5}$

15. $\sqrt{3} \cdot \sqrt{12} = \sqrt{36} = 6$

16. $\sqrt{24a} \cdot \sqrt{6b} = \sqrt{144ab} = 12\sqrt{ab}$

17. $\sqrt{48} \cdot \sqrt{32} = 4\sqrt{3}(4\sqrt{2}) = 16\sqrt{6}$

18. $\sqrt[3]{81} = \sqrt[3]{3 \cdot 27} = 3\sqrt[3]{3}$

19. $\sqrt[4]{8} \cdot \sqrt[4]{10} = \sqrt[4]{80} = \sqrt[4]{5 \cdot 16} = 2\sqrt[4]{5}$

20. $\sqrt{\frac{121}{4}} = \frac{\sqrt{121}}{\sqrt{4}} = \frac{11}{2}$

21. $\sqrt{\frac{7y}{25}} = \frac{\sqrt{7y}}{5}$

22. $\frac{6\sqrt{200}}{3\sqrt{2}} = \frac{2\sqrt{100 \cdot 2}}{\sqrt{2}} = \frac{20\sqrt{2}}{\sqrt{2}} = 20$

23. $\sqrt{\frac{5}{2}} \cdot \sqrt{\frac{3}{8}} = \sqrt{\frac{15}{16}} = \frac{\sqrt{15}}{4}$

24. $\sqrt[3]{\frac{7}{64}} = \frac{\sqrt[3]{7}}{4}$

25. $\sqrt{63x^2} = x\sqrt{63} = 3x\sqrt{7}$

26. $\sqrt{48y^3} = 4y\sqrt{3y}$

27. $\sqrt{10x^3}\sqrt{8x^2} = \sqrt{80x^5} = 4x^2\sqrt{5x}$

28. $\sqrt{\frac{7}{y^4}} = \frac{\sqrt{7}}{y^2}$

29. $\sqrt{75x^9} = \sqrt{25x^8} \cdot \sqrt{3x} = 5x^4\sqrt{3x}$

30. $\sqrt{300x^{23}} = \sqrt{100x^{22}} \cdot \sqrt{3x} = 10x^{11}\sqrt{3x}$

31. $7\sqrt{5} + 13\sqrt{5} = 20\sqrt{5}$

32. $\sqrt{50b} + \sqrt{8b} = 5\sqrt{2b} + 2\sqrt{2b} = 7\sqrt{2b}$

33. $\frac{5}{6}\sqrt{72} - \frac{3}{4}\sqrt{48} = 5\sqrt{2} - 3\sqrt{3}$

34. $2\sqrt{18} + 3\sqrt{27} - \sqrt{12} = 6\sqrt{2} + 9\sqrt{3} - 2\sqrt{3}$
$= 6\sqrt{2} + 7\sqrt{3}$

35. $\sqrt[4]{7} + 3\sqrt[3]{5} - 2\sqrt[4]{7} - \sqrt[3]{5} = -\sqrt[4]{7} + 2\sqrt[3]{5}$

36. $4\sqrt[3]{16a} + 5\sqrt[3]{2a} = 8\sqrt[3]{2a} + 5\sqrt[3]{2a} = 13\sqrt[3]{2a}$

37. $\sqrt{10}(\sqrt{5} + \sqrt{6}) = \sqrt{50} + \sqrt{60} = 5\sqrt{2} + 2\sqrt{15}$

38. $\sqrt{3a}(7\sqrt{2} + 4\sqrt{3}) = 7\sqrt{6a} + 4\sqrt{9a}$
$= 7\sqrt{6a} + 12\sqrt{a}$

39. $7\sqrt{10}(6\sqrt{2} - 3\sqrt{5}) = 42\sqrt{20} - 21\sqrt{50}$
$= 84\sqrt{5} - 105\sqrt{2}$

40. $(\sqrt{2} + \sqrt{7})(\sqrt{2} + 4\sqrt{7})$
$= 2 + 4\sqrt{14} + \sqrt{14} + 28$
$= 30 + 5\sqrt{14}$

41. $(3\sqrt{6} - 2\sqrt{5})(4\sqrt{6} + \sqrt{10})$
$= 72 + 3\sqrt{60} - 8\sqrt{30} - 2\sqrt{50}$
$= 72 + 6\sqrt{15} - 8\sqrt{30} - 10\sqrt{2}$

42. $(5\sqrt{x} - 3)^2 = 25x - 30\sqrt{x} + 9$

43. $(\sqrt{11} - \sqrt{7})(\sqrt{11} + \sqrt{7}) = 11 - 7 = 4$

44. $(2\sqrt{3} + 7\sqrt{2})(2\sqrt{3} - 7\sqrt{2}) = 12 - 98 = -86$

45. $\frac{30}{\sqrt{5}} \cdot \frac{\sqrt{5}}{\sqrt{5}} = \frac{30\sqrt{5}}{5} = 6\sqrt{5}$

46. $\frac{13}{\sqrt{50}} \cdot \frac{\sqrt{50}}{\sqrt{50}} = \frac{13\sqrt{50}}{50} = \frac{13 \cdot 5\sqrt{2}}{50} = \frac{13\sqrt{2}}{10}$

47. $\frac{7\sqrt{2}}{\sqrt{6}} = \frac{7}{\sqrt{3}} \cdot \frac{\sqrt{3}}{\sqrt{3}} = \frac{7\sqrt{3}}{3}$

48. $\sqrt{\frac{2}{3}} = \frac{\sqrt{2}}{\sqrt{3}} \cdot \frac{\sqrt{3}}{\sqrt{3}} = \frac{\sqrt{6}}{3}$

49. $\sqrt{\frac{17}{x}} = \frac{\sqrt{17}}{\sqrt{x}} \cdot \frac{\sqrt{x}}{\sqrt{x}} = \frac{\sqrt{17x}}{x}$

50. $\sqrt{\frac{5x^2}{8}} = \frac{\sqrt{5x^2}}{\sqrt{8}} = \frac{x\sqrt{5}}{2\sqrt{2}} \cdot \frac{\sqrt{2}}{\sqrt{2}} = \frac{x\sqrt{10}}{4}$

51. $\frac{11}{\sqrt{5}+2} = \frac{11}{\sqrt{5}+2} \cdot \frac{\sqrt{5}-2}{\sqrt{5}-2} = \frac{-22+11\sqrt{5}}{5-4}$
$= -22 + 11\sqrt{5} = 11(-2 + \sqrt{5})$

52. $\frac{21}{4-\sqrt{3}} = \frac{21}{4-\sqrt{3}} \cdot \frac{4+\sqrt{3}}{4+\sqrt{3}} = \frac{21(4+\sqrt{3})}{16-3}$
$= \frac{21(4+\sqrt{3})}{13}$

53. $\frac{12}{\sqrt{5}+\sqrt{3}} = \frac{12}{\sqrt{5}+\sqrt{3}} \cdot \frac{\sqrt{5}-\sqrt{3}}{\sqrt{5}-\sqrt{3}}$
$= \frac{12(\sqrt{5}-\sqrt{3})}{5-3} = 6(\sqrt{5} - \sqrt{3})$

54. $\frac{\sqrt{3}+2}{\sqrt{6}-\sqrt{3}} = \frac{\sqrt{3}+2}{\sqrt{6}-\sqrt{3}} \cdot \frac{\sqrt{6}+\sqrt{3}}{\sqrt{6}+\sqrt{3}}$

$= \frac{\sqrt{18}+3+2\sqrt{6}+2\sqrt{3}}{6-3}$

$= \frac{3+2\sqrt{3}+2\sqrt{6}+3\sqrt{2}}{3}$

55. $\sqrt{2y+3} = 5$

$2y+3 = 25$

$2y = 22$

$y = 11$

11

56. $3\sqrt{x} = \sqrt{6x+15}$

$9x = 6x + 15$

$3x = 15$

$x = 5$

5

57. $3\sqrt{z+3} = \sqrt{2z+13}$

$9(z+3) = 2z + 13$

$9z + 27 = 2z + 13$

$7z = -14$

$z = -2$

-2

58. $\sqrt{5x+1} = x+1$

$5x+1 = x^2 + 2x + 1$

$x^2 - 3x = 0$

$x = 0$ or $x = 3$

Check: $x = 0$: $1 = 1$

$x = 3$: $\sqrt{16} = 4$

The solutions are 0, 3.

59. $\sqrt{y+1} + 5 = y$

$\sqrt{y+1} = y - 5$

$y+1 = y^2 - 10y + 25$

$y^2 - 11y + 24 = 0$

$(y-3)(y-8) = 0$

$y = 3$ or $y = 8$

Check: $y = 3$: $\sqrt{4} + 5 = 7 \neq 3$

$y = 8$: $\sqrt{9} + 5 = 8 = 8$

The solution is 8.

60. $y = \sqrt{y^2 + 4y + 4}$

$y^2 = y^2 + 4y + 4$

$-4y = 4$

$y = -1$

Check: $-1 = \sqrt{1-4+4}$

$-1 = 1$ False

no solution, $\varnothing$

61. $\sqrt{x-2} + 5 = 1$

$\sqrt{x-2} = -4$

solution is $\varnothing$ since $\sqrt{x-2} \geq 0$

62. $16^{1/2} = \sqrt{16} = 4$

63. $25^{-1/2} = \frac{1}{\sqrt{25}} = \frac{1}{5}$

64. $125^{1/3} = \sqrt[3]{125} = 5$

65. $27^{-1/3} = \frac{1}{27^{1/3}} = \frac{1}{3}$

66. $64^{2/3} = (\sqrt[3]{64})^2 = 4^2 = 16$

67. $27^{-4/3} = \frac{1}{27^{4/3}} = \frac{1}{(\sqrt[3]{27})^4} = \frac{1}{3^4} = \frac{1}{81}$

68. $t = \sqrt{\frac{2s}{g}}$

$t = \sqrt{\frac{2(128)}{32}} - \sqrt{\frac{2(32)}{32}}$

$t = \sqrt{8} - \sqrt{2}$

$t = 2\sqrt{2} - \sqrt{2}$

$t = \sqrt{2} \approx 1.4$ seconds

69. $T = 2\pi\sqrt{\frac{L}{32}}$

$T \approx 2(3.14)\sqrt{\frac{8}{32}}$

$T \approx 3.14$ seconds

70. $r = \sqrt{\frac{A}{P}} - 1$

$r = \sqrt{\frac{144}{100}} - 1 = \frac{12}{10} - 1 = 0.20$ or 20%

71. perimeter $= 2(l + w) = 2(4\sqrt{20} + 2\sqrt{8})$

$8(2\sqrt{5} + \sqrt{2})$ meters

area $= lw = (4\sqrt{20})(2\sqrt{8}) = 8\sqrt{160}$

$= 32\sqrt{10}$ square meters

72. $S = 28.6A^{1/3}$

$S = 28.6(8)^{1/3}$

$S = 28.6(2)$

$S = 57.2$

57 species

73. $v = 2\sqrt{6L}$

$50 = 2\sqrt{6L}$

$25 = \sqrt{6L}$

$625 = 6L$

$104.2 \approx L$

104.2 feet

74. $t = \sqrt{\frac{2s}{g}}$

$3 = \sqrt{\frac{2s}{32}}$

$9 = \frac{2s}{32}$

$288 = 2s$

$144 = s$

144 feet

75. $r = \left(\frac{P_n}{P_0}\right)^{1/n} - 1$

$r = \left(\frac{246.7}{226.5}\right)^{1/10} - 1$

$r \approx 0.0086$

Approximately 0.86% increase per year

Chapter 9 Test

1. 7, −7, since $(7)^2 = 49$ and $(-7)^2 = 49$.

2. −8, since $(8)^2 = 64$.

3. 4, since $4^3 = 64$.

4. $\sqrt{48} = \sqrt{16 \cdot 3} = \sqrt{16}\sqrt{3} = 4\sqrt{3}$

5. $\sqrt{72x^3} = \sqrt{36 \cdot 2 \cdot x^2 \cdot x} = \sqrt{36}\sqrt{x^2}\sqrt{2x}$

$= 6x\sqrt{2x}$

6. $\sqrt{x^{29}} = \sqrt{x^{28} \cdot x} = x^{14}\sqrt{x}$

7. $\sqrt{\frac{25}{x^2}} = \frac{\sqrt{25}}{\sqrt{x^2}} = \frac{5}{x}$

8. $\sqrt[3]{\frac{5}{8}} = \frac{\sqrt[3]{5}}{\sqrt[3]{8}} = \frac{\sqrt[3]{5}}{2}$

9. $\sqrt{\frac{75}{27}} = \frac{\sqrt{75}}{\sqrt{27}} = \frac{\sqrt{25}\sqrt{3}}{\sqrt{9}\sqrt{3}} = \frac{5\sqrt{3}}{3\sqrt{3}} = \frac{5}{3}$

10. $\sqrt{\frac{64x^4}{2x^2}} = \sqrt{32x^2} = \sqrt{16}\sqrt{x^2}\sqrt{2} = 4x\sqrt{2}$

11. $\sqrt{10}\sqrt{5} = \sqrt{50} = \sqrt{25}\sqrt{2} = 5\sqrt{2}$

12. $\sqrt{6x}\sqrt{6y} = \sqrt{36xy} = 6\sqrt{xy}$

13. $\sqrt{10x^2}\sqrt{2x^3} = \sqrt{20x^5}$

$= \sqrt{4 \cdot 5 \cdot x^4 \cdot x} = 2x^2\sqrt{5x}$

14. $\sqrt{24} + 3\sqrt{54} = \sqrt{4 \cdot 6} + 3\sqrt{9 \cdot 6}$

$= 2\sqrt{6} + 9\sqrt{6} = 11\sqrt{6}$

15. $7\sqrt{8} - 2\sqrt{32} = 7\sqrt{4 \cdot 2} - 2\sqrt{16 \cdot 2}$

$= 14\sqrt{2} - 8\sqrt{2} = 6\sqrt{2}$

16. $\left(2\sqrt{2}+5\right)\left(3\sqrt{2}+4\right)$
$= 12+15\sqrt{2}+8\sqrt{2}+20$
$= 32+23\sqrt{2}$

17. $\left(\sqrt{6}+2\right)\left(\sqrt{6}-2\right) = 6-4 = 2$

18. $\left(3-\sqrt{7}\right)^2 = \left(3-\sqrt{7}\right)\left(3-\sqrt{7}\right)$
$= 9-6\sqrt{7}+7 = 16-6\sqrt{7}$

19. $\left(3\sqrt{x}+2\right)^2 = \left(3\sqrt{x}+2\right)\left(3\sqrt{x}+2\right)$
$= 9x+12\sqrt{x}+4$

20. $\frac{4}{\sqrt{5}} \cdot \frac{\sqrt{5}}{\sqrt{5}} = \frac{4\sqrt{5}}{5}$

21. $\frac{5}{4+\sqrt{3}} \cdot \frac{4-\sqrt{3}}{4-\sqrt{3}} = \frac{5\left(4-\sqrt{3}\right)}{16-3} = \frac{5\left(4-\sqrt{3}\right)}{13}$

22. $5\sqrt{3x-2}-3 = 7$
$5\sqrt{3x-2} = 10$
$\left(5\sqrt{3x-2}\right)^2 = 10^2$
$25(3x-2) = 100$
$75x = 150$
$x = 2$
Check:
$5\sqrt{3\cdot 2-2}-3 = 7$
$5\cdot 2-3 = 7$
$7 = 7$ True
The solution is 2.

23. $\sqrt{2x-1} = x-2$
$\left(\sqrt{2x-1}\right)^2 = (x-2)^2$
$2x-1 = x^2-4x+4$
$0 = x^2-6x+5$
$0 = (x-5)(x-1)$
$x = 5$ or $x = 1$
Check:
$\sqrt{2\cdot 5-1} = 5-2$
$3 = 3$ True
$\sqrt{2\cdot 1-1} = 1-2$
$1 = -1$ False
The solution is 5.

24. $9^{-1/2} = \frac{1}{\sqrt{9}} = \frac{1}{3}$

25. $8^{2/3} = \left(\sqrt[3]{8}\right)^2 = (2)^2 = 4$

26. $2\sqrt{80}+2\sqrt{45}$
$= 2\cdot 4\sqrt{5}+2\cdot 3\sqrt{5}$
$= 8\sqrt{5}+6\sqrt{5}$
$= 14\sqrt{5}$

27. Solve for d:
$3 = \sqrt{\frac{d}{16}}$
$3 = \frac{\sqrt{d}}{4}$
$12 = \sqrt{d}$
$144 = d$
Check:
$3 = \sqrt{\frac{144}{16}}$
$3 = \sqrt{9}$
$3 = 3$ True
The skydiver will fall about 144 feet in 3 seconds.

Cumulative Review Problems (Chapters 1-9)

1. $f(x) = -7.7x^3+52.7x^2-93.4x+2151$
$f(0) = -7.7(0)^3+52.7(0)^2-93.4(0) +2151$
$f(0) = 2151$
In 1985, there were 2151 thousand or 2,151,000 military personnel.

2. Possible answer: Draw a horizontal line through the point on the graph above $x = 1.5$ and $x = 4.4$. Read the y-value on the vertical axis.

$\sqrt{1.5} \approx 1.22$
$\sqrt{4.4} \approx 2.10$

3. Let w = the width of the rectangle.
$2w + 2$ = the length of the rectangle.
$2(w + 2w + 2) = 40$
$2(3w + 2) = 40$
$3w + 2 = 20$
$3w = 18$
$w = 6$
$2w + 2 = 2 \cdot 6 + 2 = 14$
dimensions: width, 6 meters; length, 14 meters

4. $8(5z - 7) - 4z = 9(4z - 6) - 3$
$40z - 56 - 4z = 36z - 54 - 3$
$36z - 56 = 36z - 57$
$-56 = -57$ false
no solution
solution: $\varnothing$

5. **a.** $2x + 3y = 7$
$3y = -2x + 7$
$y = -\frac{2}{3}x + \frac{7}{3}$

b. $y = -\frac{2}{3}x + \frac{7}{3}$
$y = -\frac{2}{3}(-4) + \frac{7}{3}$
$y = \frac{8}{3} + \frac{7}{3}$
$y = \frac{15}{3}$
$y = 5$

6. Let x = the first page.
Then $x + 1$ = the second page.
$x + x + 1 = 933$
$2x = 932$
$x = 466$
The pages are numbered 466 and 467.

7. $(6x^3 - 19x^2 + 16x - 4) \div (x - 2)$
$= 6x^2 - 7x + 2$

$$\begin{array}{r} 6x^2 - 7x + 2 \\ x - 2 \overline{) 6x^3 - 19x^2 + 16x - 4} \\ \underline{6x^3 - 12x^2} \qquad\qquad\quad \\ -7x^2 + 16x \qquad \\ \underline{-7x^2 + 14x} \qquad \\ 2x - 4 \\ \underline{2x - 4} \\ 0 \end{array}$$

8. $(2x - 3)(4x^2 + 6x + 9)$
$= 2x(4x^2 + 6x + 9) - 3(4x^2 + 6x + 9)$
$= 8x^3 + 12x^2 + 18x - 12x^2 - 18x - 27$
$= 8x^3 - 27$

9. $2x^2 + 5x = 12$
$2x^2 + 5x - 12 = 0$
$(2x - 3)(x + 4) = 0$
$2x - 3 = 0$ or $x + 4 = 0$
$x = \frac{3}{2}$ or $x = -4$
$\frac{3}{2}, -4$

10. $x^2 - 18x + 77 = (x - 7)(x - 11)$

11. $1 - \frac{3x}{2} \le x - 4$
$2\left(1 - \frac{3x}{2}\right) \le 2(x - 4)$
$2 - 3x \le 2x - 8$
$-5x \le -10$
$x \ge 2$
$\{x | x \ge 2\}$

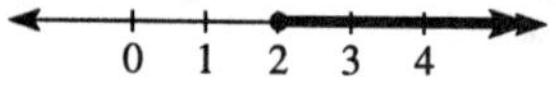

12. Graph $5x+3y\le -15$

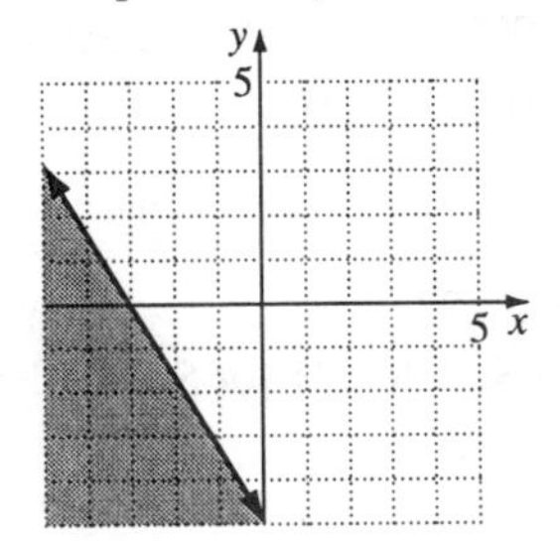

13. $8x-5y=-4$ $(\times 3)$
$2x+15y=-66$

$$\begin{aligned} 24x-15y&=-12\\ 2x+15y&=-66\\ \hline 26x&=-78\\ x&=-3\\ 2(-3)+15y&=-66\\ 15y&=-60\\ y&=-4 \end{aligned}$$

$(-3, -4)$

14. $\frac{318}{x}=\frac{56}{168}$
$56x=(318)(168)$
$x=\frac{53,424}{56}$
$x=954$ deer

15. $y=x+1$
$y=2x-1$
solution: $(2, 3)$

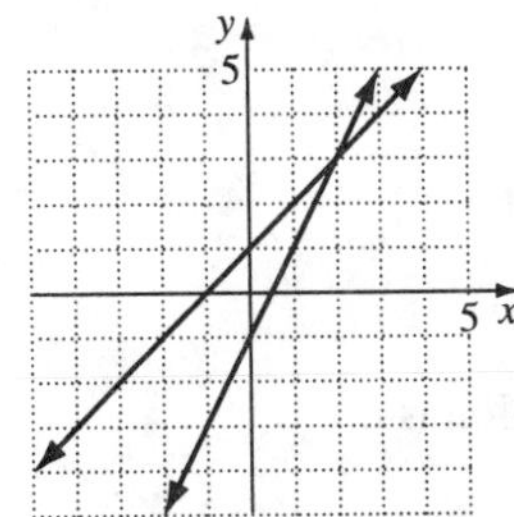

16. a. $d=45t$
$d=45(1)=45$ miles
$d=45(2)=90$ miles
$d=45(3)=135$ miles
$d=45(4)=180$ miles

b. $d=45t$

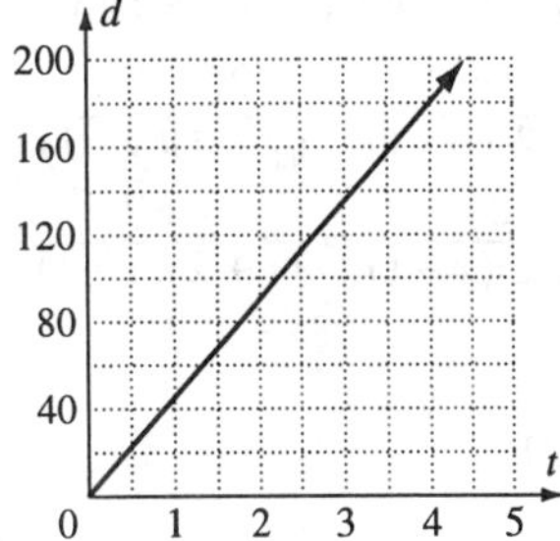

17. Points: (1995, 156.5) and (2005, 406.9)
slope $=\frac{406.9-156.5}{2005-1995}=\frac{250.4}{10}=25.04$
$25.04 billion increase each year

18. $\frac{8x^3}{-4x^7}=\frac{-2}{x^{7-3}}=-\frac{2}{x^4}$

19. $(4x+5)^2=16x^2+40x+25$

20. $x^3-25x=x(x^2-25)=x(x+5)(x-5)$

21. Let x = the smallest integer.
$x+1$ = the next consecutive integer.
$x^2+(x+1)^2=9+8x$
$x^2+x^2+2x+1=9+8x$
$2x^2-6x-8=0$
$2(x^2-3x-4)=0$
$2(x-4)(x+1)=0$
$x=4$ or $x=-1$
$x+1=5$
$x+1=0$
The integers are 4 and 5 or -1 and 0.

22. $D=\frac{n(n-3)}{2}$ $\quad D=5$
$5=\frac{n(n-3)}{2}$
$10=n^2-3n$
$n^2-3n-10=0$
$(n-5)(n+2)=0$
$n=5$ (reject $n=-2$)
The polygon with 5 sides has 5 diagonals.

23. $\frac{3y}{y^2+y-2}-\frac{2}{y+2}=\frac{3y}{(y+2)(y-1)}-\frac{2}{y+2}$
$=\frac{3y}{(y+2)(y-1)}-\frac{2(y-1)}{(y+2)(y-1)}$
$=\frac{3y-2y+2}{(y+2)(y-1)}=\frac{y+2}{(y+2)(y-1)}$
$=\frac{1}{y-1}$
$y\neq -2, 1$

24. $\frac{5x^2-6x+1}{x^2-1}\div\frac{16x^2-9}{4x^2+7x+3}$
$=\frac{(5x-1)(x-1)}{(x-1)(x+1)}\div\frac{(4x-3)(4x+3)}{(4x+3)(x+1)}$
$=\frac{5x-1}{x+1}\cdot\frac{x+1}{4x-3}=\frac{5x-1}{4x-3}$
$x\neq 1, -1, -\frac{3}{4}, \frac{3}{4}$

25. $\frac{15}{x}-4=\frac{6}{x}+3$
$x\left(\frac{15}{x}-4\right)=x\left(\frac{6}{x}+3\right)$
$15-4x=6+3x$
$-7x=-9$
$x=\frac{9}{7}$
$\frac{9}{7}$

26. $\frac{1-\frac{1}{x^2}}{1-\frac{1}{x}}$
$=\frac{\frac{x^2}{x^2}-\frac{1}{x^2}}{\frac{x}{x}-\frac{1}{x}}$
$=\frac{\frac{x^2-1}{x^2}}{\frac{x-1}{x}}$
$=\frac{(x-1)(x+1)}{x^2}\cdot\frac{x}{x-1}$
$=\frac{x+1}{x}$

27. Let x = GDP for U.S.
Then $2x - 10{,}007$ = GDP for Switzerland.
$x+2x-10{,}007=59{,}350$
$3x=69{,}357$
$x=23{,}119$
$2x-10{,}007=36{,}231$
\$23,119 for U.S., \$36,231 for Switzerland

28. $6\sqrt{75}-4\sqrt{12}$
$=6\sqrt{25\cdot 3}-4\sqrt{4\cdot 3}=6\cdot 5\sqrt{3}-4\cdot 2\sqrt{3}$
$=30\sqrt{3}-8\sqrt{3}=22\sqrt{3}$

29. $\frac{5}{6+\sqrt{11}}=\frac{5}{6+\sqrt{11}}\cdot\frac{6-\sqrt{11}}{6-\sqrt{11}}$
$=\frac{5\left(6-\sqrt{11}\right)}{36-11}=\frac{5\left(6-\sqrt{11}\right)}{25}=\frac{6-\sqrt{11}}{5}$

30. $x=\sqrt{x-2}+4$
$x-4=\sqrt{x-2}$
$(x-4)^2=\left(\sqrt{x-2}\right)^2$
$x^2-8x+16=x-2$
$x^2-9x+18=0$
$(x-6)(x-3)=0$
$x=6$ or 3
Check: $x=6$:
$6=\sqrt{6-2}+4$
$6=\sqrt{4}+4$
$6=2+4$
$6=6$ True
$x=3$:
$3=\sqrt{3-2}+4$
$3=1+4$
$3=5$ False
The only solution is 6.

Chapter 10

Problem Set 10.1

1. $x^2 = 36$ Original equation
$x = \pm\sqrt{36}$ Apply square root property
$x = 6$ or $x = -6$ Solutions
$-6, 6$

3. $y^2 = 81$ Original equation
$y = \pm\sqrt{81}$ Apply square root property
$y = 9$ or $y = -9$ Solutions
$-9, 9$

5. $x^2 = 7$
$x = \pm\sqrt{7}$
$x = \sqrt{7}$ or $x = -\sqrt{7}$
$-\sqrt{7},\ \sqrt{7}$

7. $x^2 = 50$
$x = \pm\sqrt{50} = \pm 5\sqrt{2}$
$-5\sqrt{2},\ 5\sqrt{2}$

9. $5y^2 = 20$
$y^2 = 4$ Divide both sides by 5.
$y = \pm 2$
$-2, 2$

11. $4y^2 = 49$
$y^2 = \frac{49}{4}$
$y = \pm\frac{7}{2}$
$-\frac{7}{2},\ \frac{7}{2}$

13. $y^2 - 2y = 2(3 - y)$
$y^2 - 2y = 6 - 2y$ Distributive property
$y^2 = 6$ Add $2y$ to both sides.
$y = \pm\sqrt{6}$
$-\sqrt{6},\ \sqrt{6}$

15. $2z^2 + 2z - 5 = z(z + 2) - 3$
$2z^2 + 2z - 5 = z^2 + 2z - 3$ Distributive property
$z^2 = 2$ Simplify.
$z = \pm\sqrt{2}$
$-\sqrt{2},\ \sqrt{2}$

17. $11t^2 - 23 = 4t^2 + 33$
$7t^2 = 56$
$t^2 = 8$
$t = \pm 2\sqrt{2}$
$-2\sqrt{2},\ 2\sqrt{2}$

19. $3y^2 - 2 = 0$
$3y^2 = 2$
$y^2 = \frac{2}{3}$
$y = \pm\sqrt{\frac{2}{3}} = \pm\frac{\sqrt{2}}{\sqrt{3}}\cdot\frac{\sqrt{3}}{\sqrt{3}} = \pm\frac{\sqrt{6}}{3}$
$-\frac{\sqrt{6}}{3},\ \frac{\sqrt{6}}{3}$

21. $5m^2 - 7 = 0$
$5m^2 = 7$
$m^2 = \frac{7}{5}$
$m = \pm\sqrt{\frac{7}{5}} = \pm\frac{\sqrt{7}}{\sqrt{5}}\cdot\frac{\sqrt{5}}{\sqrt{5}}$
$= \pm\frac{\sqrt{35}}{5}$
$-\frac{\sqrt{35}}{5},\ \frac{\sqrt{35}}{5}$

23. $(y - 3)^2 = 16$
$y - 3 = 4$ or $y - 3 = -4$ Square root property
$y = 7$ or $y = -1$
$-1, 7$

25. $(x+5)^2 = 121$
$x+5 = -\sqrt{121}$ or $x+5 = \sqrt{121}$
$x+5 = -11$ or $x+5 = 11$
$x = -16$ or $x = 6$
$-16, 6$

27. $(2x+1)^2 = 64$
$2x+1 = 8$ or $2x+1 = -8$
$2x = 7$ or $2x = -9$
$x = \frac{7}{2}$ or $x = -\frac{9}{2}$
$-\frac{9}{2}, \frac{7}{2}$

29. $(b+3)^2 = 5$
$b+3 = \sqrt{5}$ or $b+3 = -\sqrt{5}$
$b = \sqrt{5} - 3$ or $b = -\sqrt{5} - 3$
$\sqrt{5} - 3, -\sqrt{5} - 3$

31. $(y-2)^2 = 32$
$y-2 = 4\sqrt{2}$ or $y-2 = -4\sqrt{2}$
$y = 2+4\sqrt{2}$ or $y = 2-4\sqrt{2}$
$2-4\sqrt{2}, 2+4\sqrt{2}$

33. $(3x-1)^2 = 12$
$3x-1 = \pm 2\sqrt{3}$
$3x = 1+2\sqrt{3}$ or $3x = 1-2\sqrt{3}$
$x = \frac{1}{3} + \frac{2\sqrt{3}}{3}$ or $x = \frac{1}{3} - \frac{2\sqrt{3}}{3}$
$\frac{1}{3} - \frac{2\sqrt{3}}{3}, \frac{1}{3} + \frac{2\sqrt{3}}{3}$

35. $(6w+2)^2 = 27$
$6w+2 = \pm 3\sqrt{3}$
$6w = -2+3\sqrt{3}$ or $6w = -2-3\sqrt{3}$
$w = -\frac{1}{3} + \frac{\sqrt{3}}{2}$ or $w = -\frac{1}{3} - \frac{\sqrt{3}}{2}$
$-\frac{1}{3} - \frac{\sqrt{3}}{2}, -\frac{1}{3} + \frac{\sqrt{3}}{2}$

37. $W = 3t^2$, $W = 108$ grams
$108 = 3t^2$
$36 = t^2$
$t = 6$ (reject $t = -6$)
6 weeks

39. $d = \frac{3}{50}v^2$, $d = 150$ feet
$150 = \frac{3}{50}v^2$
$2500 = v^2$ Simplify.
$v = 50$ miles per hour

41. $A = P(1+r)^2$, $A = 121$, $P = 100$
$121 = 100(1+r)^2$
$11 = 10(1+r)$ Square root property
$1.1 = 1+r$ Divide by 10.
$r = 0.10$ or 10%

43. $A = \pi r^2$
$\pi r^2 = 49\pi$
$r^2 = 49$
$r = 7$ meters
$C = 2\pi r = 2\pi(7) = 14\pi$
radius, 7 meters; circumference, 14π meters

45. $v = 1.2 - 2000r^2$, $v = 0.7$ cm per second
$0.7 = 1.2 - 2000r^2$
$2000r^2 = 0.5$
$r^2 = 0.00025$
$r = 0.01\sqrt{2.5} \approx 0.0158$ cm
$= 0.02$ cm, rounded

47. $d = \sqrt{60^2 + 60^2} = \sqrt{60^2(2)} = 60\sqrt{2}$ feet

49. length of diagonal
$= \sqrt{30^2 + 40^2} = 50$ meters
length + width = 40 + 30 = 70 meters
distance saved = 70 – 50 = 20 meters

51. $AC = \sqrt{(5-1)^2 + (5-2)^2}$
$= \sqrt{4^2 + 3^2} = \sqrt{16+9} = \sqrt{25} = 5$

53. $N = \left(\frac{x-4}{4}\right)^2 y$

$162 = \left(\frac{x-4}{4}\right)^2 (18)$

$9 = \left(\frac{x-4}{4}\right)^2$

$\frac{x-4}{4} = -3$ or $\frac{x-4}{4} = 3$

$x - 4 = -12$ or $x - 4 = 12$

$x = -8$ or $x = 16$

Diameters are measurements of positive numbers, so diameter = 16 feet

55. Let x = length of side of larger square.

$x - 2$ = length of side of smaller square

$(x-2)^2 = 16$

$x - 2 = 4$

$x = 6$

length of side of larger square, 6 meters

57. Let x = length of side of original square.

$x + 6$ = length of side of enlarged square

$(x+6)^2 = 169$

$x + 6 = 13$

$x = 7$ feet

59. Statement b is true.

$(x+7)^2 = 0$

$x + 7 = 0$

$x = -7$

61. $(x-1)^2 - 9 = 0$

$-2, 4$

63. Students should verify answers.

65. $ax^2 - b = 0$ ($a > 0$ and $b > 0$)

$x^2 = \frac{b}{a}$

$x = \pm\sqrt{\frac{b}{a}} = \pm\frac{\sqrt{ab}}{a}$

$x = -\frac{\sqrt{ab}}{a}$ or $x = \frac{\sqrt{ab}}{a}$

$-\frac{\sqrt{ab}}{a}, \frac{\sqrt{ab}}{a}$

67. $A = p(1+r)^2$

$\frac{A}{p} = (1+r)^2$

$\sqrt{\frac{A}{p}} = 1 + r$

$r = \sqrt{\frac{A}{p}} - 1$

Review Problems

68. $6x^2 + 26x + 24$

$= 2(3x^2 + 13x + 12)$

$= 2(3x + 4)(x + 3)$

69. $\frac{x}{x^2 + 11x + 30} - \frac{5}{x^2 + 9x + 20}$

$= \frac{x}{(x+6)(x+5)} - \frac{5}{(x+5)(x+4)}$

$= \frac{x(x+4) - 5(x+6)}{(x+6)(x+5)(x+4)}$

$= \frac{x^2 + 4x - 5x - 30}{(x+6)(x+5)(x+2)}$

$= \frac{x^2 - x - 30}{(x+6)(x+5)(x+2)}$

$= \frac{(x-6)(x+5)}{(x+6)(x+5)(x+2)}$

$= \frac{x-6}{(x+6)(x+2)}$

$(x \neq -6, -5, -2)$

70. $-\sqrt{9} = -3$

$\sqrt{-9}$ is not a real number because the square of two real numbers is always positive.

Problem Set 10.2

1. x^2+12x

Add the square of one-half the coefficient of x:

$x^2+12x+\left(\frac{12}{2}\right)^2$

Simplify:

$x^2+12x+36=(x+6)^2$

3. x^2-10x

$x^2-10x+\left(-\frac{10}{2}\right)^2$

$x^2-10x+25=(x-5)^2$

5. y^2+3y

$y^2+3y+\left(\frac{3}{2}\right)^2$

$y^2+3y+\frac{9}{4}=\left(y+\frac{3}{2}\right)^2$

7. y^2-7y

$y^2-7y+\left(-\frac{7}{2}\right)^2$

$y^2-7y+\frac{49}{4}=\left(y-\frac{7}{2}\right)^2$

9. $x^2-\frac{2}{3}x$

$x^2-\frac{2}{3}x+\left(-\frac{1}{3}\right)^2$

$x^2-\frac{2}{3}x+\frac{1}{9}=\left(x-\frac{1}{3}\right)^2$

11. $y^2-\frac{1}{3}y$

$y^2-\frac{1}{3}y+\left(-\frac{1}{6}\right)^2$

$y^2-\frac{1}{3}y+\frac{1}{36}=\left(y-\frac{1}{6}\right)^2$

13. $x^2+6x=7$ Original equation

$x^2+6x+9=7+9$ Complete the square.

$(x+3)^2=16$

$x+3=\pm 4$ Square root property

$x=-7$ or $x=1$

$-7, 1$

15. $x^2-2x=2$

$x^2-2x+1=2+1$

$(x-1)^2=3$

$x-1=\pm\sqrt{3}$

$x=1+\sqrt{3}$ or $x=1-\sqrt{3}$

$1+\sqrt{3},\ 1-\sqrt{3}$

17. $y^2-6y-11=0$

$y^2-6y+9=11+9$

$(y-3)^2=20$

$y-3=\pm 2\sqrt{5}$

$x=3+2\sqrt{5}$ or $y=3-2\sqrt{5}$

$3+2\sqrt{5},\ 3-2\sqrt{5}$

19. $r^2+4r+1=0$

$r^2+4r+4=-1+4$

$(r+2)^2=3$

$r+2=\pm\sqrt{3}$

$r=-2+\sqrt{3}$ or $r=-2-\sqrt{3}$

$-2+\sqrt{3},\ -2-\sqrt{3}$

21. $x^2+3x-1=0$

$x^2+3x+\frac{9}{4}=1+\frac{9}{4}$

$\left(x+\frac{3}{2}\right)^2=\frac{13}{4}$

$x+\frac{3}{2}=\pm\frac{\sqrt{13}}{2}$

$x=-\frac{3}{2}-\frac{\sqrt{13}}{2}$ or $x=-\frac{3}{2}+\frac{\sqrt{13}}{2}$

$-\frac{3}{2}-\frac{\sqrt{13}}{2},\ -\frac{3}{2}+\frac{\sqrt{13}}{2}$

23. $y^2 = 7y - 3$

$y^2 - 7y + \left(\frac{7}{2}\right)^2 = -3 + \left(\frac{7}{2}\right)^2$

$\left(y - \frac{7}{2}\right)^2 = -3 + \frac{49}{4} = \frac{37}{4}$

$y - \frac{7}{2} = \pm\frac{\sqrt{37}}{2}$

$x = \frac{7}{2} + \frac{\sqrt{37}}{2}$ or $y = \frac{7}{2} - \frac{\sqrt{37}}{2}$

$\frac{7}{2} - \frac{\sqrt{37}}{2}, \frac{7}{2} + \frac{\sqrt{37}}{2}$

25. $2z^2 - 7z + 3 = 0$

$2\left(z^2 - \frac{7}{2}z + \frac{3}{2}\right) = 0$ (Since $2 \neq 0$, it may be canceled.)

$z^2 - \frac{7}{2}z + \left(-\frac{7}{4}\right)^2 = \left(-\frac{7}{4}\right)^2 - \frac{3}{2}$

$\left(z - \frac{7}{4}\right)^2 = \frac{49}{16} - \frac{3}{2} = \frac{25}{16}$

$z - \frac{7}{4} = \pm\frac{5}{4}$

$z = \frac{7}{4} + \frac{5}{4} = 3$ or $z = \frac{7}{4} - \frac{5}{4} = \frac{1}{2}$

$\frac{1}{2}, 3$

27. $3y^2 = 3 + 8y$

$y^2 = 1 + \frac{8}{3}y$ Divide by 3.

$y^2 - \frac{8}{3}y + \left(-\frac{4}{3}\right)^2 = 1 + \frac{16}{9} = \frac{25}{9}$

$\left(y - \frac{4}{3}\right)^2 = \frac{25}{9}$

$y - \frac{4}{3} = \pm\frac{5}{3}$

$y = \frac{4}{3} - \frac{5}{3} = -\frac{1}{3}$ or $y = \frac{4}{3} + \frac{5}{3} = 3$

$-\frac{1}{3}, 3$

29. $4y^2 - 4y - 1 = 0$

$y^2 - y = \frac{1}{4}$

$y^2 - y + \left(-\frac{1}{2}\right)^2 = \frac{1}{4} + \frac{1}{4}$

$\left(y - \frac{1}{2}\right)^2 = \frac{1}{2}$

$y - \frac{1}{2} = \pm\frac{\sqrt{2}}{2}$

$y = \frac{1}{2} + \frac{\sqrt{2}}{2}$ or $y = \frac{1}{2} - \frac{\sqrt{2}}{2}$

$\frac{1}{2} + \frac{\sqrt{2}}{2}, \frac{1}{2} - \frac{\sqrt{2}}{2}$

31. $3z^2 - 2z - 2 = 0$

$z^2 - \frac{2}{3}z = \frac{2}{3}$

$z^2 - \frac{2}{3}z + \left(-\frac{1}{3}\right)^2 = \frac{2}{3} + \frac{1}{9}$

$\left(z - \frac{1}{3}\right)^2 = \frac{7}{9}$

$z - \frac{1}{3} = \pm\frac{\sqrt{7}}{3}$

$z = \frac{1}{3} + \frac{\sqrt{7}}{3}$ or $z = \frac{1}{3} - \frac{\sqrt{7}}{3}$

$\frac{1}{3} - \frac{\sqrt{7}}{3}, \frac{1}{3} + \frac{\sqrt{7}}{3}$

33. $2t^2 = 3 - 10t$

$2t^2 + 10t = 3$

$t^2 + 5t = \frac{3}{2}$

$t^2 + 5t + \left(\frac{5}{2}\right)^2 = \frac{3}{2} + \left(\frac{5}{2}\right)^2$

$\left(t + \frac{5}{2}\right)^2 = \frac{31}{4}$

$t + \frac{5}{2} = \frac{\pm\sqrt{31}}{2}$

$t = -\frac{5}{2} - \frac{\sqrt{31}}{2}$ or $t = -\frac{5}{2} + \frac{\sqrt{31}}{2}$

$-\frac{5}{2} - \frac{\sqrt{31}}{2}, -\frac{5}{2} + \frac{\sqrt{31}}{2}$

35. $6y - y^2 = 4$
$y^2 - 6y = -4$
$y^2 - 6y + (-3)^2 = -4 + 9$
$(y-3)^2 = 5$
$y - 3 = \pm\sqrt{5}$
$y = 3 + \sqrt{5}$ or $y = 3 - \sqrt{5}$
$3 - \sqrt{5},\ 3 + \sqrt{5}$

37. $z(3z-2) = 6$
$3z^2 - 2z = 6$
$z^2 - \frac{2}{3}z = 2$
$z^2 - \frac{2}{3}z + \left(-\frac{1}{3}\right)^2 = 2 + \frac{1}{9}$
$\left(z - \frac{1}{3}\right)^2 = \frac{19}{9}$
$z - \frac{1}{3} = \pm\frac{\sqrt{19}}{3}$
$z = \frac{1}{3} + \frac{\sqrt{19}}{3}$ or $z = \frac{1}{3} - \frac{\sqrt{19}}{3}$
$\frac{1}{3} - \frac{\sqrt{19}}{3},\ \frac{1}{3} + \frac{\sqrt{19}}{3}$

39. $x^2 + x + x = x^2 + 2x$
1

41. Statement c is true.
$x^2 - 8x = -8$
$x^2 - 8x + 16 = -8 + 16$
$(x-4)^2 = 8$
$x - 4 = \pm 2\sqrt{2}$
$x = 4 \pm 2\sqrt{2}$
$4 - 2\sqrt{2},\ 4 + 2\sqrt{2}$ True

43–45. Answers may vary.

47. $x^2 + x + c = 0$
$x^2 + x + \left(\frac{1}{2}\right)^2 = -c + \frac{1}{4}$
$\left(x + \frac{1}{2}\right)^2 = \frac{1}{4} - c$
$x + \frac{1}{2} = \pm\frac{\sqrt{1-4c}}{2}$
$x = -\frac{1}{2} + \frac{\sqrt{1-4c}}{2}$ or $x = -\frac{\sqrt{1-4c}}{2}$
$-\frac{1}{2} - \frac{\sqrt{1-4c}}{2},\ -\frac{1}{2} + \frac{\sqrt{1-4c}}{2}$

Review Problems

49. $\sqrt{2x+3} = 2x - 3$
$2x + 3 = (2x-3)^2$
$2x + 3 = 4x^2 - 12x + 9$
$0 = 4x^2 - 14x + 6$
$2x^2 - 7x + 3 = 0$
$(2x-1)(x-3) = 0$
$2x - 1 = 0$ or $x - 3 = 0$
$x = \frac{1}{2}$ or $x = 3$
$\frac{1}{2},\ 3$

50. $\frac{2x+3}{x^2 - 7x + 12} - \frac{2}{x-3}$
$= \frac{2x+3}{(x-3)(x-4)} - \frac{2}{x-3} \cdot \frac{x-4}{x-4}$
$= \frac{2x+3-2(x-4)}{(x-3)(x-4)}$
$= \frac{2x+3-2x+8}{(x-3)(x-4)}$
$= \frac{11}{(x-3)(x-4)}$ $(x \neq 3, 4)$

51. $4(2x-3) + 4 = 9x + 2$
$8x - 12 + 4 = 9x + 2$
$8x - 8 = 9x + 2$
$-x = 10$
$x = -10$

Problem Set 10.3

1. $x^2+8x+15=0$

$x=\frac{-8\pm\sqrt{64-4(1)(15)}}{2(1)}$

$=\frac{-8\pm\sqrt{4}}{2}=\frac{-8\pm 2}{2}=-4\pm 1$

$x=-5$ or $x=-3$

$-5, -3$

3. $x^2+5x+3=0$

$x=\frac{-5\pm\sqrt{25-4(1)(3)}}{2(1)}=\frac{-5\pm\sqrt{13}}{2}$

$x=\frac{-5-\sqrt{13}}{2}$ or $x=\frac{-5+\sqrt{13}}{2}$

$\frac{-5-\sqrt{13}}{2}, \frac{-5+\sqrt{13}}{2}$

5. $x^2+4x-6=0$

$x=\frac{-4\pm\sqrt{16-4(1)(-6)}}{2(1)}$

$=\frac{-4\pm\sqrt{40}}{2}=\frac{-4\pm 2\sqrt{10}}{2}=-2\pm\sqrt{10}$

$x=-2-\sqrt{10}$ or $x=-2+\sqrt{10}$

$-2-\sqrt{10}, -2+\sqrt{10}$

7. $x^2+4x-7=0$

$x=\frac{-4\pm\sqrt{16-4(1)(-7)}}{2(1)}$

$=\frac{-4\pm\sqrt{44}}{2}=\frac{-4\pm 2\sqrt{11}}{2}=-2\pm\sqrt{11}$

$x=-2-\sqrt{11}$ or $x=-2+\sqrt{11}$

$-2-\sqrt{11}, -2+\sqrt{11}$

9. $x^2-3x-18=0$

$x=\frac{-(-3)\pm\sqrt{(-3)^2-4(1)(-18)}}{2(1)}$

$x=\frac{3\pm\sqrt{9+72}}{2}=\frac{3\pm\sqrt{81}}{2}=\frac{3\pm 9}{2}$

$x=6$ or $x=-3$

$-3, 6$

11. $6x^2-5x-6=0$

$x=\frac{5\pm\sqrt{(-5)^2-4(6)(-6)}}{2(6)}$

$=\frac{5\pm\sqrt{25+144}}{12}=\frac{5\pm\sqrt{169}}{12}$

$x=\frac{5\pm 13}{12}$

$x=\frac{5+13}{12}=\frac{3}{2}$ or $x=\frac{5-13}{12}=-\frac{2}{3}$

$x=\frac{3}{2}$ or $x=-\frac{2}{3}$

$-\frac{2}{3}, \frac{3}{2}$

13. $x^2-2x-10=0$

$x=\frac{2\pm\sqrt{4-4(-10)}}{2}$

$=\frac{2\pm\sqrt{44}}{2}=\frac{2\pm 2\sqrt{11}}{2}=1\pm\sqrt{11}$

$x=1+\sqrt{11}$ or $x=1-\sqrt{11}$

$1-\sqrt{11}, 1+\sqrt{11}$

15. $x^2-x=14$

$x^2-x-14=0$

$x=\frac{1\pm\sqrt{1-4(-14)}}{2}=\frac{1\pm\sqrt{57}}{2}$

$x=\frac{1+\sqrt{57}}{2}$ or $\frac{1-\sqrt{57}}{2}$

$\frac{1-\sqrt{57}}{2}, \frac{1+\sqrt{57}}{2}$

17. $6y^2+6y+1=0$

$y=\frac{-6\pm\sqrt{36-4(6)(1)}}{2(6)}$

$=\frac{-6\pm\sqrt{12}}{12}=\frac{-6\pm 2\sqrt{3}}{12}=\frac{-3\pm\sqrt{3}}{6}$

$y=\frac{-3+\sqrt{3}}{6}$ or $y=\frac{-3-\sqrt{3}}{6}$

$\frac{-3-\sqrt{3}}{6}, \frac{-3+\sqrt{3}}{6}$

19. $4x^2-12x+9=0$

$x=\dfrac{12\pm\sqrt{144-4(4)(9)}}{2(4)}=\dfrac{12\pm 0}{8}=\dfrac{3}{2}$

$x=\dfrac{3}{2}$

$\dfrac{3}{2}$

21. $y^2=2(y+1)$

$y^2-2y-2=0$

$y=\dfrac{2\pm\sqrt{4-4(1)(-2)}}{2}$

$=\dfrac{2\pm\sqrt{12}}{2}=\dfrac{2\pm 2\sqrt{3}}{2}=1\pm\sqrt{3}$

$y=1-\sqrt{3}$ or $y=1+\sqrt{3}$

$1-\sqrt{3},\ 1+\sqrt{3}$

23. $\dfrac{y^2}{4}+\dfrac{3y}{2}+1=0$

$y^2+6y+4=0$

$y=\dfrac{-6\pm\sqrt{36-4(1)(4)}}{2}$

$=\dfrac{-6\pm\sqrt{20}}{2}=\dfrac{-6\pm 2\sqrt{5}}{2}=-3\pm\sqrt{5}$

$y=-3-\sqrt{5}$ or $y=-3+\sqrt{5}$

$-3-\sqrt{5},\ -3+\sqrt{5}$

25. $2x^2-x=1$

$2x^2-x-1=0$

$(2x+1)(x-1)=0$

$x=-\dfrac{1}{2}$ or $x=1$

$-\dfrac{1}{2},\ 1$

27. $5x^2+2=11x$

$5x^2-11x+2=0$

$(5x-1)(x-2)=0$

$x=\dfrac{1}{5}$ or $x=2$

$\dfrac{1}{5},\ 2$

29. $y^2=20$

$y=\pm 2\sqrt{5}$

$y=-2\sqrt{5}$ or $y=2\sqrt{5}$

$-2\sqrt{5},\ 2\sqrt{5}$

31. $x^2-2x=1$

$x^2-2x-1=0$

$x=\dfrac{2\pm\sqrt{4+4}}{2}$

$=\dfrac{2\pm\sqrt{8}}{2}=\dfrac{2\pm 2\sqrt{2}}{2}=1\pm\sqrt{2}$

$x=1+\sqrt{2}$ or $x=1-\sqrt{2}$

$1-\sqrt{2},\ 1+\sqrt{2}$

33. $(2w+3)(w+4)=1$

$2w^2+11w+11=0$

$w=\dfrac{-11\pm\sqrt{121-4(2)(11)}}{2(2)}$

$=\dfrac{-11\pm\sqrt{121-88}}{4}=\dfrac{-11\pm\sqrt{33}}{4}$

$w=\dfrac{-11+\sqrt{33}}{4}$ or $w=\dfrac{-11-\sqrt{33}}{4}$

$\dfrac{-11-\sqrt{33}}{4},\ \dfrac{-11+\sqrt{33}}{4}$

35. $(3r-4)^2=16$

$3r-4=\pm 4$

$3r=0$ or $3r=8$

$r=0$ or $r=\dfrac{8}{3}$

$0,\ \dfrac{8}{3}$

37. $3y^2-12y+12=0$

$y^2-4y+4=0$

$(y-2)^2=0$

$y=2$

2

39. $4w^2 - 16 = 0$
$w^2 - 4 = 0$
$w^2 = 4$
$w = \pm 2$
$w = -2$ or $w = 2$
$-2, 2$

41. $\frac{3}{4}y^2 - \frac{5}{2}y - 2 = 0$
$3y^2 - 10y - 8 = 0$
$(3y + 2)(y - 4) = 0$
$y = -\frac{2}{3}$ or $y = 4$
$-\frac{2}{3}, 4$

43. $10x^2 - 11x + 2 = 0$
$x = \frac{11 \pm \sqrt{121 - 4(10)(2)}}{2(10)} = \frac{11 \pm \sqrt{41}}{20}$
$x = \frac{11 - \sqrt{41}}{20}$ or $x = \frac{11 + \sqrt{41}}{20}$
$\frac{11 - \sqrt{41}}{20}, \frac{11 + \sqrt{41}}{20}$

45. $\frac{y^2}{2} - 2y + \frac{3}{4} = 0$
$2y^2 - 8y + 3 = 0$
$y = \frac{8 \pm \sqrt{64 - 4(2)(3)}}{2(2)}$
$= \frac{8 \pm \sqrt{40}}{4} = \frac{8 \pm 2\sqrt{10}}{4}$
$y = \frac{4 + \sqrt{10}}{2}$ or $y = \frac{4 - \sqrt{10}}{2}$
$2 - \frac{\sqrt{10}}{2}, 2 + \frac{\sqrt{10}}{2}$

47. $(3x - 2)^2 = 10$
$3x - 2 = \pm\sqrt{10}$
$3x = 2 + \sqrt{10}$ or $3x = 2 - \sqrt{10}$
$x = \frac{2 + \sqrt{10}}{3}$ or $x = \frac{2 - \sqrt{10}}{3}$
$\frac{2 - \sqrt{10}}{3}, \frac{2 + \sqrt{10}}{3}$

49. $y^2 + 14y + 49 = 0$
$(y + 7)^2 = 0$
$y = -7$
-7

51. $x^2 + 9x = 0$
$x(x + 9) = 0$
$x = 0$ or $x = -9$
$-9, 0$

53. $(x - 2)^2 - 49 = 0$
$(x - 2)^2 = 49$
$x = -5$ or $x = 9$
$-5, 9$

55. $N = 0.4x^2 - 36x + 1000$
$312 = 0.4x^2 - 36x + 1000$
$0.4x^2 - 36x + 688 = 0$
$x = \frac{-(-36) \pm \sqrt{(-36)^2 - 4(0.4)(688)}}{2(0.4)}$
$x = \frac{36 \pm \sqrt{195.2}}{0.8}$
$x \approx 28$ or $x \approx 62$
About 28 years or 62 years.

57. $s = -16t^2 + 100t + 50$
$0 = -16t^2 + 100t + 50$
$8t^2 - 50t - 25 = 0$
$t = \frac{50 + \sqrt{2500 - 4(8)(-25)}}{2(8)}$
$= \frac{50 + \sqrt{2500 + 800}}{16} = \frac{50 + 10\sqrt{33}}{16}$
$t = \frac{25 + 5\sqrt{33}}{8}$ seconds ≈ 6.7 seconds, rounded

59. w = width of rectangle
$w + 3$ = length of rectangle
$w(w + 3) = 36$
$w^2 + 3w - 36 = 0$
$w = \frac{-3 + \sqrt{9 + 144}}{2} = \frac{-3 + \sqrt{153}}{2} \approx 4.7$

meters, rounded; reject negative root as it would give negative length.

$w+3=\dfrac{3+\sqrt{153}}{2}\approx 7.7$ meters, rounded

length, $\dfrac{3+\sqrt{153}}{2}\approx 7.7$ meters

width, $\dfrac{-3+\sqrt{153}}{2}\approx 4.7$ meters

61. Let b = base of triangle and h = height of triangle.

$b=2h-1$

$A=\dfrac{bh}{2}=9$

$\dfrac{(2h-1)(h)}{2}=9$

$2h^2-h=18$

$2h^2-h-18=0$

$h=\dfrac{1+\sqrt{1-4(2)(-18)}}{2(2)}=\dfrac{1+\sqrt{145}}{4}$

$b=\dfrac{1+\sqrt{145}}{2}-1=\dfrac{-1+\sqrt{145}}{2}$;

reject negative root

$h=\dfrac{1+\sqrt{145}}{4},\ b=\dfrac{-1+\sqrt{145}}{2}$

$h\approx 3.3$ inches, $b\approx 5.5$ inches, rounded

63. Let x = length of shorter leg of triangle.

$x+1$ = length of longer leg

$x^2+(x+1)^2=6^2$

$x^2+x^2+2x+1=36$

$2x^2+2x-35=0$

$x=\dfrac{-2+\sqrt{4-4(2)(-35)}}{2(2)}$

$=\dfrac{-2+\sqrt{4+280}}{4}=\dfrac{-2+2\sqrt{71}}{4}$

$x=\dfrac{-1+\sqrt{71}}{2}$ mm $\approx$ 3.7 mm, rounded

and $x+1=1+\dfrac{-1+\sqrt{71}}{2}$

$=\dfrac{1+\sqrt{71}}{2}\approx 4.7$ mm, rounded

65. $h=-0.05x^2+27$

$22=-0.05x^2+27$

$-5=-0.05x^2$

$100=x^2$

$\pm 10=x$

10 feet

67. $3x(3x+10)-x(x+4)=150$

$9x^2+30x-x^2-4x=150$

$8x^2+26x-150=0$

$4x^2+13x-75=0$

$x=\dfrac{-13\pm\sqrt{(13)^2-4(4)(-75)}}{2(4)}$

$x=\dfrac{-13\pm 37}{8}$

Lengths are positive measures, so $x = 3$.

Large rectangle is $3x$ by $3x + 10$ or 9 cm by 19 cm.

Small rectangle is x by $x + 4$ or 3 cm by 7 cm.

69. Let x = number of days for faster painter to paint house.

$x + 1$ = number of days for slower painter

Fraction painted in 1 day = $\dfrac{1}{x}$. For the painters working together,

$\dfrac{2}{x}+\dfrac{2}{x+1}=1$

$x(x+1)\left[\dfrac{2}{x}+\dfrac{2}{x+1}\right]=x(x+1)\cdot 1$

$2x+2+2x=x^2+x$

$4x+2=x^2+x$

$x^2-3x-2=0$

$x=\dfrac{3+\sqrt{3^2-4(-2)}}{2}$

$=\dfrac{3+\sqrt{17}}{2}$ days $\approx$ 3.6 days, rounded and

Slower painter takes

$x+1=1+\dfrac{3+\sqrt{7}}{2}=\dfrac{5+\sqrt{17}}{2}$ days

$\approx$ 4.6 days, rounded

71. Statement a is true.

$-2x^2 + 3x = 0$
$a = -2,\ b = 3,\ c = 0$ True

73. $d = 0.044v^2 + 1.1v,\ d = 550$ feet

$550 = 0.044v^2 + 1.1v$
$0.044v^2 + 1.1v - 550 = 0$
$v = \dfrac{-1.1 + \sqrt{1.21 - 4(0.044)(-550)}}{2(0.044)}$
$= \dfrac{-1.1 + \sqrt{98.01}}{0.088} = \dfrac{-1.1 + 9.9}{0.088}$
$v = 100$ miles per hour

75. Answers may vary.

77. $\dfrac{1}{x^2} + 3 = \dfrac{6}{x}$

$1 + 3x^2 = 6x$
$3x^2 - 6x + 1 = 0$
$x = \dfrac{6 \pm \sqrt{36 - 4(3)(1)}}{2(3)}$
$= \dfrac{6 \pm \sqrt{24}}{6} = \dfrac{6 \pm 2\sqrt{6}}{6} = \dfrac{3 \pm \sqrt{6}}{3}$
$x = \dfrac{3 + \sqrt{6}}{3}$ or $x = \dfrac{3 - \sqrt{6}}{3}$
$\dfrac{3 - \sqrt{6}}{3},\ \dfrac{3 + \sqrt{6}}{3}$

79. $\dfrac{-b + \sqrt{b^2 - 4ac}}{2a} + \dfrac{-b - \sqrt{b^2 - 4ac}}{2a}$

$= -\dfrac{2b}{2a} = -\dfrac{b}{a}$

81. $x^2 + 2\sqrt{3}x - 9 = 0$

$x = \dfrac{-2\sqrt{3} \pm \sqrt{(4)(3) - 4(1)(-9)}}{2}$
$= \dfrac{-2\sqrt{3} \pm \sqrt{48}}{2} = \dfrac{-2\sqrt{3} \pm 4\sqrt{3}}{2} = -\sqrt{3} \pm 2\sqrt{3}$
$x = \sqrt{3}$ or $x = -3\sqrt{3}$
$-3\sqrt{3},\ \sqrt{3}$

83. $(8 - x)(10 - x) = 47$

$80 - 18x + x^2 = 47$
$x^2 - 18x + 33 = 0$
$x = \dfrac{18 - \sqrt{324 - 4(33)}}{2}$
$= \dfrac{18 - \sqrt{192}}{2} = \dfrac{18 - 8\sqrt{3}}{2} = 9 - 4\sqrt{3}$
Note: The minus sign must be used, because $9 + 4\sqrt{3} > 10$.
$x = 9 - 4\sqrt{3}$ meters ≈ 2.1 meters, rounded

85. $(12.5)(8.4) - (12.5 - 2x)(8.4 - 2x) = \dfrac{(12.5)(8.4)}{2}$

$105 - (105 - 25x - 16.8x + 4x^2) = 52.5$
$4x^2 - 41.8x + 52.5 = 0$
$x = \dfrac{41.8 \pm \sqrt{(41.8)^2 - 4(4)(52.5)}}{2(4)}$
$x = \dfrac{41.8 \pm \sqrt{907.24}}{8}$
$x \approx 8.99$ or $x \approx 1.46$, but
$0 < x < 8.4$, so
$x \approx 1.46$ in.

Review Problems

86. $7(y - 2) = 10 - 2(y + 3)$

$7y - 14 = 10 - 2y - 6$
$9y = 18$
$y = 2$
2

87. $\dfrac{7}{y+2} + \dfrac{2}{y+3} = \dfrac{1}{y^2 + 5y + 6}$

$\dfrac{7(y+3) + 2(y+2)}{(y+2)(y+3)} = \dfrac{1}{(y+2)(y+3)}$
$7y + 21 + 2y + 4 = 1$
$9y = -24$
$y = -\dfrac{8}{3}$
$-\dfrac{8}{3}$

88. $x - 2y > 2$

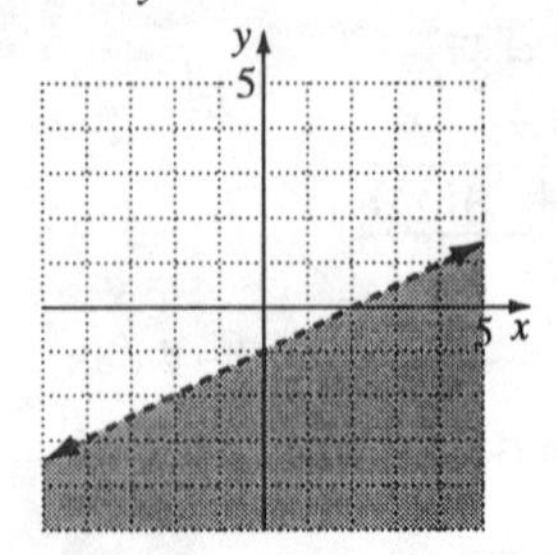

Problem Set 10.4

1. $\sqrt{-16} = \sqrt{16}\sqrt{-1} = 4i$

Product rule for radicals $i = \sqrt{-1}$

3. $\sqrt{-20} = \sqrt{20}\sqrt{-1} = \sqrt{20}i = 2i\sqrt{5}$

5. $\sqrt{-45} = \sqrt{45}\sqrt{-1} = 3i\sqrt{5}$

7. $\sqrt{-150} = \sqrt{150}i = 5i\sqrt{6}$

9. $(x-3)^2 = -9$

$x - 3 = \pm 3i$

$x = 3 - 3i$ or $x = 3 + 3i$

$3 - 3i, 3 + 3i$

11. $(x+7)^2 = -64$

$x + 7 = \pm 8i$

$x = -7 - 8i$ or $x = -7 + 8i$

$-7 - 8i, -7 + 8i$

13. $(y-2)^2 = -7$

$y - 2 = \pm\sqrt{7}i$

$y = 2 - \sqrt{7}i$ or $y = 2 + \sqrt{7}i$

$2 - i\sqrt{7},\ 2 + i\sqrt{7}$

15. $(z+3)^2 = -18$

$z + 3 = \pm 3\sqrt{2}i$

$z = -3 - 3\sqrt{2}i$ or $z = 3 - 3\sqrt{2}i$

$-3 - 3i\sqrt{2},\ -3 + 3i\sqrt{2}$

17. $x^2 + 4x + 5 = 0$

$x = \frac{-4 \pm \sqrt{16 - 20}}{2} = \frac{-4 \pm 2i}{2} = -2 \pm i$

$x = -2 - i$ or $x = -2 + i$

$-2 - i, -2 + i$

19. $x^2 - 6x + 13 = 0$

$x = \frac{6 \pm \sqrt{36 - 52}}{2} = \frac{6 \pm 4i}{2} = 3 \pm 2i$

$x = 3 - 2i$ or $x = 3 + 2i$

$3 - 2i, 3 + 2i$

21. $x^2 - 12x + 40 = 0$

$x = \frac{12 \pm \sqrt{144 - 160}}{2} = \frac{12 \pm 4i}{2} = 6 \pm 2i$

$x = 6 - 2i$ or $x = 6 + 2i$

$6 - 2i, 6 + 2i$

23. $x^2 = 10x - 27$

$x^2 - 10x + 27 = 0$

$x = \frac{10 \pm \sqrt{100 - 108}}{2}$

$= \frac{10 \pm \sqrt{-8}}{2} = 5 \pm \sqrt{2}i$

$x = 5 + \sqrt{2}i$ or $x = 5 - \sqrt{2}i$

$5 - i\sqrt{2},\ 5 + i\sqrt{2}$

25. $5y^2 = 2y - 3$

$5y^2 - 2y + 3 = 0$

$y = \frac{2 \pm \sqrt{4 - 4(5)(3)}}{2(5)}$

$= \frac{2 \pm \sqrt{-56}}{10} = \frac{1 \pm \sqrt{14}i}{5}$

$y = \frac{1}{5} + \frac{\sqrt{14}i}{15}$ or $y = \frac{1}{5} - \frac{\sqrt{14}i}{5}$

$\frac{1}{5} - \frac{i\sqrt{14}}{5}, \frac{1}{5} + \frac{i\sqrt{14}}{5}$

27. $5y^2 - y = y^2 + y - 5$

$4y^2 - 2y + 5 = 0$

$y = \frac{2 \pm \sqrt{4 - 4(4)(5)}}{2(4)}$

$= \frac{2 \pm \sqrt{-76}}{8} = \frac{1 \pm \sqrt{19}i}{4}$

$y = \frac{1}{4} + \frac{\sqrt{19}i}{4}$ or $y = \frac{1}{4} - \frac{\sqrt{19}i}{4}$

$\frac{1}{4} - \frac{i\sqrt{19}}{4}, \frac{1}{4} + \frac{i\sqrt{19}}{4}$

29. $y = -x^2 + 2x + 27$

$-x^2 + 2x + 27 = 29$

$-x^2 + 2x - 2 = 0$

$x = \frac{-2 \pm \sqrt{2^2 - 4(-1)(-2)}}{2(-1)}$

$x = \frac{-2 \pm \sqrt{-4}}{-2}$

$x = 1 \pm i$

No; there are no real solutions

31. b is true since

$(2 - i)^2 - 4(2 - i) + 5$

$= 4 - 4i + i^2 - 8 + 4i + 5 = 0.$

33. 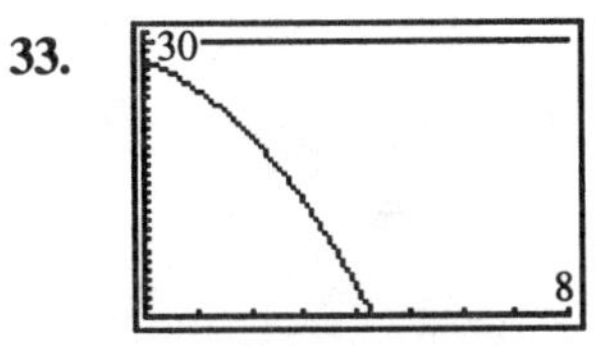

They do not intersect.

35. Answers may vary.

37. $3x(x + 3) = (x + 2)^2 - 10$

$3x^2 + 9x = x^2 + 4x + 4 - 10$

$2x^2 + 5x + 6 = 0$

$x = \frac{-5 \pm \sqrt{25 - 4(2)(6)}}{2(2)}$

$x = \frac{-5 \pm \sqrt{-23}}{4}$

$x = -\frac{5}{4} \pm \frac{i\sqrt{23}}{4}$

$-\frac{5}{4} - \frac{i\sqrt{23}}{4}, -\frac{5}{4} + \frac{i\sqrt{23}}{4}$

Review Problems

39. Let x = number of liters of 8% alcohol solution.

$x + 32$ = number of liters of 12% alcohol solution

$0.08x + 0.28(32) = 0.12(x + 32)$

$0.08x + 8.96 = 0.12x + 3.84$

$5.12 = 0.04x$

$x = 128$

128 liters of 8% alcohol solution

Let x = number of miles driven.

40. $3(39) + 0.25x = 187$

$117 + 0.25x = 187$

$0.25x = 70$

$x = 280$

280 miles

41. $(2\sqrt{3} + \sqrt{2})(2\sqrt{3} - 5\sqrt{2})$

$= (2\sqrt{3})^2 - 2\sqrt{3}(5\sqrt{2}) + \sqrt{2}(2\sqrt{3})$

$- \sqrt{2}(5\sqrt{2})$

$= 12 - 10\sqrt{6} + 2\sqrt{6} - 10$

$= 2 - 8\sqrt{6}$

Problem Set 10.5

1. $y = x^2 + 6x + 5$

x-intercepts:

$x^2 + 6x + 5 = 0$

$(x + 1)(x + 5) = 0$; $x = -1$ and -5

y-intercept: At $x = 0$, $y = 5$

Vertex:

x-coordinate: $x = -\frac{b}{2a} = -\frac{6}{2} = -3$

y-coordinate:

$y = (-3)^2 + 6(-3) + 5 = 9 - 18 + 5 = -4$

At $x = -2$, $y = 4 - 12 + 5 = -3$

Additional point: (–2, –3)

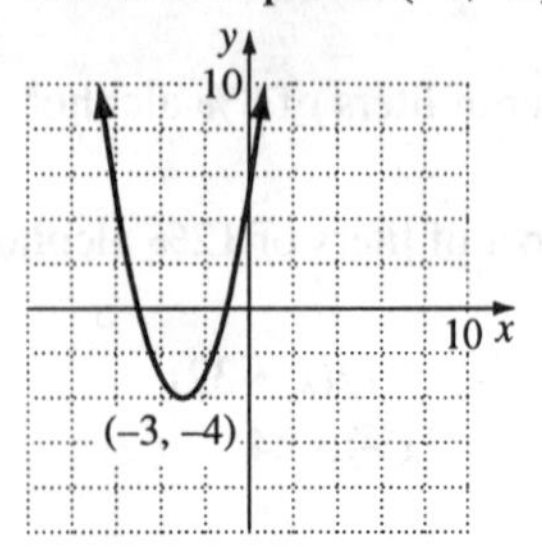

3. $f(x) = x^2 + 4x + 3$

x-intercepts:

$x^2 + 4x + 3 = (x+1)(x+3) = 0;$

$x = -1, -3$

y-intercept: At $x = 0,\ y = 3$

Vertex:

x-coordinate: $x = -\frac{b}{2a} = -\frac{4}{2} = -2$

y-coordinate:

$y = (-2)^2 + 4(-2) + 3 = 4 - 8 + 3 = -1$

At $x = -4,\ y = 16 - 16 + 3 = 3$

Additional point: (–4, 3)

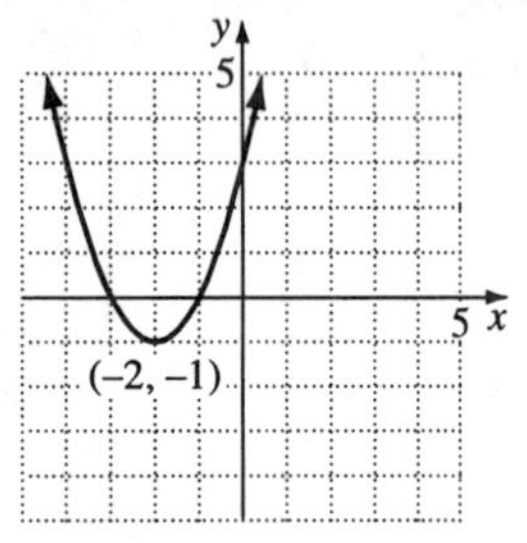

5. $y = x^2 + x$

x-intercepts: $x(x+1) = 0;\ x = 0, -1$

y-intercept: At $x = 0,\ y = 0$

Vertex:

x-coordinate: $x = -\frac{b}{2a} = -\frac{1}{2}$

y-coordinate: $y = \left(-\frac{1}{2}\right)^2 - \frac{1}{2} = -\frac{1}{4}$

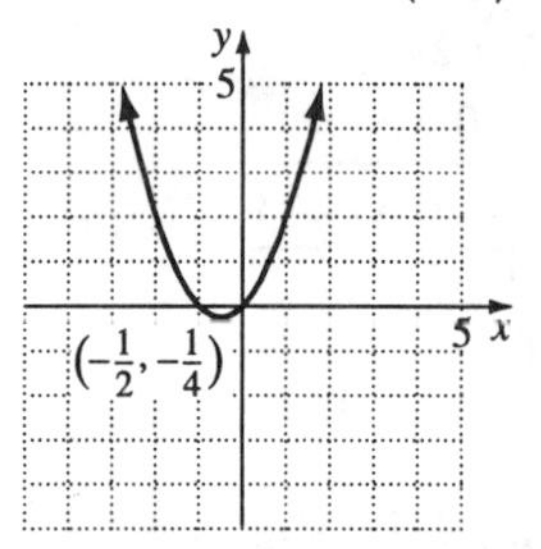

7. $f(x) = x^2 - 4$

x-intercepts: $x^2 - 4 = 0;\ (x-2)(x+2) = 0;$

$x = 2, -2$

y-intercept: At $x = 0,\ y = -4$

Vertex:

x-coordinate: $x = -\frac{b}{2a} = 0$

y-coordinate: $y = 0 - 4 = -4$

At $x = 1,\ y = -3$

Additional point: (1, –3)

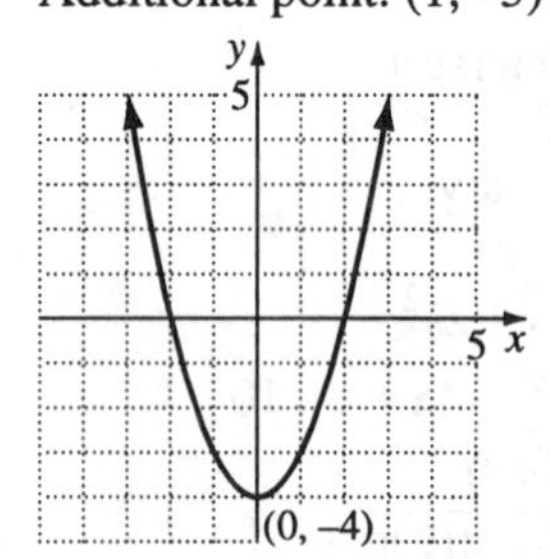

9. $y = -x^2 - 1$

No x-intercepts: $x^2 + 1 = 0$

y-intercept: At $x = 0$, $y = -1$

Vertex:

x-coordinate: $x = -\frac{b}{2a} = 0$

y-coordinate: $y = 0 - 1 = -1$

At $x = \pm 1$, $y = -2$; at $x = \pm 2$, $y = -5$

Additional points:

$(1, -2)$, $(-1, -2)$, $(2, -5)$, $(-2, -5)$

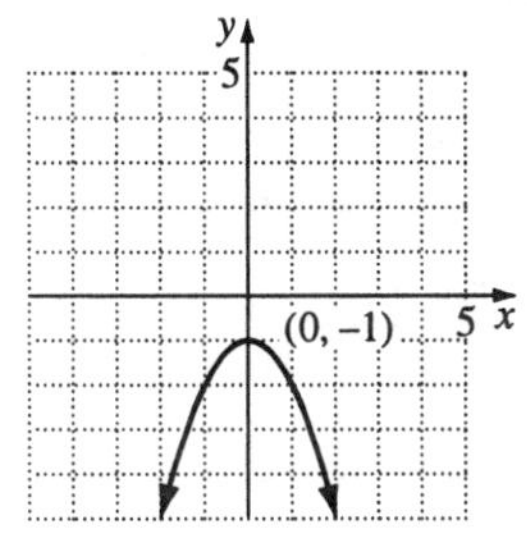

11. $y = -x^2 + 4x - 3$

x-intercepts:

$x^2 - 4x + 3 = 0$; $(x - 1)(x - 3) = 0$; $x = 1, 3$

y-intercept: At $x = 0$, $y = -3$

Vertex:

x-coordinate: $x = -\frac{b}{2a} = \frac{4}{2} = 2$

y-coordinate: $y = -2^2 + 4(2) - 3 = 1$

At $x = -1$, $y = -8$

Additional point: $(-1, -8)$

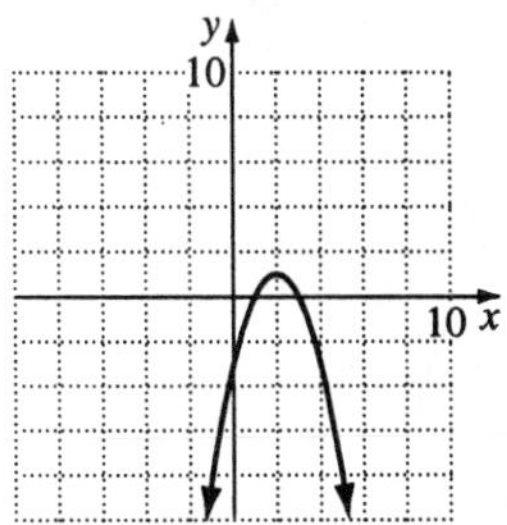

13. $f(x) = -2x^2 + 16x - 30$

x-intercepts: $x^2 - 8x + 15 = 0$

$(x - 3)(x - 5) = 0$; $x = 3, 5$

y-intercept: At $x = 0$, $y = -30$

Vertex:

x-coordinate: $x = -\frac{b}{2a} = \frac{8}{2} = 4$

y-coordinate: $y = -2(4)^2 + 16(4) - 30 = 2$

At $x = 2$, $y = -8 + 32 - 30 = -6$

Additional point: $(2, -6)$

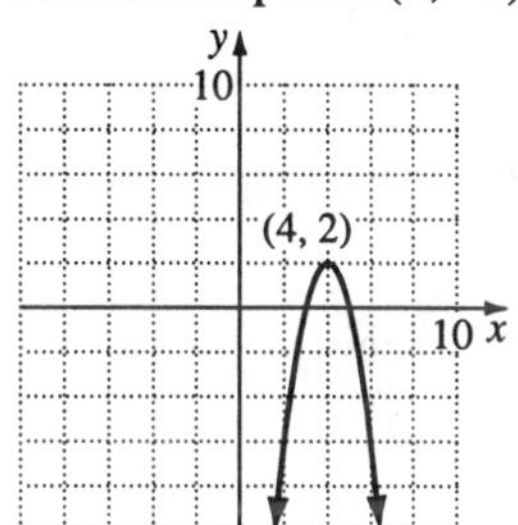

15. $y = x^2 + 4x + 4$

x-intercepts: $x^2 + 4x + 4 = 0$;

$(x + 2)^2 = 0$, $x = -2$

y-intercept: At $x = 0$, $y = 4$

Vertex:

x-coordinate: $x = -\frac{b}{2a} = -\frac{4}{2} = -2$

y-coordinate: $y = (-2)^2 + 4(-2) + 4 = 0$

At $x = 1$, $y = 1 + 4 + 4 = 9$

At $x = -1$, $y = 1$

Additional points: $(1, 9)$ and $(-1, 1)$

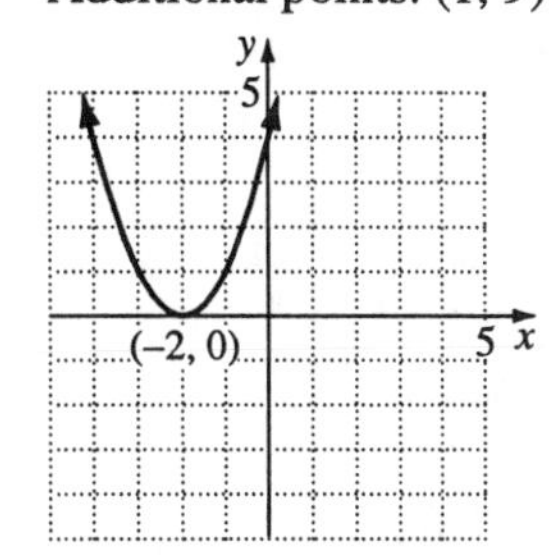

17. $g(x) = x^2 - 4x + 6$

x-intercept: $x^2 - 4x + 6 = 0$;

$b^2 - 4ac = 16 - 24 = -8 < 0$; no x-intercepts

y-intercept: At $x = 0$, $y = 6$

Vertex:

x-coordinate: $x = -\frac{b}{2a} = \frac{4}{2} = 2$

y-coordinate: $y = 2^2 - 4(2) + 6 = 2$

At $x = 1$, $y = 1 - 4 + 6 = 3$

Additional point: (1, 3)

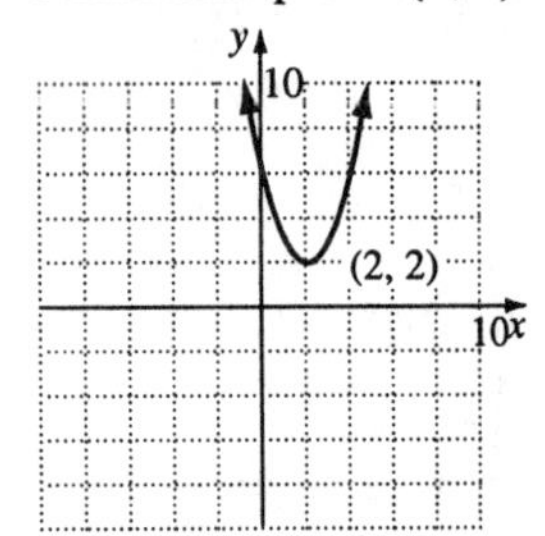

19. $y = -x^2 - 6x - 7$

x-intercepts: $x^2 + 6x + 7 = 0$;

$x = \frac{-6 \pm \sqrt{36 - 28}}{2} = -3 \pm \sqrt{2}$

y-intercept: At $x = 0$, $y = -7$

Vertex:

x-coordinate: $x = -\frac{b}{2a} = -\frac{6}{2} = -3$

y-coordinate: $y = -(-3)^2 - 6(-3) - 7 = 2$

At $x = -1$, $y = -2$

Additional point: (–1, –2)

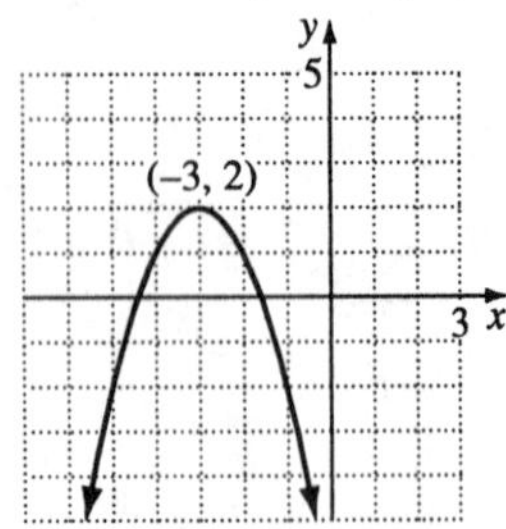

21. $f(x) = -2x^2 + 4x$

x-intercepts: $2x^2 - 4x = 0$;

$x^2 - 2x = x(x - 2) = 0$; $x = 0, 2$

y-intercept: At $x = 0$, $y = 0$

Vertex:

x-coordinate: $x = -\frac{b}{2a} = -\frac{4}{-4} = 1$

y-coordinate: $y = -2(1)^2 + 4 = 2$

At $x = 3$, $y = -2(9) + 12 = -6$

Additional point: (3, –6)

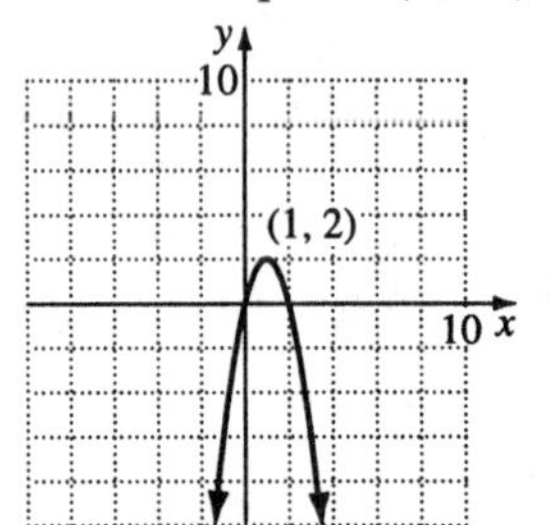

23. $y = -x^2 + 4x - 1$

x-intercepts: $x^2 - 4x + 1 = 0$;

$x = \frac{4 \pm \sqrt{16 - 4}}{2} = 2 \pm \sqrt{3}$

y-intercept: At $x = 0$, $y = -1$

Vertex:

x-coordinate: $x = -\frac{b}{2a} = \frac{4}{2} = 2$

y-coordinate: $y = -4 + 8 - 1 = 3$

At $x = 1$, $y = -1 + 4 - 1 = 2$

Additional point: (1, 2)

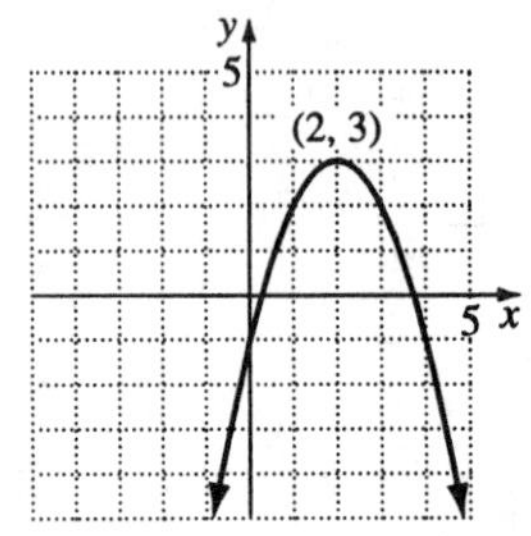

25. $h(x) = 2x^2 + 8x + 1$

x-intercepts: $2x^2 + 8x + 1 = 0$;

$x = \frac{-8 \pm \sqrt{64-8}}{4} = \frac{-4 \pm \sqrt{14}}{2}$

y-intercept: At $x = 0$, $y = 1$

Vertex:

x-coordinate: $x = -\frac{b}{2a} = -\frac{8}{4} = -2$

y-coordinate: $y = 2(-2)^2 + 8(-2) + 1 = -7$

At $x = -1$, $y = 2 - 8 + 1 = -5$

Additional point: $(-1, -5)$

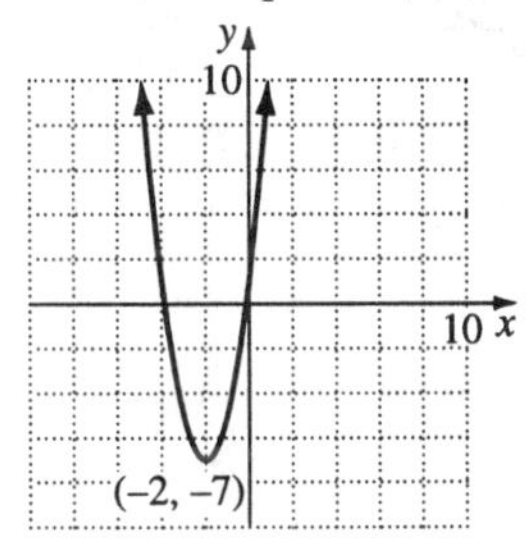

27. **a.** 2

b. 3

c. 1

29. $f(x) = -0.02x^2 + x + 1$

Vertex:

x-coordinate: $\frac{-b}{2a} = \frac{-1}{2(-0.02)} = 25$

y-coordinate:

$y = -0.02(25)^2 + 25 + 1 = 13.5$

Vertex: (25, 13.5)

In a year with 25 in. of rainfall, the tree grows 13.5 inches.

31. $P = 0.0014x^2 - 0.1529x + 5.855$

Vertex:

x-coordinate: $-\frac{b}{2a} = \frac{0.1529}{2(0.0014)} \approx 54.61$

y-coordinate:

$y = 0.0014(54.6)^2 - 0.1529(54.61) + 5.855$

≈ 1.68

Vertex: (54.61, 1.68)

\$54,610 corresponds to the minimum percent given, 1.68%

33. $h = -16t^2 + 64t + 80$

a. $0 = -16t^2 + 64t + 80$

$0 = t^2 - 4t - 5$

$0 = (t - 5)(t + 1)$

$t - 5 = 0$ or $t = 5$

5 seconds

b. $h = -16(0)^2 + 64(0) + 80$

$h = 80$

In 0 seconds, or before the person throws the ball, it is at a height of 80 ft.

c. $-\frac{b}{2a} = -\frac{64}{2(-16)} = 2$

2 seconds

d. $h = -16(2)^2 + 64(2) + 80$

$h = 144$

144 feet

e.

35. Statement c is true.

x-coordinate of vertex $= -\frac{b}{2a} = \frac{-8}{2\left(-\frac{4}{3}\right)} = 3$

y-coordinate of vertex:

$y = -\frac{4}{3}(3)^2 + 8(3) - 11$

$= -12 + 24 - 11 = 1$

The vertex is (3, 1) and is the highest point on the graph since $a < 0$. True

37. Statement d is true.
$y = -x^2 + 500x - 52,500$

39. $y = 0.011x^2 - 0.097x + 4.1$

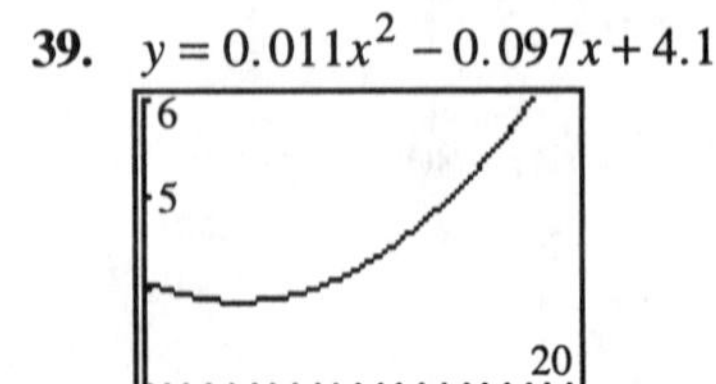

Vertex: (4.47, 3.89)
In mid 1974, about 3.89 million people held more than one job.

41. $y = (x-2)^2$
$y = x^2$
$y = (x+1)^2$

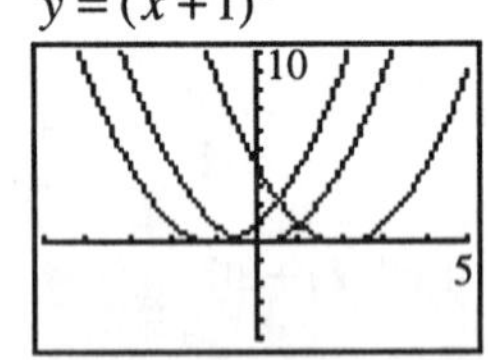

They all have the same shape, but a different y-intercept and vertex.

43–47. Answers may vary.

49. $y = 2x^2 - 8$ and $y = -2x^2 + 8$ intersect when
$2x^2 - 8 = -2x^2 + 8$
$4x^2 = 16$
$x = \pm 2,\ y = 0$
Points of intersection are (–2, 0) and (2, 0).

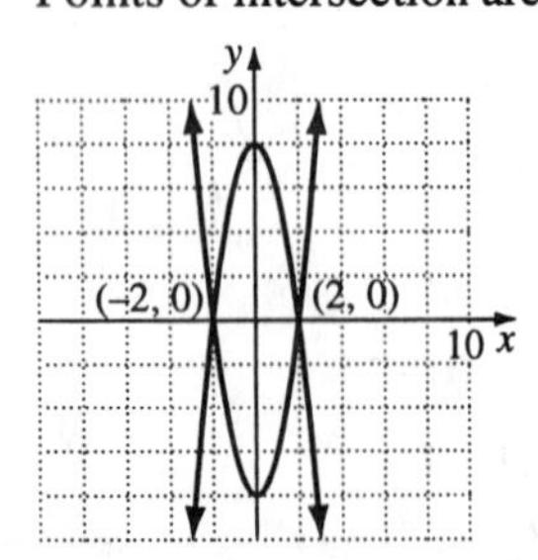

51. Vertex is maximum; opens down; 2

53. Vertex is minimum; opens up; 0

Review Problems

55. $0.00397 = 3.97 \times 10^{-3}$

56. $y = \frac{2}{3}x - 4$

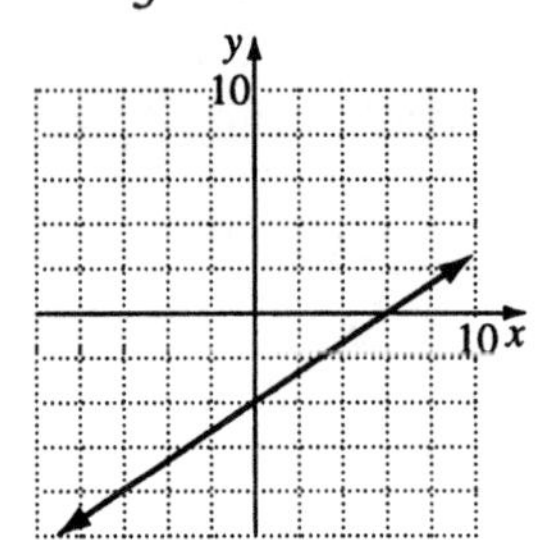

57. $3x - 4y = 8$
$-4y = -3x + 8$
$y = \frac{3}{4}x - 2$

Chapter 10 Review Problems

1. $x^2 = 64$
$x = \pm 8$
± 8

2. $y^2 = 17$
$y = \pm\sqrt{17}$
$\pm\sqrt{17}$

3. $r^2 = 75$
$r = \pm\sqrt{75} = \pm 5\sqrt{3}$
$\pm 5\sqrt{3}$

4. $(y-3)^2 = 9$
$(y - 3) = \pm 3$
$y = 3 - 3 = 0$ or $y = 3 + 3 = 6$
0, 6

5. $(x+4)^2 = 5$
$x + 4 = \pm\sqrt{5}$

$x = -4 + \sqrt{5}$ or $x = -4 - \sqrt{5}$

$-4 - \sqrt{5},\ -4 + \sqrt{5}$

6. $(2x - 7)^2 = 25$

$2x - 7 = \pm 5$

$x = \frac{1}{2}(7 - 5) = 1$ or $x = \frac{1}{2}(7 + 5) = 6$

$1, 6$

7. $(3x - 4)^2 = 18$

$3x - 4 = \pm 3\sqrt{2}$

$x = \frac{1}{3}\left(4 - 3\sqrt{2}\right) = \frac{4}{3} - \sqrt{2}$ or

$x = \frac{1}{3}\left(4 + 3\sqrt{2}\right) = \frac{4}{3} + \sqrt{2}$

$\frac{4}{3} - \sqrt{2},\ \frac{4}{3} + \sqrt{2}$

8. $x^2 = 12^2 + 8^2$

$x = \sqrt{144 + 64}$

$x = \sqrt{208} = 4\sqrt{13}$ in.

$x \approx 14.42$ in.

9. $15^2 = x^2 + 3^2$

$x = \sqrt{225 - 9}$

$x = \sqrt{216} = 6\sqrt{6}$ ft

$x \approx 14.70$ ft

10. $W = 3t^2$

$675 = 3t^2$

$t = \sqrt{\frac{675}{3}}$

$t = 15$ weeks

11. $y = \frac{1}{9000}x^2 + 5$

$45 = \frac{1}{9000}x^2 + 5$

$40 = \frac{1}{9000}x^2$

$360{,}000 = x^2$

$\pm 600 = x$

600 feet

12. $x^2 - 12x + 27 = 0$

$x^2 - 12x + \left(-\frac{12}{2}\right)^2 = -27 + \left(-\frac{12}{2}\right)^2$

$x^2 - 12x + 36 = (x - 6)^2 = 9$

$x - 6 = \pm 3$

$x = 6 - 3 = 3$ or $x = 6 + 3 = 9$

$3, 9$

13. $x^2 - 6x + 4 = 0$

$x^2 - 6x + 9 = -4 + 9$

$(x - 3)^2 = 5$

$(x - 3) = \pm\sqrt{5}$

$x = 3 - \sqrt{5}$ or $x = 3 + \sqrt{5}$

$3 - \sqrt{5},\ 3 + \sqrt{5}$

14. $3x^2 - 12x + 11 = 0$

$x^2 - 4x = -\frac{11}{3}$

$x^2 - 4x + 4 = 4 - \frac{11}{3}$

$(x - 2)^2 = \frac{1}{3}$

$x - 2 = \pm\frac{1}{\sqrt{3}}$

$x = 2 - \frac{1}{\sqrt{3}}$ or $x = 2 + \frac{1}{\sqrt{3}}$

Since

$\frac{1}{\sqrt{3}} = \frac{1}{\sqrt{3}} \cdot \frac{\sqrt{3}}{\sqrt{3}} = \frac{\sqrt{3}}{3}$

$x = 2 \pm \frac{\sqrt{3}}{3}$

$2 - \frac{\sqrt{3}}{3},\ 2 + \frac{\sqrt{3}}{3}$

15. $2x^2 + 5x - 3 = 0$

$x = \frac{-5 \pm \sqrt{25 + 24}}{4} = \frac{-5 \pm 7}{4} = \frac{1}{2},\ -3$

$x = \frac{1}{2}$ or $x = -3$

$-3,\ \frac{1}{2}$

16. $3x^2+5=9x$
$3x^2-9x+5=0$
$x=\frac{9\pm\sqrt{81-60}}{6}$
$x=\frac{9+\sqrt{21}}{6}$ or $x=\frac{9-\sqrt{21}}{6}$
$\frac{9-\sqrt{21}}{6}, \frac{9+\sqrt{21}}{6}$

17. $4y^2+2y-1=0$
$y=\frac{-2\pm\sqrt{4+16}}{8}=\frac{-2\pm2\sqrt{5}}{8}=\frac{-1\pm\sqrt{5}}{4}$
$y=\frac{-1-\sqrt{5}}{4}$ or $y=\frac{-1+\sqrt{5}}{4}$
$\frac{-1-\sqrt{5}}{4}, \frac{-1+\sqrt{5}}{4}$

18. $2x^2-11x+5=0$
$(2x-1)(x-5)=0$
$x=\frac{1}{2}$ or $x=5$
$\frac{1}{2}, 5$

19. $(3x+5)(x-3)=5$
$3x^2-9x+5x-15=5$
$3x^2-4x-20=0$
$(3x-10)(x+2)=0$
$3x-10=0$ or $x+2=0$
$x=\frac{10}{3}$ or $x=-2$
$-2, \frac{10}{3}$

20. $3x^2-7x+1=0$
$x=\frac{7\pm\sqrt{49-12}}{6}=\frac{7\pm\sqrt{37}}{6}$
$x=\frac{7+\sqrt{37}}{6}$ or $x=\frac{7-\sqrt{37}}{6}$
$\frac{7-\sqrt{37}}{6}, \frac{7+\sqrt{37}}{6}$

21. $x^2-9=0$
$x^2=9$
$x=\pm3$
$x=3$ or $x=-3$
$\{-3, 3\}$

22. $(x-3)^2-25=0$
$(x-3)^2=25$
$x-3=\pm5$
$x=3-5=-2$ or $x=3+5=8$
$\{-2, 8\}$

23. $s=-16t^2+48t+80$
If $s=0$,
$t^2-3t-5=0$
$t=\frac{3+\sqrt{9+20}}{2}=\frac{3+\sqrt{29}}{2}$ seconds
$t\approx4.2$ seconds, rounded
time to hit ground:
$\frac{3+\sqrt{29}}{2}$ seconds ≈4.2 seconds

24. Let l = length.
$l-2$ = width
$(l-2)l=16$
$l^2-2l=16$
$l^2-2l+1=16+1$
$(l-1)^2=17$
$l-1=\pm\sqrt{17}$
$l=1+\sqrt{17},\ w=-1+\sqrt{17}$
$l\approx5.1$, $w\approx3.1$, rounded

25. Let x = length of shorter leg of triangle.
$x+2$ = length of longer leg
$x^2+(x+2)^2=6^2$
$x^2+x^2+4x+4=36$
$2x^2+4x-32=0$
$x^2+2x-16=0$
$x=\frac{-2+\sqrt{4+64}}{2}=\frac{-2+2\sqrt{17}}{2}$
$=-1+\sqrt{17};\ x+2=1+\sqrt{17}$
$x\approx3.1$ feet

length of other leg ≈ 5.1 feet, rounded
lengths of legs, $-1+\sqrt{17}$ feet and $1+\sqrt{17}$ feet or rounded, 3.1 feet and 5.1 feet

26. $S=-17t^2+45t+2570$
$2370=-17t^2+45t+2570$
$17t^2-45t-200=0$
$t=\frac{45\pm\sqrt{(-45)^2-4(17)(-200)}}{2(17)}$
$t=\frac{45\pm 125}{34}$
time is a positive number
$t=5$ years
$1985+5=1990$

27. $\sqrt{-81}=9i$

28. $\sqrt{-48}=4i\sqrt{3}$

29. $\sqrt{-17}=i\sqrt{17}$

30. $(x-4)^2=-49$
$x-4=\pm 7i$
$x=4\pm 7i$
$x=4-7i$ or $x=4+7i$
$4-7i, 4+7i$

31. $(7y+1)^2=-27$
$7y+1=\pm 3\sqrt{3}i$
$y=-\frac{1}{7}+\frac{3}{7}\sqrt{3}i$ or $y=-\frac{1}{7}-\frac{3}{7}\sqrt{3}i$
$-\frac{1}{7}-\frac{3}{7}i\sqrt{3},\ -\frac{1}{7}+\frac{3}{7}i\sqrt{3}$

32. $x^2-4x+13=0$
$x=\frac{4\pm\sqrt{16-52}}{2}=\frac{4\pm\sqrt{-36}}{2}=2\pm 3i$
$x=2+3i$ or $x=2-3i$
$2-3i, 2+3i$

33. $x^2+4=3x$
$x^2-3x+4=0$
$x=\frac{3\pm\sqrt{9-16}}{2}=\frac{3\pm\sqrt{7}i}{2}$
$x=\frac{3}{2}+\frac{\sqrt{7}}{2}i$ or $x=\frac{3}{2}-\frac{\sqrt{7}}{2}i$
$\frac{3}{2}-\frac{\sqrt{7}}{2}i,\ \frac{3}{2}+\frac{\sqrt{7}}{2}i$

34. $3y^2-y+2=0$
$y=\frac{1\pm\sqrt{1-24}}{6}=\frac{1\pm\sqrt{23}i}{6}$
$y=\frac{1}{6}+\frac{\sqrt{23}}{6}i$ or $y=\frac{1}{6}-\frac{\sqrt{23}}{6}i$
$\frac{1}{6}-\frac{\sqrt{23}}{6}i,\ \frac{1}{6}+\frac{\sqrt{23}}{6}i$

35. $2y^2=3y-5$
$2y^2-3y+5=0$
$y=\frac{3\pm\sqrt{9-40}}{4}=\frac{3\pm\sqrt{31}i}{4}$
$y=\frac{3}{4}+\frac{\sqrt{31}}{4}i$ or $y=\frac{3}{4}-\frac{\sqrt{31}}{4}i$
$\frac{3}{4}-\frac{\sqrt{31}}{4}i,\ \frac{3}{4}+\frac{\sqrt{31}}{4}i$

36. $R=-2x^2+36x$
$19=-2x^2+36x$
$2x^2-36x+19=0$
$x=\frac{36\pm\sqrt{(-36)^2-4(2)(19)}}{2(2)}$
$x=\frac{36\pm\sqrt{1144}}{4}$
$x\approx 0.54$ and $x\approx 17.46$
\$0.54 or \$17.46 per pair
The applicant could be hired.

37. $y=x^2+4x-5$
x-intercepts: $x^2+4x-5=0$
$(x+5)(x-1)=0$
$x=1,-5$
y-intercept: If $x=0, y=-5$
Vertex:
x-coordinate: $-\frac{b}{2a}=-\frac{4}{2}=-2$
y-coordinate: $(-2)^2+4(-2)-5=-9$

At $x = -1$, $y = 1 - 4 - 5 = -8$
Additional point: $(-1, -8)$

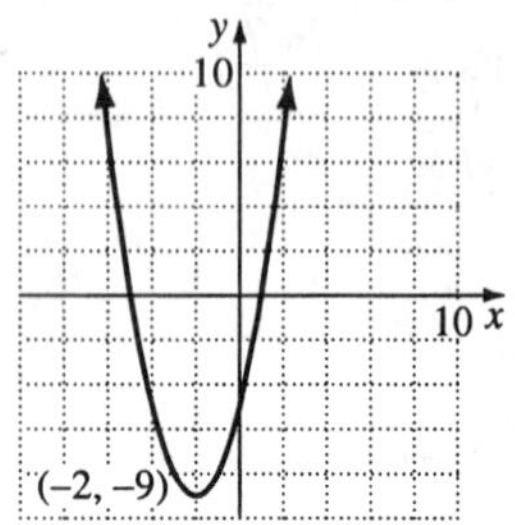

38. $y = -x^2 + 6x - 9$
$x^2 - 6x + 9 = 0$
$(x-3)^2 = 0$
x-intercept: $x = 3$
y-intercept: If $x = 0$, $y = -9$
Vertex:
x-coordinate: $-\frac{b}{2a} = \frac{6}{2} = 3$
y-coordinate: $-3^2 + 6(3) - 9 = 0$
At $x = 2$, $y = -4 + 12 - 9 = -1$
Additional point: $(2, -1)$

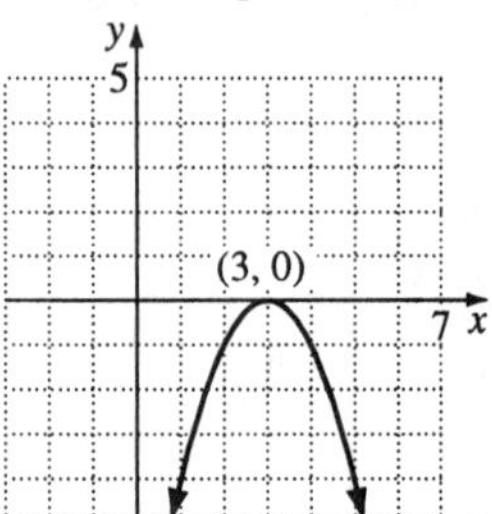

39. $y = x^2 - 6x + 7$
x-intercepts: $x^2 - 6x + 7 = 0$
$x = \frac{6 \pm \sqrt{36-28}}{2} = \frac{6 \pm 2\sqrt{2}}{2} = 3 \pm \sqrt{2}$
$x = 3 + \sqrt{2}$, $x = 3 - \sqrt{2}$
y-intercept: If $x = 0$, $y = 7$
Vertex:
x-coordinate: $-\frac{b}{2a} = \frac{6}{2} = 3$
y-coordinate: $3^2 - 6(3) + 7 = -2$
At $x = 1$, $y = 1 - 6 + 7 = 2$

Additional point: $(1, 2)$

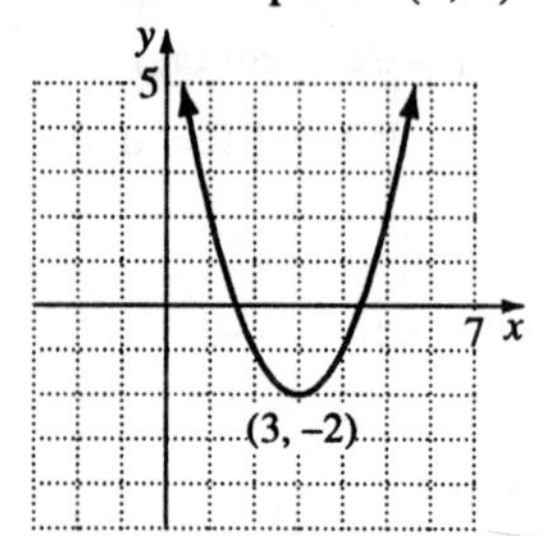

40. $y = -x^2 + 4x = x(-x + 4)$
x-intercepts: $x = 0, 4$
y-intercept: If $x = 0$, $y = 0$
Vertex:
x-coordinate: $-\frac{b}{2a} = -\frac{4}{-2} = 2$
y-coordinate: $-2^2 + 4(2) = 4$
At $x = 1$, $y = 3$
Additional point: $(1, 3)$

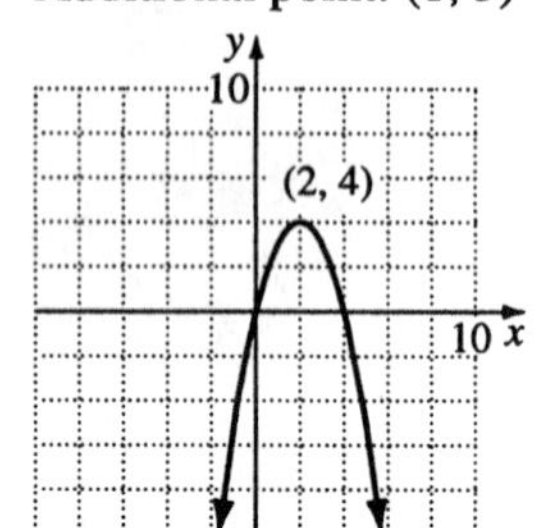

41. $y = x^2 - 4x + 10$
$x^2 - 4x + 10 = 0$
no x-intercepts:
$x = \frac{4 \pm \sqrt{16-40}}{2} = \frac{4 \pm 2\sqrt{6}i}{2} = 2 \pm \sqrt{6}i$
y-intercept: If $x = 0$, $y = 10$
Vertex:
x-coordinate: $-\frac{b}{2a} = \frac{4}{2} = 2$
y-coordinate: $2^2 - 4(2) + 10 = 6$
At $x = 1$, $y = 7$

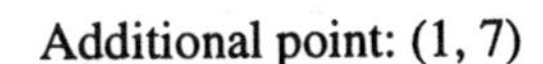

Additional point: (1, 7)

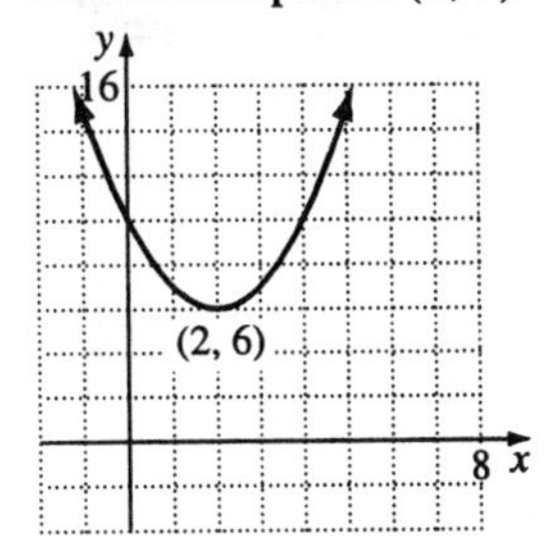

42. $y = -x^2 - 3$

$x^2 + 3 = 0$

no x-intercepts: $x = \pm\sqrt{3}i$

y-intercept: If $x = 0$, $y = -3$

Vertex:

x-coordinate: $-\frac{b}{2a} = 0$

y-coordinate: -3

At $x = 1$, $y = -4$

Additional point: $(1, -4)$

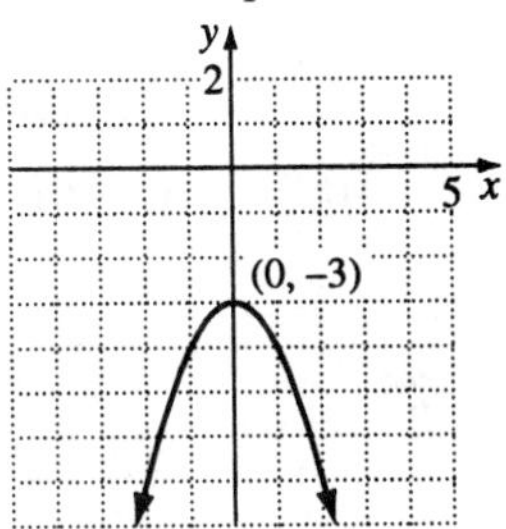

43. $d = -16t^2 + 160t$

a. $0 = -16t^2 + 160t$

$0 = -16t(t - 10)$

$t = 10$ seconds

b. $-\frac{b}{2a} = \frac{-160}{2(-16)} = 5$

5 seconds

c. $d = -16(5)^2 + 160(5) = 400$

400 feet

d.

44. $f(x) = -0.02x^2 + 0.16x + 3.95$

Vertex: x-coordinate: $-\frac{b}{2a} = \frac{-0.16}{2(-0.02)} = 4$

$f(x) = -0.02(4)^2 + 0.16(4) + 3.95 = 4.27$

Vertex: (4, 4.27)

In 1987, there were 4.27 million secretaries. Answers may vary.

Chapter 10 Test

1. $9x^2 = 54$

$x^2 = 6$

$x = \sqrt{6}$ or $x = -\sqrt{6}$

$-\sqrt{6}, \sqrt{6}$

2. $3x^2 + 5x = 0$

$x(3x + 5) = 0$

$x = 0$ or $3x + 5 = 0$

$x = -\frac{5}{3}$

$-\frac{5}{3}, 0$

3. $3x^2 + 5x + 1 = 0$

$$x = \frac{-5 \pm \sqrt{5^2 - 4 \cdot 3 \cdot 1}}{2 \cdot 3} = \frac{-5 \pm \sqrt{13}}{6}$$

$\frac{-5 - \sqrt{13}}{6}, \frac{-5 + \sqrt{13}}{6}$

4. $(x - 2)^2 = 5$

$x - 2 = \sqrt{5}$ or $x - 2 = -\sqrt{5}$

$x = 2 + \sqrt{5}$ or $x = 2 - \sqrt{5}$

$2 - \sqrt{5}, 2 + \sqrt{5}$

5. $x(x-2)=1$

$x^2-2x-1=0$

$x=\frac{2\pm\sqrt{(-2)^2-4\cdot1\cdot(-1)}}{2\cdot1}=\frac{2\pm\sqrt{8}}{2}$

$=\frac{2\pm2\sqrt{2}}{2}=1\pm\sqrt{2}$

$1-\sqrt{2},\ 1+\sqrt{2}$

6. $9x^2-6x=2$

$9x^2-6x-2=0$

$x=\frac{6\pm\sqrt{(-6)^2-4\cdot9\cdot(-2)}}{2\cdot9}=\frac{6\pm\sqrt{108}}{18}$

$=\frac{6\pm6\sqrt{3}}{18}=\frac{1\pm\sqrt{3}}{3}$

$\frac{1-\sqrt{3}}{3},\ \frac{1+\sqrt{3}}{3}$

7. $8x^2=6x-1$

$8x^2-6x+1=0$

$(4x-1)(2x-1)=0$

$4x-1=0$ or $2x-1=0$

$x=\frac{1}{4}$ or $x=\frac{1}{2}$

$\frac{1}{4},\ \frac{1}{2}$

8. $(2x+1)^2=36$

$2x+1=6$ or $2x+1=-6$

$2x=5$ $\quad$ $2x=-7$

$x=\frac{5}{2}$ $\quad$ $x=-\frac{7}{2}$

$-\frac{7}{2},\ \frac{5}{2}$

9. $3x(x-2)+1=0$

$3x^2-6x+1=0$

$x=\frac{6\pm\sqrt{(-6)^2-4\cdot3\cdot1}}{2\cdot3}=\frac{6\pm\sqrt{24}}{6}$

$=\frac{6\pm2\sqrt{6}}{6}=\frac{3\pm\sqrt{6}}{3}$

$\frac{3-\sqrt{6}}{3},\ \frac{3+\sqrt{6}}{3}$

10. $x^2-2x=-5$

$x^2-2x+5=0$

$x=\frac{2\pm\sqrt{(-2)^2-4\cdot1\cdot5}}{2\cdot1}=\frac{2\pm\sqrt{-16}}{2}$

$=\frac{2\pm4i}{2}=1\pm2i$

$1-2i,\ 1+2i$

11. $x^2+4x-3=0$

$x^2+4x=3$

$x^2+4x+4=3+4$

$(x+2)^2=7$

$x+2=\sqrt{7}$ or $x+2=-\sqrt{7}$

$x=-2+\sqrt{7}$ or $x=-2-\sqrt{7}$

$-2-\sqrt{7},\ -2+\sqrt{7}$

12. $\sqrt{-121}=\sqrt{121}\,i=11i$

13. $-\sqrt{-75}=-\sqrt{75}\,i=-5i\sqrt{3}$

14. $x^2+36=0$

$x^2=-36$

$x=\sqrt{-36}$ or $x=-\sqrt{-36}$

$x=6i$ or $x=-6i$

$-6i,\ 6i$

15. $(x-5)^2=-25$

$x-5=\sqrt{-25}$ or $x-5=-\sqrt{-25}$

$x=5+5i$ or $x=5-5i$

$5-5i,\ 5+5i$

16. $x^2-4x+7=0$

$x=\frac{4\pm\sqrt{(-4)^2-4\cdot1\cdot7}}{2\cdot1}=\frac{4\pm\sqrt{-12}}{2}$

$=\frac{4\pm2i\sqrt{3}}{2}=2\pm i\sqrt{3}$

$2-i\sqrt{3},\ 2+i\sqrt{3}$

17. $y = x^2 - 2x - 8$

1. x-intercepts
   $0 = x^2 - 2x - 8$
   $x = -2,\ x = 4$
2. y-intercept
   $y = (0)^2 - 2(0) - 8 = -8$
3. Vertex
   $x\text{-coordinate} = \frac{-b}{2a} = \frac{2}{2} = 1$
   $y\text{-coordinate} = (1)^2 - 2(1) - 8 = -9$

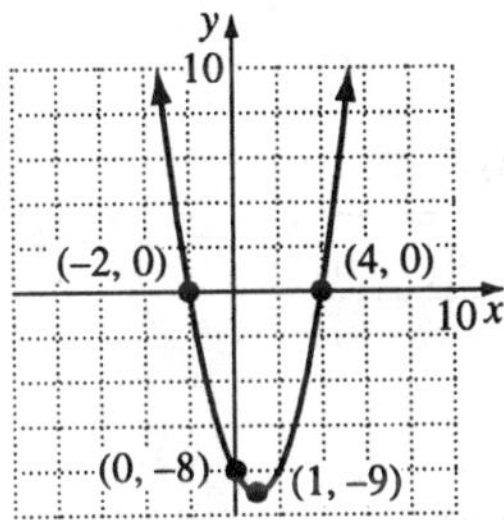

18. $y = -2x^2 + 16x - 24$

1. x-intercepts
   $0 = -2x^2 + 16x - 24$
   $x = 2$ or $x = 6$
2. y-intercept
   $y = -2(0)^2 + 16(0) - 24 = -24$
3. Vertex
   $x\text{-coordinate} = \frac{-b}{2a} = \frac{-16}{-4} = 4$
   $y\text{-coordinate} = -2(4)^2 + 16(4) - 24 = 8$

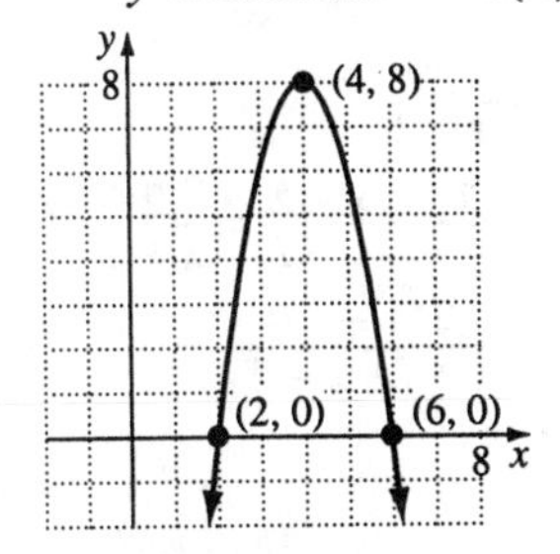

19. $(PQ)^2 + (8)^2 = (12)^2$
$(PQ)^2 = 80$
$PQ = \sqrt{80} = 4\sqrt{5}$
$4\sqrt{5}$ yards

20. $(3x + x) \cdot 3x = 72$
$12x^2 = 72$
$x^2 = 6$
$x = \sqrt{6}$
$\sqrt{6}$ inches

21. $14 = \frac{n^2 - 3n}{2}$
$28 = n^2 - 3n$
$n^2 - 3n - 28 = 0$
$(n + 4)(n - 7) = 0$
$n = -4$ or $n = 7$
n cannot be negative so $n = 7$.
There are 7 sides.

Appendix

1. $2(x-3)+5x=8(x-1)$
$2x-6+5x=8x-8$
$7x-6=8x-8$
$-x=-2$
$x=2$
2

2. $\frac{2x}{3}+\frac{1}{5}=1+\frac{3x}{5}-\frac{1}{3}$
$\frac{2x}{3}-\frac{3x}{5}=1-\frac{1}{3}-\frac{1}{5}$
$\frac{10x-9x}{15}=1-\frac{8}{15}$
$\frac{x}{15}=\frac{7}{15}$
$x=7$
7

3. $0.4(x+20)+0.5x=13.4$
$0.4x+8+0.5x=13.4$
$0.9x+8=13.4$
$0.9x=5.4$
$x=6$
6

4. $-2(y-5)+10=-3(y+2)+y$
$-2y+10+10=-3y-6+y$
$-2y+20=-2y-6$
$20=-6$ False
No solution
Solution: $\varnothing$

5. $-3(2x-4)>2(6x-12)$
$-6x+12>12x-24$
$-18x>-36$
$x<2$
$\{x|x<2\}$

1 2 3 4 5

6. $\frac{-3}{8}=\frac{x}{40}$
$-3(40)=8x$
$-\frac{120}{8}=x$
$x=-15$
-15

7. $x^2+3x=18$
$x^2+3x-18=0$
$(x+6)(x-3)=0$
$x+6=0$ or $x-3=0$
$x=-6$ or $x=3$
$-6, 3$

8. $6x^2+13x+6=0$
$x=\frac{-13\pm\sqrt{13^2-4(6)(6)}}{2(6)}$
$x=\frac{-13\pm5}{12}$
$x=-1\frac{1}{2}$ or $x=-\frac{2}{3}$
$-1\frac{1}{2}, -\frac{2}{3}$

9. $\frac{3}{y+5}-1=\frac{4-y}{2y+10}$
$\frac{3(2)}{2(y+5)}-\frac{2(y+5)}{2(y+5)}=\frac{4-y}{2(y+5)}$
$6-2y-10=4-y$
$-y=8$
$y=-8$
-8

10. $\frac{2x}{x^2-4}+\frac{1}{x-2}=\frac{2}{x+2}$
$\frac{2x}{(x-2)(x+2)}+\frac{1}{x-2}\cdot\frac{x+2}{x+2}=\frac{2}{x+2}\cdot\frac{x-2}{x-2}$
$\frac{2x+x+2}{(x-2)(x+2)}=\frac{2(x-2)}{(x+2)(x-2)}$
$3x+2=2x-4$
$x=-6$
-6

11. $x+\frac{6}{x}=-5$

$x\cdot\frac{x}{x}+\frac{6}{x}=-5\cdot\frac{x}{x}$

$\frac{x^2+6}{x}=\frac{-5x}{x}$

$x^2+6=-5x$

$x^2+5x+6=0$

$(x+2)(x+3)=0$

$x+2=0$ or $x+3=0$

$x=-2$ or $x=-3$

$-2, -3$

12. $x-5=\sqrt{x+7}$

$(x-5)^2=x+7$

$x^2-10x+25=x+7$

$x^2-11x+18=0$

$(x-9)(x-2)=0$

$x-9=0$ or $x-2=0$

$x=9$ or $x=2$, extraneous

The solution is 9.

13. $(x-2)^2=20$

$x-2=\pm\sqrt{20}$

$x-2=\pm2\sqrt{5}$

$x=2\pm2\sqrt{5}$

$2-2\sqrt{5},\ 2+2\sqrt{5}$

14. $3+x(x+2)=18$

$3+x^2+2x=18$

$x^2+2x-15=0$

$(x+5)(x-3)=0$

$x+5=0$ or $x-3=0$

$x=-5$ or $x=3$

$-5, 3$

15. $3x^2-6x+2=0$

$x=\frac{-(-6)\pm\sqrt{(-6)^2-4(3)(2)}}{2(3)}$

$x=\frac{6\pm\sqrt{36-24}}{6}$

$x=\frac{6\pm\sqrt{12}}{6}$

$x=\frac{6\pm2\sqrt{3}}{6}$

$x=\frac{3\pm\sqrt{3}}{3}$

$1-\frac{\sqrt{3}}{3},\ 1+\frac{\sqrt{3}}{3}$

16. $x^2+2x+2=0$

$x=\frac{-2\pm\sqrt{2^2-4(1)(2)}}{2(1)}$

$x=\frac{-2\pm\sqrt{4-8}}{2}$

$x=\frac{-2\pm\sqrt{-4}}{2}$

$x=\frac{-2\pm2i}{2}$

$x=-1\pm i$

$-1-i, -1+i$

17. $y=2x-3$

$x+2y=9$ (Substitute for y)

$x+2(2x-3)=9$

$x+4x-6=9$

$5x=15$

$x=3$

$y=2x-3=2(3)-3=3$

$(3, 3)$

18. $3x+2y=-2$ (1)

$-4x+5y=18$ (2)

Multiply equation 1 by 4.

Multiply equation 2 by 3.

$$\begin{array}{r} 12x+8y=-8 \\ -12x+15y=54 \\ \hline 23y=46 \end{array}$$

$y=2$

$3x+2(2)=-2$

$3x=-2-4=-6$

$x=-2$

$(-2, 2)$

19. $3x - y = 4$ (1)
$-9x + 3y = -12$ (2)

No change to equation 1.
Divide equation 2 by 3

$$\begin{aligned} 3x - y &= 4 \\ -3x + y &= -4 \\ \hline 0 &= 0 \end{aligned}$$

true for all real values of x, where $3x - y = 4$

20. $\frac{t}{a} + \frac{t}{b} = 1$

$$t\left(\frac{1}{a} + \frac{1}{b}\right) = 1$$

$$t = \frac{1}{\frac{1}{a} + \frac{1}{b}}$$

$$t = \frac{ab}{a+b}$$

21. $(0.31)(263) = 81.53$ million

22. a. 23,000

b. $60{,}000 - 8{,}000 = 52{,}000$

c. 1985, 1986, 1987

23. a. 1970, 1984, 1985; 8 per 100,000

b. 1980; 10.2 per 100,000

24. No; 15 \$10 decreases will result in maximum profit, then profit goes down.

25. a. $\frac{-196.7 - 213.1}{2} = -\$204.9$ billion

b. $-196.7 - (-73.7) = -\$123$ billion

c. $\frac{-196.4}{-3.2} = 61.375$ times

26. $\frac{10.52}{0.04} = 263$ million
Black $= 0.12(263) = 31.56$ million
White $= 0.84(263) = 220.92$ million

27. Let x = rate in Canada.
Then U.S. rate $= 4x + 91$ and
Spain rate $= x - 26$

$$x + 4x + 91 + x - 26 = 761$$
$$6x + 65 = 761$$
$$6x = 696$$
$$x = 116$$

Country	Rate per 100,000
Canada	116
U.S.	555
Spain	90
S. Africa	369
Singapore	229
Hong Kong	179
N. Ireland	126
Mexico	97
United Kingdom	93
Australia	91
Switzerland	85
France	84
Italy	80
Denmark	66
Norway	59
Netherlands	49
Japan	36

28. a. $y = 10x + 100$
where y is the debt in billions and x is the number of years after 1985.

$$320 = 10x + 100$$
$$10x = 220$$
$$x = 22$$
$$1985 + 22 = 2007$$

Year 2007

b. Fairly close

29. a. $\frac{9}{33} = \frac{3}{11}$

b. $\frac{9}{6.5}$

c. $\frac{33}{9+8.8+6.5+4.9} = \frac{33}{29.2}$

30. $d = 35t$

a. $d = 35(1) = 35$ miles
$d = 35(2) = 70$ miles
$d = 35(3) = 105$ miles
$d = 35(4) = 140$ miles

b.

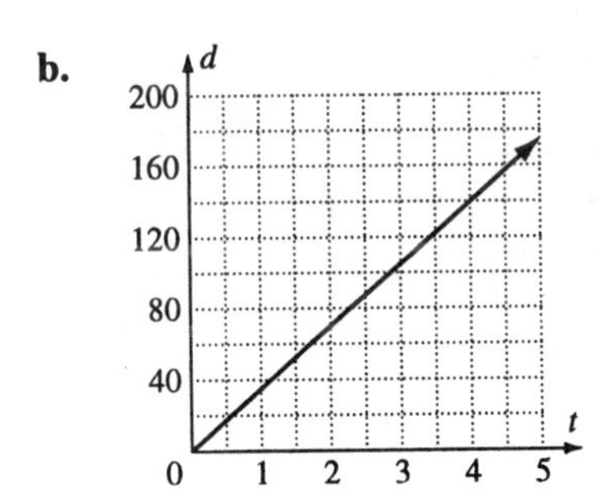

31. a. $S = 0.04x + 500$

b.

x	$S = 0.04x + 500$	(x, S)
0	$S = 0.04(0) + 500 = 500$	(0, 500)
100	$S = 0.04(100) + 500 = 504$	(100, 504)
1000	$S = 0.04(1000) + 500 = 540$	(1000, 540)
10,000	$S = 0.04(10{,}000) + 500 = 900$	(10,000, 900)

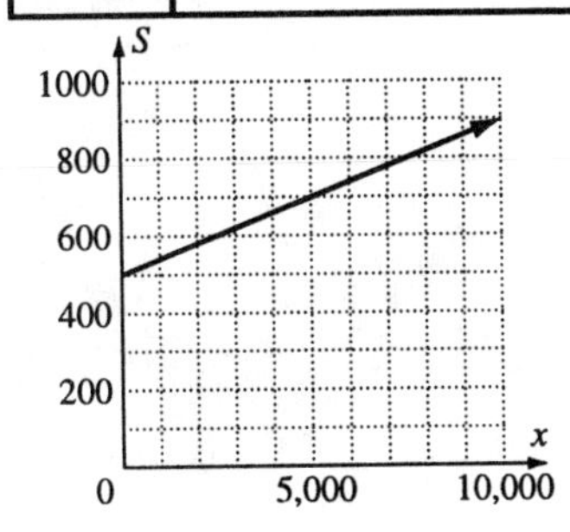

c. \$540
$S = 0.04(1000) + 500$
$S = 40 + 500$
$S = 540$

32. $f(x) = 0.3x^2 + 2.2x + 5.3$

a.

x	$f(x) = 0.3x^2 + 2.2x + 5.3$	(x, y)
0	$f(0) = 0.3(0)^2 + 2.2(0) + 5.3 = 5.3$	(0, 5.3)
1	$f(1) = 0.3(1)^2 + 2.2(1) + 5.3 = 7.8$	(1, 7.8)
2	$f(2) = 0.3(2)^2 + 2.2(2) + 5.3 = 10.9$	(2, 10.9)
3	$f(3) = 0.3(3)^2 + 2.2(3) + 5.3 = 14.6$	(3, 14.6)
4	$f(4) = 0.3(4)^2 + 2.2(4) + 5.3 = 18.9$	(4, 18.9)
5	$f(5) = 0.3(5)^2 + 2.2(5) + 5.3 = 23.8$	(5, 23.8)

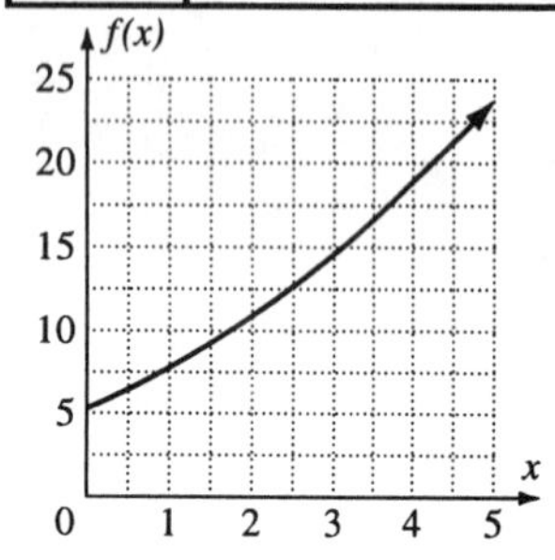

b. As time goes by, the number increases.

33. a. points (1960, 40), (1970, 25)
$\text{slope} = \frac{25 - 40}{1970 - 1960} = -1.5$
The average number is decreasing.

b. points (1980, 10), (1990, 11)
$\text{slope} = \frac{11 - 10}{1990 - 1980} = 0.1$
The average number is increasing.

c. Consistent spacing is not used between years.

34. **a.** The slope is the same, 40.

b. They will never be equal. They do not intersect.

35. $y + 3 = 0$

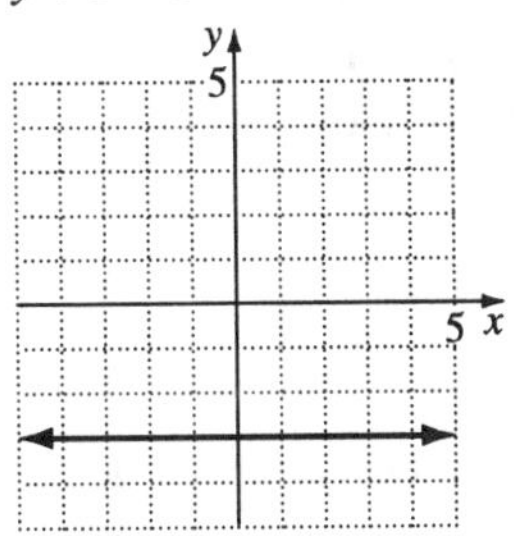

36. $3x - 2y = 6$

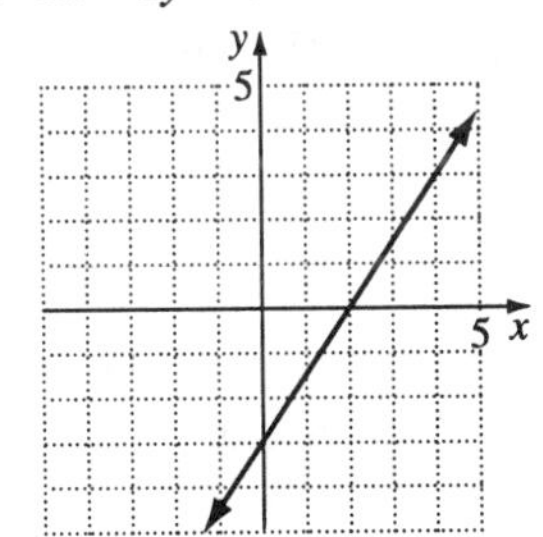

37. $y = -\frac{2}{3}x + 1$

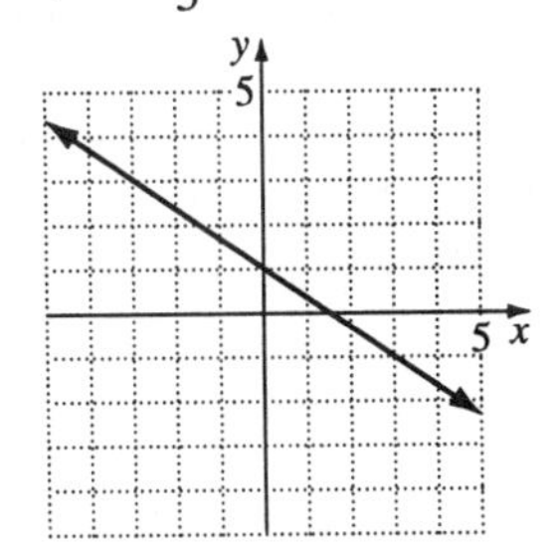

38. $5x + 2y < -10$

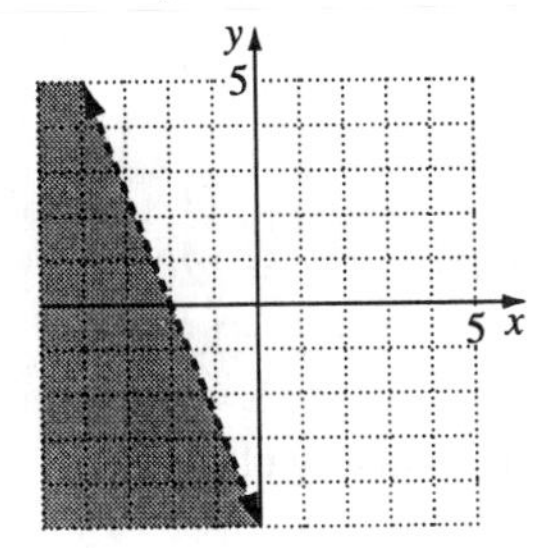

39. $y > -2x + 3$

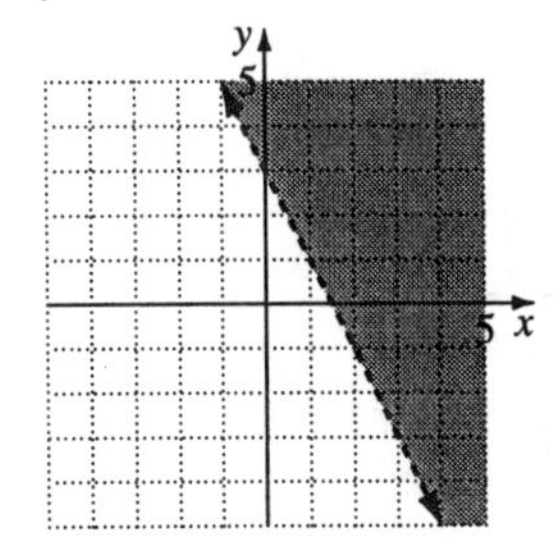

40. $y = x^3 - x$

x	–2	–1	0	1	2
y	–6	0	0	0	6

$x = -2$: $y = (-2)^3 - (-2) = -6$

$x = -1$: $y = (-1)^3 - (-1) = 0$

$x = 0$: $y = 0^3 - 0 = 0$

$x = 1$: $y = 1^3 - 1 = 0$

$x = 2$: $y = 2^3 - 2 = 6$

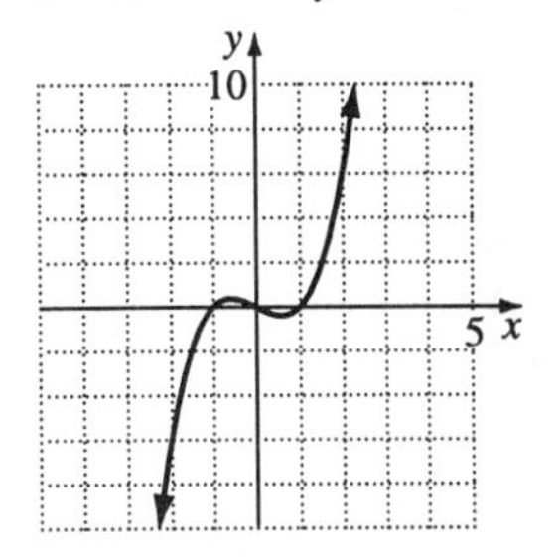

41. $y = x^2 - 2x - 8$

x-intercepts: $x^2 - 2x - 8 = 0$

$(x - 4)(x + 2) = 0$

$x = 4$ or $x = -2$

$(4, 0), (-2, 0)$

y-intercept: $y = 0 - 0 - 8 = -8$

$(0, -8)$

vertex: $x = -\frac{b}{2a} = -\frac{(-2)}{2(1)} = 1$

$y = 1^2 - 2 \cdot 1 - 8 = -9$

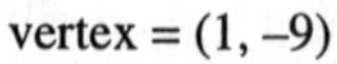
vertex = (1, –9)

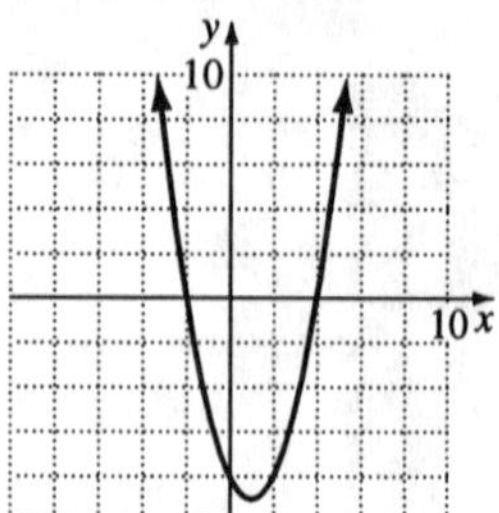

42. $2x + y = 6$
$\underline{-2x + y = 2}$
solution: (1, 4)

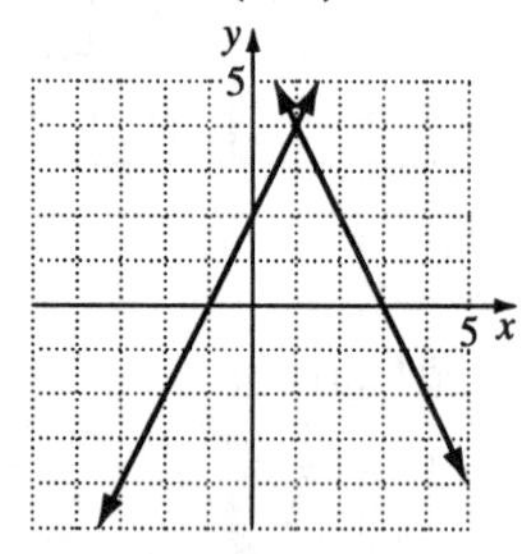

43. $2x + y < 4$
$x > 2$

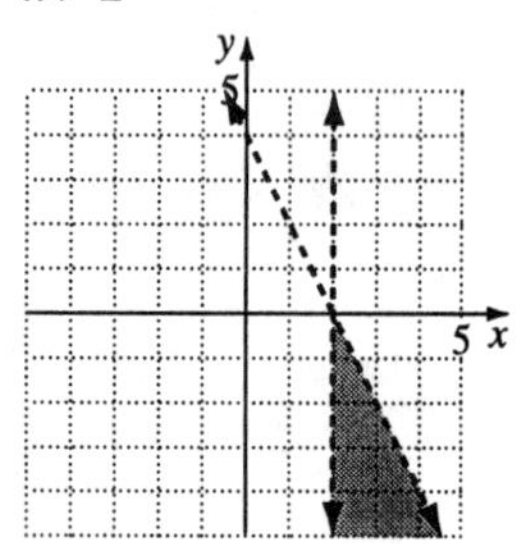

44. a. $T = D + pm$
$T - D = pm$
$p = \frac{T - D}{m}$

b. $p = \frac{1512 - 600}{16}$
$p = 57$
\$57

45. $r = 0.6(220 - a)$
$120 = 0.6(220 - a)$
$200 = 220 - a$
$a = 220 - 200$
$a = 20$
20 years

46. $y = 1200x + 6300$
where x = years of education
and y = yearly income
$19{,}500 = 1200x + 6300$
$13{,}200 = 1200x$
$11 = x$
11 years

47. $f(x) = 68.9x^2 + 1165.3x + 31{,}676$
$f(10) = 68.9(10)^2 + 1165.3(10) + 31{,}676$
$f(10) = 50{,}219$
In 1994, there were 50,219 cases commenced.

48. $A = 0.445x + 14.7$

a. 14.7; the atmospheric pressure at the surface is 14.7 pounds per square inch.

b. 0.445; the atmospheric pressure increases at 0.445 pound per square inch for each increase of 1 foot below the surface.

c. $A = 0.445(30{,}000) + 14.7$
$A = 13{,}364.7$ pounds per square inch

49. a. slope $= \frac{130 - 111}{66 - 62} = 4.75$

b. $y - 111 = 4.75(x - 62)$ or
$y - 130 = 4.75(x - 66)$

c. $y - 111 = 4.75x - 294.5$
$y = 4.75x - 183.5$

d. $y = 4.75(64) - 183.5 = 120.5$ lbs
$y = 4.75(72) - 183.5 = 158.5$ lbs

e. They are reasonably close.

50. $f(x) = 0.1x^2 - 3x + 22$
$f(90) = 0.1(90)^2 - 3(90) + 22$
$f(90) = 562$
If a plane is landing at 90 feet per second, it will need 562 feet of runway; Yes, the runway needs to be at least 562 feet.

51. **a.** $(2x + 20)(2x + 16)$
$= 4x^2 + 72x + 320$

b. $f(x) = 4x^2 + 72x + 320$

c. $f(1) = 4(1)^2 + 72(1) + 320$
$f(1) = 396$
The total area is 396 square meters with a sidewalk 1 meter wide.

52. $f(x) = 2x^2 + 22x + 320$

a. $524 = 2x^2 + 22x + 320$
$2x^2 + 22x - 204 = 0$
$2(x^2 + 11x - 102) = 0$
$2(x + 17)(x - 6) = 0$
Time is positive, so
$x = 6$ years
$1980 + 6 = 1986$

b. (6, 524)

53. $f(x) = \frac{20x}{100 - x}$

a. $f(20) = \frac{20(20)}{100 - 20} = 5;$
It costs \$5000 to remove 20%.
$f(80) = \frac{20(80)}{100 - 80} = 80;$
It costs \$80,000 to remove 80%.
$f(90) = \frac{20(90)}{100 - 90} = 180;$
It costs \$180,000 to remove 90%.

b. $x = 100$

c. The cost goes to infinity. It is impossible to remove 100% of pollutants.

d.

x	0	10	20	30	40	50	60	70	80	90	95	98	99
y	0	2	5	9	13	20	30	47	80	180	380	980	1980

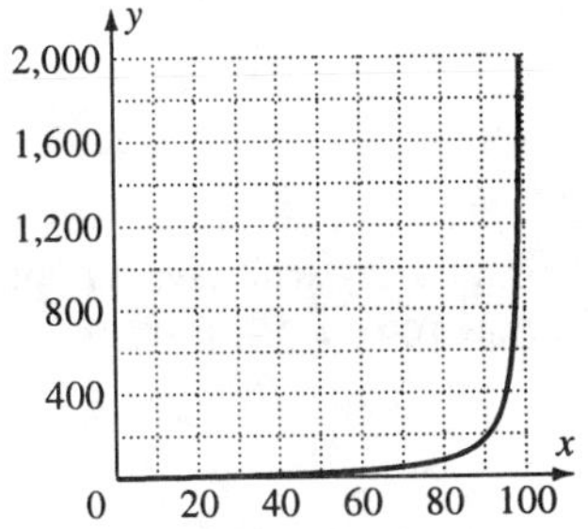

54. $I = \frac{k}{R}$
$5 = \frac{k}{22}$
$k = 110$
$I = \frac{110}{R}$
$I = \frac{110}{10}$
$I = 11$ amperes
(10, 11)

55. $v = \sqrt{\frac{Fr}{100}}$
$v = \sqrt{\frac{2000(320)}{100}}$
$v = 80$ ft per sec.

56. $S = 28.6A^{1/3}$
$S = 28.6(27)^{1/3}$
$S = 85.8$
86 species

57. points (–1, –5), (2, –2)
$d = \sqrt{(x_2 - x_1)^2 + (y_2 - y_1)^2}$
$d = \sqrt{(-2+5)^2 + (2+1)^2}$
$d = \sqrt{9+9}$
$d = \sqrt{18}$
$d = 3\sqrt{2} \approx 4.24$

58. $d^2 = \frac{4050}{I}$
$d^2 = \frac{4050}{162}$
$d = \sqrt{25}$
$d = 5$ feet

59. $f(x) = -0.004x^2 + x + 4$
vertex:
x-coordinate $= -\frac{b}{2a} = -\frac{1}{2(-0.004)} = 125$
y-coordinate =
$-0.004(125)^2 + 125 + 4 = 66.5$
vertex (125, 66.5)
When the ball has traveled 125 feet from the plate, it is 66.5 feet high.

60. $M = \frac{2x}{1-x}$
If $M = 3$, then
$3 = \frac{2x}{1-x}$
$3 - 3x = 2x$
$-5x = -3$
$x = \frac{3}{5} = 0.60$ or 60%

61. $D = \frac{n(n-3)}{2}$
If $D = 5$ diagonals, then
$5 = \frac{n(n-3)}{2}$
$n(n-3) = 10$
$n^2 - 3n - 10 = 0$
$(n-5)(n+2) = 0$
$n = 5$
number of sides, 5

62. $P = 30 - \frac{9}{t+1}$
$P = 27$ (thousands)
$27 = 30 - \frac{9}{t+1}$
$\frac{9}{t+1} = 3$
$3(t+1) = 9$
$t + 1 = 3$
$t = 2$
The community will have a population of 27,000 in 1995 + 2 = 1997.

63. $h = -16t^2 + 96t + 80$

If $h = 128$ feet, then

$128 = -16t^2 + 96t + 80$

$16t^2 - 96t + 48 = 0$

$t^2 - 6t + 3 = 0$

$t = \frac{-(-6) \pm \sqrt{(-6)^2 - 4(1)(3)}}{2(1)}$

$= \frac{6 \pm \sqrt{36 - 12}}{2}$

$= \frac{6 \pm \sqrt{24}}{2}$

$= \frac{6 \pm 2\sqrt{6}}{2}$

$= 3 \pm \sqrt{6}$

The time is $3 + \sqrt{6}$ seconds or $3 - \sqrt{6}$ seconds or rounded, 5.4 seconds or 0.6 second.

64. $4x^2 - 13x + 3 = (4x - 1)(x - 3)$

65. $4x^2 - 49 = (2x - 7)(2x + 7)$

66. $4x^2 - 20x + 25 = (2x - 5)^2$

67. $x^3 + 3x^2 - x - 3$

$= x^2(x + 3) - 1(x + 3)$

$= (x^2 - 1)(x + 3)$

$= (x + 1)(x - 1)(x + 3)$

68. $3x^2 - 75 = 3(x^2 - 25) = 3(x - 5)(x + 5)$

69. $2x^2 + 8x - 42$

$= 2(x^2 + 4x - 21)$

$= 2(x + 7)(x - 3)$

70. $-6x^2 + 7x - 2$

$= (-3x + 2)(2x - 1)$

$= -(3x - 2)(2x - 1)$

71. $x^5 - 16x$

$= x(x^4 - 16)$

$= x(x^2 + 4)(x^2 - 4)$

$= x(x^2 + 4)(x + 2)(x - 2)$

72. $6x^2 - 3x + 2$ is prime; cannot be factored over integers

73. $x^3 - 10x^2 + 25x$

$= x(x^2 - 10x + 25)$

$= x(x - 5)(x - 5)$

$= x(x - 5)^2$

74. $x^3 - 8 = (x - 2)(x^2 + 2x + 4)$

75. $14x^2y^3 - 10x^2y^2 + 4xy^2$

$= 2xy^2(7xy - 5x + 2)$

76. $x^2 + 4xy - 21y^2 = (x + 7y)(x - 3y)$

77. $6x^2 - 13xy - 28y^2 = (3x + 4y)(2x - 7y)$

78. $16x^2 - 40xy + 25y^2 = (4x - 5y)(4x - 5y)$

$= (4x - 5y)^2$

79. $24 \div 8 \cdot 3 + 28 \div (-7) = 3 \cdot 3 - 4 = 9 - 4 = 5$

80. $\frac{11 - (-9) + 6(10 - 4)}{2 + 3 \cdot 4}$

$= \frac{11 + 9 + 6(6)}{2 + 12}$

$= \frac{20 + 36}{14} = \frac{56}{14} = 4$

81. $-21 - 16 - 3(2 - 8)$

$= -21 - 16 - 3(-6)$

$= -37 + 18 = -19$

82. $-(-3y + 2) - 4(6 - 5y) - 3y - 7$

$= 3y - 2 - 24 + 20y - 3y - 7$

$= 20y - 33$

83. $(4x^2-3x+2)-(5x^2-7x-6)$
$=4x^2-3x+2-5x^2+7x+6$
$=-x^2+4x+8$

84. $(15x^2y^3-7x^2y-8x^2)$
$-(-9x^2y^3-6x^2y+5x^2-3)$
$=15x^2y^3-7x^2y-8x^2+9x^2y^3$
$+6x^2y-5x^2+3$
$=24x^2y^3-x^2y-13x^2+3$

85. $(x-2)(3x+7)$
$=3x^2+7x-6x-14$
$=3x^2+x-14$

86. $(7x+4y)(3x-5y)$
$=21x^2-35xy+12xy-20y^2$
$=21x^2-23xy-20y^2$

87. $(3x-5)^2-(2x-3)(4x+5)$
$=9x^2-30x+25-8x^2-10x+12x+15$
$=x^2-28x+40$

88. $(4y-3)(5y^2+6y-2)$
$=20y^3+24y^2-8y-15y^2-18y+6$
$=20y^3+9y^2-26y+6$

89. $(x+y)(x^2-xy+y^2)$
$=x^3-x^2y+xy^2+x^2y-xy^2+y^3$
$=x^3+y^3$

90. $\dfrac{-8x^6+12x^4-4x^2}{4x^2}=-2x^4+3x^2-1$

91. $\dfrac{20x^4y^3-5x^3y^2}{-10x^2y}=-2x^2y^2+\dfrac{xy}{2}$

92. $\dfrac{6x^2+5x-6}{2x+3}=\dfrac{(2x+3)(3x-2)}{2x+3}=3x-2$

93. $\dfrac{(4x^3)^2}{x^9}=\dfrac{16x^6}{x^9}=\dfrac{16}{x^3}$

94. $\left(\dfrac{x^4}{x^7}\right)^{-3}=\left(\dfrac{x^7}{x^4}\right)^3=\dfrac{x^{21}}{x^{12}}=x^9$

95. $\dfrac{3x^2-8x+5}{4x^2-5x+1}=\dfrac{(3x-5)(x-1)}{(4x-1)(x-1)}=\dfrac{3x-5}{4x-1}$

96. $\dfrac{y^2-y-12}{y^2-16}\cdot\dfrac{2y^2+7y-4}{y^2-4y-21}$
$=\dfrac{(y-4)(y+3)}{(y+4)(y-4)}\cdot\dfrac{(2y-1)(y+4)}{(y-7)(y+3)}=\dfrac{2y-1}{y-7}$

97. $\dfrac{15-3y}{y+6}\div(y^2-9y+20)$
$=\dfrac{-3(y-5)}{y+6}\div(y-5)(y-4)$
$=\dfrac{-3(y-5)}{y+6}\cdot\dfrac{1}{(y-5)(y-4)}$
$=\dfrac{-3}{(y+6)(y-4)}$

98. $\dfrac{x+6}{x-2}+\dfrac{2x+1}{x+3}$
$=\dfrac{(x+6)(x+3)+(2x+1)(x-2)}{(x-2)(x+3)}$
$=\dfrac{x^2+9x+18+2x^2-3x-2}{(x-2)(x+3)}$
$=\dfrac{3x^2+6x+16}{(x-2)(x+3)}$

99. $\dfrac{x}{x^2+2x-3}-\dfrac{x}{x^2-5x+4}$
$=\dfrac{x}{(x+3)(x-1)}-\dfrac{x}{(x-4)(x-1)}$
$=\dfrac{x(x-4)-x(x+3)}{(x+3)(x-1)(x-4)}$
$=\dfrac{x^2-4x-x^2-3x}{(x+3)(x-1)(x-4)}$
$=\dfrac{-7x}{(x+3)(x-1)(x-4)}$

100. $\dfrac{\frac{1}{x}-2}{4-\frac{1}{x}}=\dfrac{\frac{1-2x}{x}}{\frac{4x-1}{x}}=\dfrac{1-2x}{x}\cdot\dfrac{x}{4x-1}=\dfrac{1-2x}{4x-1}$

101. $\sqrt{50x^{11}}=\sqrt{25x^{10}}\cdot\sqrt{2x}=5x^5\sqrt{2x}$

102. $3\sqrt{20b}+2\sqrt{45b}=6\sqrt{5b}+6\sqrt{5b}=12\sqrt{5b}$

103. $2\sqrt[3]{16}-3\sqrt[3]{2}=4\sqrt[3]{2}-3\sqrt[3]{2}=\sqrt[3]{2}$

104. $\sqrt{3x}\cdot\sqrt{6x}=\sqrt{18x^2}=3x\sqrt{2}$

105. $\sqrt{5}\left(\sqrt{2}+3\sqrt{7}\right)$

$=\sqrt{5}\left(\sqrt{2}\right)+\sqrt{5}\left(3\sqrt{7}\right)=\sqrt{10}+3\sqrt{35}$

106. $\left(\sqrt{2}+3\sqrt{6}\right)\left(\sqrt{2}-\sqrt{6}\right)$

$=\left(\sqrt{2}\right)^2-\sqrt{2}\left(\sqrt{6}\right)+3\sqrt{6}\left(\sqrt{2}\right)-3\sqrt{6}\left(\sqrt{6}\right)$

$=2-\sqrt{12}+3\sqrt{12}-3(6)$

$=2+2\sqrt{12}-18$

$=-16+4\sqrt{3}$

107. $\left(2+\sqrt{5}\right)^2$

$=4+4\sqrt{5}+5$

$=9+4\sqrt{5}$

108. $\frac{2}{\sqrt{3b}}=\frac{2}{\sqrt{3b}}\cdot\frac{\sqrt{3b}}{\sqrt{3b}}=\frac{2\sqrt{3b}}{3b}$

109. $\frac{6}{\sqrt[3]{4}}$

$=\frac{6}{\sqrt[3]{4}}\cdot\frac{\sqrt[3]{2}}{\sqrt[3]{2}}$

$=\frac{6\sqrt[3]{2}}{\sqrt[3]{8}}$

$=\frac{6\sqrt[3]{2}}{2}=3\sqrt[3]{2}$

110. $\frac{\sqrt{5}}{\sqrt{5}+\sqrt{6}}$

$=\frac{\sqrt{5}}{\sqrt{5}+\sqrt{6}}\cdot\frac{\sqrt{5}-\sqrt{6}}{\sqrt{5}-\sqrt{6}}$

$=\frac{\sqrt{5}\left(\sqrt{5}-\sqrt{6}\right)}{5-6}$

$=\frac{5-\sqrt{30}}{-1}=-5+\sqrt{30}$

111. $\frac{11}{\sqrt{5}-3}$

$=\frac{11}{\sqrt{5}-3}\cdot\frac{\sqrt{5}+3}{\sqrt{5}+3}$

$=\frac{11\left(\sqrt{5}+3\right)}{5-3^2}$

$=\frac{11\sqrt{5}+33}{-4}$

112. $8^{2/3}=\left(\sqrt[3]{8}\right)^2=2^2=4$

113. **a.** Natural numbers: $\left\{6,\ \sqrt{169}\right\}$.

b. Whole numbers: $\left\{0,\ 6,\ \sqrt{169}\right\}$.

c. Integers: $\left\{-14,\ 0,\ 6,\ \sqrt{169}\right\}$.

d. Rational numbers:
$\left\{-14,\ 0,\ 0.45,\ 6,\ 7\frac{1}{5},\ \sqrt{169}\right\}$.

e. Irrational numbers: $\left\{-\pi,\ \sqrt{3}\right\}$.

f. Real numbers:
$\left\{-14,\ -\pi,\ 0,\ 0.45,\ \sqrt{3},\ 6,\ 7\frac{1}{5},\ \sqrt{169}\right\}$

114. $x^3-3x^3y+2y-5$ when $x=-3$ and $y=-4$

$=(-3)^3-3(-3)^3(-4)+2(-4)-5$

$=-27-3(-27)(-4)-8-5$

$=-27+81(-4)-13$

$=-27-324-13=-364$

115. $8+(9+5)=8+(5+9)$
Commutative Property of Addition

116. $(13\cdot7)\cdot3=13\cdot(7\cdot3)$
Associative Property of Multiplication

117. $-5\left(-\frac{1}{5}\right)=1$

Multiplicative Inverse Property

118. $6x+4(x+5)$
$=6x+4x+20$
$=10x+20$

119. a. $3x+2y=5$
$2y=-3x+5$
$y=-\frac{3}{2}x+\frac{5}{2}$

b. $y=-\frac{3}{2}(-1)+\frac{5}{2}$
$y=\frac{3}{2}+\frac{5}{2}$
$y=\frac{8}{2}$
$y=4$

120. slope: $\frac{-4-6}{3+2}=-2$
$y-6=-2(x+2)$ or
$y+4=-2(x-3)$
$y-6=-2x-4$
$y=-2x+2$

121. $\sqrt{-75}=\sqrt{(-25)(3)}=5i\sqrt{3}$

122. Let x = the number.
$5x-7=208$
$5x=215$
$x=43$

123. Let x = the smaller page number, then $x+1$ = larger page number.
$x+x+1=1097$
$2x+1=1097$
$2x=1096$
$x=548$
The pages are numbered 548 and 549.

124. Let x = length of shortest piece, then $2x+10$ = length of longest piece, and $x+17$ = length of middle-sized piece
$x+2x+10+x+17=87$
$4x+27=87$
$4x=60$
$x=15$
$2x+10=40$
$x+17=32$
The pieces are 15 in., 32 in., and 40 in.

125. Let x = height of Empire State Building
Then $2x-790$ = height of World Trade Center.
$\frac{x+2x-790}{2}=980$
$3x-790=1960$
$3x=2750$
$x=916\frac{2}{3}$
$2x-790=1043\frac{1}{3}$
The buildings are $916\frac{2}{3}$ ft and $1043\frac{1}{3}$ ft.

126. Let x = the price of VCR before reduction.
$x-0.20x=124$
$0.80x=124$
$x=\$155$

127. Let x = the number of hours worked.
$350+23x=971$
$23x=621$
$x=27$ hours

128. $\frac{23}{2}=\frac{x}{176}$
$2x=23(176)$
$x=\frac{4048}{2}$
$x=2024$ students

129. $\frac{x+76+74+78}{4}\geq 80$
$x+228\geq 320$
$x\geq 92$
92% or better

130. **a.** $21 + x < 2.5x$
$x - 2.5x < -21$
$-1.5x < -21$
$x > 14$
After 14 trips, or 15 or more trips

b. The lines intersect at $x = 14$.

131. $\frac{\$2.24}{14 \text{ oz}} = \$0.16$ per oz.

132. $\frac{4}{36} = \frac{25}{x}$
$4x = 36(25)$
$x = \frac{900}{4}$
$x = 225$ deer

133. $\frac{5}{x} = \frac{8}{72}$
$8x = 5(72)$
$x = \frac{360}{8}$
$x = 45$ feet

134. **a.** Let x = the width.
Then $x + 50$ = the length.
$2(x + x + 50) = 300$
$4x + 100 = 300$
$4x = 200$
$x = 50$
50 feet by 100 feet

b. $\frac{100}{5} = 20$
20 feet to 1 inch

135. $\frac{1}{2}(15)x = 120$
$15x = 240$
$x = 16$
16 feet

136. **a.** $2\pi r = 2\pi(5) = 10\pi \approx 31.4$ meters

b. $\pi r^2 = \pi(5)^2 = 25\pi \approx 78.5$ square meters

137. $V = \pi r^2 h$
$V = \pi(2)^2(4)$
$V = 16\pi$
$V = \pi(6)^2(4)$
$V = 144\pi$
$\frac{144\pi}{16\pi} = 9$
9 times larger

138. $96x + 378y = 1044$
$24x + 14y = 100$
where x = number of apples
and y = number of avocados

$$\begin{aligned} 96x + 378y &= 1044 \\ -96x - 56y &= -400 \\ \hline 322y &= 644 \\ y &= 2 \\ 96x + 378(2) &= 1044 \\ 96x &= 288 \\ x &= 3 \end{aligned}$$

Eat 3 apples and 2 avocados.

139. Let x = the cost of a pen.
y = the cost of a pad.

$$\begin{aligned} 10x + 12y &= 42 \\ 5x + 10y &= 29 \end{aligned}$$

Divide equation 1 by –2.
No change to equation 2.

$$\begin{aligned} -5x - 6y &= -21 \\ 5x + 10y &= 29 \\ \hline 4y &= 8 \\ y &= 2 \\ 5x + 10(2) &= 29 \\ 5x &= 9 \\ x &= \frac{9}{5} = 1.80 \end{aligned}$$

Cost of pen = \$1.80
Cost of pad = \$2

140. Let x = average income in Mississippi.
Then $2x - 1678$ = average income in Connecticut.

a. $\frac{2x-1678}{x}=1+\frac{13,216}{x}$

$2x - 1678 = x + 13,216$

$x = 14,894$

$2x - 1678 = 28,110$

Mississippi is \$14,894

Connecticut is \$28,110

b. $28,110 - 18,177 = \$9933$

$\frac{9933}{18,177} \approx 0.546$ or 55%

c. $\frac{14,894-18,177}{18,177} = -0.18$

or 18% below

141. Let x = the number.

$x+\frac{1}{x}=4$

$x^2+1=4x$

$x^2-4x+1=0$

$x=\frac{-(-4)\pm\sqrt{(-4)^2-4(1)(1)}}{2(1)}$

$=\frac{4\pm\sqrt{16-4}}{2}$

$=\frac{4\pm\sqrt{12}}{2}$

$=\frac{4\pm2\sqrt{3}}{2}$

$=2\pm\sqrt{3}$

The number is $2+\sqrt{3}$ or $2-\sqrt{3}$ or rounded, 3.7 or 0.3.

142. $x+(x+10)+(4x+20)=180$

$6x + 30 = 180$

$6x = 150$

$x = 25$

$x + 10 = 35$

$4x + 20 = 120$

25°, 35°, 120°

143. Let x = the width of the rectangle.

$2x + 1$ = the length of the rectangle.

$x(2x + 1) = 36$

$2x^2 + x - 36 = 0$

$(2x + 9)(x - 4) = 0$

$x = 4$ $\left(\text{reject } x=-\frac{9}{2}\right)$

$2x + 1 = 2(4) + 1 = 9$

dimensions:

width, 4 meters; length, 9 meters

144. $\frac{1.5\times10^8}{3\times10^5}=0.5\times10^3$ or 500 scc

145. $(24)^2=(20)^2+x^2$

$x=\sqrt{576-400}$

$x=\sqrt{176}$

$x \approx 13.3$ feet

146. $90 = k(75)$

$k = 1.2$

$x = 1.2(80)$

$x = 96$ miles/hr

147. $\frac{2}{5}x=430$

$x = 1075$

\$1075 million

148. Let x = the amount invested at 5%.

$4000 - x$ = the amount invested at 9%.

$0.05x + 0.09(4000 - x) = 311$

$0.05x + 360 - 0.09x = 311$

$-0.04x = -49$

$x = 1225$

$4000 - x = 2775$

\$1225 at 5%, \$2775 at 9%

149. Let t = the time (in hours) for boats to be 232 miles apart.
$13t + 19t = 232$
$32t = 232$
$t = 7.25$ hours

150. Let x = the amount of 80% acid solution.
$10 - x$ = the amount of 65% acid solution.
$0.80x + 0.65(10 - x) = 0.75(10)$
$0.80x + 6.5 - 0.65x = 7.5$
$0.15x = 1$
$x = 6\frac{2}{3}$
$10 - x = 3\frac{1}{3}$
$6\frac{2}{3}$ gallons of 80%, $3\frac{1}{3}$ gallons of 65%

151. x = the time for painter and assistant to work together.
$\frac{x}{4}$ = portion of job done by painter.
$\frac{x}{12}$ = portion of job done by assistant.
$\frac{x}{4} + \frac{x}{12} = 1$
$12\left(\frac{x}{4} + \frac{x}{12}\right) = 12(1)$
$3x + x = 12$
$4x = 12$
$x = 3$ days

152. Let v = the speed of the boat in still water.
c = speed of the current.
$r + c$ = the rate of boat with the current.
$r - c$ = rate of the boat against the current.
Rate × Time = Distance

$(r + c)2 = 48$
$(r - c)3 = 48$

Divide equation 1 by 2.
Divide equation 2 by 3.

$$\begin{array}{r} r + c = 24 \\ r - c = 16 \\ \hline 2r = 40 \\ r = 20 \\ 20 + c = 24 \\ c = 4 \end{array}$$

speed of boat in still water, 20 miles per hour; rate of current, 4 miles per hour

153. Let x = the rate of plane B.
$x + 50$ = the rate of plane A.

	rate × time = distance		
plane A	$x + 50$	$\frac{500}{x+50}$	500 miles
plane B	x	$\frac{400}{x}$	400 miles

$\frac{500}{x + 50} = \frac{400}{x}$
$500x = 400x + 20000$
$100x = 20000$
$x = 200$
$x + 50 = 250$
rate of plane A, 250 miles per hour; rate of B, 200 miles per hour

154.

term(n)		pattern
3	$1+2+3=6=4\cdot\frac{3}{2}$	$(3+1)\frac{3}{2}=4\cdot\frac{3}{2}$
5	$1+2+3+4+5=15=6\cdot\frac{5}{2}$	$(5+1)\frac{5}{2}=6\cdot\frac{5}{2}$
8	$1+2+3+4+5+6+7+8=36=9\cdot\frac{8}{2}$	$(8+1)\frac{8}{2}=9\cdot\frac{8}{2}$
10	$1+2+\cdots+9+10=55=11\cdot\frac{10}{2}$	$(10+1)\frac{10}{2}=11\cdot\frac{10}{2}$
70	$1+2+\cdots+68+69+70=?$	$(70+1)\frac{70}{2}=71\cdot\frac{70}{2}=2485$
n	$1+2+\cdots+(n-2)+(n-1)+n=?$	$(n+1)\frac{n}{2}=\frac{n^2+n}{2}$

155. w, x, y, z represent natural numbers.
$x > w$
$y = z - 1$
$z = x + 4$
$x > w$
$x + 4 > w + 4$
$z > w + 4$

156. **a.** $10 + 4 - 8 = 6$
$x = 10,\ y = 4,\ z = 8$

b. $4 \cdot 8 - 10 = 22$
$a = 4,\ b = 8,\ c = 10$

c. $10 \div 2 + 8 = 13$
$r = 10,\ s = 2,\ w = 8$

157. c is true. Let X = any number.
$X > A$
$B < X$
$C = A$
$X > C$ True

158.

	first term $n = 1$	second term $n = 2$	third term $n = 3$	fourth term $n = 4$	fifth term $n = 5$	rule for the nth term
a.	1	3	5	7	9	$2n - 1$
b.	5	8	11	14	17	$3(n + 1) - 1$
c.	1	4	9	16	25	n^2
d.	0	3	8	15	24	$n^2 - 1$
e.	1	8	27	64	125	n^3

159. There are five sizes of square possible.

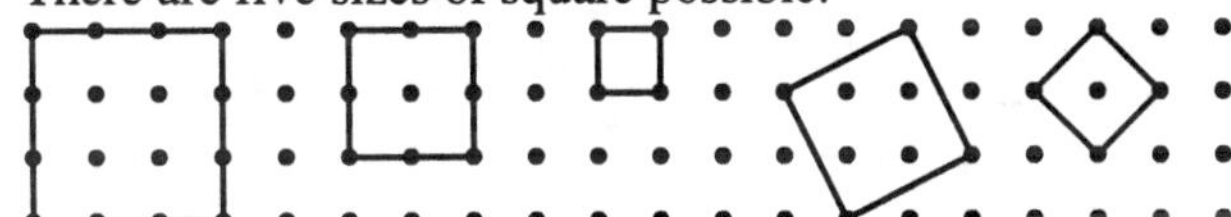

160. Answers may vary. Sample given.

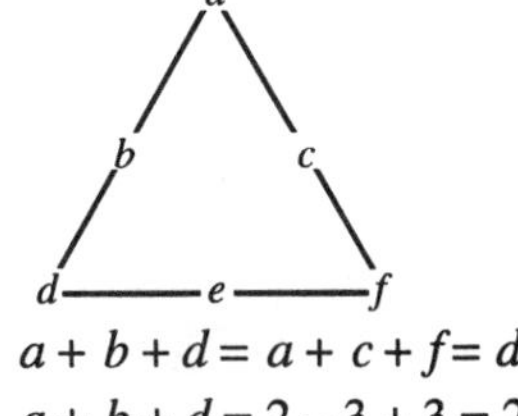

$a + b + d = a + c + f = d + e + f$
$a + b + d = 2 - 3 + 3 = 2$
$a + c + f = 2 - 1 + 1 = 2$
$d + e + f = 3 - 2 + 1 = 2$
$a = 2,\ b = -3,\ c = -1,\ d = 3,\ e = -2,\ f = 1$

161. **a.** possible two digit odd number: 11, 13, 15, 17, …, 97, 99

b. divisible by 3: 15, 21, 27, 33, 39, 45, 51, 57, 63, 69, 75, 81, 87, 93, 99

c. divisible by 5: 15, 45, 75

d. digits sum to an odd number:
15: $1 + 5 = 6$
45: $4 + 5 = 9$
75: $7 + 5 = 12$

e. only number is 45

162. change for a quarter:

no. of pennies	no. of nickels	no. of dimes
	1	2
5		2
	3	1
5	2	1
10	1	1
15		1
	5	0
5	4	0
10	3	
15	2	
20	1	
25		

12 ways

163. $\frac{120}{12} = 10$ rows

First row has 10 cans, second row has 9 cans, etc.

$10 + 9 + 8 + 7 + 6 + 5 + 4 + 3 + 2 + 1 = 55$

55 cans

164. $\frac{1}{2}$

165. $64 - 24 = 40$ cubes